Biomedical
TRANSDUCERS
and INSTRUMENTS

Tatsuo Togawa
Toshiyo Tamura
P. Åke Öberg

CRC Press
Boca Raton   New York

Acquiring Editor:	Marsha Baker
Project Editor:	Sarah Fortener
Cover design:	Dawn Boyd
Manufacturing:	Carol Royal

Library of Congress Cataloging-in-Publication Data

Togawa, Tatsuo, 1937-
  Biomedical transducers and instruments / by Tatsuo Togawa, Toshiyo
Tamura, P. Åke Öberg.
      p. cm.
  Includes bibliographical references and index.
  ISBN 0-8493-7671-8
  1. Biomedical engineering.   2. Physiology--Measurement-
-Instruments.   I. Tamura, Toshiyo.   II. Öberg, P. Åke.   III. Title.
  [DNLM: 1. Transducers.   2. Electronics, Medical--instrumentation.
3. Biomedical Engineering--instrumentation.   QT 26 T645b   1997]
R856.T64   1997
610′.28--dc21
DNLM/DLC
for Library of Congress   96-40279
                                                                                                      CIP

No claim to original U.S. Government works
International Standard Book Number 0-8493-7671-8
Library of Congress Card Number 96-40279
Printed in the United States of America 1 2 3 4 5 6 7 8 9 0
Printed on acid-free paper

Preface

In biomedical research, as well as in clinical medicine, the acquisition of physiological data from the living body is important not only to increase our understanding of basic biological mechanisms but also to support diagnostic procedures. The quality of such measurements largely depends on the performance of the transducers and instruments in the measurement system. The living body is a "difficult" object to measure; accurate measurements of physiological signals require transducers and instruments that have high specificity and selectivity and do not interfere with the systems under study.

In order to overcome the many challenges of this field, a great variety of transducers and instruments have been designed and presented. To be able to select the "best" method for studying new problems, detailed knowledge of sensor and instrument properties is required. This book is written to provide up-to-date information for such requirements.

The first chapter presents a brief summary of key concepts required to understand measurement systems and instruments. The remaining chapters are organized around important quantities studied in the life sciences: pressure, flow, temperature, and motion, for example. This method of organizing the material is inspired by Richard S. C. Cobbold's well-known monograph, *Transducers for Biomedical Measurements: Principles and Applications* (John Wiley & Sons, 1974). The popularity of this book for over 20 years is partly related to its clear structure, which makes it convenient to use as a reference book.

Our book does not require a professional background in engineering or medicine. A background in basic science is sufficient. Clinical engineers can use the book as a reference book in their daily work in hospitals.

For the biomedical engineer, this book will facilitate the systematic understanding of transducers and instruments used in the health sciences. The researcher can use our book as a reference text, as alternative approaches to the solution of measurement problems are presented.

The application of sensor and instrument principles to problems in life also has a practical perspective. It is important to take into account problems such as biocompatibility, electromagnetic interference, and ways of attaching probes to the biological system under study. We have tried to cover this part by giving direct guidance in the text and also by providing many references to practical papers and books.

Many of the illustrations in the book are borrowed from other authors in original or modified form. We would like to thank these authors and copyright owners for their permission to republish the figures. We would also like to thank Per Sveider, Yuichi Kimura, Kimio Otsuka, Hirokazu Saito, and Mitsuihiro Ogawa, who helped us prepare the figures, and Britt Axelsson and Lindy Powell Gustavsson, who edited the manuscript and revised the English text, respectively.

Tatsuo Togawa
Toshiyo Tamura
P. Åke Öberg

The Authors

Tatsuo Togawa, Ph.D., received his Ph.D. degree in applied physics in 1965 at the University of Tokyo. He spent 3 years at the Institute of Medical Electronics, University of Tokyo, as a research assistant. He then moved to the Institute for Medical and Dental Engineering, Tokyo Medical and Dental University, and since 1972 he has served as Professor of the Division of Instrumentation Engineering. From 1993 to 1994, he was Director of that institute. He is a past president of the Japan Society of Medical Electronics and Biological Engineering. Throughout his career, he has been involved in research in biomedical measurements and instrumentation. His scientific works have been published in various books and as more than 90 articles in international journals. He is an editorial member of several international journals. Since 1973, he has been cooperating with Professor P. Åke Öberg, Department of Biomedical Engineering, Linköping University, Sweden. In 1994, he received the degree of Doctor *honoris causa* from Linköping University.

Toshiyo Tamura, Ph.D., received his Ph.D. degree in 1980 from Tokyo Medical and Dental University. From 1980, he has served as a research associate in the Institute for Medical and Dental Engineering, Tokyo Medical and Dental University. In 1984 he spent a year and a half at the Department of Biomedical Engineering, Linköping University, Linköping, Sweden, under Professor P. Åke Öberg. During this period, he was involved in research regarding peripheral blood flow measurement. From 1991 to 1993, he was an Associate Professor, Department of Electrical Engineering, Faculty of Engineering, Yamaguchi University, Ube, Yamaguchi. Since 1993, he has served as an Associate Professor at the Institute for Medical and Dental Engineering, Tokyo Medical and Dental University. His research and teaching activities have focused on biomedical transducers involving noninvasive apparatus and biosignal analysis. His scientific work is represented by more than 60 published articles in international journals. He is also an active member in several national and international societies and has presented numerous lectures at international meetings.

P. Åke Öberg, Ph.D., received his MSEE degree from the Chalmers University of Technology, Göteborg, Sweden, in 1964 and his Ph.D. in Biomedical Engineering from Uppsala University in 1971. From 1963 to 1972 he worked in the Department of Physiology and Medical Biophysics, Uppsala University. In 1972, he joined Linköping University, where he is currently a Professor of Biomedical Engineering and Director of the Department of Clinical Engineering, University Hospital. From 1982 to 1983, he spent a year at the University of Washington in their Bioengineering Center, where he worked with laser-Doppler blood flow measurements.

Dr. Öberg is the founding chairman of the Clinical Engineering Division of the International Federation for Medical and Biological Engineering and is now a board member of the division, His research interests include biomedical instrumentation, transducers, and clinical engineering. He has published over 300 scientific papers and books in the fields of biomedical and clinical engineering, physiology, and bio-optics. Dr. Öberg is a past president of the Swedish Society of Medical Physics and Medical Engineering. He is now the deputy editor of the *Journal of Medical and Biological Engineering and Computing*. Dr. Öberg is a fellow of the Swedish Academy of Engineering Sciences (1980), Royal Swedish Academy of Sciences (1987), and an honorary member of the Hungarian Academy of Engineering Sciences.

Contents

Chapter 1

Fundamental Concepts

This chapter provides brief definitions and explanations of some fundamental concepts in biomedical measurement techniques. We recommend the use of this chapter as a checklist of the minimum knowledge required when doing research or practicing medicine that involves measurements employing transducers and measurement systems.

Different definitions for the same technical terms often appear in different fields. In such cases, the content of this chapter is limited to the situation where physiological measurements are concerned. More general as well as more detailed explanations can be found in standard textbooks on measurement techniques.

1.1 SIGNALS AND NOISE IN THE MEASUREMENT

1.1.1 MEASUREMENT

A *measurement* is a procedure by which an observer determines the quantity which characterizes the property or state of an object. The quantity to be determined is the object quantity of the measurement. In this book, physical or chemical quantities that contain physiological information are considered to be the object quantities. Sometimes such quantities can be estimated by the human senses, for instance through visible observations. But, to obtain objective, reproducible, and quantitative results, instruments should be used for which the results are given as the output of the measurement system.

The physical characteristics of the output depends on the type of instrument used. When electronic instruments are used, the output will be in the form of an electric potential. This output can also be converted into digital values if desired. In any case, the original physical or chemical quantity is converted into convenient forms such as an electric potential or a numerical value. To describe the object quantity correctly from the output of an instrument, the relation between the output of the instrument and the object quantity must be defined. To achieve this, good calibration procedures are required (see Section 1.3.1). The term *measurement* always implies the whole procedure by which the object quantity is correctly determined. Further details about the theory and philosophy of measurement will be found in standard textbooks on measurement science (for example, see Finkelstein[1]).

1.1.2 SIGNALS AND NOISE

In a measurement, the *signal* is defined as the component of a variable which contains information about the object quantity, whereas *noise* is defined as a component unrelated to the object quantity. Thus, the signal is the desired component, and noise is the unwanted component in a measurement.

Signals and noise are not uniquely defined by their range or variation pattern but depend on the intention of the observer. For example, in biopotential measurements the electromyogram (EMG), which measures the potential generated by muscles, gives information about the muscle activity; thus, the EMG can be regarded as a signal. But the EMG is an unwanted component for another observer who is interested in obtaining nerve action potentials. In this situation, the EMG component is considered to be a kind of noise. In other words, the definition of signals and noise depends on the intention of the observer and is quite arbitrary in general measurement situations. That situation is different in communication technology, where a sender sends signals to a receiver, so that the signal is fully determined by the intention of the sender.

In actual measurement situations, there are no general rules by which signals and noise can be distinguished. Only detailed knowledge about the nature of the measurement object and possible disturbances in the system can help to distinguish signals from noise. Typical sources of noise and their characteristics are discussed in Section 1.1.6.

1.1.3 AMPLITUDE AND POWER

In biomedical measurements, object quantities are usually time-varying. Usually, unwanted time-varying components are superimposed on the signal. If an electronic instrument is used to measure the object quantity, it provides an electric potential output containing the signal and noise components which both have to be regarded as time-varying.

To describe the range of variation of a time-varying signal, the concepts *amplitude* and *power* are commonly used. Sinusoidal variation is considered as a fundamental pattern of variation, because any time-varying variable in a definite time interval can be expressed as a sum of sinusoidal variations as described below. When sinusoidal variations are concerned, the difference between the maximal positive and the maximum negative peaks is a measure of the amplitude, which is called the *peak-to-peak value*. The peak-to-peak value can also be defined for any time-varying signal that is not sinusoidal or periodic. The *root-mean-square amplitude* is a convenient measure of the variability of the signal. If the signal is denoted $x(t)$, then the root-mean-square amplitude is defined as $\sqrt{\overline{x(t)^2}}$. Both peak-to-peak values and root-mean-square amplitudes have the same dimensions as that of the variable itself.

The power is a quantity defined as the time average of the square of the variable. For a variable $x(t)$, the power is given as $\overline{x(t)^2}$. If $x(t)$ corresponds to the electric potential developed over a resistor, R, the power dissipation in the resistor is proportional to $\overline{x(t)^2}/R$. In this case, $\overline{x(t)^2}/R$ has a physical meaning of power which is defined as the amount of energy dissipation per unit time. But the concept of power defined as $\overline{x(t)^2}$ is used more widely even though it does not directly correspond to the physical concept of power. As described in Section 1.1.4, the quantity $\overline{x(t)^2}$ is useful when frequency components of the variable are considered.

1.1.4 POWER SPECTRUM

The *power spectrum* is the distribution of power corresponding to frequency components of the variable. As rigorously treated by the theory of Fourier transform, any periodic function of time $x(t)$ having a period T and time average equal to zero can be expressed as the sum of sinusoidal components of different frequencies as

$$x(t) = \sum_{n=1}^{\infty} \left(A_n \cos(n\omega_0 t) + B_n \sin(n\omega_0 t) \right) \tag{1.1}$$

where $\omega_0 = 2\pi/T$, and A_n and B_n are coefficients called the Fourier coefficients and defined as

$$A_n = \frac{2}{T} \int_{-T/2}^{+T/2} x(t) \cos(n\omega_0 t) dt \tag{1.2}$$

$$B_n = \frac{2}{T} \int_{-T/2}^{+T/2} x(t) \sin(n\omega_0 t) dt . \tag{1.3}$$

Then the total power is given as

$$\overline{x(t)^2} = \frac{1}{2}\sum_{n=1}^{\infty}\left(A_n^2 + B_n^2\right).$$

(1.4)

This expression implies that the power can be expressed as the sum of all the powers of the sinusoidal components of different frequencies.

A similar expression can be given for nonperiodic function using the Fourier transform. Similar to Equation (1.1) for a periodic function, the nonperiodic function $x(t)$ can be written as

$$x(t) = \frac{1}{2\pi}\int_{-\infty}^{+\infty} X(\omega)e^{-i\omega t}d\omega$$

(1.5)

where $X(\omega)$ is the Fourier transform of $x(t)$ defined as

$$X(\omega) = \int_{-\infty}^{+\infty} x(t)e^{i\omega t}dt.$$

(1.6)

It can be shown that the power is given as

$$\overline{x(t)^2} = \frac{1}{2\pi}\int_{-\infty}^{+\infty} |X(\omega)|^2 d\omega .$$

(1.7)

In this equation, the sum in Equation (1.1) is replaced by an integral, and Fourier coefficients are replaced by a continuous function of the angular frequency. But the function $|X(\omega)|^2$ can be understood as a component of power corresponding to the angular frequency ω. This function is called the *power density*, and power spectrum is expressed by the power density as a function of the angular frequency ω.

Further details of mathematical treatments and their physical meanings will be found in ordinary textbooks on Fourier transforms (for example, Champeney[2]).

1.1.5 SIGNAL-TO-NOISE RATIO

The *signal-to-noise ratio* is generally defined as the ratio of the value of the signal to that of the noise and is often denoted simply by S/N (or SNR). Commonly used is the ratio of the power of the signal to that of the noise, but the ratio of peak-to-peak values or also root-mean-square amplitudes of the signal to that of the noise may be used as the signal-to-noise ratio, if it is previously defined.

In practical situations, the signal-to-noise ratio is considered in limited frequency ranges, and its value is always different for different frequency ranges because the signal and the noise always have different frequency spectra.

The ratio is often expressed in decibels (db), defined as

$$10 \log_{10} S/N$$

(1.8)

when the ratio of power of the signal to that of the noise is used, while it is

$$20 \log_{10} S/N$$

(1.9)

when the ratio of amplitude of the signal to that of the noise is used.

1.1.6 DIFFERENT TYPES OF NOISE

Different types of noise arising from different souces may appear in actual measurement situations and can be characterized by their power spectra and treated theoretically. Some types of noise are discussed briefly in the following sections, but more detailed treatment can be found in adequate textbooks (for example, van der Ziel[3]). In addition to these types of noise, all unwanted signals superimposed on the signal are classified as noise, irrespective of whether the sources are identified or not.

1.1.6.1 Thermal Noise

Thermal noise is a kind of noise which is caused by the random thermal agitation. The power density of the thermal noise is uniformly distributed through the whole frequency range, and its power is proportional to the temperature. For example, across the terminals of a resistor of resistance R, the noise potential $v(t)$ appears in a frequency range Δf, so that it obeys a relation

$$\overline{v(t)^2} = 4kTR\Delta f \tag{1.10}$$

where k is the Boltzmann constant (1.38×10^{-23} J/K), and T is the absolute temperature. This relation is known as Nyquist's equation.

1.1.6.2 1/f Noise

1/f noise is a kind of noise characterized by its power spectrum so that the power density is inversely proportional to the frequency over the lower frequency range. The $1/f$ noise may have different origins. When a current is passing through a semiconductor device, the $1/f$ noise is caused by the fluctuation of carriers in the semiconductor. This type of noise is also called *flicker noise*. Flicker noise is also generated in a resistor when a current is flowing through it.

In many unstable quantities, which may change in a long time interval of days, months, or even years, it is often observed that the power density of the fluctuation in the quantity is almost inversely proportional to the frequency, and thus it is considered to be the $1/f$ noise. The drift is also considered as a very low frequency component of the $1/f$ noise.

Fluctuations in which the power density is inversely proportional to the frequency are found in physiological quantities such as fluctuation in the heart rate, which is considered to reflect physiological activities.[4] If this measurement is used to study physiological activities, it cannot be considered as a noise but rather as a signal.

1.1.6.3 Interference

Interference is a kind of noise caused by physical or chemical events outside the object and the measurement system. The interference is sometimes caused by natural phenomenon such as lightning, but more commonly by artificial sources. Power-lines often cause interference by electromagnetic coupling to the object of measurement. Sources other than electromagnetic ones can cause interference. For example, fluorescent lamps may cause noise in optical measurement systems.

The power spectrum of noise due to interference depends on the source. The power-line frequency (50/60 Hz) and its harmonics often appear in the power spectrum of the noise, when electric instruments supplied from the power-line are used near the object and for the measurement system. Electronic instruments which utilize pulse or switching operation generate noise in a wider frequency range. Machines having mechanical moving parts may generate vibration, which interferes with mechanical measurement systems. The power spectrum of the mechanical interference may have peaks corresponding to mechanical resonance frequencies of the machine itself and other materials excited mechanically. Details of actual situations of interference and practical techniques for reducing noise can be found in appropriate textbooks (for example, Morrison[5]).

1.1.6.4 Artifact

The term *artifact* usually implies a component of noise superimposed on the object quantity and caused by external influences such as movement. The motion artifact often appears in biopotential measurement when using skin surface electrodes. It is considered that this is due in part to the potential generated by the epidermal layers of the skin and in part to the change in electrode potentials generated at the interfaces of the electrolyte and the metal. The wave form of the artifact depends on the nature of the external influence. Sometimes, the motion artifact resembles biopotential signals such as those of the electroencephalogram (EEG) and electrocardiogram (ECG), and thus it is difficult to remove the noise from the signal by simple methods such as the use of a band-pass filter.

Artifacts can be reduced by suppressing the process which couples the interference to the object or the measurement system. For example, motion artifacts at the recording electrodes can be reduced by

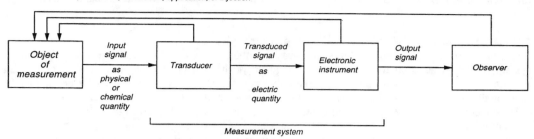

Figure 1.1 The general structure of a measurement system.

the use of nonpolarizing electrodes. While the electrode potential varies due to the variation of the ionic concentration near the electrode surface, that variation is suppressed in nonpolarizing electrodes by employing adequate dissociation equilibrium. The abrasion or puncture of the epidermal part of the skin also reduces motion artifacts; because of that, the potential developed in the skin is suppressed by short-circuiting the stratum corneum.[6-8] (See Section 6.3.1.4.)

1.2 CHARACTERISTICS OF THE MEASUREMENT SYSTEM

1.2.1 TRANSDUCER AND MEASUREMENT SYSTEM

In a measurement procedure, the observer obtains information about the object by using an appropriate measurement system. Usually, a measurement system consists of the transducer and the electric instrument as shown in Figure 1.1. The physical or chemical quantity that characterizes the object is detected by the transducer and is converted into an electric quantity, which is displayed by using an appropriate electronic instrument that will transfer the results to the observer.

Sometimes, measurements require active procedures to be applied to the object, such as excitation, illumination, irradiation, stimulation, application, or injection. Such procedures are considered to be a part of the measurement process and are provided by the transducer or other parts of the measurement system.

Although an active procedure is unavoidable in some measurements, its influence on the object should be minimized for two reasons, i.e., to minimize hazard and to minimize the change in the object quantity due to the active procedure. On the other hand, a situation often occurs in which measurement becomes much easier and more accurate if the applied energy or strength of the active procedure is increased. The level of the active procedure should then be determined as a compromise between minimized influence to the object and maximized performance of the measurement system.

The transducer is an essential part of the measurement system, because the quality of the measurement system is determined mostly by the transducer used. Signal-to-noise ratio, for example, is always determined mainly by the transducer, as long as adequately interfaced electronic circuits are employed.

Different types of object quantities require different kind of transducers. Also, different kinds of transducers are required according to the requirements of different measurement situations, such as different signal amplitudes and frequency ranges; accuracy requirements; limitations in size, shape, or materials; or invasiveness of the measurement procedure. In biomedical measurements, transducers designed for other purposes are commonly unsuitable even if fundamental characteristics such as type of object quantities, measurement ranges, or frequency responses are acceptable. Actually, most transducers used in biomedical measurements are designed so that they can be applied to the body with minimum subsidiary effects to obtain the desired biological information correctly. Most parts of this book concern transducers applicable to biomedical measurements.

1.2.2 STATIC CHARACTERISTICS

In most measurement systems, the output of the measurement system at every moment can be fully determined by its input as the object quantity at that moment, if the change of the object quantity is slow enough. In such a situation, the input-output relationship of the measurement system can be uniquely determined without depending on the time course. The object quantity and the characteristics which represent the relation between the object quantity and the output of the measurement system are called the *static characteristics*. Fundamental features of a measurement system can be characterized by the static characteristics.

1.2.2.1 Sensitivity, Resolution, and Reproducibility

The term *sensitivity* is always used in such a way that the sensitivity of a transducer or a measurement system is high when a small change in object quantity causes a large change in its output. But this quantitative definition of sensitivity is not the only one. In some fields, the sensitivity is defined as the ratio of the output to the input. In this definition, the numerical value which represents the sensitivity is large when the sensitivity is high. In other fields, the sensitivity is defined as the ratio of the input to the output. This factor corresponds to the amount of change in the object quantity which produces a unit change in the output. By this definition, the numerical value is small when the sensitivity is high. The sensitivity has a dimension when the dimension of the object quantity and that of the output are different. Sensitivities for quantities of different objects are represented in different units, such as mV/kPa, μA/K, mV/pH, etc.

The sensitivity can be a constant value when the change in the output is linearly related to the change in the object quantity, whereas it cannot be constant when the response is nonlinear. In such a case, sensitivity depends on the absolute value of the object quantity.

The *resolution* is the least value of the object quantity that can be distinguished at the output of the measurement system. A change in the object quantity that is smaller than the resolution of the measurement system will not produce a detectable change in its output which can be distinguished from noise. The numerical value of the resolution is small when the resolution is high. The resolution has the same dimension as that of the object quantity.

The *reproducibility* describes how close to one another repeated outputs are when the same quantity is measured repeatedly. Quantitatively, the reproducibility of a measurement system is defined as a range in the object quantity so that the results of successive measurements for the same quantity fall into that range with a given probability. If the probability level is not specified, it is usually understood to be 95%. When the range is narrow, the reproducibility is high. The term *repeatability* is also used to express the similar concept of reproducibility, but repeatability is understood to be the reproducibility in a short time interval when these two terms are distinguished.

1.2.2.2 Measurement Range

The *measurement range* is the total range of the object quantity within which the measurement system works so as to meet the nominal performance of the measurement system. Thus, the measurement range depends on other performance requirements such as sensitivity, resolution, or reproducibility. If the requirements are high, the measurement range will be narrow. Sometimes different measurement ranges are specified for different requirements. For example, in a thermometer, the measurement range is from 30 to 40°C for reproducibility of 0.1°C, and from 0 to 50°C for reproducibility of 0.5°C.

The measurement range states the maximum allowable change of the object quantity as long as the nominal performance of the measurement system is expected. On the other hand, minimum detectable change of the object quantity is stated by the resolution. The ratio of measurement range to resolution is called *dynamic range*. The dynamic range is a nondimensional value and is sometimes expressed in decibels (db).

The dynamic range has to be considered when the signal is converted into a digital quantity. For example, the number of bits of an analog-to-digital converter and the data format or number of digits of the digital display are determined so that the maximum digital number usable by these devices is large enough compared to the dynamic range.

1.2.2.3 Linearity or Nonlinearity

The *linearity* describes how close the input-output relationship of the measurement system is to an appropriate straight line. Different definitions of linearity are used depending on which straight line is considered. The straight line defined by the least square fit to the input-output relation can be used, whereas other straight lines defined by the least square fit positioned to pass either the origin or the terminal point, or both, can also be used. When the straight line passing through the origin is used, the linearity of this specific definition is sometimes called *zero-based linearity* or *proportionality*.

As a quantitative measure of the linearity, the maximum deviation of the input-output relation curve from a straight line is used. However, the term *nonlinearity* is conventionally used to indicate this value, because the numerical value when using this definition is large when the deviation of the input-output relation curve from the straight line is significant.

When the linearity is high (or the nonlinearity is small), the input-output relationship can be regarded as a straight line, and thus the sensitivity can be regarded as constant. On the other hand, when the linearity is low (or nonlinearity is large), the sensitivity depends on the input level.

Although higher linearity is desired in most measurement systems, accurate measurement is possible even if the response is nonlinear, as long as the input-output relation is fully determined. By using a computer, the input can be estimated at every sampling interval from the output and knowledge of the input-output relation.

1.2.2.4 Hysteresis

Hysterisis is a kind of phenomenon in which different output values appear corresponding to the same input no matter how slow the speed of change of the input is. If a measurement system has hysteresis, the input-output relation curve is not unique but depends on the direction change as well as the range of successive input values.

There are different causes of hysteresis, such as backlash in mechanical coupling parts, viscoelasticity or creep of mechanical elements, magnetization of ferromagnetic materials, or adsorption and desorption of materials on electrochemical devices. While hysteresis due to backlash is independent of the range of variations of the input, hysteresis attributable to other causes depends on the variations of the input, so that large variations cause large hysteresis. To minimize hysteresis, larger inputs appearing at a transient or by the artifacts beyond the normal variation range of the signal should be avoided. The transducer design in which the input is limited within the normal measurement range is advantageous not only to protect sensing elements in the transducer from destruction but also to reduce hysteresis. For example, the stopper for the diaphragm or beam in mechanical transducers is employed for this reason.

1.2.3 DYNAMIC CHARACTERISTICS

The *dynamic characteristics* of a measurement system describe a transient input-output relation, whereas the static characteristics describe the relation when the input remains constant or changes slowly. Dynamic characteristics are required when the response to time-varying inputs is of concern. The time-varying pattern of the object quantity is observed as a wave form, but true wave forms will not be reproduced unless the dynamic characteristics are excellent.

The dynamic characteristics are particularly important when the transducer is part of a control system. Instability or oscillation may occur due to a poor dynamic response of the transducer.

The most common cause affecting the dynamic characteristics of the measurement system is the presence of elements that store and release energy when the object quantity varies. For example, inertial elements, such as masses and inductances, and compliant elements, such as springs and electric and heat capacitances, are such system components. If the displacements of mechanical parts and fluids cause significant time delays, they will also affect the dynamic characteristics of the system. Besides the measurement system, the object matter of the measurement and the interfacing media may also affect the dynamic characteristics of the whole measurement process. In such a situation, the dynamic characteristics should be discussed, including the object or the interfacing media, as in the case of the catheter-transducer systems in pressure measurements (see Section 2.2.2).

Although some important concepts and terms regarding the dynamic characteristics of the measurement systems are explained briefly in the following sections, we also recommend the study of standard textbooks.[9,10]

1.2.3.1 Linear and Nonlinear Systems

The term *linear system*, or the concept that a system is linear, always refers to a system or a system condition that can be represented by a linear differential equation. In the linear system, the response to simultaneous inputs is the sum of their independent inputs. Sometimes, the linear system is defined by this condition. A system which does not meet this condition is called a *nonlinear system*.

In the linear systems, the dynamic characteristics are the same regardless of the amplitude of the input. The amplitude of the response is simply proportional to the input amplitude, because a large input can be regarded as a sum of small inputs; hence, the response corresponding to the large input is the sum of the responses corresponding to the small inputs. This property is important because many convenient parameters that characterize the system can be defined regardless of the signal amplitude.

Real systems cannot always be linear when the input increases far beyond the measurement range. On the other hand, most measurement systems can be regarded as linear if the variations of the input are small. Even in a nonlinear system, an appropriate linear system can be assumed to be a result of the approximation in a small measurement range. In a linear system, the response to a sinusoidal input is also sinusoidal with the same frequency as that of the input, while other frequency components such as higher harmonics may appear in a nonlinear system.

1.2.3.2 Frequency Response

The *frequency response* refers to the distribution of the amplitude and the phase shift of the output to sinusoidal inputs of unit amplitude over the whole frequency range in which the dynamic characteristics are considered. Usually, the frequency response is defined only for linear systems.

The output of a linear system can be described as the sum of the responses corresponding to sinusoidal inputs having different frequencies, because the input is expressed as the sum of sinusoidal functions, such as Equation (1.1) or (1.5). Therefore, the frequency response provides complete characteristics about the output of the system for any input.

When the input-output relation of a system is described by a constant-coefficient, first-order differential equation, the system is called a *first-order system*. The differential equation which describes a first-order system is written as

$$a_1 \frac{dy(t)}{dt} + a_0 y(t) = x(t) \tag{1.11}$$

where $x(t)$ and $y(t)$ are the input and the output of the system and a_0 and a_1 are constants. Then the frequency response of this system can be represented as shown in Figures 1.2(a) and (b), where f_c is given as $a_0/2\pi a_1$ and is called the *cut-off frequency*.

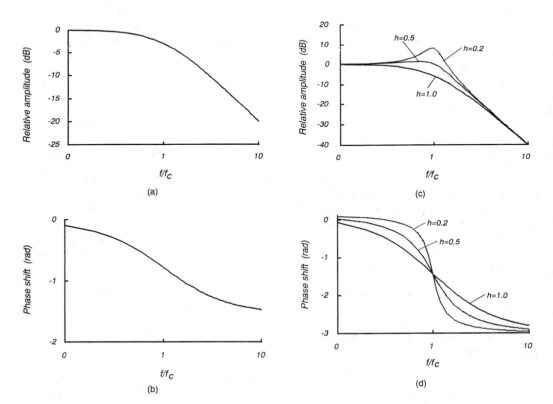

Figure 1.2 Frequency responses of first- and second-order linear systems. Amplitude and phase responses of the first-order system, **(a)** and **(b)**, and those of the second-order system, **(c)** and **(d)**.

The *second-order system* is a system that can be described by a second-order constant-coefficient differential equation written as

$$a_2 \frac{d^2 y(t)}{dt^2} + a_1 \frac{dy(t)}{dt} + a_0 y(t) = x(t) \tag{1.12}$$

where a_0, a_1, and a_2 are constants. The frequency response of this system can be represented as shown in Figures 1.2(c) and (d), where

$$f_0 = \frac{1}{2\pi} \sqrt{\frac{a_0}{a_2}} \tag{1.13}$$

which is called the *natural frequency*, and

$$h = \frac{a_1}{2\sqrt{a_0 a_2}} \tag{1.14}$$

which is called the *damping coefficient*.

1.2.3.3 Time Constant, Response Time, Rise Time, and Settling Time

When the input of a system changes abruptly from one level to another, the behavior of the output can be characterized by some specific parameters according to the type of the system. Such parameters are defined as a unit step input in which the input is zero before a specific time and unity after that time

The *time constant* is defined in the first-order system. As shown in Figure 1.3(a), the response of a first-order system to a unit step input is a process approaching the final value exponentially, and the time constant τ is defined as the time required for the output to reach $1 - 1/e \approx 0.673$ of the final value and is given as a_1/a_0 for the system represented by Equation (1.11).

In the second-order system, the response to a unit step input varies with the damping coefficient, as shown in Figure 1.3(b). Some parameters are used to express how fast the system can follow the input. The *response time* is usually defined as the time to reach 90% of the final value, and the *rise time* is usually defined as the time that the output changes from 10 to 90% of the final value to a unit step input. The *settling time* is defined as the time required for the output to settle within a definite range near the final value; for example, the range is defined as ±5% of the final value to a unit step input.

1.3 DETERMINATION OF ABSOLUTE QUANTITY

Usually, measurements are performed to determine the absolute values of the object quantities in physical or chemical units, while only relative values are required in some situations. The measurement of absolute quantities requires calibration of the measurement system unless the measurement system has the standard in itself. The accuracy of the measurement is assessed by the amount of error, which is the deviation of the measured value from the true value. There are many sources of error, such as the use of unreliable standards, inadequate calibration procedures, noise contamination of the signal, poor static and dynamic characteristics of the measurement system, and unsuitable data processing. Errors are sometimes classified according to their nature, such as random, systematic, quantization, and dynamic errors.

1.3.1 STANDARD AND CALIBRATION

The fundamental units of physical and chemical quantities are defined referring to the standards so that the unit quantities are derived from the material properties. In the SI (Systéme International d'Unités), seven base units are defined, i.e., length in meters, mass in kilograms, time in seconds, electric current in amperes, thermodynamic temperature in Kelvins, amount of substance in moles, and luminous intensity in candela. All other units can be derived from the base units (*ISO Standard Handbook*[11]).

The base units are universal because they are defined by the properties of the atoms or molecules. For example, the thermodynamic temperature is defined as the fraction of 1/273.16 of the temperature

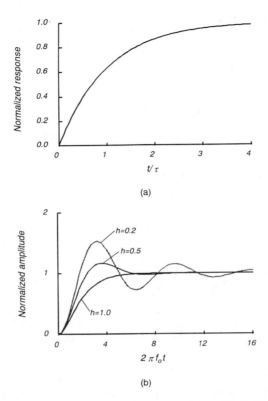

(a)

(b)

Figure 1.3 Responses to a step input in the first-order system **(a)** and that of the second-order system **(b)**.

of the triple point of water. However, other convenient material properties can also be used as the standard as long as the quantity determining the material is reproducible and stable enough. For example, the melting point of gallium, which is 29.771°C, is a convenient standard in the biomedical temperature range, and it was confirmed that the temperature is reproducible within a deviation of 0.002°C.[12]

To calibrate a measurement system, the object quantities produced by standards are applied to the input, and the relationship between the object quantity and the output is determined quantitatively through the whole measurement range. A measurement system which is stable enough and correctly calibrated by the standards can also be a standard instrument, so that other measurement systems can be calibrated compared with that standard instrument. For example, the standard thermometer having a crystal-resonator temperature transducer can be used to determine a temperature absolutely within a deviation of 0.01°C.

1.3.2 ACCURACY AND ERROR

The term *accuracy* describes how close the measured value is to the true value. The difference between the values is termed the *error*. While the errors in successive measurements may not be the same even though the object quantity remains unchanged, quantitative measure of error can be given by appropriate definition, such as the range within which 95% of measurements fall.

The error may depend on the level of the object quantity, especially when the measurement range includes from small to large values of the object quantity. The error may be small when the object quantity is small, while it may be large when the object quantity is large. In such a situation, the relative error, which is defined as the ratio of the error to the true value, can be a convenient figure to evaluate the performance of the measurement system.

When the measurement system is calibrated adequately, the error will be reduced to a limit determined by the reproducibility. Sometimes, the accuracy of the measurement system becomes poor due to the drift in a long period of time, even though the reproducibility has not changed so much. In such a situation, the accuracy can be recovered to that of the initial level by a recalibration. By repeated calibration procedures, the accuracy can be maintained within a definite range for a long time period.

1.3.3 TYPES OF ERROR

In actual measurement situations, there are always many sources of errors. The errors arising from different sources are classified into different types. The natures of typical types of error are discussed briefly in the following sections, but more detailed explanations can be found in standard texts.[13,14]

1.3.3.1 Random Error

The *random error* is a kind of error that appears unpredictably in repeated measurements. Random noise superimposing on the signal and short-term fluctuations of the measurement system may cause random errors. Deviations of the measured values are distributed on positive and negative sides of the true value with equal probabilities, and the mean of the deviations is zero. Thus, if the measurements are repeated while the object quantity is unchanged, the average of measured values approaches the true value.

The statistical property of random errors can be determined from the distribution of measurement values when the same object quantity is measured repeatedly. In practice, it is often assumed that random errors are distributed according to a normal distribution. That assumption is secured theoretically when the errors are considered to be the result of a large number of small uncorrelated component errors. According to the central limit theorem, it is expected that the sum of many uncorrelated random values forms a normal distribution.

For the normal distribution, the standard deviation of the mean of n measurements, which is the standard error, is reduced inversely proportional to the square root of n. Consequently, repeated measurement combined with signal averaging is an effective way to reduce random errors.

1.3.3.2 Systematic Error

The *systematic error* is the bias from the true value appearing equally in repeated measurements of the same object quantity. Systematic errors have different origins, such as direct current (d.c.) or very low frequency components of noise contaminating the signal, drift of the measurement system, inadequate calibration, uncorrected nonlinearity, or rounding down in digital data processing. When the same object quantity is measured by two different measurement systems, the difference between averages of measurement values of these two systems indicates the presence of a systematic error.

Systematic errors cannot be eliminated by repeated measurements and averaging. It is also difficult to identify all causes of systematic errors. One practical way of reducing systematic errors is recalibration of the measurement system under the measurement conditions over the required measurement range. As long as recalibrations are performed repeatedly at adequate intervals, the systematic errors can be significantly reduced for a long period of time even though the drift of the measurement system is significant.

Sometimes, when a systematic error is suspected due to the influence of environments such as climatic change, two identical measurement systems are used, so that the object and a standard are measured simultaneously. If the systematic errors in both outputs are the same, the error will be cancelled out in the difference between these two outputs, and then the object quantity can be measured correctly as the deviation from the standard value.

1.3.3.3 Quantization Error

The *quantization error* is an error caused by conversion of an analog value to a corresponding digital value, as in an analog-to-digital converter. The error is the difference between the original analog value and the converted digital value.

When an analog value is converted into a digital value, an ambiguity corresponding to 1 in the least-significant bit will remain. The maximum value of the digital output is limited by the digital word length or the number of bits of the analog-to-digital converter. If the word length is 8, the maximum value of the digital data is 255, and the ambiguity is about 0.4% of the full range of the digital value. However, the measurement range in an actual measurement situation does not always just fit the full range of the digital data format. Thus, the relative portion of the quantization error in the measurement range will become serious if the measurement range occupies only a small part of the range of the digital data.

1.3.3.4 Dynamic Error

The *dynamic error* is a kind of error occurring from imperfect dynamic response of the measurement system. It occurs when the object quantity varies so quickly that the output of the measurement system does not follow the change of the input. The difference between the output value and the object quantity is the dynamic error. When the dynamic error exists, the instantaneous output of the measurement system cannot be regarded as a function of the input of only that moment, but it depends on the past history of the input.

The possible dynamic error of a system is estimated from the response corresponding to a stepwise input. As shown in Section 1.2.3.3 and Figure 1.2, the response curve for a stepwise input can be determined by a few parameters in simple systems such as the first- or second-order linear systems, and the dynamic error for a stepwise input is the deviation of the response curve from the final steady-state value. In higher order systems, some parameters, such as the settling time, can still be defined in the response curve corresponding to a stepwise input. It can be expected that the dynamic error will remain in a definite range if a time longer than the settling time passes after abrupt change in the object quantity.

REFERENCES

1. Finkelstein, L., Theory and philosophy of measurement, in *Handbook of Measurement Science,* Vol. 1, Sydenham, P. H., Ed., John Wiley & Sons, Chichester, 1982, chap. 1.
2. Champeney, D. C., *Fourier Transforms and their Physical Applications,* Academic Press, London, 1973.
3. van der Ziel, A., *Noise in Measurements,* John Wiley & Sons, New York, 1976.
4. Musha, T., Takeuchi, H., and Inoue, T., 1/f fluctuations in the spontaneous spike discharge intervals of a giant snail neuron, *IEEE Trans. Biomed. Eng.,* BME-30, 194, 1983.
5. Morrison, R., *Noise and other Interfering Signals,* John Wiley & Sons, New York, 1992.
6. Tam, H. W. and Webster, J. G., Minimizing electrode motion artifact by skin abrasion, *IEEE Trans. Biomed. Eng.,* BME-24, 134, 1977.
7. Ödman, S., Potential and impedance variations following skin deformation, *Med. Biol. Eng. Comput.,* 19, 271, 1981.
8. Webster, J. G., Reducing motion artifacts and interference in biopotential recording, *IEEE Trans. Biomed. Eng.,* BME-31, 823, 1984.
9. Doebelin, E. O., *Measurement Systems: Application and Design,* 4th ed., McGraw-Hill, New York, 1990.
10. Palls-Areny, R. and Webster, J. G., *Sensors and Signal Conditioning,* John Wiley & Sons, New York, 1991.
11. *ISO Standard Handbook, Quantity and Units,* International Organization for Standards, Geneva, 1993.
12. Sostman, H. E., Melting point of gallium as a temperature calibration standard, *Rev. Sci. Instrum.,* 48, 127, 1977.
13. Hofman, D., Measurement errors, probability and information theory, in *Handbook of Measurement Science,* Vol. 1, Sydenham, P. H., Ed., John Wiley & Sons, Chichester, 1982, 24.
14. Finkelstein, L., Errors and uncertainty in measurements and instruments, in *System & Control Encyclopedia,* Singh, M. G., Ed., Pergamon Press, Oxford, 1987, 1553.

Pressure Measurement

2.1 OBJECT QUANTITIES

2.1.1 UNITS OF PRESSURE

Pressure is defined as the force exerted per unit area. In the SI system, the unit of pressure is Pa (pascal), which, by definition, is equal to one newton per square meter (N/m^2).

Physiological pressures have usually been expressed in millimeters of mercury (mmHg) or centimeters of water (cmH_2O). These units will be replaced by the SI unit pascal.

Because the pascal, Pa, is too small to use for physiological pressures, kPa ($=10^3 Pa$) is generally used. The conversions from mmHg and cmH_2O to kPa are

$$1 \text{ mmHg} = 133.322 \text{ Pa} = 0.133322 \text{ kPa}$$

$$1 \text{ cmH}_2\text{O} = 98.0665 \text{ Pa} = 0.098066 \text{ kPa.}$$

Physiological pressures are normally measured and expressed relative to the atmospheric pressure. Since the pressure is measured as a differential pressure in relation to the atmosphere, it is normally unnecessary to specify the absolute atmospheric pressure. However, measurements with certain types of transducers, such as an implantable transducer which measures absolute pressure, are affected by changes in atmospheric pressure. A change in atmospheric pressure can cause significant errors unless the pressure being measured is properly corrected by means of the actual atmospheric pressure.

Atmospheric pressure under standard conditions is 1 atm and is defined as

$$1 \text{ atm} = 101.325 \text{ kPa.}$$

Atmospheric pressure variation can be measured by a barometer. The standard scale of the barometer is mbar, and the conversion to Pa and mmHg is

$$1 \text{ mbar} = 0.1 \text{ kPa.}$$

Another unit of pressure which is widely used in engineering fields is kgf/cm^2, and the conversion to Pa is

$$1 \text{ kgf/cm}^2 = 98.0665 \text{ kPa.}$$

Torr is sometimes used and is equivalent to mmHg.

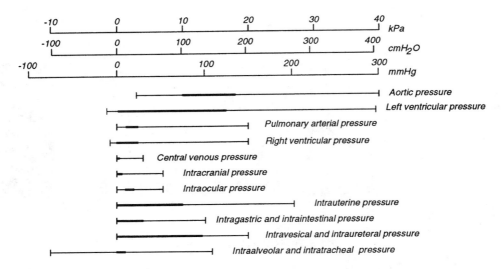

Figure 2.1 Variable ranges of pressures in the body cavities in physiological conditions (thick lines) and unphysiological conditions (thin lines).

2.1.2 REQUIREMENTS FOR PRESSURE MEASUREMENT
2.1.2.1 Physiological Pressure Ranges and Measurement Sites

Pressures in the human body are measured as a part of clinical examinations and for physiological studies. Figure 2.1 shows ranges of pressures in normal and abnormal situations.

Pressures in the cardiovascular system can be measured in many different ways. Arterial blood pressure is routinely measured in most patients and is accepted as an index of circulatory condition. The quantities normally measured are systolic and diastolic pressures. In the systolic phase, the aortic valve of the heart is open, and the arterial pressure consequently reflects the mechanical activity of the heart ventricle. On the other hand, in the diastolic phase, the aortic valve is closed, and then the time course of the arterial pressure reflects the movement of blood from the aorta to the peripheral vascular system. The arterial pulse pressure, which is defined as the difference between systolic and diastolic pressures, is also an important quantity relating to the characteristics of the heart and the arterial system.

The mean arterial pressure is the average arterial pressure during an entire cardiac cycle. It is usually used when the characteristics of the cardiovascular system as a whole are discussed. For example, the ratio of the mean arterial pressure and the cardiac output constitutes an approximate estimate of peripheral blood flow resistance. The mean arterial pressure can be estimated from the systolic and diastolic pressures. To calculate it exactly, it must be obtained from the whole time course of arterial pressure.

The left ventricular pressure reflects the pumping action of the ventricle. In particular, the slope of the ascending part of the ventricular pressure curve indicates the force developed by the ventricle at the initiation of a contraction and is usually denoted as dp/dt, which has been used for assessing the cardiovascular function.[1] In diastole, the left ventricular pressure in the normal heart is below 1 kPa (8 mmHg), and end diastolic pressure is an important quantity which represents ventricular filling just before an initiation of ejection.

Right ventricular and pulmonary arterial pressures are generated by the contraction of the right ventricle. In normal circulation, these pressures are lower than systemic arterial pressure, since the resistance of the pulmonary circulation is about one fourth that of the systemic circulation. Exceptionally high pulmonary arterial pressure, however, is sometimes observed in the patient with severe pulmonary disease due to pulmonary arterial stenosis or a ventricular septal defect. Pressure measurements via cardiac catheterization are necessary to diagnose these diseases.

Pulmonary wedge pressure is also commonly measured. It is the pressure observed when the catheter introduced into the pulmonary artery is wedged at a branch of the pulmonary artery. The pulmonary wedge pressure is usually in between the true capillary pressure and the pressure in the left atrium, and it is used as an estimation of left atrial pressure.[2]

Central venous pressure is the pressure close to the right atrium and is a sum of the pressures developed by venous elasticity and intrapleural pressure. The absolute pressure in the intrapleural space is normally

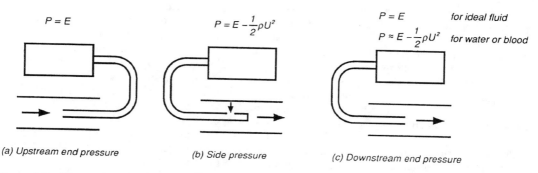

$P = E$

$P = E - \dfrac{1}{2}\rho U^2$

$P = E$ *for ideal fluid*

$P \approx E - \dfrac{1}{2}\rho U^2$ *for water or blood*

(a) Upstream end pressure   (b) Side pressure   (c) Downstream end pressure

Figure 2.2 Effects of kinetic energy in pressure measurements with catheters having different opening sites relative to flow direction.

below 1 kPa (10 cmH$_2$O). Intrapleural pressure is normally almost equal to atmospheric pressure, and the central venous pressure can be an index of the blood volume in the venous system and the elasticity of the veins. As long as total blood volume and venous elasticity are unchanged, venous pressure varies according to changes in cardiac function. Since central venous pressure rises when cardiac performance deteriorates, it is an important pressure to monitor in the patient with cardiac failure.[3]

In the larger vessels, where blood flow velocity is high, the pressure observed in the vessel is affected to a varying degree by the kinetic energy of the fluid, which is dependent on the direction of the catheter tip with respect to the direction of the blood stream. According to the Bernoulli's theorem, total energy per unit volume of fluid is shown as

$$E = P + \rho g h + \frac{1}{2}\rho U^2 = \text{const.} \tag{2.1}$$

in a steady flow of ideal fluid (incompressible with a viscosity of zero), where U is flow velocity, P is the static pressure, ρ is the density, g is acceleration of gravity, and h is height. The energy, E, is constant along a stream line. The first term represents static pressure, the second term the gravitational potential energy, and the third term the kinetic energy. As long as the height, h, is unchanged, a change in the kinetic energy causes a change in pressure.

The kinetic energy varies with the linear velocity of the fluid at the opening of the catheter. If $h = 0$ and if the opening of the catheter is directed upstream, as shown in Figure 2.2(a), flow velocity U at the opening is zero, and the pressure is thus equal to E. When the opening of the catheter is directed at a right angle to the stream, as shown in Figure 2.2(b), the observed pressure is usually called side pressure and is equal to $E - \rho U^2/2$ because the flow velocity at the opening is U. If the opening faces downstream, as shown in Figure 2.2(c), the pressure will be the same as (a) because the flow velocity at the opening will be zero, provided the fluid is ideal. However, in actual fluid such as blood or water, a vortex appears downstream from the catheter, and the flow velocity at the opening facing downstream is not zero. In this situation, the observed pressure is somewhat unstable, but a pressure near the side pressure is usually observed.

The effect of kinetic energy in the actual circulatory system is very different in different parts of the vascular tree. In the aorta, the maximum flow velocity is roughly 100 cm/s with a kinetic energy of 0.5 kPa (4 mmHg). If the systolic blood pressure is 16 kPa (120 mmHg), the contribution from kinetic energy is not more than 3%, the same amount as the maximum expected error. A larger error, however, can occur in the pulmonary·arterial pressure measurement. Since the maximum flow velocity is roughly 90 cm/s, the maximum contribution from kinetic energy will be 0.4 kPa (3 mmHg). The pulmonary arterial pressure is approximately 2.7 kPa (20 mmHg), and the contribution from kinetic energy can consequently be about 15% of the total pressure. If the blood flow increases, kinetic energy is increased in proportion to the square of the blood velocity, and the error will increase. In the clinical situation, however, the catheter introduced into the right atrium and pulmonary artery is facing downstream, and the observed pressure is thus considered to be close to the side pressure. In a central vein, the blood velocity is normally 30 cm/s or less, and the contribution of kinetic energy is not more than 0.05 kPa (0.35 mmHg).[4]

In addition to the cardiovascular system, there are many cavities, conduits, and spaces where pressures are measured. Intracranial pressure is considered as the pressure in the intracranial space, and it is usually measured as the pressure in the cerebrospinal space, ventricles of the brain, or intradural or extradural spaces. These pressures are below 1 kPa (8 mmHg) in normal subjects and are sometimes elevated in the patient with a brain tumor, edema, or hemorrhage. Continuous monitoring of intracranical pressure is consequently recommended for patients in whom such a condition is suspected.

The intraocular pressure is the inner pressure of the eye ball, and it is usually measured indirectly at the corneal surface. In the normal subject, intraocular pressure is in the range of 1.3 to 2.6 kPa (10 to 20 mmHg); in a patient with glaucoma, it is elevated above 3.3 kPa (25 mmHg).

The intrauterine pressure is the pressure of the amniotic fluid in the amniotic cavity. It is considered as the sum of pressures developed by the uterine contraction and intraperitoneal pressure. The elevation of the pressure caused by the uterine contraction is about 5 to 11 kPa (40 to 80 mmHg).

The intragastric and intraintestinal pressures are measured in the alimentary canal. These pressures are considered as the sum of the pressures developed by the peristaltic motions of the stomach or intestine and the pressure in the peritoneal cavity. The maximum pressure developed by the motion of the gastric organs is about 4 kPa (30 mmHg).

The urinary bladder pressure is caused by the fluid pressure in the bladder. It is measured to examine urinary reflex and contraction of the urinary bladder. The urinary reflex occurs at about 5 kPa (50 cmH$_2$O), and the urinary bladder pressure increases to 10 to 15 kPa (100 to 150 cmH$_2$O) during micturition. The urethra is always closed by the sphincter urethrae, except during micturition, and the pressure required to introduce fluid into the urethra is measured to examine the strength of contraction of the sphincter urethrae.

Intrapleural pressure and intratracheal pressure are measured to estimate lung compliance, airway resistance, and the mechanical activity of ventilation. Intrapleural pressure is difficult to define as a fluid pressure because there is no appreciable amount of fluid in the intrapleural space. Thus, intrapleural pressure is usually determined by balloon pressure when a balloon is placed in the esophagus. Mouth pressure is sometimes substituted for intratracheal pressure. The variation in intrapleural pressure is always below 0.5 kPa (5 cmH$_2$O) in normal respiration and can be varied from −10 to +15 kPa (−100 to +150 cmH$_2$O) during maximum voluntary ventilation.

2.1.2.2 Reference Point for Pressure Measurement

Most physiological pressure measurements are performed in order to study physiological functions such as heart and muscle activity. An observed pressure, however, is composed not only of the pressure produced by organ and tissue activity but also the pressure generated by gravitational force and atmospheric pressure. Thus, it is sometimes necessary to differentiate the physiological pressure component from the components of gravitational and atmospheric origin.

Atmospheric pressure is applied uniformly to the human body. Thus, the transducer which measures pressure relative to the atmosphere is not affected by a change in the atmospheric pressure. However, when a transducer which measures absolute pressures is used, variations in atmospheric pressure should be considered. In such cases, it is necessary either to calibrate frequently in regard to the atmospheric pressure or to make corrections in atmospheric pressure as measured by another transducer.

The effect of gravitational force is somewhat complicated. The circulatory system can be depicted as in Figure 2.3. If the pressure drop due to flow resistance and kinetic energy is negligible, the pressure difference between two points in a vessel will be equal to the difference in the gravitational potential, which is estimated as ρgh, where ρ is the density of fluid between these two points, h is the difference in altitude, and g is the acceleration of gravity.

Due to gravitational force, the pressure at a specific site may change when there is a change in posture. To avoid any ambiguity in this regard, it is postulated that most clinical pressure measurements are performed with the patient in a well-defined posture. However, even if the posture is the same, some ambiguity still remains due to the level at which the transducer is placed. The reference point at which the pressure is zero is used in determining the appropriate level to place the transducer.

There is a site in the cardiovascular system where the pressure remains almost constant regardless of posture. It has been shown that right atrial pressure is the most stable pressure in relation to posture changes. This characteristic is important in maintaining stable circulation while a person is moving around.[5,6]

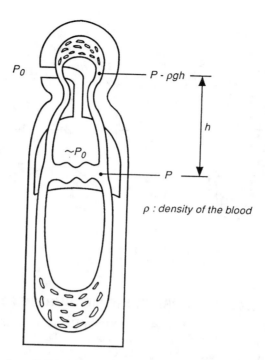

Figure 2.3 The effect of gravitational force on pressures at different sites in the circulatory system.

The central vein is located in the thorax close to the heart, and the pressure in the intrathoracic space is close to atmospheric pressure regardless of posture. This means that the external pressure applied to the central vein and the right atrium is stable, and the pumping action of the heart consequently remains stable. Cardiac function is also very sensitive to right atrial pressure.

As long as pressure is measured at the level of the right atrium, variation in the pressure caused by posture changes will be small. The blood pressure measurement ordinarily taken on the upper arm is quite adequate in this sense because a person's upper arm is usually kept almost at the same level as the right atrium. When it is necessary to measure a pressure at a height which is different from that of the right atrium, it may be helpful to make a correction for the difference in height so that consistent data can be obtained when different observations are made.

Thus, the right atrium is accepted as the reference point for pressure measurements. Figure 2.4 shows a convenient way to determine the reference point.[6] The reference level is located at the center of the fourth intercostal space and costal cartilage junction anteriorly and about 10 cm from the back, or half of the antero-posterior diameter of the chest on the longitudinal axis. The anterior chest wall to mid-left atrium distance assessed by echocardiography is also considered as an accurate determination of the reference level.[7]

When a fluid-filled catheter is used to transmit internal pressure in the body to an external transducer, the pressure at the tip of the catheter can be measured by placing the transducer at the same level as the tip. Practically, it is more convenient to place the transducer at the reference level so that the pressure at the reference level can be measured, regardless of the exact level of the tip of the catheter.

However, this is not precisely accurate, since the densities of the fluid in the catheter (for saline, the density is about 1.009 g/cm^3) and in the conduit between the measuring site and the reference point (for blood, the density is about 1.055 g/cm^3) are usually different, so that the contribution of gravity to the pressure, ρgh, is not the same.

For central venous pressure monitoring in patients whose body movements are not restricted, it is difficult to keep the transducer at the level of the reference point. To compensate for the shift of the reference point, a differential pressure measurement has been proposed in which an additional catheter filled with water is attached to the patient so that the meniscus in the catheter is close enough to the reference point so the difference in the pressures of the venous and reference catheters gives true venous pressure.[8,9]

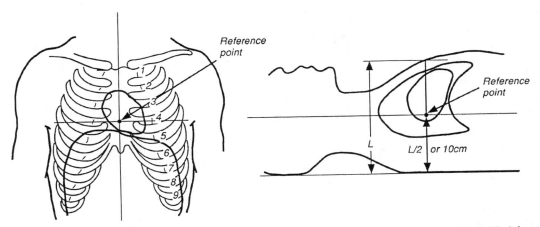

Figure 2.4 A convenient way to determine the reference point for pressure measurement. (Modified from Guyton, A. C. and Greganti, F. P., *Am. J. Physiol.*, 185, 137, 1956.)

2.2 DIRECT PRESSURE MEASUREMENT

2.2.1 CATHETERS AND THE DIAPHRAGM-TYPE PRESSURE TRANSDUCER

In order to measure a pressure inside the body cavity directly, a catheter filled with saline, with a pressure transducer connected to it, is commonly used. According to Pascal's principle, the pressure change at the tip of the catheter inside the body can be mediated to the transducer outside the body.

2.2.1.1 Catheters for Pressure Measurements

Catheters and needles of different sizes can be used for pressure measurements. A typical setup for cardiovascular pressure measurement is a flexible plastic catheter with a luer-lock connector at the proximal end, so that it can be connected to a stop cock or to other instruments. The size of the catheter is sometimes denoted in French scale (Fr or F, each unit being equivalent to a 0.33-mm outer diameter).

Catheters of different designs are available. Some have openings only at the tip, some have an open or closed tip with one or many side holes, and others have a double lumen with one opening at the tip and the other at some distance from the tip. A catheter with a balloon near the tip is used for pulmonary arterial measurement. The balloon is carried forward naturally by the blood flow, and this makes it possible to place the catheter at the pulmonary arteries.[10]

X-ray monitoring is sometimes necessary during catheter insertion. For this reason the catheter should be X-ray opaque. When the catheter is placed in a blood vessel, care should be taken to prevent blood coagulation, which not only disturbs pressure measurement but also can cause serious thromboembolism. Although catheters used for these purposes are specially made, no material is completely anticoagulant. Thus, a slow infusion of an anticoagulant agent in saline is recommended for long-term measurements. Typically, saline containing about 2000 units of heparin per liter can be continuously infused at a rate of about 3 to 6 ml per hour.

2.2.1.2 Diaphragm Displacement Transducer

Most pressure transducers for direct pressure measurements have an elastic diaphragm, and its displacement or strain is detected by a sensing element such as the strain gauge or a variable capacitance. Although the amount of deformation of the diaphragm due to the applied pressure is nonlinear, it can be regarded as linear when the diaphragm is thin and the deformation is small compared with the thickness of the diaphragm. In a circular flat diaphragm with clamped edges, displacement of the diaphragm at a distance r from the center is given as

$$z(r) = \frac{3\left(1 - \mu^2\right)\left(R^2 - r^2\right)^2 \Delta P}{16\, Et^3}$$

(2.2)

where μ is Poisson's ratio, R is diaphragm radius, t is diaphragm thickness, ΔP is pressure difference, and E is Young's modulus. The displacement is maximum at the center, which can be written as

$$z(0) = \frac{3\left(1-\mu^2\right)R^4 \Delta P}{16\, Et^3}.$$

(2.3)

Strain of the diaphragm in a radial component, ε_r, and a tangential component, ε_t, is expressed as

$$\varepsilon_r(r) = \frac{3\Delta P\left(1-\mu^2\right)}{8t^2 E}\left(R^2 - 3r^2\right)$$

(2.4)

$$\varepsilon_t(r) = \frac{3\Delta P\left(1-\mu^2\right)}{8t^2 E}\left(R^2 - r^2\right).$$

(2.5)

These strain components are equal at the center, i.e.,

$$\varepsilon_r(0) = \varepsilon_t(0) = \frac{3\Delta P\left(1-\mu^2\right)R^2}{8t^2 E}.$$

(2.6)

Figure 2.5 shows distributions of diaphragm displacement and two components of the strain.

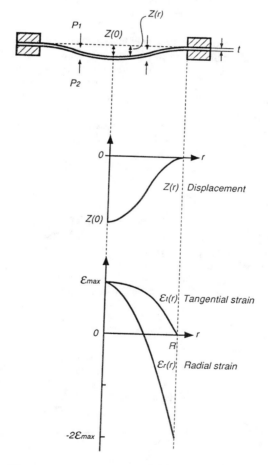

Figure 2.5 Deformation of a thin circular diaphragm with clamped edge; distribution of diaphragm displacement; $Z(r)$, and that of tangential and radial stress components, $\varepsilon_t(r)$, $\varepsilon_r(r)$.

Volume displacement, which is defined as the volume change caused by the deformation of the diaphragm, is given as

$$V = \frac{\pi(1-\mu^2)R^6 \Delta P}{16\,Et^3}.$$

(2.7)

In Equations (2.2) to (2.6), the dimension of ΔP and E cancel out each other, and all remaining variables — R, r, and t — have a dimension of length.

These relations provide the fundamental characteristics for a transducer design. The displacement and strains of the diaphragm for a given pressure depend on the geometry of the diaphragm. As seen in Equations (2.2) to (2.6), the sensitivity of the transducer is determined when the geometry and the components of the diaphragm material are given. However, if the thickness changes in proportion to the radius, Equation (2.6) does not change. The ratio of displacement at the center and radius, $Z(0)/R$, is also unchanged as long as R/t remains constant. In other words, for geometrically similar diaphragms of different sizes, the strains and the portion of displacement relative to the radius are unchanged, assuming that the material of the diaphragms is the same. Thus, pressure transducers of different sizes which have equal sensitivity can be constructed using the same material and a geometrically similar design. As long as the sensitivity is the same, a smaller diaphragm is advantageous, since the volume displacement is reduced in proportion to the third power of the radius if R/t is unchanged, as seen in Equation (2.7). Very small pressure transducers have been made by means of silicon micromachining technology. The lower limit of size is determined by the noise level due to the Brownian motion of molecules. This effect, however, is insignificant in physiological pressure ranges, even for a diaphragm with a diameter of 0.1 mm.[11]

Different principles can be used to detect displacement or strains of the diaphragm. Strain-gauge and capacitive methods are most commonly used in pressure transducers used for physiological measurements. In catheter tip pressure transducers, optical methods are also used.

The strain-gauge type has been widely used. This principle utilizes metal and semiconductor elements in which electrical resistance varies with strain. Although the relation between electrical resistance and strain is nonlinear, the relation can be regarded as linear when the strain is less than 0.5%. If the length and its change are L and ΔL, and the electrical resistance and its change are R and ΔR, their ratio, G, is defined by

$$G = \frac{\Delta R/R}{\Delta L/L}$$

(2.8)

and is constant and called the gauge factor. The gauge factor, G, for metals is about 2.0 and is caused only by the dimensional change, while semiconductors have larger gauge factors, between -100 to $+140$, in which resistance change due to the piezoresistive effect is added.

Figure 2.6 shows an example of the pressure transducer with wire strain gauges. The displacement of the diaphragm is transmitted to a platform to which wire strain gauges are connected, so that a displacement of the platform causes expansion in two gauges and compression in two other gauges.

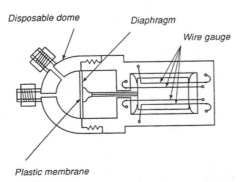

Figure 2.6 An example of the pressure transducer with wire strain gauges and a disposable dome.

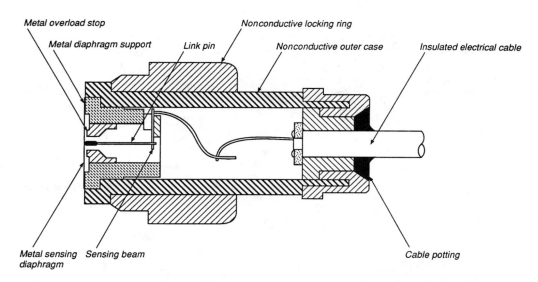

Metal overload stop   Nonconductive locking ring

Metal diaphragm support   Link pin   Nonconductive outer case   Insulated electrical cable

Metal sensing   Sensing beam   Cable potting
diaphragm

Figure 2.7 A pressure transducer using a beam-type silicon strain gauge (Spectramed P1OEZ; Oxnard, CA).

As mentioned earlier, the sensitivity of the diaphragm type of pressure transducer can be estimated from its geometry and the materials used in it. For example, if a steel diaphragm is used, Young's modulus is roughly 2×10^{11} N/m^2, and Poisson's ratio is 0.3; if the radius is 5 mm and the thickness of the diaphragm is 0.1 mm, then from Equation (2.3) the displacement at the center is estimated at approximately 0.007 mm for an applied pressure of 13 kPa (100 mmHg). When a metal strain-gauge 10 mm in length is used to detect displacement, the strain is about 0.07%, and if the gauge factor is 2, then the ratio of change in electrical resistance is 0.14%. A bridge with these gauges gives an output voltage of about 7 mV for 13 kPa (100 mmHg) at 5-V excitation. The volume displacement is also estimated at about 0.2 mm^3 for the pressure change of 13 kPa (100 mmHg) from Equation (2.7).

When a catheter and transducer are connected to the patient, all surfaces exposed to body fluids should be sterile. However, it is not convenient to have to sterilize the whole transducer assembly frequently. To avoid having to do so, a disposable dome is used in which a thin plastic membrane separates the diaphragm of the transducer from the body fluids inside the dome.

Figure 2.7 shows an example of a cardiovascular pressure transducer (Spectromed P1OEZ; Oxnard, CA) in which a beam-type semiconductor strain-gauge element is used to detect displacement of the metal diaphragm.

The semiconductor strain-gauge is also used in miniature clinical pressure transducers. The major advantage of this strain-gauge type is that the gauge pattern can be fabricated on the silicon substrate using ordinary integrated circuit processing technology. In addition, the elastic beam or diaphragm can be fabricated in the same silicon substrate. Using silicon micromachining technology, very small sensing elements can be realized. This technique also diminishes mechanical or thermal instabilities caused by the bonding of the strain-gauge on an elastic material.

The silicon diaphragm on which strain-gauges are fabricated is also used in clinical pressure transducers such as disposable transducers and catheter-tip transducers. Figure 2.8 shows an example of the disposable pressure transducer (Cobe Lab CDXIII, Inc.; Lakewood, CA) with cross-sections of the silicon diaphragm and the tube with the sensing element.[12]

To detect diaphragm displacement, a capacitive method can be used. If two plate electrodes are arranged in parallel and close enough to each other so that the effect of fringing electric fields is negligible, the capacitance is expressed as

$$C = \frac{\varepsilon A}{d} \qquad (2.9)$$

where ε is the dielectric constant of the medium ($8.85 \times 10{-}12$ F/m in air), A is the area, and d is the separation. If one electrode is attached to the diaphragm and the other is fixed to the housing as shown in Figure 2.9(a), the distance d varies according to the displacement of the diaphragm. Although the capacitance is a nonlinear function of d in Equation (2.9), the linear output can be obtained by a simple

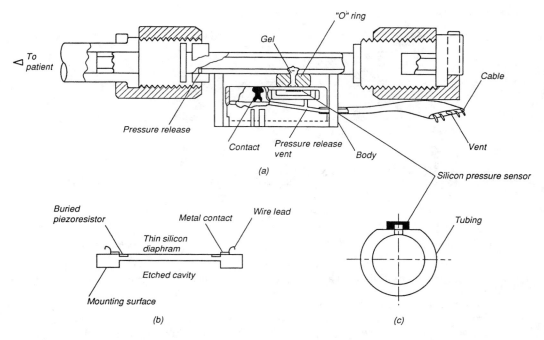

Figure 2.8 A disposable pressure transducer (Cobe CDXIII; Lakewood, CA). **(a)** The silicon-diaphragm type pressure sensor. **(b)** Its attachment to tubing **(c)**. (Modified from Spotts, E. L. and Frank, T. P., *J. Clin. Eng.*, 7, 197, 1982.)

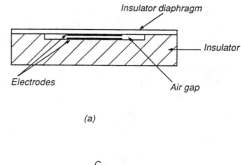

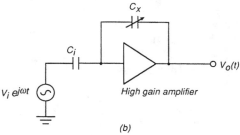

Figure 2.9 A capacitive pressure transducer **(a)**, and a circuit which provides linear output to the displacement **(b)**.

Table 2.1 Examples of Specifications of Commercial Pressure Transducers

	Transducer Type					
	Gould P23ID	Spectramed PIOEZ	Nor-Art AE840	Cobe CDXIII	Hewlett-Packard 1290A	Spectramed DX-300
Pressure range (mmHg)	−50 to +300	−50 to +300	−30 to +300	−50 to +300	−30 to +300	−50 to +300
Maximum overpressure (mmHg)	5000	10,000	4500	4000	6000	10,000
Volume displacement (mm³/100 mmHg)	.04	.04	.25	.001[a]	.2	
Sensing element	Wire gauge	Silicon beam diffused gauge	Silicon beam diffused gauge	Silicon diaphram diffused gauge	Quartz diaphragm capacitive type	Silicon beam diffused gauge
Remarks			Disc type	Disposable	Disc type	Disposable

[a] Without dome.

circuit as shown in Figure 2.9(b). When a sinusoidal excitation voltage, $V_i e^{j\omega t}$, is applied and the gain of the amplifier is large enough, the sum of currents at the input part of the amplifier should be zero and its potential becomes almost zero due to the negative feedback operation. Thus,

$$j\omega t C_i V_i e^{j\omega t} + j\omega C_x V_0(t) = 0 \tag{2.10}$$

where ω is angular frequency of excitation. Then, the output voltage, $V_0(t)$, is given as

$$V_0(t) = \frac{C_i}{C_x} V_i e^{j\omega t}. \tag{2.11}$$

Substituting Equation (2.9) so that $C_x = C$, output is expressed as

$$V_0(t) = \frac{d}{\varepsilon A} C_i V_i e^{j\omega t} \tag{2.12}$$

which is a linear function of the distance, d.

The capacitive method is utilized in a physiological pressure transducer (for example, Hewlett-Packard 1290A; Andover, MA). It has a quartz diaphragm which is mechanically stable and can tolerate being brushed with detergent.

In Table 2.1, the performances of some commercially available physiological transducers are compared. These transducers are designed for blood pressure measurement using hydraulic coupling. Detailed evaluations of commercially available physiological pressure transducers have been published by the Emergency Care Research Institute (ECRI).[13,14] These evaluations show that most commercial transducers, both reusable and disposable, are stable enough so that the drift during an 8-hr period is less than ±0.13 kPa (±1 mmHg). However, it has been reported that larger drifts are sometimes observed in reusable transducers when a disposable dome is attached and filled with saline, while the transducer without the dome is stable.[15] Thus, it is recommended that drift be evaluated under the same conditions as those under which the transducer is used during actual measurements.

2.2.2 DYNAMIC RESPONSE OF CATHETER-TRANSDUCER SYSTEMS

In order to observe pressure wave forms precisely using a fluid-filled catheter and pressure transducer, the catheter-transducer system should be prepared so the pressure wave at the tip of the catheter is transmitted to the transducer fast enough and without significant distortions of its waveform. Usually the time lag due to the pressure wave propagation along the catheter is insignificant, even though the transmission velocity of the pressure wave is not equal to the velocity of sound in free water, which is about 1599 m/s, but is sometimes reduced to around 400 m/s.[16] The time lag for a 1-m catheter will be only 2.5 ms or less.

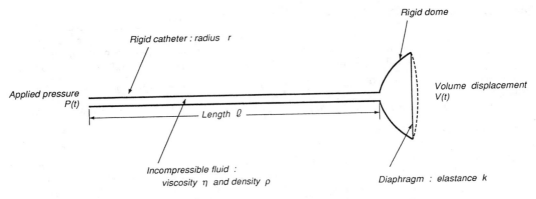

Figure 2.10 A simplified model of the catheter-transducer system.

Distortion of the pressure waveform, which is more serious, is commonly observed. What commonly occurs is that the elasticity of the transducer, the flow resistance of the catheter, and the inertance of the fluid in the system compose a resonant system. If the resonant frequency of the system is located in the frequency range of the measurement, the observed wave form will be distorted significantly. This situation can be analyzed theoretically and experimentally by a simple mechanical model. The model provides an adequate formulation for most of the characteristics of the catheter-transducer system, as discussed in the following sections.

2.2.2.1 Evaluation of Dynamic Response of the Catheter-Transducer System

The catheter-transducer system is considered to be approximated by a simplified model, as shown in Figure 2.10, in which a rigid tube is connected to a compliant chamber and is filled with an incompressible viscous fluid. The input to this system is the pressure at the tip of the catheter, $P(t)$, and the output is the volume displacement, $V(t)$, which can be detected by some sensing elements in the transducer.

Response characteristics of this simplified model have been analyzed in detail.[17,18] The kinetic equation of the fluid in the catheter for the fluid movement along the catheter x-axis is expressed as

$$F = m\frac{d^2x}{dt^2} + c\frac{dx}{dt} + kx \tag{2.13}$$

where F is external force, m is mass of the fluid in the catheter, c is flow resistance, and k is elastance. Parameters in the equation are given as

$$m = \rho\pi r^2 l \tag{2.14}$$

$$c = 8\eta\pi l \tag{2.15}$$

$$k = \pi^2 r^4 K \tag{2.16}$$

where r and l are radius and length of the catheter, ρ and η are density and viscosity of the fluid, and K is the elastance of the pressure transducer.

Then, as described in Section 1.2.3.2, natural frequency and damping factor are given as

$$f_0 = \frac{1}{2\pi}\sqrt{\frac{k}{m}} = \frac{r}{2}\sqrt{\frac{K}{\pi\rho l}} \tag{2.17}$$

$$h = \frac{c}{2\sqrt{mk}} = \frac{4\eta l}{r^3\sqrt{\pi\rho K}}. \tag{2.18}$$

When a sinusoidal pressure wave of constant amplitude is applied as input to this system, relative amplitude, γ, which is defined as the ratio of the output amplitude at the given frequency and that for steady pressure input, and the phase angle, θ, can be written as

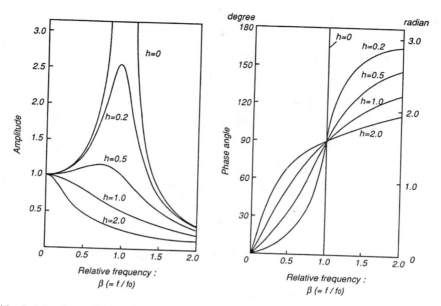

Figure 2.11 Relation between relative amplitude and relative frequency (left), and that between phase angle and relative frequency (right) in the simplified model shown in Figure 2.10.

$$\gamma = \frac{1}{\sqrt{\left(1-\beta^2\right)^2 + \left(2h\beta\right)^2}} \tag{2.19}$$

$$\theta = \tan^{-1}\frac{2h\beta}{1-\beta^2} \tag{2.20}$$

where β is relative frequency defined as

$$\beta = \frac{f}{f_0}. \tag{2.21}$$

Figure 2.11 shows the relation between relative amplitude and relative frequency (left), and that between phase angle and relative frequency (right). These results imply that as long as the simplified model shown in Figure 2.10 is applicable, the response of the catheter-transducer system is completely determined by two parameters, i.e., natural frequency, f_0, and damping factor, h.

From these results, the natural frequency, f_0, and the damping factor, h, can be calculated theoretically from Equations (2.17) and (2.18), and then frequency response and phase shift characteristics are determined, as long as the inner radius and length of the catheter and elastance of the transducer with the dome are known. For example, when a 1.25-m long 5-F standard wall catheter which has an internal diameter of about 0.66 mm is connected to a transducer with $K = 3.3 \times 10^4$ Pa m^{-3} (corresponding to a volume displacement of about 0.04 mm³/100 mmHg), the natural frequency, f_0, is estimated to be about 48 Hz and the damping factor, h, is estimated to be about 0.085. As another example, when a 0.5-m 8-F standard wall catheter which has an internal diameter of about 1.42 mm is connected to a transducer having the same elastance, f_0 is estimated to be about 163 Hz and h is estimated to be about 0.0054.

In the actual catheter-transducer system, f_0 and h can be determined from the step response, i.e., the output when a step change in pressure is applied. Figure 2.12 shows a simple method for doing this.[18,19] The catheter-transducer system is filled with water and connected to the bulb of a sphygmomanometer and a rubber balloon. After the balloon is inflated, it is punctured by a flame. A sudden drop in pressure at the tip of the catheter will occur. If the damping factor, h, is smaller than unity, the output of the

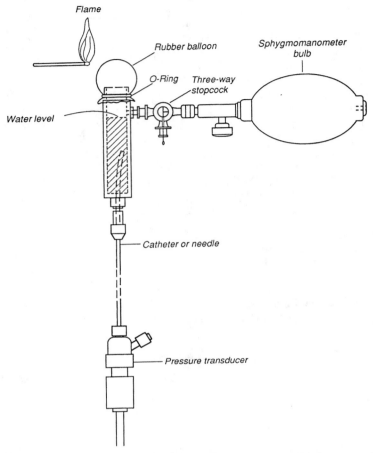

Figure 2.12 A simple measurement method of the step response of the catheter-transducer system. (Modified from Yang, S. S. et al., in *Cardiac Catheterization Data to Hemodynamic Parameters,* 2nd ed., F. A. Davis, Philadelphia, 1978.)

system shows a damping oscillation, as shown in Figure 2.13. By measuring amplitudes of successive peaks, the damping factor, h, is calculated as

$$h = \sqrt{\left(\ln \frac{P_{i+1}}{P_i}\right)^2 \Big/ \left[\pi^2 + \left(\ln \frac{P_{i+1}}{P_i}\right)^2\right]} \tag{2.22}$$

and the natural frequency, f_0, is calculated as

$$f_0 = \sqrt{1-h^2}\Big/T \tag{2.23}$$

where T is the time interval between successive peaks.

The frequency response of the catheter-transducer system can also be determined by using a sinusoidal pressure generator.[20,21] By scanning the driving frequency, the frequency response of the system can be determined directly from the ratio of driving and observed pressure amplitudes. Although this method is not as simple as the step response measurement, it is sometimes required nevertheless, as direct measurements can be made even though the system is nonlinear, while the step response analysis is only applicable in linear systems.

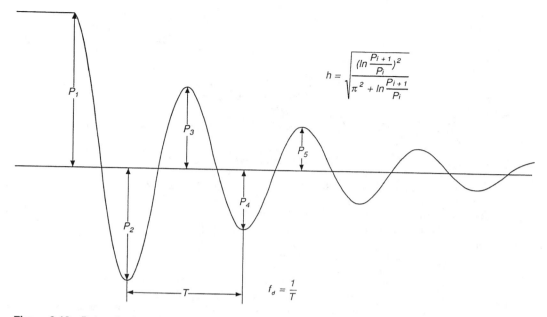

$$h = \sqrt{\frac{(ln\frac{P_{i+1}}{P_i})^2}{\pi^2 + ln\frac{P_{i+1}}{P_i}}}$$

$$f_d = \frac{1}{T}$$

Figure 2.13 Determination of the damping factor, *h*, and the resonant frequency, f_d, from a step response.

As long as these measurements are performed carefully, the natural frequency and the damping factor of an actual catheter-transducer system are always close to the theoretical estimations. However, an unreasonably low natural frequency can be observed occasionally. In such cases, it is likely that air bubbles have been trapped in the transducer dome or in the lumen of the catheter.[22] Air is a compressible gas, and it reduces the elastance, *K*, of the system and causes a decrease in f_0, as seen in Equation (2.17). Air bubbles in a transparent dome can be found via visual inspection, while air remaining in the catheter is more difficult to detect. There are some recommended procedures for displacing air in the catheter system, such as initially filling the catheter system with boiled saline or 50% alcohol in water, or flushing initially with carbon dioxide for a few minutes.[19,23,24]

2.2.2.2 Improvement of Dynamic Response

The catheter-transducer system should have a flat frequency response over the range in which signal frequency components are contained. This requirement is satisfied if the natural frequency is high enough compared to the highest frequency component in the signal and the damping is adequate, i.e., *h* = 0.6–0.7. In arterial pressure measurements, the pressure pulse contains at least up to 6th, but some times up to 20th, significant harmonics of the fundamental frequency, that is, the heart cycle.[25] If the maximum heart rate is 200 beats per minute, the required frequency range will be at least up to 20 and sometimes up to 66 Hz.

It is not difficult to obtain a natural frequency in the actual catheter-transducer systems that is higher than the frequency postulated above. However, when the usual cardiovascular catheter is connected directly to the transducer, the damping coefficient is always far below 0.6, and sharp resonance consequently appears. There are various methods available for improving this situation, such as using a (1) catheter of adequate size, (2) mechanical damper, (3) filter or electronic compensation circuit, or (4) numerical calculation by a computer.

An adequate combination of catheter size and transducer elastance can be determined using the previously mentioned theory. However, to obtain an appropriate f_0 and *h*, a very fine, short catheter should be used. For example, using a 5-cm long catheter with an internal diameter of 0.2 mm with a transducer having an elastance of 3.3×10^{14} Pa m^{-3} (volume displacement of 0.04 mm^3/100 mmHg), $f_0 = 72$ Hz and *h* = 0.6 will be obtained at 37°C. This size is impractical for most cardiovascular measurements, while it is appropriate for taking measurements via direct puncture of the cavity or organ with a fine needle.

The mechanical damper is commonly used to obtain adequate damping. It is essentially a variable flow resistance which is inserted between the catheter and the transducer. As long as the diameter of the

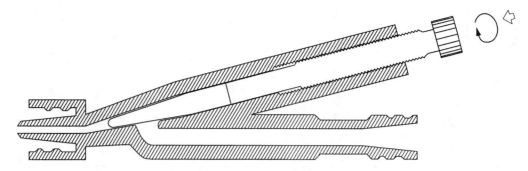

Figure 2.14   An example of a mechanical damping device.

catheter and the elastance of the transducer are unchanged, f_0 and K are unchanged, as seen in Equations (2.16) and (2.17). Thus, $h = c/c_0$ can be adjusted by c, which can be adjusted by the flow resistance in the catheter.

To adjust the flow resistance without changing the inertial component of the fluid in the catheter, a screw clamp can be used. This is attached to a small elastic tube inserted between the catheter and the transducer.[26] More convenient disposable devices are available commercially. Figure 2.14 shows an example. In this device, flow resistance can be adjusted by a needle valve.

Electronic filters or compensating circuits are also used to improve the dynamic response. The simplest method involves the use of the low-pass filter. Even if a flat frequency response up to the natural frequency cannot be realized, it is still acceptable provided the natural frequency is high enough. For example, a natural frequency above 100 Hz can be obtained with an 8-F catheter with a transducer with higher elastance. By using a low-pass filter with a cutoff frequency of about 20 to 22 Hz, the maximum rate of the left ventricular pressure buildup (max. dp/dt) can be estimated within an error of 10%.[27]

Further improvement of response characteristics can be obtained by using a resonance circuit. It has been shown that an almost flat frequency response up to the natural frequency can be obtained with typical catheter-transducer systems by using a compensating amplifier which has an adjustable resonance circuit.[28]

Computer analysis can also be used to compensate for the catheter-transducer characteristics. For example, Fourier analysis is applicable for almost periodic wave forms, such as those of intracardiac or arterial pressure. As long as the system can be regarded as linear, its response can be determined by the power amplitude and phase shift to a sinusoidal input in the whole frequency range of the signal. On the other hand, when a pressure wave form of one cardiac cycle is given, power amplitude and phase shift for every harmonic frequency component can be calculated using Fourier analysis. By correcting every harmonic component by the response characteristics of the system, original input wave form consequently will be reconstructed by inverse Fourier transformation.

This procedure was applied to left ventricular pressure measurement in dogs using an ordinary catheter-transducer system without a mechanical damper and was compared with intracardiac measurement by a catheter-tip transducer. The results showed that the reconstructed pressure wave forms were close to those obtained by intracardiac measurement, and when harmonic components up to the 15 to 24th harmonics were used in computation, maximum dp/dt estimated from the reconstructed wave forms did not differ more than 5% from original wave forms.[29] Figure 2.15 shows an example of left ventricular pressure and its time-derivative wave forms obtained by an intracardiac measurement, by using a catheter-transducer system, and by a reconstruction using up to 20 harmonic components.

2.2.3 CATHETER-TIP PRESSURE TRANSDUCER

The catheter-tip pressure transducer, which has been developed primarily for accurate measurement of pressure wave forms in the cardiovascular system, has a pressure-sensing element at the tip of a catheter. Although an adequately damped catheter-transducer system is acceptable for most clinical cardiovascular pressure measurements, the catheter-tip pressure transducer has many advantages. It has no time delay, it has a flat frequency response up to several kHz, frequent or continuous saline injection is not necessary in order to avoid blood clotting in the fluid-filled catheter, and it is less affected by mechanical motion

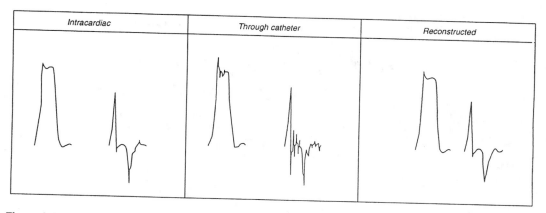

Intracardiac	Through catheter	Reconstructed

Figure 2.15 An example of left ventricular pressure and its time-derivative wave forms, those obtained through a catheter-transducer system, and wave forms reconstructed using harmonic components. (From Futamura, Y. et al., *Jpn. J. Med. Electr. Biol. Eng.*, 4, 214, 1975. With permission.)

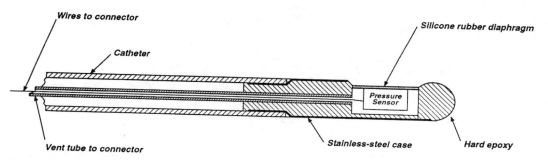

Figure 2.16 A catheter-tip pressure transducer with a beam-type silicone strain-gauge. (Courtesy of Millar Instruments, Inc.; Houston, TX).

of the catheter. The catheter-tip pressure transducer, however, also has some disadvantages, such as the inconvenience of recalibration and the fact that it is fragile and expensive.

Many different principles can be used in detecting pressure at the tip of a catheter. Examples are semiconductor strain gauges and capacitive and optical methods. In earlier attempts, the metal wire strain-gauge, the bulk silicone gauge, and the differential transformer were used.[30,31]

Progress in integrated circuit technology has made possible the fabrication of very small strain gauge elements on the silicone tip. Micromachining can be used to fabricate small, thin diaphragm beams or cantilevers of different shapes which have been applied to catheter-tip pressure transducers.

Catheter-tip pressure transducers with beam-type silicon strain gauges which detect the side pressure via a silicone-rubber diaphragm have been used extensively.[32] Figure 2.16 shows a cross-sectional view of the commercial model of the transducer (Mikro-Tip® Catheter Pressure Transducer, Millar; Houston, TX). The catheter has a vent tube which connects the rear side of the diaphragm to the atmosphere so that it can measure pressure relative to the atmospheric pressure. The catheters are manufactured in different sizes from 3-F to 8-F (o.d. 1.0 to 2.67 mm), the nominal pressure range is from −6.5 to +40 kPa (−50 to +300 mmHg), and the resonance frequency is from 35 to 50 kHz.

The silicone diaphragm with diffused strain gauges has also been used in catheter-tip transducers for side pressure or end pressure measurements.[33-35] Figure 2.17(a) shows a cross-section of a silicone diaphragm, and (b) shows the catheter tip on the end of which the diaphragm is mounted.[33]

A different type of strain-gauge, based on the transverse piezoresistive effect, is also considered applicable to the catheter-tip transducer. The transverse piezoresistive effect is a phenomenon in which a strain causes electrical potential perpendicular to the direction of the applied current.[36] As an example, a rectangular silicone diaphragm in the (001) plane was fabricated with a resistor strip directed at a 45° angle to one side.[37] The current is applied along the resistor strip, and the output voltage is detected

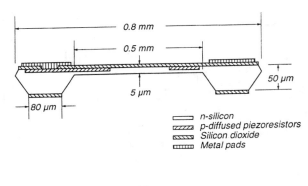

(a)

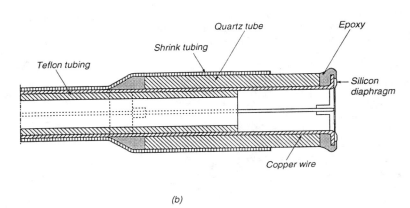

(b)

Figure 2.17 Cross-section of a silicone diaphragm pressure sensor **(a)**, and a catheter-tip transducer having the diaphragm in its end **(b)**. (From Samaun, T., Wise, K. D., and Angell, J. B., *IEEE Trans. Biomed. Eng.*, BME-20, 101, 1973. With permission.)

from the electrode pair arranged on both sides of the strip. This method is considered advantageous because the output is always zero at zero strain, and thus the zero drift is small.

Use of capacitive type pressure sensors for the catheter-tip transducers has been reported by Clark and Wise[38] and Chau and Wise.[39,40] A catheter size as small as 0.5-mm o.d. was used. The construction of this transducer is shown in Figure 2.18. The silicone diaphragm with its rim is fabricated using precision silicon micromachining technology and is bonded with an electrostatic seal to a glass support. A diaphragm size of $290 \times 550 \times 2$ μm, a capacitor plate separation of 2.2 μm, and pressure accuracy of 0.13 kPa (1 mmHg) were attained.[40] In the ultra-miniature pressure transducer, it can be shown theoretically that, to obtain higher sensitivity, the capacitive type is better than the piezoresistive type, while to obtain a wider pressure range, the piezoresistive type is to be preferred.[39]

Fiberoptic catheters for pressure measurement have been studied by many investigators.[41-45] Optical fibers transmit light through so-called cladded fibers made by transparent material such as glass or plastic in which fibers with a core with a high refractive index are surrounded by a shell with a low refractive index. The light transmits by total reflection at the boundary of these two media.

In fiberoptic catheters, light from a source is transmitted to the tip through the optical fiber. The reflected light from a reflector, which moves in relation to the applied pressure, enters the fiber and the intensity of the reflected light is measured by a photo detector. There are two different types: one measures the end-pressure, while the other measures the side pressure.

An example of construction of the end-pressure type of fiberoptic catheter is shown in Figure 2.19. The ends of the illuminating and detecting fibers are in a common plane, and the reflecting diaphragm is placed near the fiber end. If the distance between the fiber end and the reflector, d, is smaller than the separation between the center of the illuminating fiber and that of the detector fiber, a, reflected light intensity increases as a increases; if $d \ll a$, then the reflected light intensity is proportional to $(d/a)^{3/2}$.

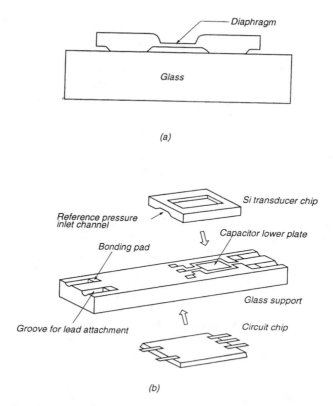

(a)

(b)

Figure 2.18 Construction of a capacitive-type pressure transducer. **(a)** Silicon chip with a diaphragm bonded to a glass support using electrostatic seal. **(b)** Assembly for the catheter-tip transducer. (From Chau, H. L. and Wise, K. D., *Transducers '87*, The Institute of Electrical Engineers of Japan, Tokyo, 1987. With permission.)

On the other hand, if d is larger than a, the reflected light intensity decreases as a increases; if $d \gg a$ then the reflected light intensity is proportional to d^{-2}, as shown in Figure 2.20.[43]

In the actual design, a catheter of 1.5-mm o.d. with a diaphragm 6 μm thick was fabricated. It could measure in a pressure range of −6.7 to +27 kPa (−50 to +200 mmHg) with nonlinearity of less than 2.5%. A commercial catheter of similar construction has been produced (Model 110/140, Camino Laboratories; San Diego, CA). The size of the catheter is 4-F (1.3-mm o.d.), and the pressure range is −6.3 to 31.5 kPa (−10 to 250 mmHg). A thermodilution catheter with a fiberoptic pressure transducer is also available. Its size is 8-F and it is designed for monitoring pulmonary wedge pressure and cardiac output using the thermo-dilution technique (see Section 3.3.3).

Fiberoptic catheters for side pressure measurement have also been produced.[44-45] A cross-section of an example of this type is shown in Figure 2.21.[45] This catheter consists of a single plastic fiber and a cantilever glass-beam directed to a reflecting corner. Illuminating and reflecting lights are separated by a beam splitter. Using micro-machining technology, the transducer tip has been fabricated small enough to be assembled in a stainless tube of 0.45 mm o.d., and the size of the catheter is 0.5 mm o.d.

Compared with other catheter-tip pressure transducers, the fiberoptic catheter has the advantages of safety in terms of current leakage and being unaffected by electromagnetic interference. In addition, the structure of the optical fiber system is simpler and may become inexpensive enough to be disposable.

2.2.4 IMPLANTABLE PRESSURE TRANSDUCERS AND PRESSURE TELEMETERING CAPSULES

For long-term monitoring of pressure in the body cavities, it is preferable to use a pressure transducer that is separated from external instruments so that it can either be surgically implanted (as in the case of the intracranial pressure measurements) or swallowed (for pressure measurements in the gastrointestinal tract). In these applications, absolute pressure transducers are preferable, although relative pressure transducers with vent tubes or snorkels were used in some earlier attempts at intracranial pressure

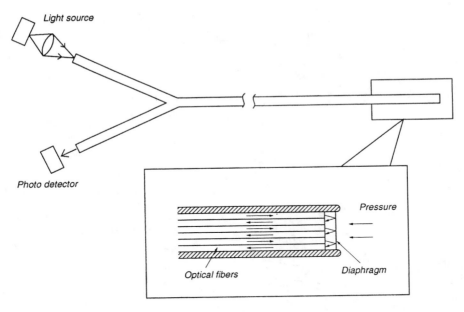

Figure 2.19   A fiberoptic end-pressure transducer.

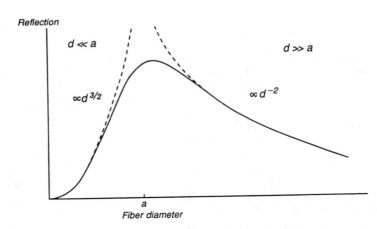

Figure 2.20   Variation in the reflected light intensity when the distance between the fiber-end surface and the reflecting diaphragm varies. (Modified from Lindström, L. H., *IEEE Trans. Biomed. Eng.*, BME-17, 207, 1970.)

monitoring. When a radio transmitter and an absolute pressure transducer are combined, the transducer unit can be isolated and placed in the body without any connections.

2.2.4.1   Implantable Pressure Transducers

Many kinds of implantable pressure transducers have been developed, primarily for intracranial and cardiovascular pressure measurements. In some earlier attempts at intracranial pressure measurement using implantable transducers, displacement of the diaphragm was detected by strain-gauges[46,47] or capacitance changes.[48] In these transducers, the rear side of the diaphragm was connected to the atmosphere via a vent tube. Figures 2.22(a) and (b) are constructions of such transducers. Both were designed to be implanted in a hole bored in the skull. The transducer shown in Figure 2.22(a) has a diaphragm with strain-gauges and an outer chamber made of thin plastic film. By inflating the outer chamber and also applying the same pressure to the rear side of the diaphragm, the baseline reading can be checked *in vivo*.[47] In the transducer shown in Figure 2.22(b), pressure is detected by a capacitive

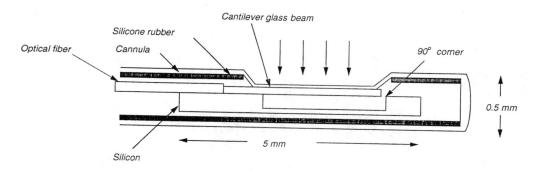

Figure 2.21 A fiberoptic side-pressure transducer. (From Tenerz, L. and Hök, B. H., *Transducers '87,* The Institute of Electrical Engineers of Japan, Tokyo, 1987, 312. With permission.)

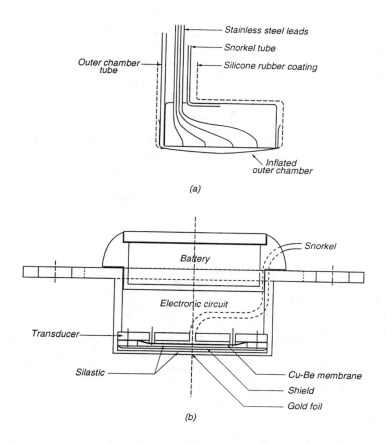

Figure 2.22 Implantable intracranial pressure transducers. **(a)** Strain-gauge-type transducer with an outer chamber for *in vivo* calibration. **(b)** Capacitive-type transducer with a transmitter for telemetry. (Part (a) from Eversden, I. D., *Med. Biol. Eng.,* 8, 159, 1970. Part (b) from Brock, M. and Diefenthäler, K., in *Intracranial Pressure,* Brock, M. and Dietz, H., Eds., Springer-Verlag, Berlin, 1972, 21. With permission.)

change, and it has a transmitter and battery for telemetering. The capacitance chamber is designed so that displacement of the diaphragm is limited by a contact to the opposite wall. By applying negative pressure through the vent tube just until the limit of the displacement is reached, the output can be calibrated *in vivo*.[48]

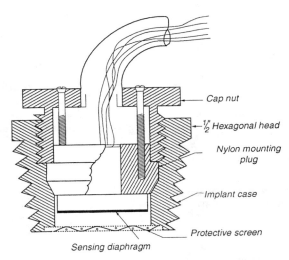

Cap nut

$\frac{1''}{2}$ Hexagonal head

Nylon mounting plug

Implant case

Protective screen

Sensing diaphragm

Figure 2.23 An implantable intracranial pressure transducer measuring absolute pressure. (From Tindall, G. T. et al., in *Intracranial Pressure*, Brock, M. and Dietz, H., Eds., Springer-Verlag, Berlin, 1972, 9. With permission.)

Figure 2.23 shows an example of the intracranial pressure transducer which measures absolute pressure.[49] A disc-shape absolute pressure transducer is assembled in a stainless-steel housing which fits to the drilled hole in the skull. A similar transducer including a telemetry transmitter has also been reported.[50]

Some attempts at passive telemetry of intracranial pressure have been reported. In these systems the implanted part of the system has a resonant circuit, and the resonant frequency, which varies with the pressure, is detected from outside the body. Figure 2.24(a) shows an example in which the inductance of the coil is altered by the displacement of a ferrite core connected to the diaphragm.[51] Figure 2.24(b) shows an example in which capacitance is altered by a displacement of the nitrogen-filled bellows.[52]

By using micromachining and integrated circuit technology, very small and highly stable pressure transducers have been developed for implantable devices. Most of them are designed for absolute pressure measurement, so that applied pressure is measured against a pressure in the reference chamber, which is usually a vacuum.

Figure 2.25(a) shows an example of a piezoresistive absolute pressure transducer designed for biomedical applications.[53] In this transducer, a silicone diaphragm with diffused gauges is fabricated in a silicon substrate, and the rear side of the diaphragm is capped by another silicon substrate so as to form a reference chamber. To connect two silicone substrates, an eutectic seal with an alloy of 78% gold and 22% tin is used. The size of the transducer is $1.25 \times 3.75 \times 0.125$ mm, and the square-shaped diaphragm is 0.8×2.3 mm in size with a thickness of between 16 and 24 µm. The ultimate stability for temperature changes is about 0.13 kPa/K (1 mmHg/K), and a long-term stability of 0.9 kPa/month (6 mmHg/month) can be attained. To form the reference chamber, electrostatic sealing between silicone and glass is used. The electrostatic sealing is formed using a combination of heat and electrostatic potential. The silicone and glass are placed in contact, and a voltage of 400 to 500 V is applied during heating to about 450°C.[54] In the electrostatically sealed absolute pressure transducer of a configuration similar to Figure 2.25(a), very small long-term drift of less than 0.13 kPa/month (1 mmHg/month) was attained.[53]

Capacitive pressure transducers are also fabricated by silicon micromachining technology. Figure 2.25(b) shows an example in which silicon direct bonding was used to form the reference chamber.[55] The bonding is formed by heating in an oxygen atmosphere to 1100°C. The oxygen in the reference chamber is consumed to oxidize the silicon, and a vacuum cavity is thus formed.

These silicon-to-glass or silicon-to-silicon bonding techniques are useful not only for forming the reference chamber, but also for encapsulation of the electronic circuit which is fabricated on the same silicon chip. An example of such a structure is shown in Figure 2.26.[54] With this structure, the electronic circuit can be protected from penetration of the body fluid when it is used in the implantable device. This structure has been employed in implantable pressure transducers.[55-57] Integration of the detecting circuit on the same chip of the capacitor will be advantageous in reducing the effect of stray capacitance components. To convert capacitance change into an electrical signal, a reference capacitor C_o is sometimes used in addition to the pressure-sensitive capacitor C_x, so that C_x-C_o is pressure sensitive. Figure 2.27(a)

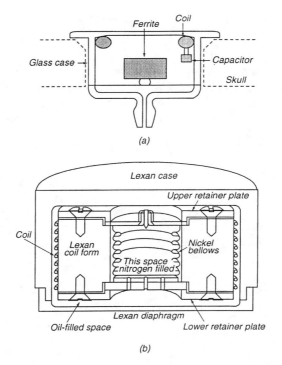

(a)

(b)

Figure 2.24 Passive telemetering intracranial pressure transducers. **(a)** Using inductance change. **(b)** Using capacitance change. (Part (a) from Hill, D. G. and Allen, K. L., *Med. Biol. Eng. Comput.*, 15, 666, 1977. Part (b) from Viernstein, L. J., *Johns Hopkins API Tech. Dig.*, 1, 135, 1980. With permission.)

shows an example of the formation of reference capacitor C_o on the same chip as C_x. The ring structure is shown in Figure 2.27(b), in which the center area is used for C_x and the periphery consists of C_o. Figure 2.27(c) shows a pressure transducer in which two ring structure capacitors and a signal-detecting circuit are formed on a rectangular silicon chip 1.5 × 3.8 mm in size and bonded to the glass plate. This transducer has an absolute stability of 0.04 kPa/month (0.3 mmHg/month).[56]

In the application to clinical or experimental pressure measurements, these transducers are used either by being inserted directly into the body cavity where pressure is to be measured or by being connected to a fluid-filled tube which communicates the pressure being measured.

Figure 2.28 shows an example of the transducer used in the intracranial pressure telemetry system.[58] In this transducer, the silicon diaphragm is electrostatically bonded to a Pyrex glass tube, and the pressure being measured is applied from inside the tube. The tube is shielded in a titanium package, so that the outside of the tube is a reference chamber for absolute pressure measurement. To use this transducer for intracranial pressure measurements, a fluid-filled catheter is connected to the lateral cerebral ventricle. The radio transmitter is provided for the intracranial pressure telemetry. In animal experiments, baseline drift was investigated during 180 days of implantation, and an absolute stability of 0.13 kPa/month (1 mmHg/month) was attained.[58]

2.2.4.2 Pressure Telemetering Capsules

The swallowable capsule for gastrointestinal pressure telemetry has been studied by many investigators.[59-65] Most of the capsules consist of a pressure transducer, amplifier, radio transmitter, and battery. Passive telemetry, in which the electric power is supplied from outside the body by electromagnetic induction, has also been attempted.[62] Sizes of pressure measurement capsules were 0.7 to 10 mm in diameter, and 19 to 30 mm in length. Construction of two examples of pressure capsules are shown in Figure 2.29. Pressure is detected by the change in inductance in a resonant circuit, and thus it modulates the oscillation frequency. These capsules can measure intestinal pressure from 0 to 200 kPa (0 to 200 cmH$_2$O) for up to about 5 days.

The major problem in the clinical use of the capsule for gastrointestinal measurements is the difficulty in recovering it for repeated use. This problem will be solved when an inexpensive single-use capsule

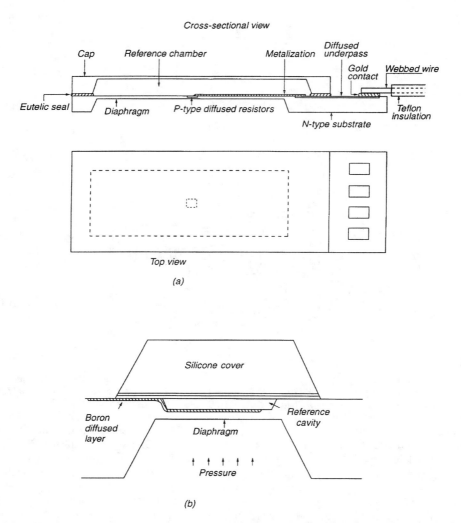

Figure 2.25   Absolute pressure transducers for implantable devices fabricated by integrated circuit technology. **(a)** Piezoresistive type. **(b)** Capacitive type. (Part (a) from Ko, W. H. et al., *IEEE Trans. Electron Dev.,* ED-26, 1896, 1979. Part (b) from Shoji, S. et al., *Transducers '87,* The Institute of Electrical Engineers of Japan, Tokyo, 1987, 305. With permission.)

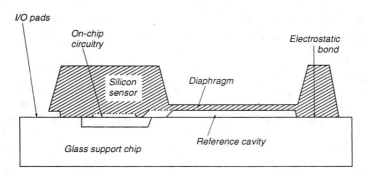

Figure 2.26   A capacitive-type absolute pressure transducer with the electronic circuit encapsulated in a closed chamber using silicon-to-glass bonding technique. (From Wise, K. D., in *Implantable Sensor for Closed-Loop Prosthetic Systems,* Ko, W. H., Ed., Futura, Mount Kisco, NY, 1985, 3. With permission.)

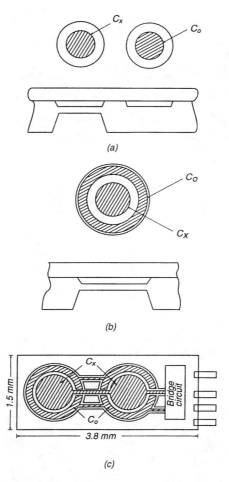

Figure 2.27 Constructions of capacitive-type pressure transducers having a pressure-sensitive capacitor, C_x, and reference capacitor, C_o. **(a)** Two-circular structure, **(b)** ring structure, **(c)** two ring structures with a bridge circuit formed on a rectangular silicon chip. (Modified from Ko, W. H. et al., *IEEE Trans. Electron Dev.*, ED-29, 48, 1982.)

becomes available. The pressure measurement capsule has also been used for uterine contraction and fetal heart-sound measurements by introducing it into the uterine cavity during labor.[63]

2.2.5 PRESSURE MEASUREMENTS IN SMALL VESSELS

In microvascular studies, pressure measurements in small vessels are required. To measure intravascular pressure directly, the small bevelled-glass micropipet is inserted into the vessel by direct puncture under microscopic view, and the pressure transducer is connected to it. In principle, the pipet-transducer system can be considered similar to the catheter-transducer system discussed in Section 2.2.2.1. However, the natural frequency will become lower when the internal radius of the pipet is reduced, as long as the elastance of the transducer is unchanged.

To improve dynamic performance of the pipet-transducer system, the use of a highly rigid transducer is preferable. More sophisticated methods have also been used, in which a servo-control technique is applied to compensate for the volume displacement.

2.2.5.1 Highly Rigid Transducer System

To obtain a natural frequency in an acceptable range for micropipet systems, the transducer should be highly rigid. For this purpose different transducers, both specially designed and commercial ones, have been examined. In an earlier attempt, a capacitive transducer was designed to measure internal pressure

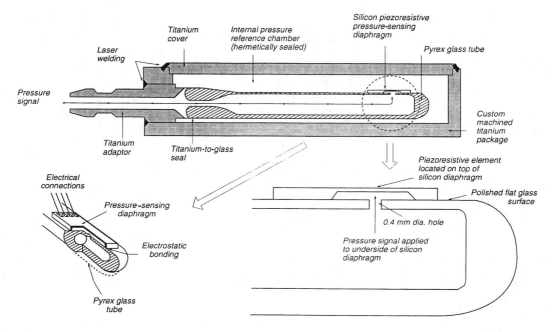

Figure 2.28   Construction of an implantable transducer for intracranial pressure telemetry. (Modified from Leung, A. M. et al., *IEEE Trans. Biomed. Eng.,* BME-33, 386, 1986.)

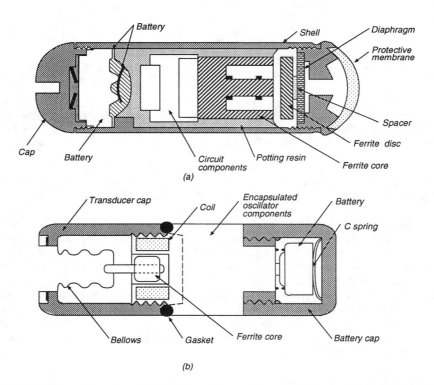

Figure 2.29   Two pressure capsules. (Modified from Farrar, J. T. et al., *Science,* 126, 975, 1957; Horowitz, L. and Farrar, J. T., *Gastroenterology,* 42, 455, 1962.)

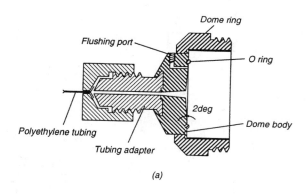

(a)

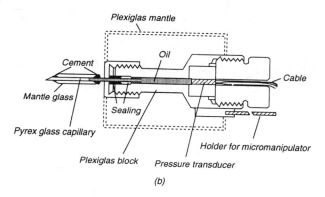

(b)

Figure 2.30 Two transducer domes for microvascular pressure measurements. **(a)** For Statham-Gould P23Gb pressure transducer. **(b)** For Kulite CPL-070-4 pressure transducer. (Part (a) modified from Levasseur, J. E. et al., *J. Appl. Physiol.*, 27, 422, 1969. Part (b) from Wunderlich, P. and Schnermann, J., *Pflügers Arch.*, 313, 89, 1969. With permission.)

of the arteriole, and a natural frequency of 25 Hz was attained when a micropipet 30 μm in diameter was used. However, natural frequency was reduced to 10 Hz when a 20-μm pipet was used.[66]

To obtain a higher natural frequency, it is important to minimize elastic components in the transducer system. Thus, besides minimizing volume displacement of the diaphragm, care should be taken to increase the rigidity of the dome and connecting parts, to minimize fluid volume in the dome to reduce the effect of compressibility of the fluid, and to remove air bubbles carefully.

Commercial pressure transducers with small volume displacement can also be used by applying specially designed domes. Figure 2.30(a) shows a transducer system with a Statham P23Gb pressure transducer (Gould; Clevelend OH).[67] The fluid volume in this dome is only 6% of that in the standard dome. When connected to a micropipet 25 μm in outer diameter and 15 μm in inner diameter, a natural frequency of 45 Hz was obtained.

Figure 2.30(b) shows another transducer system with a Kulite CPL-070-4 pressure transducer.[68] The transducer has a silicon diaphragm 1.2 mm in diameter and 1.8 μm in thickness. However, the frequency range in which distortion is insignificant was up to 3 Hz for a micropipet 12 μm in diameter at the tip, and up to 4 Hz when the diameter at the tip was 15 μm.

To record arterial pulse pressure, a frequency response of up to 6 Hz is usually required. It would thus be difficult to use a micropipet smaller than 15 μm in diameter as long as it is connected directly to a conventional pressure transducer, even when a specially designed dome is employed.

2.2.5.2 Servo-Controlled Pressure Measuring System

Although arterial pulse pressure measurement via a pipet smaller than 15 μm in diameter is difficult when using a pressure transducer in direct connection, it is still possible by connecting the pipet to a servo-controlled pressure measuring system, even when the diameter of the pipet is 1 μm or less.

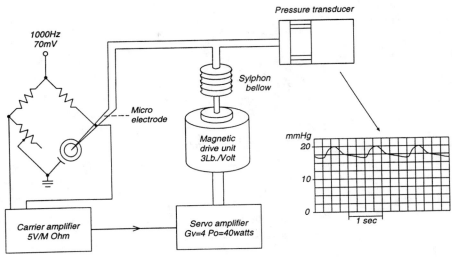

Figure 2.31 Block diagram of the servo-controlled pressure measuring system. (From Wiederhielm, C. A. et al., *Am. J. Physiol.*, 207, 173, 1964. With permission.)

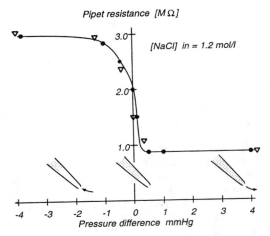

Figure 2.32 An example of the relationship between pipet resistance and pressure difference. (From Fox, J. R. and Wiederhielm, C. A., *Microvasc. Res.*, 5, 317, 1973. With permission.)

Figure 2.31 shows the block diagram of the servo-controlled pressure measuring system proposed by Wiederhielm and his colleagues.[69,70] Essential parts of the system consist of a Wheatstone bridge which detects the change in pipet resistance and a counter pressure drive unit. The pipet is filled with concentrated sodium chloride solution, typically 0.5 to 2 mol/l, which has a conductivity 3 to 6 times greater than that of physiological saline or plasma. For visual observation, a green dye is sometimes added to the inner solution. When the tip of the pipet is placed in the body fluid, the fluid at the tip can flow inward or outward according to the pressure difference between the outside and the inside of the pipet. Inward flow causes an increase in electrical resistance between the inside and outside fluid, and outward flow causes a decrease in the electrical resistance. Figure 2.32 shows the relationship between pipet electrical resistance and pressure difference of a pipet with a diameter of 0.3 μm.[70] The unit is controlled so as to generate counter pressure which compensates for the imbalance of the Wheatstone bridge. The servo-control system maintains a constant pipet resistance. Thus, constant pressure difference between the inside and outside of the pipet will be maintained, and the resulting pressure difference can be modified by adjusting a resistor on one side of the Wheatstone bridge.

In actual operation, inside pressure is maintained at about 0.013 kPa (0.1 mmHg) greater than the outside pressure. As long as the pressure difference between the outside and the inside of the pipet is maintained constant, the outside pressure can be determined by the inside pressure, which can be recorded by connecting an ordinary pressure transducer to the counter-pressure drive unit.

Due to the servo-controlled operation, almost the same frequency responses can be obtained in a wide range of pipet sizes, even though flow resistance of the pipet changes greatly. It can be applied with 0.5 to 5 μm pipets, and a frequency response up to 35 Hz is attained.[71,72] For pressure measurements in vessels of capillary dimensions, a 1- to 2-μm tip is used, and insertion of such pipets causes little or no mechanical interference with flow in vessels as small as 10 to 12 μm.[73]

The dynamic properties of pipets smaller than 0.1 μm were also investigated. In smaller pipets, flow resistance of the internal fluid is greatly increased, and thus the sensitivity in electrical resistance of the pipet decreases. On the other hand, due to the small bulk flow through the tip and short diffusion distance, the time required to reach equilibrium is reduced. As a result, a better frequency response can be realized. It was shown that a response up to 60 Hz was obtained with a 0.06-μm pipet. However, it also should be pointed out that the small pipets are sensitive to resistivity and temperature changes in the external solution; they are affected by contact potentials and are more likely to become plugged.[70] The complete system for microvascular pressure measurement is commercially available (Servo-nulling Pressure Measuring System, Instruments for Physiology and Medicine; San Diego, CA). The application techniques of the servo-controlled pressure measuring system are well established[73-77] and have even been used in coronary arterial pressure measurements in the beating heart.[78,79]

2.2.6 PRESSURE MEASUREMENTS IN COLLAPSIBLE VESSELS AND INTERSTITIAL SPACES

Since pressure is a quantity defined in fluid, the space where pressure is measured has to be filled with fluid. However, there is no appreciable space filled with fluid in sites such as collapsed vessels or tubules and interstitial spaces. Even in these sites, pressure measurements are sometimes required.

To measure pressures in these sites, the fluid at the object site is kept at an equilibrium with a fluid in the measurement system to which a pressure transducer can be connected. To do this, many different techniques have been developed.

2.2.6.1 Pressure Measurements in Collapsible Vessels

The esophagus, the anus, and the urethra are usually collapsed by the contraction of the sphincter muscles. If there is an appreciable amount of fluid inside, the strength of the sphincter muscle contraction will be estimated by the internal pressure. However, it is not adequate simply to insert an open-tipped catheter, because elastic walls or surrounding tissues are deformed to such an extent that the external force developed by the sphincter is balanced by the elastic force, and the pressure is thus blocked.

One solution to this difficulty is the use of a balloon catheter. When a balloon catheter is placed in a collapsible vessel and adequately inflated by air, the pressure in the balloon will be equilibrated to the surrounding tissue pressure. If the volume of the air in the balloon is inadequate, the error increases greatly. Figure 2.33 shows an example of the pressure-volume characteristics of a balloon catheter system for esophageal pressure measurement.[80] The volume of the balloon should be adjusted into the working range.

More commonly, a constantly perfused side-hole catheter has been used. A small amount of water is continuously perfused through a side-hole of a catheter by a low compliance perfusion pump, and the pressure is measured by a pressure transducer connected to the flow system. This method has been applied to anal,[81] esophageal,[82,83] and urethral[84] sphincter pressure measurements. In these measurements, the flow rate is typically 0.3 to 2 ml/min.

The dynamic response of the perfused catheter is mainly determined by the perfusion flow rate and the compliance of the perfusion system.[85] The response to a rapid increase in the pressure is limited by the rate of pressure rise in the perfusion system when the side hole of the catheter is suddenly closed. The rate of pressure rise is almost proportional to the perfusion flow rate and the rigidity of the perfusion system. However, an increase of the perfusion rate may increase the pressure artifact caused by flow variation. Thus, the use of a low-compliance perfusion system is preferable.

A pump commonly used for this purpose is the syringe pump, in which the syringe piston is propelled by a mechanical drive unit. To obtain lower compliance, the use of a glass syringe is recommended. The compliance of a glass syringe is about one sixth that of a plastic syringe of the same size.[86]

Although syringe pumps are convenient and have been used in the perfusion system for intraluminal pressure measurements, compliance of the pump still limits the frequency response characteristics and

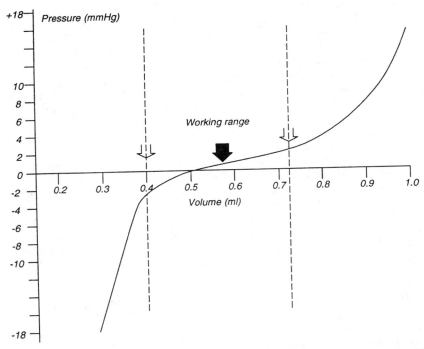

Figure 2.33 An example of the pressure-volume characteristics of a balloon-catheter system. (From Maxted, K. J. et al., *Med. Biol. Eng. Comput.*, 15, 398, 1977. With permission.)

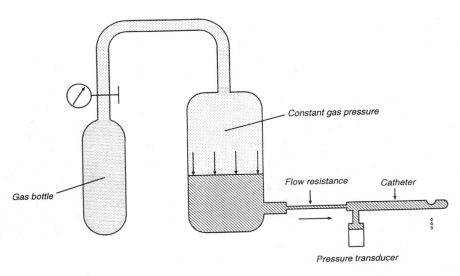

Figure 2.34 A pneumo-hydraulic perfusion system. (Modified from Ask, P., *Med. Biol. Eng. Comput.*, 16, 732, 1978.)

causes signal distortions. Moreover, imperfections in the mechanical propulsion of the piston in conventional syringe pumps cause flow variation and subsequent artifacts in the pressure signal.

To obtain a low-compliance, stable perfusion system, pneumohydraulic pumps have been introduced.[87,88] The pneumohydraulic pump consists of a high-pressure chamber and a flow resistance, as shown in Figure 2.34. The gas pressure comes from a compressed air bottle and is regulated by a pressure regulator applied to the infusion fluid, which is usually sterile water. The perfusion rate is determined

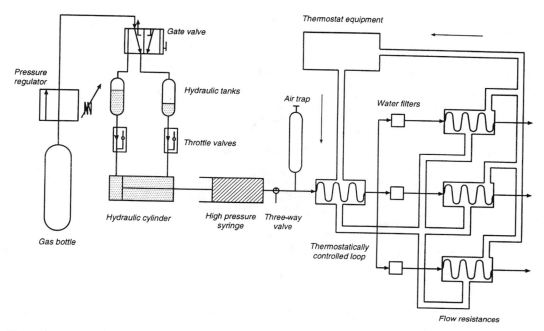

Figure 2.35 Block diagram of a three-channel, low-compliance perfusion system. (Modified from Ask, P., *Med. Biol. Eng. Comput.*, 16, 732, 1978.)

by this pressure and the flow resistance in the high pressure chamber. To obtain a flow rate of about 1 ml/min, a stainless steel tube of 0.1-mm internal diameter and 630-mm length was used when the applied pressure was about 1.100 kPa (10.8 atm). While a counter pressure affects the perfusion flow rate, a pressure increase of 13.3 kPa (100 mmHg) causes only a 1.3% reduction of the flow when the applied pressure is 1000 kPa. An evaluation using the esophageal model showed that the almost flat frequency response over the range required for esophageal measurements was obtained by this perfusion system with a 1.5-m catheter, an internal diameter of 0.9 to 1.7 mm, and a perfusion rate of 0.5 to 1.0 ml/min.[88]

Figure 2.35 shows a block diagram of the three-channel perfusion system for simultaneous pressure measurements at different sites. In this system, the temperatures of flow resistances are controlled by a thermostat, so that flow rates are not changed due to the temperature coefficients of viscosity.

To measure sphincter pressure, a catheter with a side hole is inserted beyond the site of the sphincter and then withdrawn gradually. With this procedure, an axial pressure profile is obtained, and the sphincter pressure is given as peak pressure. Continuous monitoring of sphincter pressure is sometimes required, especially for the upper and lower esophageal sphincters. For such measurements, a catheter with a sleeve is proposed.[89] In this catheter, an approximately 5-cm long sleeve made from a silicone rubber sheet is glued to the flat side of a catheter at each side of the sheet except the distal side. One lumen in the catheter is opened under the sleeve for perfusion. Using this catheter, maximal esophageal sphincter pressure can be monitored as long as the sphincter lies under the sleeve, whereas measurements using a side-hole catheter require accurate positioning so that the side hole is located at the center of the sphincter. Figure 2.36 shows a sleeve catheter designed for pressure measurement in the human upper esophageal sphincter. This catheter has proximal and distal side holes which allow for positioning of the sleeve in the high-pressure zone of the upper esophageal sphincter.[90]

2.2.6.2 Interstitial Pressure Measurements

There is little or no free fluid in the interstitium and most of the interstitial fluid is entrapped within the gel-like material. The pressure to be measured is the fluid equilibration pressure, which is the pressure in saline solution in equilibrium with the gel-phase.[91,92] Thus, the essential technique required for interstitial pressure measurement is to realize and maintain an equilibrium between the fluid in the interstitium and saline solution at the pressure transducer.

The major techniques used for interstitial pressure measurements involve chronically implanted perforated capsules, chronically implanted porous capsules, micropipets, and wick-in-needles. In early

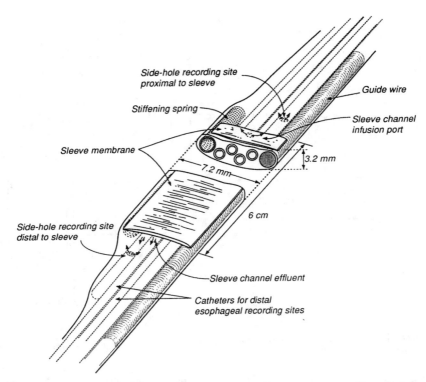

Figure 2.36 A sleeve catheter for esophageal sphincter pressure measurement. (From Kahrilas, P. J. et al., *Dig. Dis. Sci.*, 32, 121, 1987. With permission.)

studies, attempts were made to use intratissue balloons and hypodermic needles. However, it was found that pressures with these techniques were always inconsistent with pressures obtained with fluid equilibration, and the techniques are consequently considered inadequate for interstitial pressure measurement.[91]

Chronically implanted capsules can be used only in animal experiments. Chronically implanted perforated capsules are hollow plastic cylinders or balls. The diameters of these capsules are typically 0.8 to 3 cm, with 100 to 250 holes about 1 mm in diameter. The capsules are implanted subcutaneously, and about 4 weeks after implantation the tissue and vascular system grow into the inside of the capsule, as shown in Figure 2.37.[93] To take measurements, a hypodermic needle is inserted into the free fluid in the capsule, and fluid pressure is measured by an ordinary pressure transducer.

The chronically implanted porous capsule consists of a porous polyethylene cylinder about 3 mm in diameter and about 5 mm in length, with a pore size of about 60 μm.[94] A polyethylene tube 10 to 15 cm long is connected to the capsule, and at implantation the tube is capped with a plug and is left in the body of the experimental animal. After 4 weeks, the pressure in the capsule is measured by either an ordinary pressure transducer or a servo-controlled counter pressure system, described in Section 2.2.4.2.

The micropipet technique is the same technique used in microcirculation studies, as described in Section 2.2.4.2. A glass micropipet, 1 to 3 μm in diameter, is connected to a servo-controlled counter-pressure system, and pressure is measured by a direct puncture.[95]

In the wick-in-needle technique, a hypodermic needle is provided with a side hole 2 to 4 mm long and is filled with a multifilamentous nylon thread as shown in Figure 2.38.[96,97] A polyethylene catheter connects the needle to the pressure transducer. The wick-in-needle is inserted in the tissue, and fluid communication between needle and tissue fluid is confirmed by compression or decompression of the catheter operating a screw clamp. When fluid communication is satisfactory, the pressure change due to a displaced volume returns to control level within 1 to 5 min.[96]

These different methods for conducting interstitial pressure measurements were compared in cats[98] and in dogs.[92] Both evaluations showed that all four techniques (i.e., the chronically implanted perforated and porous capsules, the wick-in-needle, and the micropipets) gave similar interstitial pressures of about −0.26 kPa (−2 mmHg) for the subcutis under control conditions. When Ringer's solution was infused

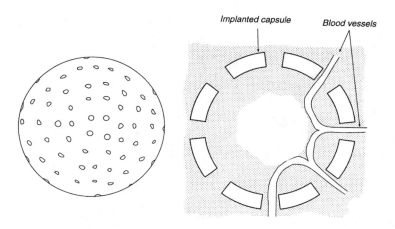

Figure 2.37 A perforated capsule for interstitial fluid pressure measurement. The tissue of the vascular system grows into the inside of the capsule, as shown on the right. (Modified from Guyton, A. C., *Circ. Res.,* 12, 399, 1963.)

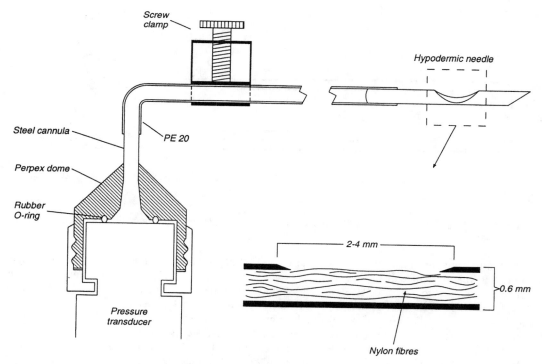

Figure 2.38 The wick-in-needle technique for interstitial pressure measurement. (From Fadnes, H. O. et al., *Microvasc. Res.,* 14, 27, 1977. With permission.)

intravenously, marked pressure rises were observed in both capsules, whereas such a change was not observed in the wick-in-needle and the micropipets. All methods gave similar pressures 120 to 210 min after the end of infusion. In acute dehydration induced by peritoneal dialysis with hypertonic glucose, only capsular pressures changed accordingly. Based on these observations, it is suspected that capsule measurements are easily influenced by a change in colloid osmotic pressure of the surrounding fluid, and thus capsular techniques are inadequate for measurements of changes in interstitial fluid pressure that take place in less than a few hours.[98]

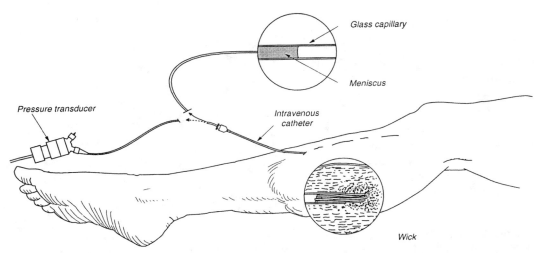

Figure 2.39 Compartment pressure measurement using the wick technique. (From Hargen, A. R. et al., *Microvas. Res.,* 14, 1, 1977. With permission.)

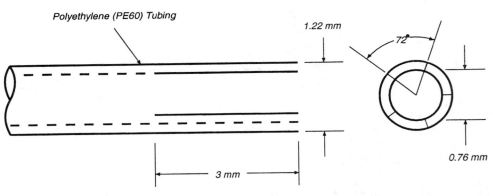

Figure 2.40 A slit catheter for interstitial pressure monitoring. (Modified from Castle, G. S. et al., *J. Clin. Eng.,* 6, 219, 1981.)

Interstitial pressure measurements were also used for the clinical diagnosis of compartment syndromes. In this type of patient, interstitial pressure that is slightly negative in the normal subject is sometimes elevated to over +4 kPa (30 mmHg). Figure 2.39 shows a method using the wick technique.[99] Polyglycolic acid fibers with a diameter of 20 mm were used as a wick material. The wick was 4 cm in length and was pulled 1 cm into a polyethylene catheter of 0.9-mm o.d. and 0.6-mm i.d. The wick catheter was introduced into the muscle using an intravenous catheter positioning unit. The pressure was measured by either a pressure transducer or a glass capillary tube, by which meniscus movement was followed microscopically so that equilibrium would yield interstitial fluid pressure.

A slit catheter, shown in Figure 2.40, was also used for clinical monitoring of interstitial pressure.[100] An experiment involving the compartment in a dog's leg showed that the slit catheter could measure intracompartment pressure in a range of 0 to 10.6 kPa (0 to 80 mmHg) with a standard deviation of 0.15 kPa (1.1 mmHg) and was more stable than wick-in-needle methods.

Colloid osmotic pressure of the interstitial fluid is measured to estimate protein concentration. Conventionally, colloid osmotic pressure is measured for sampled fluids obtained from the implanted wick.[96,101,102] To record osmotic pressure continuously, implantable colloid osmometers were developed. Figure 2.41 shows an example.[103] It consists of a needle with a side hole covered with a semipermeable hollow fiber (Amicon UM-10). The polyethylene catheter is connected to a pressure transducer, and osmotic pressure is recorded as a negative hydrostatic pressure. For *in vitro* and *in vivo* measurements, equilibration times were about 10 min.[103]

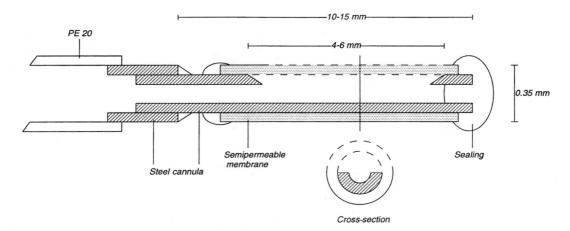

Figure 2.41 An implantable catheter for colloid osmotic pressure measurement. (From Reed, R. K., *Microvasc. Res.*, 18, 83, 1979. With permission.)

2.2.7 DIFFERENTIAL PRESSURE MEASUREMENTS

Measurements of pressure differences are sometimes required, such as to evaluate a pressure drop in the respiratory airway or in the cardiovascular system. In flow measurements using small flow resistances, measurements of small pressure differences under varying pressure levels are necessary (see Sections 3.2.8 and 3.4.1.2).

In principle, a pressure difference can be measured by two independent pressure transducers. However, if the pressure difference is small compared with the variation of each pressure, small changes in sensitivity and the zero level of each transducer may cause large measurement errors in differential pressures. In such situations, direct measurement of differential pressure is recommended.

For measuring small differential pressures in gases, there are many kinds of differential pressure transducers based on various principles (such as strain-gauge, capacitive, inductive, and optical), and most of them are commercially available. Commonly, the differential pressure transducer has a diaphragm to which the differential pressure to be measured is applied, and displacement or strain is detected at one side of the diaphragm. If the detecting element does not tolerate moisture, only dry gas can be used, at least in one part of the differential pressure input, and this is generally not acceptable for respiratory measurements. Most commercial differential pressure transducers of the capacitive type are used only for noncorrosive, noncondensable gases and are not appropriate for respiratory measurements, even though highly sensitive transducers are available.

By employing two diaphragms, it is possible to keep the detecting element from the applied gases. Figure 2.42 shows an example of such construction.[104] This transducer has two diaphragms connected to each other by a rigid spacer. The force acting on the connected diaphragm system is determined by the differential pressure, and the displacement is detected by capacitance electrodes at the rear sides of the diaphragms.

The number of available transducers for differential pressure measurements in liquid is limited. Among them, the variable inductance differential pressure transducer[105] is one of the most convenient transducers. Its construction is shown in Figure 2.43(a). The diaphragm is made by magnetic stainless-steel, and inductances of two coils placed close to the diaphragm are changed by a displacement of the diaphragm. Figure 2.43(b) shows an example of a simple demodulation circuit which provides the d.c. output.

Variable inductance type transducers are available commercially (MP45, DP45, DP103, Validyne Engineering Co.; Northridge, CA). Interchangeable diaphragms are supplied for these transducers so that different pressure measurement ranges can be obtained by replacing the diaphragm. For the diaphragm of highest sensitivity, a measurement range of ± 0.14 kPa (± 1.4 cmH$_2$O) is obtained in the MP45, ± 0.20 kPa (± 2.0 cmH$_2$O) in the DP45, and ± 0.056 kPa (± 5.6 mmH$_2$O) in the DP 103. Using a carrier frequency of 3 to 5 kHz, a flat frequency response up to 1000 Hz can be obtained. Nominal nonlinearity is less than 0.5% of full scale, and the hysteresis error is less than 0.1%.

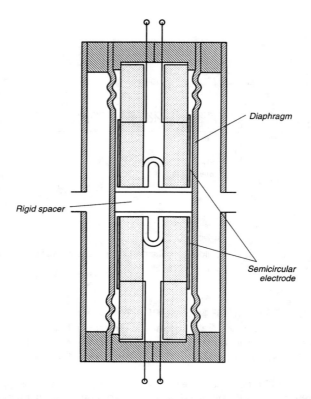

Figure 2.42   A capacitive-type differential pressure transducer with two diaphragms connected to each other by a rigid spacer. (Modified from Rabek, J. W., *U.S. Patent 3,* 965, 746, 1976.)

2.3 INDIRECT PRESSURE MEASUREMENT

The most successful clinical application of indirect pressure measurement is the occluding cuff technique for blood pressure measurement. This technique has been used as the conventional auscultatory blood pressure measurement and has also been applied to the automated blood pressure recording system. Instantaneous arterial pressure can also be measured by a cuff with a servo-controlled pressure driving system.

Internal pressures in some body cavities can be measured indirectly by reaction force measurements. This technique can be applied to intraocular pressure, intra-amniotic pressure, and intracranial pressure measurements.

2.3.1   INDIRECT MEASUREMENT OF SYSTOLIC, DIASTOLIC, AND MEAN BLOOD PRESSURE

All occluding cuff techniques for indirect blood pressure measurement employ the inflatable cuff which is wrapped around an extremity, typically the upper arm. When the cuff is inflated, full cuff pressure is transmitted to the tissue around the artery, as long as the cuff size is adequate. The lumen of the artery will open and close following the positive and negative transmural pressure, which is the pressure difference between the inside and outside of the vessel. Thus, by varying the cuff pressure gradually and detecting the pressure point at which the lumen is just opened or closed, intravascular pressure can be measured from the cuff pressure.

To detect the lumen opening, the auscultatory techniques as well as the automated technique based on the Korotkoff sounds principle have been widely used, while other techniques such as ultrasound and oscillometric methods have also been employed. In these indirect arterial pressure measurements, the accuracy of measurements is affected mainly by the design of the occluding cuff and the procedure used for detecting vascular opening.

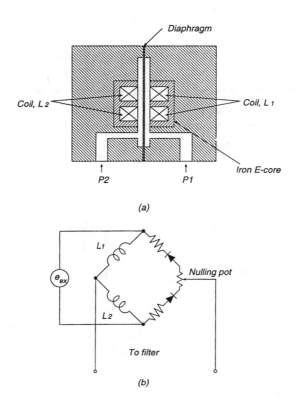

Figure 2.43 **(a)** Variable-inductance-type differential pressure transducer. **(b)** Simple demodulation circuit. (From Doebelin, E. O., *Measurement System Application and Design,* rev. ed., McGraw-Hill, New York, 1975, 387. With permission.)

2.3.1.1 Cuff Design for Indirect Blood Pressure Measurements

The conventional sphygmomanometer cuff currently used for measuring indirect blood pressure in the upper arm consists of an inflatable bladder within a restrictive cloth sheath, as shown in Figure 2.44.

The size of the inflatable bladder should be large enough so that the pressure in the bladder fully transmits into the underlying tissue containing the large artery. For clinical sphygmomanometers, the bladder size has been empirically determined. According to the recommendations on blood pressure measurement provided by the British Hypertension Society, the width of the bladder should be at least 40% of the circumference of the upper arm, and the length should be at least 80% of the circumference.[106] The American Heart Association also provides recommendations for human blood pressure determination by sphygmomanometers. According to these recommendations, the width of the bladder should be 40% of the circumference of the midpoint of the limb (or 20% wider than the diameter), and the length should be twice the recommended width.[107] Recommendations are also given for bladder diameters for blood pressure cuffs as shown in Table 2.2. These recommendations are consistent regarding bladder width, which should be about 1.2 times the diameter of the arm.

The postulations for cuff size are also supported by a model analysis.[108] In this analysis, it was assumed that the arm was a rotationally symmetric cylinder of an elastic, noncompressible material, and a pressure field was applied for the cylinder corresponding to a cuff with a width of $2b \times r$, as shown in Figure 2.45(a), where b is the ratio of the cuff width to the arm diameter, r. Then the radial stress at the center of the arm was calculated analytically. The results are shown in Figure 2.45(b). It was found that for cuff widths less than 1.2 times the diameter of the arm (i.e., $b < 1.2$), bladder pressure is not accurately transmitted to the center of the arm even at the midpoint of the cuff, and thus erroneously high blood pressures will be obtained. For cuff widths greater than 1.2 times the diameter of the arm (i.e., $b \geq 1.2$), the pressure is accurately transmitted to all depths of the arm, at least under the midpoint of the cuff, as well as some distance away from the midpoint.

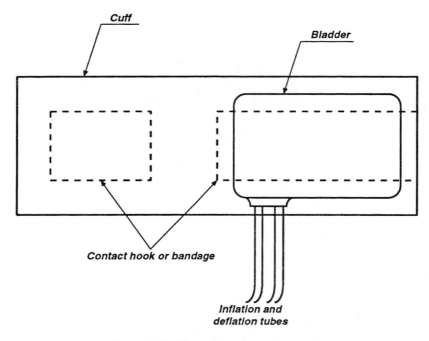

Figure 2.44   The sphygmomanometer cuff.

Table 2.2   Recommended Bladder Dimensions
for Blood Pressure Cuff[107]

Arm Circumference at Midpoint[a] (cm)	Cuff Name	Bladder Width (cm)	Bladder Length (cm)
5–7.5	Newborn	3	5
7.5–13	Infant	5	8
13–20	Child	8	13
17–26	Small adult	11	17
24–32	Adult	13	24
32–42	Large adult	17	32
42–50[b]	Thigh	20	42

[a]   Midpoint of arm is defined as half the distance from the acro-
mion to the olecranon.
[b]   In persons with very large limbs, the indirect blood pressure
should be measured in the leg or forearm.

Effects of cuff size on the accuracy of the auscultatory blood pressure measurement were studied experimentally by making comparisons with direct measurement.[108-110] These studies showed that smaller cuffs gave higher blood pressure, as predicted from the above model analysis, and the standard cuff bladder which is 12 cm wide and 23 cm long was sometimes inadequate. Simpson and his colleagues[109] showed that differences in pressures measured with a 12 × 23-cm cuff bladder and a 14 × 35-cm cuff bladder were 1 kPa (7.3 mmHg) for systolic and 0.9 kPa (6.9 mmHg) for diastolic pressure in 24 subjects with arm circumferences of 21 to 46.5 cm. van Montfrans and his colleagues[110] also studied rectangular cuff bladders from 12 × 23 cm to 14 × 38 cm and a conical cuff for 18 normal sized arms (circumference < 34 cm) and 19 obese arms (circumference ≥ 34 cm). Statistically significant differences were observed for different cuff sizes only in systolic pressures in the obese arm group; these authors recommended a 12 × 23-cm cuff bladder for normal arms, and a 14 × 38-cm cuff bladder for obese arms.

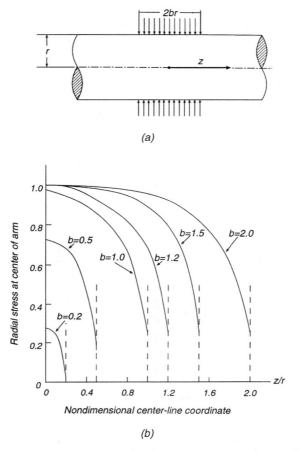

(a)

(b)

Figure 2.45 A simplified model of the arm **(a)**, and calculated radial stress distribution at the arm axis **(b)**. (From Alexander, H. et al., *Med. Biol. Eng. Comput.*, 15, 2, 1977. With permission.)

2.3.1.2 Detection of Korotkoff Sounds

The auscultation method of indirect blood pressure measurement is based on the fact that when the occluding cuff pressure is slowly reduced from a pressure above systolic pressure to a pressure below diastolic pressure, acoustic waves called Korotkoff sounds are generated while the cuff pressure remains between the systolic and diastolic pressures.

The origins of Korotkoff sounds are not completely understood, though many different mechanisms for sound production have been proposed such as turbulence, flutter of the artery or relaxation oscillation, water hammer effect, and sudden stretching of the arterial wall.[111] Despite the lack of a theoretical basis, Korotkoff's method has been widely accepted and the techniques have been standardized.

In blood pressure measurement using the conventional sphygmomanometer, a stethoscope is placed on the brachial artery close to the distal side of the cuff. The cuff pressure is raised to about 4 kPa (30 mmHg) above the estimated systolic pressure and then reduced at a rate of 0.26 to 0.4 kPa/s (2 to 3 mmHg/s). The systolic pressure is given as the cuff pressure when clear tapping sounds first appear, and the diastolic pressure is given as the cuff pressure when the sounds disappear. This technique was first proposed by Korotkoff in 1905, and it was generally accepted after its reliability was confirmed by many comparative studies with direct measurement and other indirect techniques which had been used before Korotkoff's method was proposed.[112]

Korotkoff sounds can be detected by microphones, instead of a stethoscope, placed at the distal side of the cuff, but more conveniently the microphone can be placed beneath the cuff or even in the cuff bladder.[113] Figure 2.46 shows typical data for relative sound intensities and fundamental frequencies in

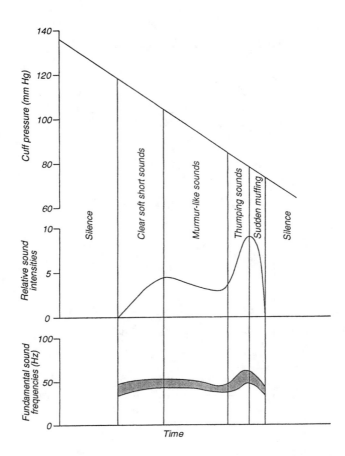

Figure 2.46 Characteristics of Korotkoff sounds in the typical normotensive subject. (From Greatorex, C. A., in *Noninvasive Physiological Measurements,* Vol. 1, Rolfe, P., Ed., Academic Press, London, 1979, 193. With permission.)

normotensive subjects when the cuff pressure was reduced according to the conventional auscultatory technique.[114] As seen in the figure, principal components of Korotkoff sounds are low in frequency, with a peak at about 45 Hz, but higher harmonic components are also present, and the major frequency components fall in a band from about 20 to 300 Hz.[112]

As long as characteristics of Korotkoff sounds are known, automatic detection algorithms can be designed. In actual measurement situations, noises superpose on the signal, so that a narrow bandpass filter should usually be used. When the cuff pressure is reduced at a normal rate, changes in amplitude are different for different frequency bands and are also different for the systole and diastole. According to a study by Golden et al.,[115] a maximum increase in amplitude at the systolic transition occurred in a frequency band of 18 to 26 Hz, and a maximum decrease in amplitude at the diastolic transition occurred in a frequency band of 40 to 60 Hz.

In automated detection systems based on Korotkoff sounds, the effect of noise is a serious problem. While the effect of ambient noise can be reduced by using an appropriate bandpass filter or covering the microphone with the cuff, motion artifact is more difficult to avoid because it can exist in the same frequency band as the major component of Korotkoff sounds. Motion artifacts can be induced by movements of the whole body, arm, hand, or even finger and are difficult to avoid completely. To reduce motion artifacts, taping a microphone to the upper arm using micropore surgical tape has been recommended. By doing so it has been reported that systolic and diastolic pressures can be determined even during jogging on a treadmill.[116] When an ECG signal is available, further reduction of noise can be attained by using a time window which opens about 80 to 250 ms after the *R* wave in ECG.[117]

In the blood pressure measurement using Korotkoff's method, evaluation of accuracy is a difficult problem. There are many studies in which indirect blood pressure measurements using Korotkoff's method have been compared with direct measurements.[108,110,118-125] From these reports it appears that direct and indirect measurements result in almost the same pressures, at least in certain types of subjects. For example, Ragan and Bordley[118] reported that in 138 observations in 51 adult subjects, auscultatory systolic readings were within ±1.3 kPa (±10 mmHg) of the intraarterial pressure in 83% of the comparisons. However, many authors have stated that large discrepancies, greater than ±4 kPa (±30 mmHg), have occasionally been observed which cannot be explained by the materials being used for the measurements, such as using a cuff of the wrong size or inadequate damping or natural frequency in the direct pressure measurement system. Bruner et al.[123,124] have pointed out that reduction of blood flow in the brachial artery causes an underestimation in indirect measurements, especially for systolic pressures. That situation can easily occur when there is peripheral vasoconstriction, where decreases in blood flow will cause degradation of Korotkoff sounds and indirect measurement may underestimate systolic pressure. In addition, actual peak pressure sometimes elevates when the peripheral resistance is increased, and discrepancies between indirect and direct measurements of systolic pressure will become even greater.

As long as systolic and diastolic pressures are determined by Korotkoff sounds, these problems also appear in automated systems. However, if similar criteria are used for determining the systolic and diastolic pressures, based on signal processing of Korotkoff sounds in automated systems as in conventional auscultatory techniques, it may be expected that both measurements will result in almost the same pressure, even if this value differs from the one obtained by direct measurement. Actually, Wolthuis et al.[117] showed that their automated system can detect Korotkoff sound signals corresponding to systolic and diastolic pressure within 2 beats from the auscultatory determination, when the cuff pressure was reduced at a rate of about 0.4 kPa/s (3 mmHg/s). Thus, discrepancies in systolic and diastolic pressure for both measurements will be within 0.8 kPa (6 mmHg) when the heart rate is 60 beats per minute.

2.3.1.3 Mean Blood Pressure Measurements by the Oscillometric Method

The oscillometric method is an indirect method of measurement of mean arterial pressure based on a principle in which the oscillation in the occluding cuff pressure due to the arterial volume pulsation has the maximum amplitude when the cuff pressure is close to the mean arterial pressure. Historically, the oscillation in the occluding cuff pressure was used for blood pressure measurements even before Korotkoff's method was proposed, and in earlier investigations the cuff pressure for maximum oscillation was sometimes confused with the diastolic pressure. Many comparative studies with direct arterial pressure measurements in animal and human subjects, as well as model experiments, have shown that the pressure for maximum oscillation is always very close to the mean arterial pressure.[126]

The principle of the oscillometric method was demonstrated by simplified model experiments. Posey and his colleagues[126] used a saline-filled compression chamber as a model of the arm. An arterial segment was placed in it and perfused by blood from the carotid artery of the animal. Oscillation of the chamber pressure was then observed. Results from this experiment showed that the lowest chamber pressure for maximum oscillations, when the amplitude remains constant, was only few percent below true mean arterial pressure.

Yamakoshi and his colleagues[127] also tried a similar experiment but used an air-filled chamber. They observed volumetric oscillation and concluded that the maximum volume pulsation occurred when the chamber pressure was equal to the mean arterial pressure within an error of ±0.4 kPa (±3 mmHg). They also compared the pressure for maximum oscillation in amplitude of photo-plethysmographic pulsation at the human finger to the directly measured mean brachial arterial pressure and found that indirectly measured finger pressures were approximately 5% lower than mean brachial pressures.[128] Figure 2.47 shows an example of the volume oscillation signal detected by transmission-type photo-plethysmography at the index finger. In this example, cuff pressure for maximum amplitude of the plethysmographic signal was only 0.4 kPa (3 mmHg) below directly measured mean pressure in the brachial artery, and the point of disappearance was 0.26 kPa (2 mmHg) below systolic pressure.

Ramsey[129] used a standard cuff and automatic determination system of the maximum oscillation in the cuff pressure, and pressures for maximum oscillation were compared with direct arterial pressure measurements. The results showed that the overall error was only 0.03 kPa (0.23 mmHg) with a standard deviation of 0.56 kPa (4.21 mmHg) in 28 measurements in 17 subjects. These observations indicate that mean arterial pressure measurement by oscillometric methods is reliable and accurate enough for most clinical applications.

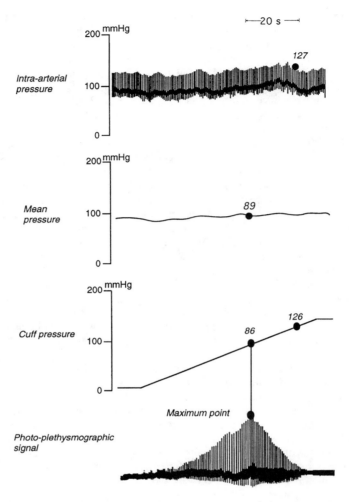

Figure 2.47 Examples of instantaneous and mean intra-arterial pressures, cuff pressure, and photo-plethys-mographic signals obtained from the index finger of a normotensive subject. (From Yamakoshi, K. et al., *Med. Biol. Eng. Comput.*, 20, 314, 1982. With permission.)

Since the auscultatory method became a standard technique of clinical blood pressure measurement, the oscillometric method has seldom been used clinically. However, mean blood pressure measurement by the oscillometric method has some advantages which are worth noting: (1) The oscillation can be detected by either cuff pressure or the plethysmogram, using the mechanical, photoelectric, or impedance method; (2) the maximum amplitude of the oscillation can be determined objectively and easily, because a sharp maximum in amplitude is usually observed; and (3) the procedure for detecting maximum amplitude can easily be automated, thus the oscillometric method is more convenient than the auscultatory method for continuous blood pressure monitors. The difficulty in applying the oscillometric method clinically is due to the fact that it gives systolic and mean blood pressures, while the auscultatory method gives systolic and diastolic pressures. Also, clinicians are very familiar with systolic and diastolic pressure, but not with arterial mean pressure.

To obtain diastolic pressure, an estimation method was proposed in which diastolic pressure is calculated from systolic and mean pressures and volume plethysmographic wave forms.[130] The method is based on the assumption that the arterial pressure wave form is similar to the volume plethysmographic wave form. Figure 2.48 shows an example of these wave forms. This means that the pressure-volume relation of the artery can be regarded as linear in the range of a pressure pulse amplitude. If these two wave forms are similar, ratios of mean and the full amplitudes of these pulse waves are the same, i.e.,

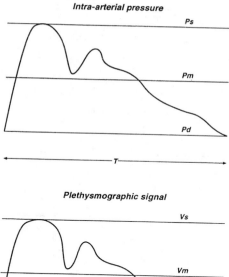

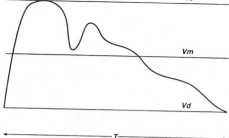

Figure 2.48 An example of intra-arterial pressure and plethysmographic wave forms in one cardiac cycle. If two wave forms are similar, the diastolic pressure can be calculated from the systolic and mean pressures using corresponding values in the plethysmographic wave form. (Modified from Shimazu, H. et al., *Med. Biol. Eng. Comput.*, 24, 549, 1986.)

$$\frac{P_m - P_d}{P_s - P_d} = \frac{V_m - V_d}{V_s - V_d} = k \tag{2.24}$$

where P_s, P_d, and P_m are systolic, diastolic, and mean pressures; V_s, V_d, and V_m are systolic, diastolic, and mean arterial volumes. Thus, diastolic pressure, P_d, can be obtained as

$$P_d = P_m - \frac{k}{1 - k}\left(P_s - P_m\right). \tag{2.25}$$

P_m can be measured by the maximum oscillation method, and P_s can also be determined as the pressure obtained when the cuff pressure is decreased gradually and the plethysmographic signal appears. Thus, by determining k from the recorded plethysmographic wave forms, diastolic pressure can be estimated.

Shimazu and his colleagues[131] studied this method using transmission-type photo-plethysmography in the human finger and compared the results with direct measurements. Their findings showed that the error in diastolic pressure estimation was about ±0.7 kPa (±5 mmHg). They also applied this method in the case of using reflection-type photo-plethysmography in the forearms and ankles of babies and children. The estimation error for diastolic pressure was about ±0.9 kPa (±7 mmHg).

2.3.1.4 Blood Pressure Measurements by Doppler Ultrasound

In indirect blood pressure measurements with the occluding cuff, the Doppler ultrasound measurement can be used to determine systolic and diastolic pressures instead of Korotkoff sounds. Two different methods have been proposed. The first involves detecting arterial wall motion, and the second involves detecting arterial blood velocity under the occluding cuff.

The principle of the first method is shown in Figure 2.49.[132] The transmitter and receiver crystals are incorporated inside the occluding cuff, so that the reflection signal from the arterial wall can be received.

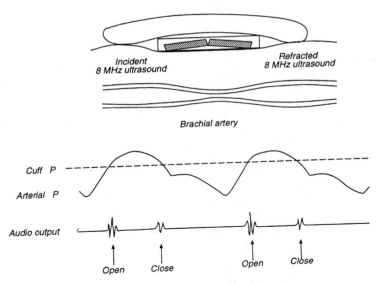

Incident
8 MHz ultrasound

Refracted
8 MHz ultrasound

Brachial artery

Cuff P

Arterial P

Audio output

Open   Close   Open   Close

Figure 2.49 Indirect blood pressure measurement using Doppler ultrasound which detects arterial wall motion. When cuff pressure remains in between the systolic and diastolic pressures, the artery opens and closes in each cardiac cycle. (From Stegall, H. F. et al., *J. Appl. Physiol.*, 25, 793, 1968. With permission.)

As the cuff pressure is reduced, the arterial segment under the cuff opens when the intravascular pressure exceeds the cuff pressure. The arterial wall motion at the opening and closing of the segment can be detected as the Doppler shift signal in the received ultrasound. When 8-MHz ultrasound is used, the Doppler shift at the opening is ordinarily observed in a frequency range of 200 to 500 Hz, and that at the closing is 30 to 100 Hz.[132]

The cuff pressure at the first opening signal equals systolic pressure. As the cuff pressure is reduced, the closing signal falls later and later in the cycle until it finally merges with the opening signal of a subsequent cardiac cycle. At this point, the cuff pressure equals diastolic pressure.

Comparative studies have shown that systolic and diastolic pressures obtained by the Doppler measurements of arterial wall motion are always close enough to those of direct measurements.[132-135] They have also pointed out that the Doppler method can be used even when patients are in shock and Korotkoff sounds are inaudible.

A problem with the Doppler ultrasound method is that the ultrasound beam is relatively narrow, and a slight movement of the probe thus causes a failure of Doppler signal detection. This problem is serious when the Doppler is used for continuous blood pressure monitoring. In a commercial unit (Arteriosonde®, Kontron Instruments; Everett, MA), this difficulty has been solved by using several different crystals for emitters and receivers.[135]

The second method for indirect blood pressure measurement by Doppler ultrasound involves detecting arterial blood velocity by Doppler shift of the backscattered ultrasound from the blood in the artery. The technique is the same as the ultrasonic Doppler flowmeter which is described in detail later (see Section 3.2.2). To measure arterial pressure, the conventional occluding cuff for the upper arm is used, and the ultrasound probe is applied to the skin over the radial artery.[136] The cuff is inflated above systolic pressure and gradually deflated. Systolic pressure is regarded as the cuff pressure when the blood flow signal is first recognized. As the cuff pressure is reduced further, the Doppler signal increases in amplitude, but as long as the cuff pressure is above diastolic pressure, the Doppler signal reaches zero near the end of each pulse, and when the cuff pressure reaches diastolic pressure, the Doppler signal remains above zero.

Systolic and diastolic pressures measured by this method were close to those obtained by direct measurements in patients with normal circulation. In patients in shock, the systolic pressures were in accordance with direct measurements, even when Korotkoff sounds were inaudible. However, diastolic pressure could not be measured in patients with clinical hypotension and shock, possibly caused by severe vasoconstriction and loss of reactive hyperemic response.[136]

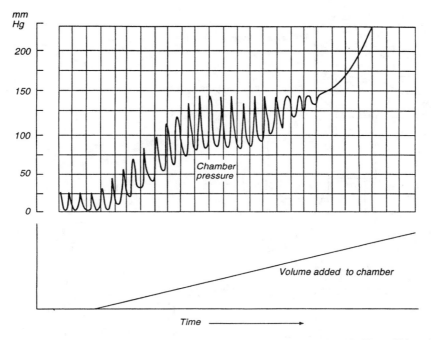

Figure 2.50 Pressure in a rigid chamber in which a pulsating vessel is enclosed. (From Shirer, H. W., *IRE Trans. Bio-Med. Electron.,* 9, 116, 1962. With permission.)

2.3.2 INDIRECT MEASUREMENTS OF INSTANTANEOUS ARTERIAL PRESSURE

Although instantaneous arterial pressure has usually been measured directly by inserting a catheter into the artery, it can be measured indirectly under certain conditions. It has been pointed out that when the vessel is enclosed in a rigid chamber filled with an adequate amount of liquid and if all tension on the vessel wall is removed so that the vessel is a sack tube, then the chamber pressure will equal the endovascular pressure.[25]

Figure 2.50 shows the relationship between the chamber pressure and the amount of liquid gradually added into the chamber. Intravascular pressure will appear in the chamber when a certain amount of liquid is added. This technique is known as the vessel wall unloading method (or vascular unloading method).

In the actual limb, however, the conditions for vessel wall unloading cannot be realized satisfactorily by simply enclosing the skin surface with a rigid chamber, because the deformation of the tissue between the chamber and the vessel makes it impossible to keep the chamber noncompliant. This problem was solved by introducing a hydraulic servo-control system.

Figure 2.51 shows a continuous recording system for arterial pressure wave form in which the finger cuff pressure is driven by the photo-plethysmographic signal.[137,138] In a closed-loop operation, when the blood volume in the finger tissue is going to increase, light absorption will increase and photo-plethysmograph output corresponding to the transmitted light will decrease. If this signal causes an increase in the cuff pressure, extravascular pressure is increased, and further increase in the vascular volume will be restrained. When the blood volume in the tissue is going to decrease, a similar response will be expected so that further volume change is restrained. Thus, the closed loop operation reduces vascular volume change, and if the loop gain is high enough the tissue under the cuff cannot change its volume, i.e., the situation is the same as being surrounded by a noncompliant chamber. By adjusting the amount of fluid in the cuff to an appropriate level, cuff pressure will equal intravascular pressure. In the actual system, finger cuff pressure was controlled by an electropneumatic transducer, and it was demonstrated that instantaneous arterial pressure could be recorded continuously, and the obtained wave forms were close to those recorded directly.[138]

A similar system, but one which used a water-filled compression chamber, was also constructed, and its performance was studied in detail (Figure 2.52).[139-141] The compression chamber surrounding the

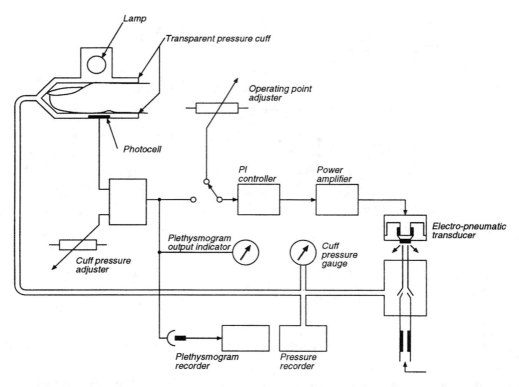

Figure 2.51 A continuous blood pressure recording system using a pneumatic-driven finger cuff. (Modified from Peñáz, J., *Dig. 10th Int. Conf. Med. Biol. Eng.,* Dresden, 1973, 104; Peñáz, J. et al., *Zschr. Med. Jahrg.,* 31, 1030, 1976.)

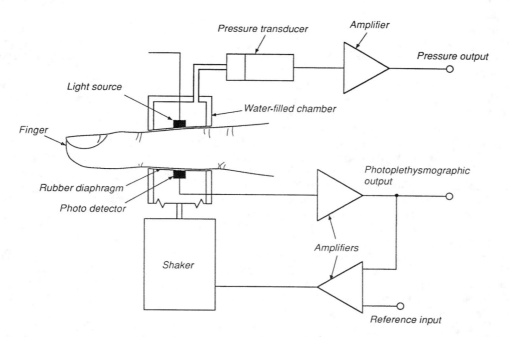

Figure 2.52 A continuous blood pressure recording system using a water-driven cuff. (Modified from Yamakashi, K. et al., *IEEE Trans. Biomed. Eng.,* BME-27, 150, 1980.)

finger has a diaphragm with an effective diameter area of 11 cm^2 which is connected to an electromagnetic shaker and has a natural frequency of 80 Hz and a maximum generating force of about 49 N (5 kg$_f$). Before measurement, open-loop operation is carried out by adjusting the d.c. level of the cuff pressure controller so that plethysmographic output allows the maximum amplitude. In such conditions, the arterial wall is unloaded. Then, closed-loop operation is started. If the loop gain in closed-loop operation is sufficiently high, any variation in vascular volume due to a change in arterial pressure is instantaneously compensated for, so that the vascular volume is clamped at the level at which the vascular wall is unloaded.

According to a study of four normotensive and six hypertensive patients undergoing arterial catheterization, it was found that observed cuff pressure wave forms in the closed-loop operation were similar to the wave forms recorded directly from the brachial arteries. Figure 2.53 shows an example of recordings of brachial arterial pressure, finger cuff pressure, photo-plethysmographic output, and volume displacement at the diaphragm. Systolic and diastolic pressures obtained from the cuff pressure records were highly correlated to those of the brachial artery, with correlation coefficients of 0.992 for systolic and 0.978 for diastolic pressures, while indirectly measured pressures in the finger were 1 to 1.7 kPa (8 to 13 mmHg) lower than the brachial arterial pressures, possibly due to the difference in intravascular pressure at the different recording sites.[140]

An additionally improved system has also been developed in which the finger cuff pressure is controlled by air.[142] In this system, an electropneumatic transducer for servo-control of air pressure, a manometer, and the plethysmographic preamplifier are mounted in a small box which is strapped to the back of the patient's hand to keep connections to the finger cuff as short as possible. Simultaneous measurements with indirect and direct methods in 17 patients showed that the average differences in each patient between indirectly measured finger blood pressure and directly measured pressures in the radial artery ranged from –0.6 to 1.1 kPa (–4.6 to 7.9 mmHg), with a mean of 0.1 ± 0.5 kPa (0.8 ± 3.8 mmHg). It was also demonstrated that the arterial pressure could be measured even during cardiopulmonary bypass when it was non-pulsatile. This type of instrument has been supplied commercially (Finapress®, Ohmeda; Madison, WI).

The vascular unloading method can also be applied to the upper arm using a standard cuff. Figure 2.54 shows the system studied by Aaslid and Brubakk.[143] Because of the difficulty in detecting arterial volume changes by photo-plethysmography in the upper arm, a Doppler ultrasound blood velocimetry was introduced. A range-gated Doppler measurement (see Section 3.2.2.4) was used to detect blood velocity in the brachial artery. An ultrasound probe that produces an ultrasound beam 20 mm wide is attached near the axilla to detect blood velocity in the brachial artery 2 to 5 cm proximal to the cuff. When the cuff is not inflated, the mean blood velocity in the brachial artery is normally 5 to 10 cm/s. The cuff pressure is controlled by an electropneumatic system, and in closed loop operation it is controlled so that the brachial blood velocity is maintained at a reference level, which is normally 3 cm/s. In this situation, the artery is partially occluded, and thus the vascular wall will be unloaded. It was shown that when the loop gain of the servo-system is sufficiently high, arterial pressure wave forms can be recorded successfully.[144]

A detailed study in 23 patients in which indirect measurements were compared with intraarterial pressures showed that deviations between pressures determined by the two methods were 0.07 ± 0.04 kPa (0.5 ± 3.2 mmHg) for the systolic, –0.08 ± 0.4 kPa (–0.6 ± 2.2 mmHg) for the mean, and –0.03 ± 0.4 kPa (–0.2 ± 2.7 mmHg) for the diastolic pressures.

Compared with the vascular unloading measurement on the finger, the measurement on the upper arm is advantageous in that the measurement site is closer to the central artery and blood pressure values, and wave forms consequently can be expected to be closer to those of the central artery. However, the measurement in the upper arm restricts blood flow to the lower arm, so that it cannot be carried out for long periods of time. It is suggested that the cuff be deflated after a flow restriction period of 2 minutes. When a fully continuous recording is required it is suggested that cuffs and ultrasound probes should be applied to both arms and that flow restriction and rest should be alternated between arms.[143]

2.3.3 INDIRECT PRESSURE MEASUREMENTS BY REACTION FORCES

When a rigid material is pressed against the skin surface or the surface of the organ, a reaction force will be exerted from the body to the applied material. A part of the reaction force is derived from elastic forces due to the deformation of the tissue, but internal pressure in the body space beneath the material also contributes to the reaction force. Under certain conditions in which elastic forces can be reduced to zero,

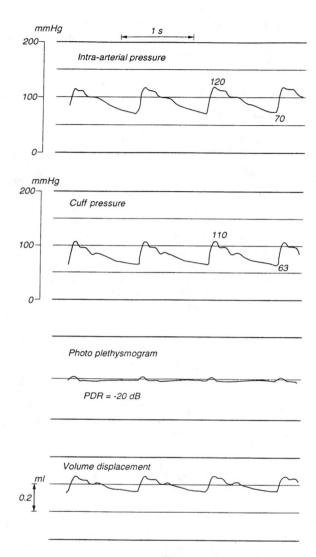

Figure 2.53 Examples of the recording of bracheal arterial pressure, cuff pressure, photo-plethysmogram, and volume displacement at the diaphragm obtained by the unit shown in Figure 2.52. (Modified from Yamakashi, K. et al., *IEEE Trans. Biomed. Eng.*, BME-27, 150, 1980.)

the internal pressure can be measured quantitatively from the reaction force. This principle has been used for noninvasive measurements of intraoccular, intra-amniotic, intra-abdominal, and intracranial pressures.

2.3.3.1   Applanation Method

When a flat surface is pressed against an object which has a flexible boundary at the outside and is filled with fluid inside that has a positive pressure, as shown in Figure 2.55(a), the internal pressure, P, may be measured by the force, F, exerted on the plane and the area of contact, a, as,

$$P = \frac{F}{a} \tag{2.26}$$

as long as the elastic force due to deformation of the surface layer at the edge of the contact surface is negligibly small. This is known as the applanation method.

The guard ring technique shown in Figure 2.55(b) has also been used. In this technique, the force is measured by the central part of the contact surface, and the surface of the guard ring is maintained in

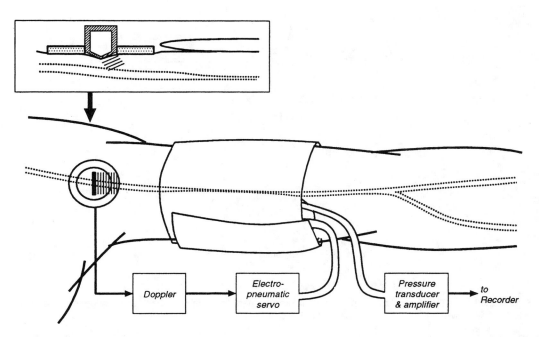

Figure 2.54 A blood pressure recording system using a pneumatic driven upper arm cuff and the Doppler ultrasound blood velocity probe. (From Aaslid, R. and Brubakk, A. O., *Circulation*, 64, 751, 1981. With permission.)

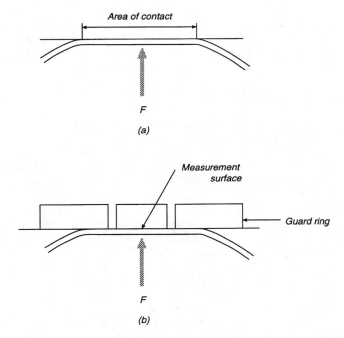

Figure 2.55 **(a)** Applanation method for indirect internal pressure measurement. Internal pressure can be estimated from the external force and the area of contact. **(b)** To eliminate the effect of the bending force at the edge of the contact surface, the guard ring technique is advantageous.

the same plane as the measurement surface. The major advantage of the guard ring technique is that although a deformation at the edge of the contact surface causes elastic force, it does not affect the force on the measurement surface located at the center.

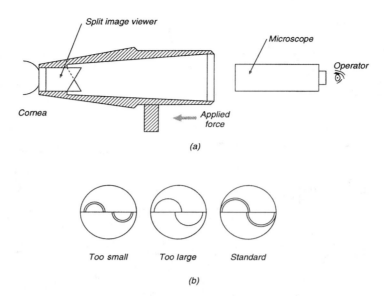

(a)

Too small   Too large   Standard

(b)

Figure 2.56 **(a)** The Goldmann applanation tonometer, and **(b)** patterns observed through the microscope. (Modified from Moses, R. A., *Am. J. Ophthalmol.*, 86, 376, 1971.)

In the applanation method with a guard ring, the surface of the guard ring and the measurement surface should be coplanar throughout the whole measurement range. To realize that, no displacement of the measurement surface relative to the guard ring is allowed. This means that the force applied to the measurement surface should be measured without displacement, which requires a force transducer of zero compliance. In principle, force measurement without displacement can be realized by compensating estimated displacement in the force transducer by using a servo-control drive system as shown later. However, force transducers with an extremely small compliance are available (see Section 4.3), so that as long as such transducers are used, servo-controlled compensation of the displacement in the force measurement is practically unnecessary for the most physiological pressure measurements.

2.3.3.2 Intraocular Pressure Measurements

Using applanation method, intraocular pressure measurement can be performed indirectly across the cornea. Although the indentation method, in which a rod is applied to the cornea and the indentation depth is measured, has been used for estimating intraocular pressure, more accurate measurements can be realized using the applanation method. A tonometer of this type is usually called the applanation tonometer.

Many different applanation tonometers have been described. The Goldmann applanation tonometer employs force measurement without a guard ring.[145,146] For this measurement, a flat plate is pressed against the cornea and the force is adjusted so that the contact diameter is a definite value, usually 3.06 mm. This adjustment is performed by viewing the contact surface through a special prism assembly as shown in Figure 2.56(a). The boundary of the contact area can be observed by instilling fluorescein in the tear liquor and illuminating with blue light. Due to the prism assembly, the split image shown in Figure 2.56(b) is seen through the microscope. When the inner edges of the meniscus in the two fields are matched, a standard contact area is obtained, and then the force is measured by a spring balance. A further increase in the contact area causes an increase in intraocular pressure, and the force increases again. The diameter of the contact area during the force measurement was determined from an observation that if the diameter of the contact area remains between 3 and 3.5 mm, the elastic force and surface tension balance each other in normal human corneas. In this amount of applanation, an increase in intraocular pressure caused by a compressed volume can be disregarded. The measurement error was estimated as less than ±0.13 kPa (±1 mmHg).[145]

The Halberg tonometer is similar in principle to the Goldmann tonometer.[147,148] The difference is that in the Halberg tonometer, a constant force is applied gravitationally, and the diameter of the applanated area is measured optically on a scale divided into 0.1 mm. This tonometer has the advantage of being simple, sturdy, and portable, but it is less accurate than the Goldmann tonometer.

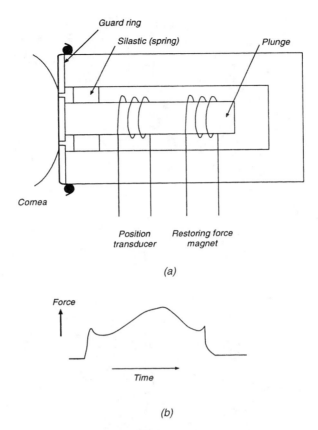

Figure 2.57 **(a)** The Mackay-Marg applanation tonometer, and **(b)** a typical record. (Modified from Walker, R. E. and Litovitz, T. L., *Exp. Eye Res.,* 13, 14, 1972.)

The applanation tonometer with a guard ring was proposed by Mackay and Marg.[149] It has a plunger 1.5 mm in diameter and a coplanar guard ring; approximately a 3-mm diameter area of the cornea is flattened by the plate. To measure applied force caused by maintaining surfaces of the plunger and the guard ring in a coplanar position, a servo-control unit was introduced. As shown in Figure 2.57(a), a deviation of the plunger surface from the guarding surface is detected by a differential transducer (see Section 4.2.1), and the plunger is driven so that it is pushed back to the correct coplanar position.

The measurement is made by recording the time course of the sustained force of the plunger, which is obtained as the current supplied to the driving coil. A typical record is shown in Figure 2.57(b). At the contact of the probe to the cornea, the force increases as the area of applanation increases, until this area completely covers the plunger surface. A further increase of the bending force in the cornea exerted at the boundary of the applanation becomes sustained by the guard ring, and the restoring force of the plunger drops. Further advance of the applanation area causes an increase in intraocular pressure, and the force increases again. Thus, the force of the first dip corresponds to the intraocular pressure. The total time of application of the probe to the cornea is only about one second, and an anesthetic is not required for the measurement.

Instead of using the servo-control system to maintain coplanarity of the plunger and the guard ring, a low compliance force transducer can be used.[150] Figure 2.58 shows a tonometer with a quartz crystal transducer. Force is detected as a charge induced by the piezoelectric effect of the quartz crystal (see Section 4.3). The deflection of the quartz transducer is a few angstroms in the measurement range, so that coplanarity can be maintained.

The pneumatic applanation tonometer has a thin rubber diaphragm which is pressurized by gas, and when the diaphragm is applied to the cornea it is applanated automatically.[151-153] The construction of the probe is shown in Figure 2.59. A constant gas flow is supplied from a jet tube located at the center of the probe, and the gas leaks through the gap between the tip of the jet tube and the thin silastic membrane

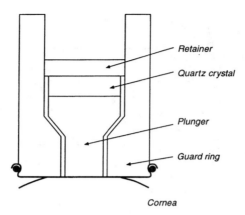

Retainer

Quartz crystal

Plunger

Guard ring

Cornea

Figure 2.58 A quartz crystal tonometer. (Modified from Mackay, R. S. et al., *IRE Trans. Bio-Med. Electron.*, 9, 174, 1962.)

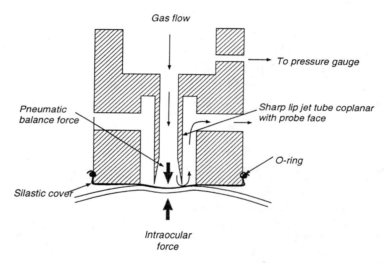

Gas flow

To pressure gauge

Pneumatic balance force

Sharp lip jet tube coplanar with probe face

O-ring

Silastic cover

Intraocular force

Figure 2.59 A pneumatic applanation tonometer. (Modified from Walker, R. E. et al., *Exp. Eye Res.*, 13, 187, 1972.)

which covers the probe. Pressure in the jet tube depends on the flow resistance of the gap, which varies with displacement of the membrane and, as a consequence, the membrane remains at a position where pneumatic force exerted on the area at the tip of the jet tube balances intraocular force exerted on the area inside the contact point of the membrane and the guard ring. Under this condition of equilibrium, it was confirmed that pressure in the jet tube is almost proportional to intraocular pressure. Thus, intraocular pressure can be measured by measuring gas pressure in the jet tube.

Pneumatic applanation tonometer equipment is available commercially (Pneumatonograph®, Alcon Laboratories; Fort Worth, TX). A clinical evaluation of 200 eyes in 100 patients showed that pneumatic tonometer readings agreed with Goldmann tonometer readings within ±0.26 kPa (±2 mmHg) in 75% of eyes, and ±0.4 kPa (±3 mmHg) in 85% of eyes.[154]

A noncontact applanation tonometer has also been developed.[155] The principle involved in this tonometer is shown in Figure 2.60. Applanation is produced by a controlled air pulse of linearly increasing force impinging on the cornea. A monitoring system senses light that is reflected from the corneal surface and records a maximal signal at the instant of applanation. Special alignment is performed visually by an optical spherometer so that the optical spherometer's axis is normal to the local corneal vertex, and the distance from the pneumatic orifice to the cornea is adjusted — 11 mm, for example, for a cornea with a radius of curvature of 7.9 mm. An air pulse is produced by a cylinder and a piston driven by a

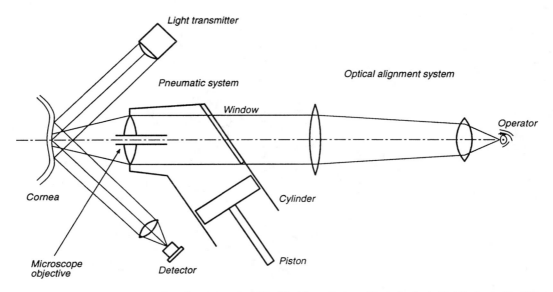

Figure 2.60 A noncontact applanation tonometer. (Modified from Forbes, M. et al., *Arch. Ophthalmol.*, 91, 134, 1974.)

solenoid, and the force exerted by the air pulse increases linearly for the first 8 ms. The curvature of the corneal surface is monitored by a collimated light beam, and when applanation is achieved, the cornea acts as a plane mirror and causes a maximum signal at the detector. At the instant when applanation is detected, the solenoid is immediately shut off to minimize air pulse force impinging upon the cornea. In a comparative study with the Goldmann tonometer in 570 eyes, both measurements agreed well within a standard deviation of 0.38 kPa (2.86 mmHg).

2.3.3.3 Intra-amniotic and Intra-abdominal Pressure Measurements

Intra-amniotic and intra-abdominal pressures can be measured noninvasively by applanation transducers with a guard ring. An applanation transducer for intra-amniotic pressure measurement for the diagnosis and controlled treatment of dystocia was designed by Smyth.[156] Figure 2.61 shows the construction of the transducer. It has a pressure-sensing area of 5 cm^2 in the center and a guard ring 7.5 cm in outer diameter. The pressure-sensing area is supported by a spring, which is mounted upon the guard ring; the deflection of the spring is measured by strain-gauges adherent to the spring. The deflection of the pressure sensing area with respect to the guard ring surface is within 0.025 cm for 11.8 kPa (120 g/cm^2), which corresponds to an intra-amniotic pressure developed by a very strong uterine contraction.

The transducer is applied to the patient over a fluid-filled part of the uterus and is held in place by hand or by a stiff elastic belt passed around the patient, avoiding fetal back and breach. The pressure of application must be sufficient to flatten the abdominal surface into contact with the guard ring.

According to Smyth,[156] the accuracy of noninvasive intra-amniotic pressure was ±0.098 kPa (±1 cmH$_2$O), as long as the deflection of the pressure-sensing area was within ±0.02 cm and the effect of hydrostatic forces was corrected adequately.

By applying the same transducer to the flank, over the abdominal cavity but not over the uterus, intra-abdominal pressure can be measured. Simultaneous measurement of intra-amniotic and intra-abdominal pressures can be made, and the pressure developed by true uterine effort can consequently be measured, since the intra-amniotic pressure is the sum of the pressures due to uterine muscle tone and the intra-abdominal pressure. A similar transducer was also used for a study of patients with symptomatic gastroesophageal reflux in whom intra-abdominal pressure was monitored over 24 hours.[157]

2.3.3.4 Intracranial Pressure Measurement in Newborn Infants

Although intracranial pressure has to be measured by direct insertion of an intraventricular catheter or by implantation of a pressure transducer in adults (as mentioned in Section 2.2.4), it can be measured indirectly and noninvasively in the newborn infant via the fontanelle. The fontanelle is the opening in the skull due to incomplete development of the bone, and the anterior fontanelle is normally present in

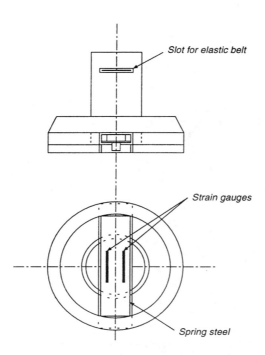

Figure 2.61   An applanation transducer for intra-amniotic and intra-abdominal pressure measurement. (Modified from Smyth, C. N., *J. Obstet. Gynaec. Br. Emp.*, 64, 59, 1957.)

the newborn and early periods of infancy. The anterior fontanelle of the newborn is about 2 cm in size, although it is quite variable in size and shape.[158] Intracranial pressure can be roughly estimated by palpation of the anterior fontanelle. Quantitative estimation of intracranial pressure via the fontanelle has been studied in two different ways, i.e., the oscillometric method and the applanation method.

The oscillometric method is based on the assumption that the volume pulsation amplitude reaches a maximum when external and internal pressures are balanced. This principle is similar to that of the oscillometric method for mean arterial blood pressure measurement (see Section 2.3.1.3). To apply external pressure on the fontanelle, Purin[159] used a circular chamber 5 cm in diameter, with a thin rubber diaphragm, which is called Marey's tambour. The pressure in the tambour was increased and decreased gradually, and volume pulsation was recorded by a kymograph. If the assumption of the oscillometric method is valid, pulsation amplitude will be maximized when the pressure in the tambour equals the intracranial pressure. Purin compared this method with measurements by lumbar or ventricular puncture in 13 infants and found that both measurements were identical. That observation was confirmed in a study using a tambour with a diameter of 4.3 cm;[160] however, a study using a 3-cm latex-covered tambour showed that peak pulsation amplitude was identified in only one of the six patients.[161]

The applanation method (see Section 2.3.3.1) can also be used for intracranial measurement via the fontanelle. Wealthall and Smallwood[161] used an applanation transducer (APT-16, Hewlett Packard; Andover, MA), which consisted of a plunger and a guard ring. The diameter of the plunger was about 6.4 mm, and the plunger movement was 1.1 μm for an applied pressure of 0.13 kPa (1 mmHg). The movement of the plunger was measured by a differential transducer. To apply the transducer to the scalp over the fontanelle, a flange was attached to the scalp by means of adhesive. A comparative study in five patients with the direct intracranial pressure measurements by ventricular puncture showed that the applanation transducer faithfully measured intracranial pressure over a wide range in each patient, and the correlation coefficient between both measurements was 0.98.

Robinson and his colleagues[162] also used the same transducer (APT-16, Hewlett Packard; Andover, MA), and they pointed out that the applanation transducer underestimated intracranial pressure when it was above 0.98 kPa (10 cmH$_2$O), but this error could be reduced by advancing the plunger to 30 μm relative to the guard ring surface, so that the plunger surface would be in the same plane as the guard ring surface when a pressure of 2.2 kPa (22.4 cmH$_2$O) was exerted on the plunger surface. By this modification, they obtained reproducible results over the range 0.5 to 59 kPa (5 to 60 cmH$_2$O).

A similar applanation transducer was reported by Edwards[163] that consisted of a plunger, a guard ring, and a differential transducer, and the plunger movement for an exerted pressure of 0.08 kPa (10 cmH$_2$O) was about 7 μm.

REFERENCES

1. Mason, D. T., Usefulness and limitations of the rate of rise of intraventricular pressure (dp/dt) in the evaluation of myocardial contractility in man, *Am. J. Cardiol.*, 23, 516, 1969.
2. Buchbinder, N. and Ganz, W., Hemodynamic monitoring: invasive techniques, *Anesthesiology*, 45, 146, 1976.
3. Longerbeam, J. K., Vannix, R., Wagner, W., and Joergenson, E., Central venous pressure monitoring: a useful guide to fluid therapy during shock and other forms of cardiovascular stress, *Am. J. Surg.*, 110, 220, 1965.
4. Burton, A. C., Kinetic energy in the circulation, in *Physiology and Biophysics of the Circulation*, Year Book Medical Publishers, Chicago, 1965, 102.
5. Greganti, F. and Guyton, A. C., Right ventricule as physiological level for reference of pressures in circulatory system, *Am J. Physiol.*, 183, 622, 1955.
6. Guyton, A. C. and Greganti, F. P., A physiologic reference point for measuring circulatory pressures in the dog — particularly venous pressure, *Am J. Physiol.*, 185, 137, 1956.
7. Chandraratna, P. A. N., Determination of zero reference level for left atrial pressure by echocardiography, *Am. Heart J.*, 89, 159, 1975.
8. Blackburn, J. P., Self-leveling venous pressure transducer, *Br. Med. J.*, 4, 825, 1968.
9. Corbett, G. A., Preston, T. D., and Bailey, J. S., A self-leveling central venous electromanometer, *Med. Biol. Eng.*, 12, 366, 1974.
10. Swan, H. J. C., Ganz, W., Forrester, J., Marcus, H., Diamond G., and Chonette, D., Catheterization of the heart in man with use of a flow-directed balloon-tipped catheter, *New Engl. J. Med.*, 283, 447, 1970.
11. Chau, H. L. and Wise, K. D., Noise due to Brownian motion in ultrasensitive solid-state pressure sensors, *IEEE Trans. Electron Dev.*, ED-34, 859, 1987.
12. Spotts, E. L. and Frank, T. P., A disposable blood pressure transducer system, *J. Clin. Eng.*, 7, 197, 1982.
13. Emergency Care Research Institute (ECRI), Physiological pressure transducer, *Health Dev.*, 8, 199, 1979.
14. Emergency Care Research Institute (ECRI), Disposable pressure transducer, *Health Dev.*, 13, 268, 1984.
15. Gordon, V. L., Welch, J. P., Carley, D., Teplick, R., and Newbower, R. S., Zero stability of disposable and reusable pressure transducers, *Med. Instrum.*, 21, 87, 1987.
16. Latimer, K. E., The transmission of sound waves in liquid-filled catheter tubes used for intravascular blood-pressure recording, *Med. Biol. Eng.*, 6, 29, 1968.
17. Fry, D. L., Physiological recording by modern instruments with particular reference to pressure recording, *Physiol. Rev.*, 40, 753, 1960.
18. Yang, S. S., Bentivoglio, L. G., Maranhão, V., and Goldberg, H., Basic measurements and calculations, in *Cardiac Catheterization Data to Hemodynamic Parameters*, 2nd ed., F. A. Davis, Philadelphia, 1978.
19. Shapiro, G. G. and Krovetz, L. J., Damped and undamped frequency responses of underdamped catheter manometer systems, *Am. Heart J.*, 80, 226, 1970.
20. Stegall, H. F., A simple, inexpensive, sinusiodal pressure generator, *J. Appl. Physiol.*, 22, 591, 1967.
21. Foreman, J. E. K. and Hutchison, K. J., Generation of sinusoidal fluid pressures of relatively high frequency, *J. Appl. Physiol.*, 29, 511, 1970.
22. Henry, W. L., Wilner, L. B., and Harriso, D. C., A calibration for detecting bubbles in cardiac catheter-manometer systems, *J. Appl. Physiol.*, 23, 1007, 1967.
23. Fry, D. L., Noble, F. W., and Mallos, A. J., An evaluation of modern pressure recording systems, *Circ. Res.*, 5, 40, 1957.
24. Dear, H. D. and Spear, A. F., Accurate method for measuring *dP/dt* with cardiac catheters and external transducers, *J. Appl. Physiol.*, 30, 897, 1971.
25. Shirer, H. W., Blood pressure measuring methods, *IRE Trans. Bio-Med. Electron.*, 9, 116, 1962.
26. Sutterer, W. F. and Wood, E. H., Strain-gauge manometers: application to recording of intravascular and intracardiac pressures, in *Medical Physics*, Vol. 3, Glasser, O., Ed., Year Book Medical Publishers, Chicago, 1960, 641.
27. Gould, L. K., Trenholme, S., and Kennedy, J. W., *In vivo* comparison of catheter manometer systems with the catheter-tip micromanometer, *J. Appl. Physiol.*, 34, 263, 1973.
28. Noble, F. W. and Barnett, G. O., An electric circuit for improving the dynamic response of the conventional cardiac catheter system, *Med. Electron. Biol. Eng.*, 1, 537, 1963.
29. Futamura, Y., Tachibana, T., Ichie, Y., Takeuchi, S., Yasui, S., Mizuno, Y., and Hisada, S., Analysis of left ventricular pressure pulses and characteristics of catheter-manometer system, *Jpn. J. Med. Electr. Biol. Eng.*, 4, 214, 1975.
30. Allard, E. M., Sound and pressure signals obtained from a single intracardiac transducer, *IRE Trans. Biomed. Electron.*, 9, 74, 1962.
31. Piemme, T. E., Pressure measurement: electrical pressure transducers, *Prog. Cardiovasc. Dis.*, 5, 574, 1963.
32. Millar, H. D. and Baker, L. E., A stable ultraminiature catheter-tip pressure transducer, *Med. Biol. Eng.*, 11, 86, 1973.
33. Samaun, T., Wise, K. D., and Angell, J. B., An IC piezoresistive pressure sensor for biomedical instrumentation, *IEEE Trans. Biomed. Eng.*, BME-20, 101, 1973.

34. Okino, H., Kitano, T., Igarashi, I., Inagaki, H., and Mizuno, M., Miniaturized blood pressure transducer, *Biomedizinische Technik*, 24, 56, 1979.

35. Esashi, M., Komatsu, H., Matsuo, T., Takahashi, M., Takishima, T., Imabayashi, K., and Ozawa, H., Fabrication of catheter-tip and sidewall miniature pressure sensor, *IEEE Trans. Electron Dev.*, ED-29, 57, 1982.

36. Pfann, W. G. and Thurston, R. N., Semiconducting stress transducers utilizing the transverse and shear piezoresistance effects, *J. Appl. Physics*, 32, 2008, 1961.

37. Bao, M., Wang, Y., and Qi, W., New development on design method of piezoresistive pressure sensor, *Transducer '87*, The Institute of Electrical Engineers of Japan, Tokyo, 1987, 299.

38. Clark, S. K. and Wise, K. D., Pressure sensitivity in anisotropically etched thin-diaphragm pressure sensors, *IEEE Trans. Electron Dev.*, ED-26, 1887, 1979.

39. Chau, H. L. and Wise, K. D., Scaling limits in batch-fabricated silicon pressure sensors, *IEEE Trans. Electron Dev.*, ED-34, 850, 1987.

40. Chau, H. L. and Wise, K. D., An ultraminiature solid-state pressure sensor for a cardiovascular catheter, *Transducer '87*, The Institute of Electrical Engineers of Japan, Tokyo, 1987, 344.

41. Polanyi, M. L., Medical application of fiberoptics, in *Digest of 6th Int. Conf. on Med. Electron. Biol. Eng.*, Tokyo, 1965, 598.

42. Ramirez, A., Hood, Jr., W. B., Polanyi, M., Wagner, R., Yankopolos, N. A., and Abelmann, W. H., Registration of intravascular pressure and sound by a fiberoptic catheter, *J. Appl. Physiol.*, 26, 679, 1969.

43. Lindström, L. H., Miniaturized pressure transducer intended for intravascular use, *IEEE Trans. Biomed. Eng.*, BME-17, 207, 1970.

44. Saito, K. and Matsumoto, H., Development and evaluation of fiberoptic pressure catheter, in *Digest of 11th Int. Conf. Med. Electron. Biol. Eng.*, Ottawa, 1976, 690.

45. Tenerz, L. and Hök, B. H., 0.5-mm diameter pressure sensor for biomedical applications, *Transducers '87*, The Institute of Electrical Engineers of Japan, Tokyo, 1987, 312.

46. Nornes, H. and Serck-Hanssen, F., Miniature transducer for intracranial pressure monitoring in man, *Acta Neurol. Scand.*, 46, 203, 1970.

47. Eversden, I. D., Modifications to a miniature pressure transducer for the measurement of intracranial pressure, *Med. Biol. Eng.*, 8, 159, 1970.

48. Brock, M. and Diefenthäler, K., A modified equipment for the continuous telemetric monitoring of epidural or subdural pressure, in *Intracranial Pressure*, Brock, M. and Dietz, H., Eds., Springer-Verlag, Berlin, 1972, 21.

49. Tindall, G. T., McGraw, C. P., and Iwata, K., Subdural pressure monitoring in head-injured patients, in *Intracranial Pressure*, Brock, M. and Dietz, H., Eds., Springer-Verlag, Berlin, 1972, 9.

50. Ikeyama, A., Furuse, M., Nagai, H., Maeda, S., Inagaki, H., Igarashi, I., and Kitano, T., Epidural measurement of intracranial pressure by newly-developed pressure transducer, *Jpn. J. Med. Elect. Biol. Eng.*, 20, 1, 1982.

51. Hill, D. G. and Allen, K. L., Improved instrument for the measurement of c.s.f. pressures by passive telemetry, *Med. Biol. Eng. Comput.*, 15, 666, 1977.

52. Viernstein, L. J., Intracranial pressure monitoring, *Johns Hopkins API Tech. Dig.*, 1, 135, 1980.

53. Ko, W. H., Hynecek, J., and Boettcher, S. F., Development of a miniature pressure transducer for biomedical applications, *IEEE Trans. Electron Devices*, ED-26, 1896, 1979.

54. Wise, K. D., Solid-state sensor technology, in *Implantable Sensor for Closed-Loop Prosthetic Systems*, Ko, W. H., Ed., Futura, Mount Kisco, NY, 1985, 3.

55. Shoji, S., Nisase, T., Esashi, M., and Matsuo, T., Fabrication of an implantable capacitive type pressure sensor, *Transducers '87*, The Institute of Electrical Engineers of Japan, Tokyo, 1987, 305.

56. Ko, W. H., Bao, M-H., and Hong, Y-D., A high-sensitivity integrated-circuit capacitive pressure transducer, *IEEE Trans. Electron Dev.*, ED-29, 48, 1982.

57. Smith, M. J. S., Bowman, L., and Meindl, J. D., Analysis, design, and performance of a capacitive pressure sensor IC, *IEEE Trans. Biomed. Eng.*, BME-33, 163, 1986.

58. Leung, A. M., Ko, W. H., Spear, T. M., and Bettice, J. A., Intracranial pressure telemetry system using semicustom integrated circuits, *IEEE Trans. Biomed. Eng.*, BME-33, 386, 1986.

59. Farrar, J. T., Zworykin, V. K., and Baum, J., Pressure-sensitive telemetering capsule for study of gastrointestinal motility, *Science*, 126, 975, 1957.

60. Farrar, J. T. and Bernstein, J. S., Recording of intraluminal gastrointestinal pressures by a radiotelemetering capsule, *Gastroenterology*, 35, 603, 1958.

61. Mackay, R. S., Radio telemetering from within the human body, *IRE Trans. Med. Electron.*, 6, 100, 1959.

62. Farrar, J. T., Berkley, C., and Zworykin, V. K., Telemetering of intraenteric pressure in man by an externally energized wireless capsule, *Science*, 131, 1814, 1960.

63. Smyth, C. N. and Wolff, H. S., Application of endoradiosound or "wireless pill" to recording of uterine contractions and foetal heart sounds, *Lancet*, 2, 412, 1960.

64. Horowitz, L. and Farrar, J. T., Intraluminal small intestinal pressures in normal patients and in patients with functional gastrointestinal disorders, *Gastroenterology*, 42, 455, 1962.

65. Rowlands, E. N. and Wolff, H. S., The radio pill. Telemetering from the digestive tract, *Br. Comm. Electron.*, 598, 1960.

66. Rappaport, M. B., Bloch, E. H., and Irwin, J. W., A manometer for measuring dynamic pressures in the microvascular system, *J. Appl. Physiol.*, 14, 651, 1959.

67. Levasseur, J. E., Funk, F. C., and Patterson, Jr., J. L., Physiological pressure transducer for microchemocirculatory studies, *J. Appl. Physiol.*, 27, 422, 1969.
68. Wunderlich, P. and Schnermann, J., Continuous recording of hydrostatic pressure in renal tubules and blood capillaries by use of a new pressure transducer, *Pflügers Arch.*, 313, 89, 1969.
69. Wiederhielm, C. A., Woodbury, J. W., Kirk, S., and Rushmer, R. F., Pulsatile pressures in the microcirculation of frog's mesentery, *Am. J. Physiol.*, 207, 173, 1964.
70. Fox, J. R. and Wiederhielm, C. A., Characteristics of the servo-controlled micropipet pressure system, *Microvasc. Res.*, 5, 324, 1973.
71. Intaglietta, M., Pressure measurements in the microcirculation with active and passive transducers, *Microvasc. Res.*, 5, 317, 1973.
72. Intaglietta, M., Pawula, R. F., and Tompkins, W. R., Pressure measurements in the mammalian microvasculature, *Microvasc. Res.*, 2, 212, 1970.
73. Zweifach, B. W., Quantitative studies of microcirculatory structure and function. 1. Analysis of pressure distribution in the terminal vascular bed in cat mesentry, *Circ. Res.*, 34, 843, 1974.
74. Lipowsky, H. H. and Zweifach, B. W., Methods for the simultaneous measurement of pressure differentials and flow in single unbranched vessels of the microcirculation for rheological studies, *Microvasc. Res.*, 14, 345, 1977.
75. Bhattacharya, J. and Staub, N. C., Direct measurement of microvascular pressures in the isolated perfused dog lung, *Science*, 210, 327, 1980.
76. Raj, J. U. and Chen, P., Microvascular pressures measured by micropuncture in isolated perfused lamb lungs, *J. Appl. Physiol.*, 61, 2194, 1986.
77. Fike, C. D., Lai-Fook, S. J., and Bland, R. D., Microvascular pressures measured by micropuncture in lungs of newborn rabbits, *J. Appl. Physiol.*, 63, 1070, 1987.
78. Tillmanns, H., Steinhausen, M., Leinberger, H., Thederan, H., and Kübler, W., Pressure measurements in the terminal vascular bed of the epimyocardium of rats and cats, *Circ. Res.*, 49, 1202, 1981.
79. Nellis, S. H. and Liedtke, A. J., Pressure and dimensions in the terminal vascular bed of the myocardium determined by a new free-motion technique, in *Microcirculation of the Heart*, Tillmanns, H., Kübler, W., and Zebe, H., Eds., Springer-Verlag, Berlin, 1982, 61.
80. Maxted, K. J., Shaw, A., and MacDonald, T. H., Choosing a catheter system for measuring intra-oesophageal pressure, *Med. Biol. Eng. Comput.*, 15, 398, 1977.
81. Harris, L. D., Winans, C. S., and Pope, II, C. E., Determination of yield pressures: a method for measuring anal sphincter competence, *Gastroenterology*, 50, 754, 1966.
82. Pope, II, C. E., A dynamic test of sphincter strength: its application to the lower esophageal sphincter, *Gastroenterology*, 52, 779, 1967.
83. Winans, C. S. and Harris, L. D., Quantitation of lower esophageal spincter competence, *Gastroenterology*, 52, 773, 1967.
84. Harrison, N. W. and Constable, A. R., Urethral pressure measurement: a modified technique, *Br. J. Urol.*, 42, 229, 1970.
85. Ask, P., Öberg, P. Å., and Tibbling, L., Static and dynamic characteristics of fluid-filled esophageal manometry systems, *Am. J. Physiol.*, 233, E389, 1977.
86. Shaw, A., Baillie, A. D., and Runcie, J., Oesophageal manometry by liquid-filled catheters, *Med. Biol. Eng. Comput.*, 18, 488, 1980.
87. Arndorfer, R. C., Stef, J. J., Dodds, W. J., Linehan, J. H., and Hogan, W. J., Improved infusion system for intraluminal esophageal manometry, *Gastroentorology*, 73, 23, 1977.
88. Ask, P., Low-compliance perfusion pump for esophageal manometry, *Med. Biol. Eng. Comput.*, 16, 732, 1978.
89. Dent, J., A new technique for continuous sphincter pressure measurement, *Gastroenterology*, 71, 263, 1976.
90. Kahrilas, P. J., Dent, J., Dodds, W. J., Hogan, W. J., and Arndorfer, R. C., A method for continuous monitoring of upper esophageal sphincter pressure, *Dig. Dis. Sci.*, 32, 121, 1987.
91. Guyton, A. C., Granger, H. J., and Taylor, A. E., Interstitial fluid pressure, *Physiol. Reviews*, 51, 527, 1971.
92. Wig, H., Reed, R. K., and Aukland, K., Measurement of interstitial fluid pressure in dogs: evaluation of methods, *Am J. Physiol.*, 253 (*Heart Cir. Physiol.*, 22), H283, 1987.
93. Guyton, A. C., A concept of negative interstitial pressure based on pressures in implanted perforated capsules, *Circ. Res.*, 12, 399, 1963.
94. Ott, C. E., Cluche, J-L., and Knox, F. G., Measurement of renal interstitial fluid pressure with polyethylene matrix capsules, *J. Appl. Physiol.*, 38, 937, 1975.
95. Wig, H., Reed, R. K., and Aukland, K., Micropuncture measurement of interstitial fluid pressure in rat subcutis and skeletal muscle: comparison to wick-in-needle technique, *Microvasc. Res.*, 21, 308, 1981.
96. Fadnes, H. O. and Aukland, K., Protein concentration and colloid osmotic pressure of interstitial fluid collected by the wick technique: analysis and evaluation of the method, *Microvasc. Res.*, 14, 11, 1977.
97. Fadnes, H. O., Reed, R. K., and Aukland, K., Interstitial fluid pressure in rats measured with a modified wick technique, *Microvasc. Res.*, 14, 27, 1977.
98. Wig, H., Comparison of methods for measurement of interstitial fluid pressure in cat skin/subcutis and muscle, *Am. J. Physiol.*, 249 (*Heart Circ. Physiol.*, 18), H929, 1985.
99. Hargens, A. R., Mubarak, S. J., Owen, C. A., Garetto, L. P., and Akeson, W. H., Interstitial fluid pressure in muscle and compartment syndromes in man, *Microvasc. Res.*, 14, 1, 1977.

100. Castle, G. S. P., Lorgan, J. G., Rorabeck, C. H., and Hardie, R., The slit catheter system: a new diagnostic method for measurement of limb compartmental pressure, *J. Clin. Eng.*, 6, 219, 1981.
101. Kramer, G. C., Sibley, L., Aukland, K., and Renkin, E. M., Wick sampling of interstitial fluid in rat skin: further analysis and modifications of the method, *Microvasc. Res.*, 32, 39, 1986.
102. Joles, J. A., Koomans, H. A., and Berckmans, R. J., Colloid osmotic pressure of subcutaneous wick fluid in rats, *Microvasc. Res.*, 35, 139, 1988.
103. Reed, R. K., An implantable colloid osmometer: measurements in subcutis and skeletal muscle of rats, *Microvasc. Res.*, 18, 83, 1979.
104. Rabek, J. W., Pressure transducer, *U.S. Patent 3*, 965, 746, 1976.
105. Doebelin, E. O., *Measurement System Application and Design*, rev. ed., McGraw-Hill, New York, 1975, 387.
106. Petrie, J. C., O'Brien, E. T., Littler, W. A., and de Swiet, M., Recommendations on blood pressure measurement, *Br. Med. J.*, 293, 611, 1986.
107. Kirkendall, W. M., Feinleib, M., Freis, E. D., and Mark, A. L., Recommendations for human blood pressure determination by sphygmomanometers, *Circulation*, 62, 1146A, 1980.
108. Alexander, H., Cohen, M. L., and Steinfeld, L., Criteria in the choice of an occluding cuff for the indirect measurement of blood pressure, *Med. Biol. Eng. Comput.*, 15, 2, 1977.
109. Simpson, J. A., Jamieson, G., Dickhaus, D. W., and Grover, R. F., Effect of size of cuff bladder on accuracy of measurement of indirect blood pressure, *Am. Heart J.*, 70, 208, 1965.
110. van Montfrans, G. A., van der Hoeven, G. M. A., Karemaker, J. M., Wieling, W., and Dunning, A. J., Accuracy of auscultatory blood pressure measurement with a long cuff, *Br. Med. J.*, 295, 354, 1987.
111. Ur, A. and Gordon, M., Origin of Korotkoff sounds, *Am. J. Physiol.*, 218, 524, 1970.
112. Geddes, L. A., Hoff, H. E., and Badger, A. S., Introduction of the auscultatory method of measuring blood pressure — including a translation of Korotkoff's original paper, *Cardiovasc. Res. Cent. Bull.*, 5, 57, 1966.
113. Geddes, L. A. and Moore, A. G., The efficient detection of Korotkoff sounds, *Med. Biol. Eng.*, 6, 603, 1968.
114. Greatorex, C. A., Noninvasive blood pressure measurement, in *Noninvasive Physiological Measurements*, Vol. 1, Rolfe, P., Ed., Academic Press, London, 1979, 193.
115. Golden, Jr., D. P., Wolthuis, R. A., Hoffler, G. W., and Gowen, R. J., Development of a Korotkov sound processor for automatic idenfication of auscultatory events. Part 1. Specification of preprocessing bandpass filters, *IEEE Trans. Biomed. Eng.*, BME-21, 114, 1974.
116. Kantrowitz, P., A review of noninvasive blood pressure measurement using a cuff, with particular respect to motion artifact, *Biomed. Eng.*, 8, 480, 1973.
117. Wolthuis, R. A., Golden, Jr., D. P., and Hoffler, G. W., Development of a Korotkov sound processor for automatic identification of auscultatory events. Part 2. Decision logic specifications and operational verification, *IEEE Trans. Biom. Eng.*, BME-21, 119, 1974.
118. Ragan, C. and Bordley, III, J., The accuracy of clinical measurements of arterial blood pressure, *Bull. Johns Hopkins Hosp.*, 49, 504, 1941.
119. Roberts, L. N., Smiley, J. R., and Manning, G. W., A comparison of direct and indirect blood-pressure determinations, *Circulation*, 8, 232, 1953.
120. van Bergen, F. H., Weatherhead, D. S., Treloar, A. E., Dobkin, A. B., and Buckley, J. J., Comparison of indirect and direct methods of measuring arterial blood pressure, *Circulation*, 10, 481, 1954.
121. Holland, W. W. and Humerfelt, S., Measurement of blood-pressure: comparison of intra-arterial and cuff values, *Br. Med. J.*, 2, 1241, 1964.
122. Nagle, F. J., Naughton, J., and Balke, B., Comparisons of direct and indirect blood pressure with pressure-flow dynamics during exercise, *J. Appl. Physiol.*, 21, 317, 1966.
123. Bruner, J. M. R., Krenis, L. J., Kunsman, J. M., and Sherman, A. P., Comparison of direct and indirect methods of measuring arterial blood pressure, Part I, *Med. Instrument.*, 15, 11, 1981.
124. Bruner, J. M. R., Krenis, L. J., Kunsman, J. M. and Sherman, A. P., Comparison of direct and indirect methods of measuring arterial blood pressure, Part II, *Med. Instrument.*, 15, 97, 1981.
125. Bruner, J. M. R., Krenis, L. J., Kunsman, J. M., and Sherman, A. P., Comparison of direct and indirect methods of measuring arterial blood pressure, Part III, *Med. Instrument.*, 15, 182, 1981.
126. Posey, J. A., Geddes, L. A., Williams, H., and Moore, A. G., The meaning of the point of maximum oscillations in cuff pressure in the indirect measurement of blood pressure, Part 1, *Cardiovasc. Res. Cent. Bull.*, 8, 15, 1969.
127. Yamakoshi, K., Shimazu, H., Shibata, M., and Kamiya, A., New oscillometric method for indirect measurement of systolic and mean arterial pressure in the human finger. Part 1. Model experiment, *Med. Biol. Eng. Comput.*, 20, 307, 1982.
128. Yamakoshi, K., Shimazu, H., Shibata, M., and Kamiya, A., New oscillometric method for indirect measurement of systolic and mean arterial pressure in the human finger. Part 2. Correlation study, *Med. Biol. Eng. Comput.*, 20, 314, 1982.
129. Ramsey, III, M., Noninvasive automatic determination of mean arterial pressure, *Med. Biol. Eng. Comput.*, 17, 11, 1979.
130. Shimazu, H., Ito, H., Kobayashi, H., and Yamakoshi, Y., Idea to measure diastolic arterial pressure by volume oscillometric method in human fingers, *Med. Biol. Eng. Comput.*, 24, 549, 1986.

131. Shimazu, H., Kobayashi, H., Ito, H., and Yamakoshi, K., Indirect measurement of arterial pressure in the limbs of babies and children by the volume oscillometric method, *J. Clin. Eng.*, 12, 297, 1987.

132. Stegall, H. F., Kardon, M. B., and Kemmerer, W. T., Indirect measurement of arterial blood pressure by Doppler ultrasonic sphygmomanometry, *J. Appl. Physiol.*, 25, 793, 1968.

133. Kirby, R. R., Kemmerer, W. T., and Morgan, J. L., Transcutaneous Doppler measurement of blood pressure, *Anesthesiology*, 31, 86, 1969.

134. Zahed, B., Sadove, M. A., Hatano, S., and Wu, H. H., Comparison of automated Doppler ultrasound and Kortkoff measurements of blood pressure of children, *Anesthesia Analgesia*, 50, 699, 1971.

135. Gundersen, J. and Ahlgren, I., Evaluation of an automatic device for measurement of the indirect systolic and diastolic blood pressure, Arteriosonde 1217, *Acta Anaesth. Scand.*, 17, 203, 1973.

136. Kazamias, T. M., Gander, M. P., Franklin, D. L., and Ross, Jr., J., Blood pressure measurement with Doppler ultrasonic flowmeter, *J. Appl. Physiol.*, 30, 585, 1971.

137. Peñáz, J., Photoelectric measurement of blood pressure, volume and flow in the finger, *Dig. 10th Int. Conf. Med. Biol. Eng. Dresden*, 1973, 104.

138. Peñáz, J., Voigt, A., and Teichmann, W., Beitrag zur fortlaufenden indirekten Blutdruckmessung, *Zschr. Med. Jahrg.*, 31, 1030, 1976.

139. Yamakoshi, K., Shimazu, H., and Togawa, T., Indirect measurement of instantaneous arterial blood pressure in the rat, *Am. J. Physiol.*, 237, H632, 1979.

140. Yamakoshi, K., Shimazu, H., and Togawa, T., Indirect measurement of instantaneous arterial blood pressure in the human finger by the vascular unloading technique, *IEEE Trans. Biomed. Eng.*, BME-27, 150, 1980.

141. Yamakoshi K., Kamiya, A., Shimazu, H., Ito, H., and Togawa, T., Noninvasive automatic monitoring of instantaneous arterial blood pressure using vascular unloading technique, *Med. Biol. Eng. Comput.*, 21, 557, 1983.

142. Smith, N. T., Weseling, K. H., and de Wit, B., Evaluation of two prototype devices producing noninvasive, pulsatile, calibrated blood pressure measurement from a finger, *J. Clin. Monit.*, 1, 17, 1985.

143. Aaslid, R. and Brubakk, A. O., Accuracy of an ultrasound Doppler servo method for noninvasive determination of instantaneous and mean arterial blood pressure, *Circulation*, 64, 75, 1981.

144. Brubakk, A. O. and Aaslid, R., A new method for indirect measurement of pulsatile arterial blood pressure, *ISAM 1979*, 1979, 468.

145. Goldmann, H. and Schmidt, Th., Über Applanationstonometrie, *Ophthalmologica*, 134, 221, 1957.

146. Moses, R. A., The Goldmann applanation tonometer, *Am. J. Ophthalmol.*, 46, 865, 1958.

147. Halberg, G. P., Hand applanation tonometer, *Trans. Am. Acad. Ophthalmol. Otolaryngol.*, 72, 112, 1968.

148. François, J., Vancea, P. and Vanderkerckhove, R., Halberg tonometer. An evaluation, *Arch Ophthalmol.*, 86, 376, 1971.

149. Mackay, R. S. and Marg, E., Fast, automatic, electronic tonometers based on an exact theory, *Acta Ophthalmol.*, 37, 495, 1959.

150. Mackay, R. S., Marg, E., and Oechsli, R., Quartz crystal tonometer, *IRE Trans. Bio-Med. Electron.*, 9, 174, 1962.

151. Langham, M. E. and McCarthy, E., A rapid pneumatic applanation tonometer. Comparative findings and evaluation, *Arch Ophthalmol.*, 79, 389, 1968.

152. Walker, R. E. and Litovitz, T. L., An experimental and theoretical study of the pneumatic tonometer, *Exp. Eye Res.*, 13, 14, 1972.

153. Walker, R. E., Litovitz, T. L., and Langham, M. E., Pneumatic applanation tonometer studies. II. Rabbit cornea data, *Exp. Eye Res.*, 13, 187, 1972.

154. Quigley, H. A. and Langham, M. E., Comparative intraocular pressure measurements with the pneumatonograph and Goldmann tonometer, *Am. J. Ophthalmol.*, 80, 266, 1975.

155. Forbes, M., Pico, G., Jr., and Grolman, B., A noncontact applanation tonometer. Description and clinical evaluation, *Arch. Ophthalmol.*, 91, 134, 1974.

156. Smyth, C. N., The guard-ring tocodynamometer. Absolute measurement of intra-amniotic pressure by a new instrument, *J. Obstet. Gynaec. Br. Emp.*, 64, 59, 1957.

157. Wernly, J. A., DeMeester, T. R., Bryant, G. H., Wang, C.-I, Smith, R. B., and Skinner, D. B., Intra-abdominal pressure and manometric data of the distal esophageal spincter, *Arch Surg.*, 115, 534, 1980.

158. Popitch, G. A. and Smith, D. W., Fontanels: range of normal size, *J. Ped.*, 80, 749, 1972.

159. Purin, V. R., Measurement of the cerebrospinal fluid pressure in the infant without puncture. A new method, *Pediatriya*, 43, 82, 1964

160. Hayashi, T., Studies on early diagnosis of infantile central nervous system disorders, especially, for hydrocephalic infant — measurement of intracranial pressure via the anterior fontanelle and analysing the fontanelle "pulse wave", *Bull. Kurume Med. Coll.*, 38, 401, 1975.

161. Wealthall, S. R. and Smallwood, R., Methods of measuring intracranial pressure via the fontanelle without puncture, *J. Neurol. Neurosurg. Psychiat.*, 37, 88, 1974.

162. Robinson, R. O., Rolfe, P., and Sutton, P., Noninvasive method for measuring intracranial pressure in normal newborn infants, *Develop. Med. Child Neurol.*, 19, 305, 1977.

163. Edwards, J., An intracranial pressure tonometer for use on neonates: preliminary report, *Develop. Med. Child Neurol.*, 32, Suppl. 38, 1974.

Chapter 3

Flow Measurement

3.1 OBJECT QUANTITIES

3.1.1 UNITS OF FLOW MEASUREMENTS

The amount of flow is usually measured as a volume flow rate, which is the rate at which a volume crosses a surface. In the SI system, the unit of volume flow rate is m³/s, but in physiological measurements, the units l/s, l/min, or ml/min are commonly used. The mass flow rate can also be used, which is the rate at which a mass crosses a surface. The unit of mass flow rate in the SI system is kg/s.

When a fluid, such as blood, is supplied to or drained from the tissue uniformly, the amount of flow is commonly measured as volume flow rate per unit mass of the tissue. The SI unit is m³/s·kg, but ml/min·100 g is commonly used in physiological measurements. Similarly, a volume flow rate per unit volume of tissue can also be considered that represents the number of turnovers of the fluid in the unit time interval; thus, the units are s⁻¹, min⁻¹, etc. Flow can also be characterized in terms of flow velocity. The SI unit for flow velocity is m/s.

If flow velocity is not uniform, a velocity gradient, which is the derivative of velocity in respect to a specific coordinate, exists. In viscous fluid, a tangential force called shear stress appears in proportion to the velocity gradient. In other words, when the flow velocity $U(z)$ has a velocity gradient, $dU(z)/dz$, in the z direction, the shear stress τ can be represented as

$$\tau = \mu \frac{dU(z)}{dz} \tag{3.1}$$

where μ is the viscosity of the fluid that is constant in Newtonian fluid. The unit of shear stress in the SI system is pascal (Pa) and that of viscosity is Pa·s. In physiological measurements, P (poise) and cP (centipoise) are commonly used for viscosity and are converted to Pa·s as

$$1\,\text{P} = 0.1\,\text{Pa·s} \tag{3.2}$$

$$1\,\text{cP} = 10^{-3}\,\text{Pa·s}. \tag{3.3}$$

For a flow having a mean flow velocity, $\overline{U}$, in a circular cross-sectional tube with an internal diameter, d, a quantity, Re, defined as

$$\text{Re} = \frac{\rho \overline{U} d}{\mu} \tag{3.4}$$

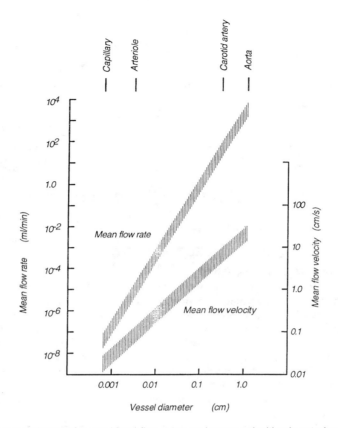

Figure 3.1 A rough estimate of mean blood flow rates and mean velocities in arteries of different sizes.

is called the Reynold's number, where ρ is the density of the fluid. Reynold's number is a nondimensional quantity.

For gas flow, the volume flow rate can be defined in a way similar to that for a liquid. However, gas is compressible, and thermal expansion is not negligible under many physiological measurement conditions. Thus, the exact molar content of gas or each gas content across a surface cannot be defined simply by its volume. For accurate measurements, pressure and temperature should at least be identified. Changes in water vapor pressure may also affect the relative content of the gas composition.

Generally, any volume or volume flow rate can be converted to the quantity measured at the standard temperature and pressure, dry (STPD), which corresponds to 0°C, 101.325 kPa (1 atm), and zero water vapor pressure. This conversion can be made as long as temperature, pressure, and water vapor content are known whenever the measurement is made.

In respiratory measurements, conditions which differ from standard ones are sometimes used. Typical body temperature conditions are ambient pressure and saturated with water vapor (BTPS), and ambient temperature and pressure saturated with water vapor (ATPS). These conditions are assigned to each unit, such as l/s STPD, l/s BTPS, etc.

3.1.2 REQUIREMENTS FOR MEASUREMENT RANGES
3.1.2.1 Blood Flow in a Single Vessel
The blood flow rate, as well as mean flow velocity in a blood vessel, can be roughly estimated by the size of the blood vessel, because the vessel size can vary adaptively with the blood flow rate.[1] In an actual arterial system, a correlation exists between blood flow rate or velocity and arterial diameter, as shown in Figure 3.1. Blood flow rate is roughly proportional to the third power of vessel diameter, while mean flow velocity is roughly proportional to the diameter. The ratio of flow rates in a larger artery about 2 cm in diameter and a capillary of 6 μm is more than 10^9, and that of flow velocity is about 2000. No one measurement method is available which is applicable to the whole range of flow rate or velocity. Thus, different methods should be used for different ranges of flow or velocity.

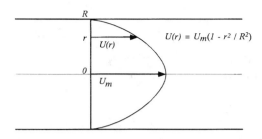

Figure 3.2 Parabolic velocity profile realized in steady laminar flow in a long circular conduit.

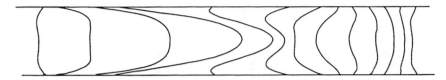

Figure 3.3 Schematic of typical velocity profiles in the artery in one cardiac cycle.

In a blood vessel, or in any conduit having a flowing fluid, flow velocity is never uniform over a cross-section but instead has a velocity distribution. If a conduit is assumed to be a long straight tube having a circular cross-section and flow is assumed to be steady and laminar, a parabolic velocity profile as shown in Figure 3.2 will develop. The velocity, $U(r)$, at a point at distance, r, from the center of the tube is expressed as

$$U(r) = U_m\left(1 - \frac{r^2}{R^2}\right) \tag{3.5}$$

where R is the internal radius of the tube, and U_m is maximum velocity. Thus, flow rate, Q, is obtained as

$$Q = \int_0^R U(r) \cdot 2\pi r dr = \frac{1}{2}\pi R^2 U_m. \tag{3.6}$$

The flow rate, Q, divided by the cross-sectional area, πr^2, is mean velocity and is $U_m/2$; i.e., the mean velocity is just a half of the maximum velocity when the velocity profile is parabolic.

Where the blood flow is not steady but pulsatile, the velocity profile differs from the parabolic one. In a large artery, very high velocity can occur temporarily so that turbulent flow appears. If some obstructions exist in the artery, turbulence will occur more easily. Even in turbulent flows, the flow rates can be measured by appropriate methods, but an exact velocity profile cannot be identified.

When a flow is pulsatile, the velocity profile varies in shape from time to time. In Figure 3.3, typical velocity profiles in a large artery during a heart beat are shown. The backward velocity component is commonly observed even in small arteries, and the discrimination of flow direction is usually required in instantaneous velocity measurements. Backward flow occurs at the end of systole and is substantial in patients with aortic valvular insufficiency.

3.1.2.2 Tissue Blood Flow

Tissue blood flow differs significantly for various tissues and physiological conditions. Figure 3.4 shows normal ranges of tissue blood flow rates of different organs in humans.[2] Tissue blood flow is usually represented as the volume flow rate per unit mass of the tissue. If a region of a tissue is perfused uniformly, the tissue blood flow is estimated by the total blood flow perfused into the region divided by the weight of the tissue in the perfused region. If the flow is not uniformly distributed, the average tissue blood flow can be defined as a rough estimate of the circulatory condition. For example, the average tissue blood flow in a segment of a limb can be regarded as an index of the peripheral circulation.

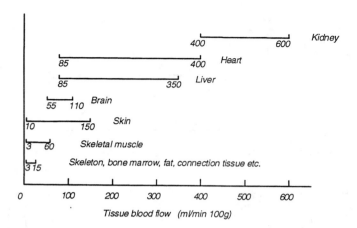

Figure 3.4 Rounded figure of tissue blood flow in different organs. Each range roughly indicates variation from rest to maximal vasodilatation.

The local tissue blood flow in an organ is also an important quantity for clinical diagnosis and physiological studies. If the local tissue blood flow can be measured at every part of an organ, the complete flow distribution will be obtained. Such a technique, called flow imaging, is valuable in clinical diagnoses and physiological studies.

3.1.2.3 Respiratory Gas Flow
The ventilation of the lungs can be assessed by studying a gas volume and its variations in the lung. Inspiratory and expiratory gas volumes or gas flow measurements provide data which characterize the ventilation of the lung, while actual gas volumes in the lung cannot be determined simply, because some amount of gas always remains in the lung even at maximal expiratory effort.

The gas flow at the airway is almost equal to the time derivative of the gas volume in the lung, as long as the temperature, pressure, and water vapor content of the ventilating gas are unchanged. Thus, flow measurement at the airway can be substituted by instantaneous lung volume measurement, and lung volume measurement can be substituted by instantaneous flow measurement at the airway as long as the initial volume is known.

In clinical spirometry, measurement ranges of flow should cover peak flow rate at a maximal expiratory effort. According to the standard presented by the American Thoracic Society (ATS), the required ranges for flow and volume measurements are 0 to 12 l/s and 0 to 7 l.[3]

In respiratory measurement, gas composition may change significantly. Oxygen and carbon dioxide contents in expired air vary depending on the gas transfer rate in the lung. Oxygen content is increased when pure oxygen is added to the inspired air. Tracer gases such as helium and argon are sometimes used in the pulmonary function tests. In anesthetic monitors, flow measurements are required for air containing an anesthetic gas.

A difference in the relative composition of gases may affect flow measurements due to the difference in the physical properties of these gases. Although a flow measurement method that is unaffected by gas composition is desirable, existing methods are more or less affected by some physical properties of gas, and thus corrections are sometimes required when gas composition varies widely. The physical properties of mixed gas are usually assumed to be a linear relation of the value of each gas, although small deviations from this assumption exist in some gas mixtures.[4]

3.2 BLOOD FLOW MEASUREMENTS IN SINGLE VESSELS

In this section, the different methods for blood flow measurements in single vessels are discussed according to the measurement principles, while versatile applications are considered for each method.

3.2.1 ELECTROMAGNETIC FLOWMETER
3.2.1.1 Principle
The electromagnetic flowmeter is based on the principle that when a fluid containing electric charges flows in a magnetic field, an electromotive force is generated. If a particle having a charge, q, moves

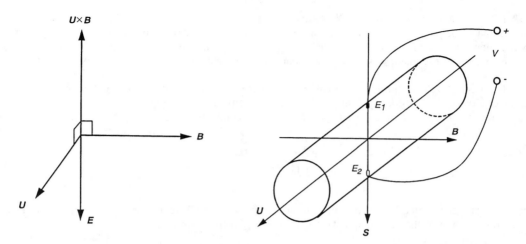

Figure 3.5 Relationship between flow velocity, U; magnetic flux density, B; developed electric field, E; and el80 ectromotive force, V.

with a velocity $\mathbf{U}$ in a magnetic field of magnetic flux density, $\mathbf{B}$, then a force, $\mathbf{F}$, will be exerted on the particle, which is expressed in vector form as

$$\mathbf{F} = q(\mathbf{U} \times \mathbf{B}). \tag{3.7}$$

In an electrolyte solution, such as blood flowing across a magnetic field, ions of positive and negative charges will move in opposite directions; consequently, an electric field, $\mathbf{E}$, will be generated, so that $\mathbf{F}$ is balanced with the electric force, $q\mathbf{E}$, i.e.,

$$q\mathbf{E} + q(\mathbf{U} \times \mathbf{B}) = 0. \tag{3.8}$$

Thus, if two electrodes are placed along this electric field, a potential difference

$$V = \mathbf{S} \cdot \mathbf{E} = -\mathbf{S} \cdot (\mathbf{U} \times \mathbf{B}) \tag{3.9}$$

will appear between this electrode pair, where $\mathbf{S}$ is a vector corresponding to a segment connecting the locations of two electrodes. These relations are shown in Figure 3.5. If $\mathbf{U}$ and $\mathbf{B}$ are perpendicular, the electromotive force, V, can be obtained as

$$V = d \cdot U \cdot B \tag{3.10}$$

where d is the distance between two electrodes, $U = |\mathbf{U}|$, and $B = |\mathbf{B}|$.

In the actual blood vessel, the blood velocity is not uniform. Nevertheless, it can be shown that Equation (3.10) is still valid by taking U as the mean velocity as long as the velocity profile is axisymmetric about the longitudinal axis of the vessel, as mentioned later. This is an advantage of the electromagnetic flowmeter. Using mean velocity, the flow rate, Q, is expressed as

$$Q = \frac{\pi d^2 U}{4} = \frac{\pi d V}{4B} \tag{3.11}$$

as long as the flow is axisymmetric. In SI terminology, Q is expressed in m³/s, B in tesla, d in m, and V in volts.

3.2.1.2 Factors Affecting the Measurements

While the principle of the electromagnetic blood flowmeter is rather simple, there are many factors which may affect sensitivity. At the very least, velocity profile, magnetic field distribution, electric conductivities of the inside vessel wall, and outside media may affect the electromotive force to some extent.

The effect of velocity profile was analyzed by Shercliff.[5] If the vessel is assumed to be circular, the conductivity of the outside medium is zero; the velocity profile is represented as $U(r, \theta)$; and the electromotive force, V, at an electrode pair arranged perpendicular to the magnetic field is given as

$$V = \frac{2B}{\pi a} \int_0^{2\pi} \int_0^a U(r,\theta) \cdot W(r,\theta) r \; dr \; d\theta \qquad (3.12)$$

where B is magnetic flux density, a is the radius of the vessel, r and θ are the variables of polar coordinates in a cross-section of the vessel, and $W(r,\theta)$ is a weight function expressed as

$$W(r,\theta) = \frac{a^4 + a^2 r^2 \cos 2\theta}{a^4 + r^4 + 2a^2 r^2 \cos 2\theta} \qquad (3.13)$$

where the angle, θ, is zero at the direction of the magnetic field. If the velocity profile is axisymmetric, $U(r,\theta) = U(r)$. While

$$\int_0^{2\pi} W(r,\theta) d\theta = 2\pi, \qquad (3.14)$$

the expression of the electromotive force, V, for a flow having an axisymmetric velocity profile, can be reduced as

$$V = \frac{2B}{\pi a} \cdot 2\pi \int_0^{2\pi} U(r,\theta) r dr = 2a \cdot U \cdot B, \qquad (3.15)$$

which is exactly Equation (3.10).

The weight function $W(r, \theta)$ has a distribution as shown in Figure 3.6. The weight is the contribution of the velocity component at a point (r, θ) to the electromotive force induced at the electrode. It is recognized from this figure that the weight function is greater at points near the electrodes. Therefore, if a velocity profile is not axisymmetric but flow is concentrated near the electrode, induced voltage will be higher even though the flow rate is the same. Actually, it was confirmed experimentally that the apparent sensitivity was changed from 0.5 to 2 times when a fluid jet was injected into various positions in the cross-section of the tube.[6,7] In an actual vascular system, similar situations may occur when the electrode is located close to a vascular branch.

Inhomogeneity of the magnetic field also affects the sensitivity. In actual electromagnetic blood flowmeter probes, an electrically excited magnet or a pair of coreless coils is placed close to the vessel wall, and because the size of the magnet or coil is comparable to the vessel diameter, magnetic field inhomogeneity is unavoidable. However, in certain situations, the effect of the velocity profile on the sensitivity can be reduced using an inhomogenous magnetic field.[8] It was shown that when a pair of coreless exciting coils (as shown in Figure 3.7) is used, field distribution near the electrodes can be adjusted by the angle, θ, so that the effect of the velocity profile is minimized.[9] It was shown that the sensitivity does not change for a flow with a Reynold's number between 500 and 16,000 when $\theta = 55°$.

The effects of conductivity of the wall and inside and outside media were studied in detail.[10-14] A model, as shown in Figure 3.8 (left), is considered, where a and b are the internal and external radii of the vessel, and σ_1 and σ_2 are the electric conductivities of the internal medium and the wall, respectively. Electric conductivity outside the vessel is assumed to be zero. Then, the electromotive force induced at the electrodes E_1, E_2 is written as

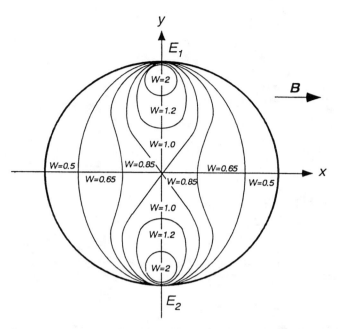

Figure 3.6 The weight function corresponding to Equation (3.13), which represents the relative flow contribu-tions to the electromotive force induced across the electrodes E_1 and E_2 when a uniform magnetic field is applied perpendicular to the electrode pair. (From Shercliff, J. A., *J. Appl. Physiol.*, 25, 817, 1954. With permission.)

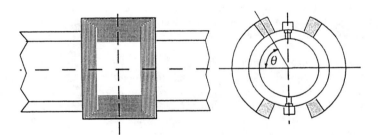

Figure 3.7 An electromagnetic flowmeter configuration using a pair of symmetric coreless exciting coils. (Modified from Clark, D. M. and Wyatt, D. G., in *New Findings in Blood Flowmetry*, Cappelen, C. H. R., Ed., Universitetsforlaget, Oslo, 1968, 49.)

$$V = 2sbBU \qquad (3.16)$$

where s is the relative sensitivity, which is given as

$$s = \frac{\dfrac{2\sigma_1}{\sigma_2}}{\dfrac{\sigma_1}{\sigma_2}\left(1 - \dfrac{a^2}{b^2}\right) + 1 - \dfrac{a^2}{b^2}} \cdot \qquad (3.17)$$

The relation between s and a/b is shown in Figure 3.8 (right).

According to Ferguson and Landahl,[11] the electric conductivity of a fresh sample of the aortic wall is about 0.32 to 0.47 mho m^{-1}, while that of the blood is about 0.61 to 0.70 S (siemens) m^{-1}, thus σ_1/σ_2 is estimated to be 1.4 to 1.9, and if $a/b = 0.8$ then the deviation of relative sensitivity from unity is less than 15%. However, if physiological saline is perfused instead of blood, σ_1 increases to about 2 mho

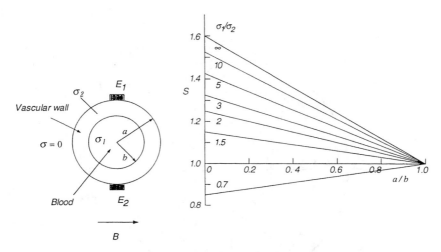

Figure 3.8 Cross-section of the vessel (left) and the relationship between the relative sensitivity, *s*, and the relative internal diameter, *a/b*, for various conductivity ratios, σ_1/σ_2 (right). (From Wyatt, D. G., *Phys. Med. Biol.*, 13, 529, 1968. With permission.)

m^{-1}, σ_1/σ_2 becomes 4 to 7, and relative sensitivity may increase by more than 20%. Therefore, for sensitivity calibrations, the use of a solution having similar electric conductivity to that of the fluid to be measured is recommended.

When the medium outside the vessel has a higher electric conductivity, sensitivity is reduced. In extreme situations where the vessel is surrounded by physiological saline, sensitivity will be reduced to about 50%. However, by insulating the outside of the vessel wall over a distance 1.3 times the vascular diameter on both sides of the electrode along the axis, a reduction in sensitivity can be avoided.[11] Further details about the factors affecting the sensitivity and accuracy in electromagnetic blood flow measurements are discussed in a review article by Wyatt.[15]

3.2.1.3 Methods of Magnetic Field Excitation

In earlier studies of electromagnetic blood flowmeters, static magnetic fields were used. However, due to a large electrode polarization potential superimposed on the blood flow-induced signal, it is difficult to obtain stable records even if nonpolarized electrodes (see Section 7.2.1.1) are used. To eliminate the effect of the polarization potential and to simplify the probe design, the alternating magnetic field was introduced.[16]

Using a.c. field excitation, a blood flow-induced signal is obtained as an a.c. potential and can be separated from the d.c. polarization potential. However, an alternating magnetic field also induces a large a.c. potential directly in a loop circuit composed by the lead wires and the electrolyte between two electrodes. This potential is called the transformer component.

Because the transformer component is an induced potential proportional to the time derivative of the magnetic flux across a loop, its amplitude is proportional to the excitation frequency, and the phase differs by 90° from both the excitation field and blood flow-induced signal. In principle, the flow signal can be separated from the transformer component by its phase relations. However, in the actual situation of blood flowmetry, it is difficult to eliminate a transformer component that is always larger than the blood flow signal by several orders of magnitude.

To detect the blood flow component selectively, different techniques have been proposed, as shown in Figure 3.9. Figure 3.9(a) shows a technique called gated sine wave.[17,18] A sine wave excitation is used, but a gate is opened for a short interval at the peak of the excitation where the blood flow signal is at its maximum while the transformer component is close to zero, because its phase differs 90° from the excitation field.

In the method shown in Figure 3.9(b), a square wave excitation is used. A large transformer component is generated just after inverting the field, and this transient potential attenuates in a short time if the time constant of the input circuit is short. Thus, by opening a gate near the end of each square wave excitation, the blood flow signal can be obtained selectively.[19-23] Figure 3.9(c) shows a technique similar to the square wave excitation. By using a trapezoidal wave, the amplitude of the transformer component can be reduced.[24,25] Figure 3.9(d) shows the method of sawtooth wave excitation.[26] In sawtooth wave excitation,

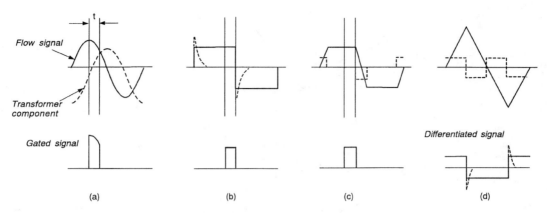

Figure 3.9 Flow signals and transformer components with gated or differentiated signals corresponding to sine wave **(a)**, square wave **(b)**, trapezoidal wave **(c)**, and sawtooth wave **(d)** excitations.

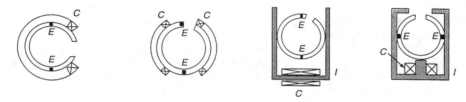

Figure 3.10 Different configurations of coreless and cored probes; *C*: coil, *E*: electrode, *I*: iron core.

a magnetic field varies at a constant rate and reverses periodically. Hence, the transformer component becomes a square wave. By differentiating the combined signal, a square wave flow signal with a transformer component will be obtained which is the same as the signal in the square wave excitation system.

In order to minimize the effect of the transformer component, the excitation frequency should be chosen carefully. Because the amplitude of the transformer component is proportional to the rate of change in magnetic flux, a lower frequency may be preferable to reduce the transformer component. On the other hand, the excitation frequency should be sufficiently higher than the highest frequency component in the flow signal. The excitation frequency is determined as a compromise between these postulates. In commercial instruments, an excitation frequency of 100 to 1000 Hz has been used, while the excitation frequency is selective in some flowmeters (for example, the MFV-2100, Nihon Koden Kogyo Co., Japan).

3.2.1.4 Perivascular Probes

Figure 3.10 shows the different configurations of perivascular electromagnetic flowmeter probes. In the coreless probe, the excitation field strength is proportional to the excitation current and the number of turns of the coil. By using an iron core, the field strength can be increased about twice.[27] In the iron-core probe, the temperature increase can be reduced significantly. To obtain the same magnetic flux, the total cross-sectional area of the wires in the coil can be increased and total wire length can be reduced in the core configuration; hence, ohmic losses in the coil can be reduced significantly.

In the smaller vessels, the amplitude of the signal is reduced roughly in proportion to the square of the vascular radius as long as the magnetic flux density remains the same. As has been mentioned earlier (Equation 3.16), the electromotive force is proportional to the vascular diameter and the blood velocity, which is roughly proportional to the diameter (see Section 3.1.2.1). To increase the signal amplitude, the magnetic flux density should be increased, thus the iron-core configuration is preferred. The magnetic flux density of a typical probe for a vessel diameter of 10 mm is typically about 3×10^{-3} T, while that for a vessel diameter of 1 mm is 3×10^{-2} T or more.[28,29] Even the magnetic flux density is increased about 10 times as much; the amplitude of the flow signal for a 1-mm vessel is about 1/10 that for a 1-cm vessel.

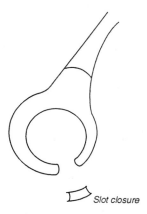

Figure 3.11 A typical perivascular probe with the slot closure.

When a small flow signal is measured during strong magnetic field excitation, the polarization potential due to the eddy current concentrated on the electrode surface can cause unwanted potentials. Wyatt[30,31] and Hognestad[32] showed that the effect can be reduced by using recessed electrodes.

Leakage currents from the exciting coils also cause serious noise if the insulation is insufficient. The excitation voltage may sometimes be 10^6 or more times greater than the flow signal amplitude. This means that careful encapsulation of the exciting coil is required. Capacitive leakage may also be a problem; however, in most situations this effect can be eliminated by electrostatic shielding.[27,30,31] Perivascular probes for vessels from 35 mm down to 0.5 mm in diameter are available commercially.

The inner diameter of the probe must fit the outer diameter of the vessel. Some authors[33] recommend that a perivascular probe should have a diameter about 90% of the vessel diameter to get good contact between electrode and vessel. Actually, one manufacturer provides perivascular probes for smaller diameter ranges from 0.5 to 4.0 mm in 0.05-mm increments and for larger diameter ranges of 10.0 to 35.0 mm in 1.0-mm increments (Skakar Medical; Delft, The Netherlands).

Figure 3.11 shows a typical configuration of the perivascular probe. The blood vessel is passed through a slot and allowed to expand into the lumen so that the external surface of the vessel is in direct contact with the inside wall of the probe. Usually the slot is closed by a plug, which not only keeps the vessel in the lumen of the probe but also reduces the extravascular current component.

Clinical use of perivascular probes is usually limited to surgical procedures. They are, however, often used in animals for chronic measurements. For this purpose, the probe is implanted in the body with a transcutaneous connector to which the signal and excitation cables are connected. For long-term measurements, instabilities in sensitivity and baselines are serious problems. For example, an observation of the carotid loop in sheep for 9 to 19 days showed that sensitivity varied from 67 to 93% that of *in vitro*, and offset errors were –100 to +370 ml/min at zero flow.[34] The authors suggested leak current as a possible cause of such instability.

The shape and the material of the electrodes having direct contact with the vascular wall are important items in a probe design for chronic implantation. Folts and Rowe[35,36] showed a perivascular probe design for measurements at the ascending aorta. This probe could be used for up to one year. This probe had a form which made it suitable for implanting in the major aorta, but other probes have caused a rupture of the aorta within a month.

Although chronic use of perivascular probes in patients is rare, Williams et al.[37,38] described a removable electromagnetic flow probe. The probe had a pair of coils with a pliable core and two gold electrodes. All components were enclosed in a slender silastic moulding of uniform cross-sectional area. The probe was positioned around the ascending aorta, as shown in Figure 3.12.[37] The probe shape was maintained by using a nylon probe. The nylon probe could be removed easily from the aorta due to the flexibility of the moulding. The probe has been used in 20 patients, and cardiac output was measured for up to 8 days.

3.2.1.5 Intravascular Probes

Catheter-type electromagnetic velocity probes have been developed and used clinically for monitoring blood velocity in the large arteries and veins. The electromagnetic velocity probe using a coreless coil

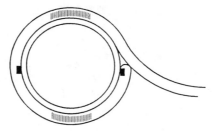

Figure 3.12 A removal electromagnetic flow probe positioned around the aorta. A pair of coils with pliable cores and electrodes was enclosed in a flexible silastic molding. (From Williams, B. T. et al., *Rev. Surg.,* 26, 227, 1969. With permission.)

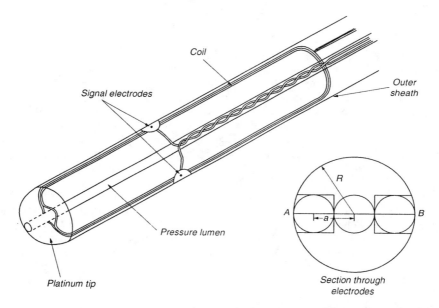

Figure 3.13 An electromagnetic velocity probe using a coreless coil. (From Mills, C. J. and Shillingford, J. P., *Cardiovasc. Res.,* 1, 263, 1967. With permission.)

was first developed by Mills.[39] The design of a modified probe[40] is shown in Figure 3.13. A catheter 3 mm in diameter contains a coil of 30 turns with signal electrodes on its surface. The sensitivity was 13.6 μV/m·s when a 975-Hz sine-wave excitation and a probe current of 1 A were used.

Because the width of the excitation coil in this type of probe is small compared to the vessel diameter, the magnetic field generated by the coil is not uniform over the entire cross-sectional area of the vessel. The field is concentrated in a region near the surface of the probe, causing blood flow near the probe surface to contribute more to the flow signal as compared to the flow in the rest of the vessel. As shown in Figure 3.14, the weight function for Mills' probe is very different from that of the perivascular probe.[41] An estimate showed that 90% of the flow signal is generated in an annulus 1.5 mm thick.[15] Due to these restrictions, only a small part of the flow in the cross-section contributes to the signal, and this type of probe is not regarded as a flowmeter but rather is called a velocity probe.

Nevertheless, it has also been shown that the intravascular electromagnetic catheter can be used as a flow monitor, because the signal of the velocity probe is highly correlated to the flow rate under certain conditions. A simultaneous measurement of the modified velocity probe using an iron-core magnet with the perivascular probe showed that in the measurements of ascending aortic flow in dogs, standard error for a single estimate of flow by the catheter for a 5-ms time interval during systolic ejection was ± 4.14% of the peak flow rate.[42]

In a clinical study, the cardiac output was calculated from the product of mean velocity measured by a velocity probe and the radiologically measured cross-sectional area of the ascending aorta or pulmonary

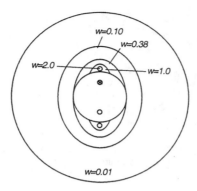

Figure 3.14 Normalized weight function of Mills' probe. (From Bevir, M. K., *Phys. Med. Biol.*, 16, 229, 1971. With permission.)

artery. The data obtained from this study were compared with cardiac output determined by the indicator-dilution methods, resulting in a correlation coefficient of 0.73.[43]

The sensitivity of the intravascular velocity probe increases with increasing excitation current. A limiting factor is, however, the temperature increase in the probe. An observation in Mills' probe showed that for the excitation current of 0.6 A, the power consumption was about 0.45 W, resulting in a surface temperature rise of up to about 10°C.[44] The heat production in leads seems to be greater than that in the tip of the probe.[45]

The typical commercial intravascular electromagnetic catheter has the following specifications: catheter diameter, 2.0 to 2.7 mm (6 to 8 Fr); measurement range, –200 to +200 cm·s^{-1}; sensitivity, 0.3 μVm^{-1}s with 0.6-A excitation current; heat production at the sensor tip, 0.2 W; and leakage current, 2.5 μA for 500 V (Mikro-Tip®, SVPC-6XXX, Millar; Houston, TX).

Various designs of electromagnetic velocity probes have been attempted. Warbasse et al.[46] used a U-shaped, iron-core electromagnet. The electrodes were located at the end of the core. In this configuration, the sensitive part can be located very near to the end of the catheter. The flow was found to be disturbed at the catheter tip, and the magnetic field varied at the end of the catheter.[15] Stein and Schuette[47] described a catheter probe having a short cylindrical tube at the catheter tip so that the blood can flow through it when the tube is wedged into a branch of a vessel. The magnetic field is generated by a solenoid coil with a hollow iron core, and the flow signal is detected by a pair of electrodes attached to the inner surface of the cylindrical tube.

Kolin et al.[48,49] designed an electromagnetic probe which measures the flow in a side branch of the aorta. It had a lumen located some distance from the tip at a right angle to the catheter axis and was sensitive to the flow through the lumen. The catheter was inserted into the aorta, and after the lumen was positioned in the opposite side branch the catheter was arched by means of a pull-wire so that the lumen was placed firmly against the branch entrance.

Kolin[50] also described a radial field electromagnetic intravascular probe. As shown in Figure 3.15, it had a magnet at the center of the probe, and a radial magnetic field was generated. A circular current, as shown by arrows in the figure, will be induced if the probe is surrounded by a conductive media. However, by adding a radial septum the circular current is blocked and an electromotive force is induced that can be detected by a pair of electrodes at both sides of the septum. In the actual probe, a perforated septum is added (as shown in the figure) to accomplish centering of the transducer in the blood vessel.

Although the configuration seems to be inconvenient for practical use, it has a particular property. The magnetic flux density, *B*, of an ideal radial field can be represented as

$$B = \frac{B_0}{r} \tag{3.18}$$

where *r* is the radius. The electromotive force, *V*, for a flow velocity, *U*, is obtained by integrating *U·B* over the whole circle:

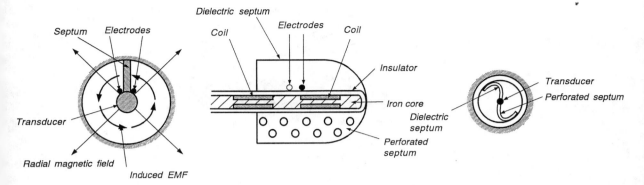

Figure 3.15 A radial field intravascular velocity probe. (From Kolin, A., *IEEE Trans. Biomed. Eng.*, BME-16, 220, 1969. With permission.)

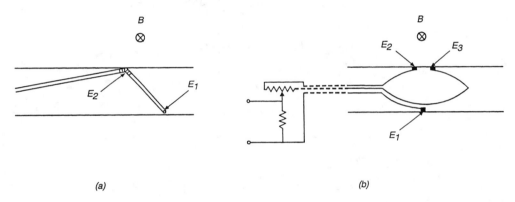

(a) (b)

Figure 3.16 Intravascular electrode systems for detecting flow signals using externally applied magnetic field: **(a)** use of a bending catheter and **(b)** use of springs with electrodes. (Part (a) modified from Kolin, A., *Proc. Nat. Acad. Sci.*, 59, 808, 1968. Part (b) modified from Biscar, J. P., *IEEE Trans. Biomed. Eng.*, BME-20, 62, 1973.)

$$V = \int_{0}^{2\pi} \frac{B_0}{r} r d\theta = 2\pi U B_0. \qquad (3.19)$$

Because the last expression is independent of the radius of the integration path, it is suggested that the dimension of the probe can be miniaturized without a loss of sensitivity. In Mills' probe, though, the sensitivity is reduced in proportion to its geometrical dimension.

3.2.1.6 Use of External Field Excitation

Two different methods have been studied. In the first, a magnetic field is applied from the outside of the body, and the flow signal is detected by intravascular electrodes. In the second, a magnetic field is also applied from outside the body, but the flow signal is detected on the body surface, making the measurement noninvasive.

Only a few attempts have been made to use intravascular flow detection with externally applied magnetic fields. As shown in Figure 3.16(a), Kolin[51] used a catheter that is bent near the tip and has two electrodes at both sides of the bent section. These electrodes are placed at the opposite wall sections of the blood vessel. Biscar[52] reported an intravascular electrode system, as shown in Figure 3.16(b). The flow signal is detected by two electrodes, E_1 and E_2, and the additional electrode, E_3, is used for the reduction of the transformer component. Although the construction of the catheter is simplified by the use of an external field, it still requires a surgical procedure and a large coil is required for excitation.

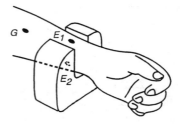

E₁, E₂: Recording electrodes
G: ground electrodes

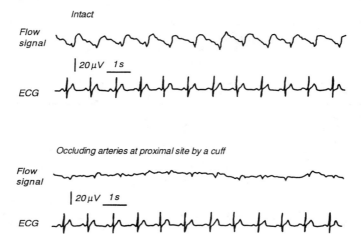

Figure 3.17 Blood flow measurement at the wrist using a permanent magnet (above) and an example of the record (below). (Modified from Okai, O. et al., *Jpn. Heart J.,* 15, 469, 1974.)

Noninvasive measurement of the electromagnetic flow signal was first demonstrated in rabbits positioned in a static magnetic field of 1 T. The field was directed from front to back, and the flow-related signal was recorded between the forelimbs.[53] A similar recording was made in the human thorax using a 0.7-T magnet and also in the human wrist with a 0.1-T magnet.[54,55] Figure 3.17 shows the method of measurement at the wrist and an example of the recording.[55] In animal studies, the observed signal was compared with the recording obtained by a perivascular electromagnetic flowmeter probe. The authors found a linear relationship between the amplitudes of the signals.[56] Using an averaging technique, pulsatile flow signals can be obtained in weaker magnetic fields. Lee et al.[57] recorded the femoral, popliteal, and posterior tibial arterial flow pulsatiles in densities of 0.03 T by averaging 32 to 64 successive wave forms. Boccalon et al.[58-60] also recorded the flow signals in a static magnetic field of 0.06 T using an averaging technique.

The signal amplitude in the noninvasive method can be analyzed in simplified models.[58-60] The simplest case is a hypothetical cylindrical limb, which has only one artery at the center of the cylinder, and the tissue has uniform conductivity. If an electrode pair is placed on the limb perpendicular to the magnetic field, this model corresponds to the same situation of the perivascular probe being attached onto the vessel with a thick wall. The sensitivity factor, *s*, will be represented by Equation (3.17), but in this case, the inner radius, *a*, is much smaller than the outer radius, *b*, i.e., $a \ll b$, thus Equation (3.17) can be simplified as

$$s = \frac{2\sigma_1/\sigma_2}{1+\sigma_1/\sigma_2} \qquad (3.20)$$

where σ_1 and σ_2 are the electric conductivity of the blood and the tissues, respectively. Thus, the expected signal, V, is

$$V = \frac{2BQ}{\pi b}\frac{2\sigma_1/\sigma_2}{1+\sigma_1/\sigma_2}.$$ (3.21)

This expression implies that the observed potential is proportional to the blood flow rate and does not relate to the vessel diameter. More complex models were also analyzed.[60,61]

Noninvasive measurement using alternating magnetic fields was also reported.[62] The authors used a magnetic field of about 0.03 T at a depth of 1 cm below the skin and detected the flow signal by an electrode pair separated by 2 cm. The excitation frequency was 400 Hz. They demonstrated that the flow signal from the cubital and brachial arteries could be recorded. However, the transformer component was extremely large, thus they had to use a feedback system for transformer component suppression.

3.2.2 ULTRASONIC BLOOD FLOWMETERS

A sound wave propagating in a moving medium is affected by the velocity of the medium, and the sound scattered by a moving object is also affected by the velocity of the scattering object. Both phenomena can be used to measure blood flow velocity or flow rate in a blood vessel. The fundamental properties of sound wave propagation are summarized in the first part of this section, but more details will be found in ordinary textbooks.[33,63,64] In the later part, two types of ultrasonic flowmeters, the transit time or phase shift flowmeter and the Doppler flowmeter, are described.

3.2.2.1 Propagation of Ultrasound in the Tissue

Ultrasound is defined as sound having a frequency above the human audible range. Ultrasound propagates in the soft tissue with a velocity of about 1500 m/s. The sound velocity, c, frequency, f, and wavelength, λ, are related as

$$c = f\lambda.$$ (3.22)

As an example, the frequency of ultrasound of 1 MHz is about 1.5 mm in tissue.

A sound at a point in a medium is characterized by the sound pressure and the sound particle velocity (or medium velocity). If a sound wave propagates in one direction, the amplitude of sound pressure, p, is proportional to that of sound particle velocity, U, so that

$$p = \rho cU$$ (3.23)

where ρ is the density of the medium. The ratio of amplitudes of sound pressure and particle velocity is called a characteristic impedance (or a characteristic acoustic impedance) of the medium, that is

$$Z = p/U = \rho c.$$ (3.24)

Characteristic impedances of actual media are listed in Table 3.1. At the boundary between two media having different characteristic impedances, the sound wave will be partially reflected. If a sound wave with the sound pressure amplitude, p_i, impinges perpendicularly on a boundary between two media having characteristic impedances, Z_1 and Z_2, then the sound pressure of the reflected wave, p_r, is given as

$$p_r = \left|\frac{Z_1 - Z_2}{Z_1 + Z_2}\right|p_i.$$ (3.25)

Thus, when the difference between characteristic impedances of two media is small, the reflection is small. On the other hand, if $Z_1 \gg Z_2$ or $Z_1 \ll Z_2$, then $p_r \cong p_i$, which means that the impinged wave is fully reflected. Sound energy (or sound power), I, is expressed as

Table 3.1   Typical Values for Ultrasonic Attenuation Coefficients

	Velocity (m/s)	Impedance (10^6 Pa·s·m^{-1})	Attenuation (dB/cm at 1 MHz)	Half Value Layer at 1 MHz (cm)
Air (STP)	330	0.0004	12	0.25
Water	1480	1.48	0.002	1500
Fat	1450	1.38	0.63	4.76
Blood	1570	1.61	0.18	16.67
Kidney	1560	1.62	1.0	3.00
Soft tissue	1540	1.63	0.70	4.29
Liver	1550	1.65	0.94	3.19
Muscle				
Along fiber	1580	1.70	1.3	2.31
Across fiber	—	—	3.3	0.91
Bone	4080	7.80	15	0.20
Ceramic crystal	5100	3.0	2.3	—
Plexiglas	2670	3.2	2.3	—

Source: From Goldstein, A., in *Encyclopedia of Medical Devices and Instrumentations,* Webster, J. G., Ed., John Wiley & Sons, New York, 1988. With permission.

$$I = \frac{1}{2}pu = \frac{1}{2}\frac{p^2}{Z} \ . \tag{3.26}$$

For an incident sound wave with sound energy, I_i, the sound energy of the reflected wave, I_r, is obtained by

$$I_r = \left(\frac{Z_1 - Z_2}{Z_1 + Z_2}\right)^2 I_i. \tag{3.27}$$

Thus, the sound energy of transmitted sound, I_t, is given as

$$I_t = I_i - I_r = \frac{4Z_1 Z_2}{\left(Z_1 + Z_2\right)^2} I_i. \tag{3.28}$$

Substituting characteristic impedance values of media, the transmittance of the boundary of air and water is obtained as about 0.13%, and that of a ceramic crystal and water is about 20%.

When ultrasound propagates in tissue, it is attenuated by absorption, reflection, and scattering. If a plane wave propagates in the x direction in an absorbing medium, and if it has a sound pressure $p(0)$ at $x = 0$, then sound pressure at x is represented as

$$p(x) = p(0)e^{-\alpha x} \tag{3.29}$$

where α is the absorption coefficient. The absorption coefficients for the ultrasound of 1 MHz are also listed in Table 3.1. The absorption coefficient varies almost in proportion to the sound frequency. Because sound energy is proportional to a square of sound pressure, sound energy at x, $I(x)$, is expressed as

$$I(x) = I(0)e^{-2\alpha x}. \tag{3.30}$$

To generate and detect ultrasound, a piezoelectric transducer is commonly used. The active element of the transducer is a plate of piezoelectric crystal with thin metallic electrodes on both sides. When an alternating potential is applied to the electrodes, the thickness of the piezoelectric crystal changes alternately, and a sound wave is generated into the surrounding medium. When a sound wave penetrates a piezoelectric crystal, it causes a mechanical strain in the material. Electric polarization appears at the electrodes, thus generating a signal related to the sound pressure.

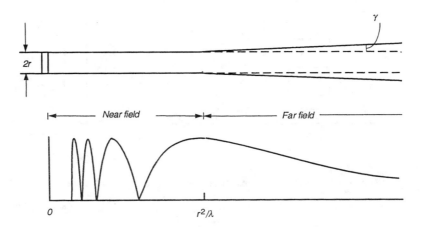

Figure 3.18 The ultrasound beam pattern in front of a circular transducer (above) and sound pressure amplitude profile along the axis (below).

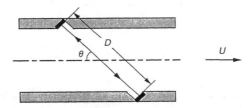

Figure 3.19 Arrangement of crystals for transit time or phase shift measurement.

The piezoelectric crystal of lead zirconate titanate (PZT) is commonly used for generating and detecting ultrasound. In this material, about 50% of the applied electric energy can be converted into mechanical energy, when the transducer is operated at below mechanical resonant frequency, while the efficiency may be as high as 90% when it is operated at the resonance frequency.[65] Although mechanical resonance is advantageous for increasing efficiency in a continuous wave operation, it is undesirable for pulsed operation because it lengthens the pulse duration. To dampen the mechanical resonance, an absorbing medium is attached to the rear side of the transducer as the backing material.

When a transducer is operated in a continuous wave, a specific beam pattern is generated in the medium. The beam pattern can be determined by the geometry of the transducer. By applying Huygens' principle, it can be assumed that the transducer surface consists of a number of point sources emitting a spherical wave. If a flat, disk-shaped transducer is operated in the thickness-expanding mode, the beam pattern is characterized by a near field (or Fresnel zone) and a far field (or Fraunhofer zone). In the near field, diffraction can occur among sound waves generated at different sites on the transducer surface, and consequently a spatially varying sound pressure amplitude appears. In the far field, the sound pressure amplitude is spatially uniform. Figure 3.18 shows the beam pattern when the transducer radius, r, is about 10 λ. The beam pattern of the near field is cylindrical, and its length is about r^2/λ. The beam pattern of the far field is conical, and the half-angle, γ, of divergence of the cone is given by

$$\sin\gamma = 0.6\frac{\lambda}{r}. \tag{3.31}$$

3.2.2.2 Transit Time and Phase Shift Ultrasound Flowmeters

When a sound wave travels upstream or downstream in a flowing fluid, the apparent sound velocity differs from the sound velocity in resting fluid, because a sound wave travels at a constant velocity relative to the medium. When a transmitting and a receiving crystal are arranged as shown in Figure 3.19 and the fluid velocity is uniform over the whole cross-section of the conduit, transit time, T, can be expressed as

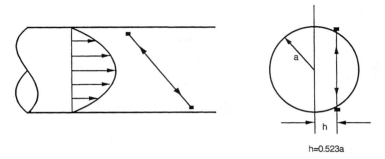

h=0.523a

Figure 3.20   A method to minimize the influence of the velocity profile.

$$T = \frac{D}{c \pm U \cos\theta} \tag{3.32}$$

where D is the distance between transmitting and receiving elements, U is the fluid velocity, and θ is the beam angle with respect to the axis of the conduit; a plus or minus sign corresponds to traveling downstream or upstream. The difference in transit time of the downstream and the upstream sound waves can then be obtained as

$$\Delta T = \frac{2DU \cos\theta}{c^2 - U^2 \cos^2\theta} \approx \frac{2DU \cos\theta}{c^2} \tag{3.33}$$

if $U \ll c$. The phase difference, $\Delta\phi$, caused by this transit time difference, is

$$\Delta\phi = \omega\Delta T = \frac{2\omega DU \cos\theta}{c^2} \tag{3.34}$$

where ω is the angular frequency of the sound wave. As an example, if $D = 2$ cm, $U = 10$ cm/s, and $\theta = 45°$, then ΔT is about 1.3×10^{-9} s. For a sound wave of 1 MHz, $\omega = 2\pi \times 10^6$ rad/s, and $\Delta\phi$ is about 8×10^{-3} rad $\approx 0.46° \approx 28'$.

The velocity profile is not uniform in the actual fluid flow, such as blood flow in the vessel, so the above estimation should be corrected. When the flow is laminar and stationary, a parabolic velocity profile is expected. Under such conditions, the actual flow will be 33% greater than the flow estimated by the transit time, assuming a uniform velocity profile.[66] However, a simple calculation shows that if the path of the ultrasound beam is shifted about 52.3% of the radius, as shown in Figure 3.20, the error in the flow estimation can be minimized.

To detect transit time differences or phase differences, various techniques have been proposed. Block diagrams of these techniques are shown in Figure 3.21. In this figure, (a) is a technique for transit time measurement.[67-69] At the first clock pulse, a short ultrasound pulse is emitted from one crystal and detected by another crystal. The time detector generates output proportional to transit time. At the next clock pulse, two crystals are switched, and a transit time corresponding to the reverse direction is detected. By alternating these operations and by performing synchronous rectification, a signal proportional to the transit time difference can be obtained.

Figure 3.21(b) shows a method for detecting the phase difference between upstream and downstream signals.[70] The detected signal is amplified and heterodyned so that it is converted to an audio frequency signal having the same phase information as the original signal. Then, the phase change relative to a reference signal is detected by a phase detector. By switching the transmitting and receiving crystals with a chopper and performing synchronous rectification to the phase detector output, the phase difference between upstream and downstream signals can be obtained.

Figure 3.21(c) is also a method of detecting the phase differences in which each crystal is simultaneously operated as the transmitter and the receiver.[71] Two sound waves in different frequencies (f_1 and f_2) are transmitted upstream and downstream. Each received signal is mixed with oscillator output having

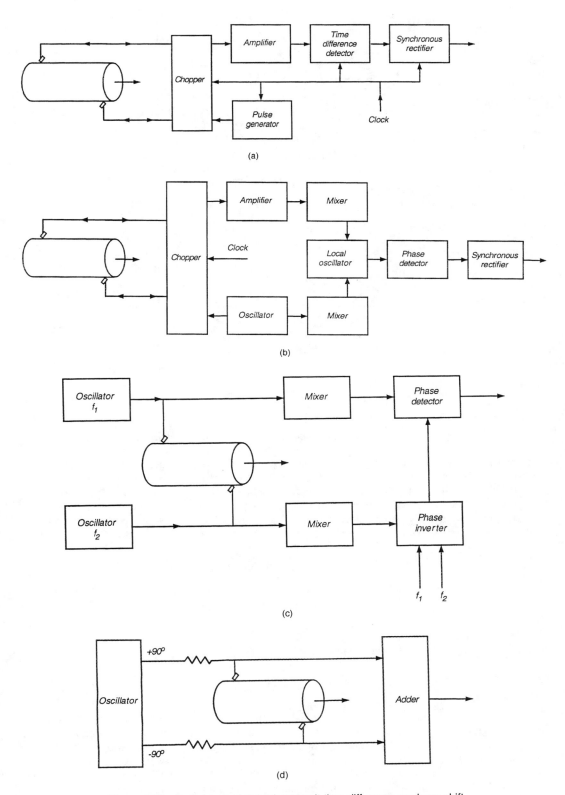

Figure 3.21 Techniques of detecting transit time difference or phase shift.

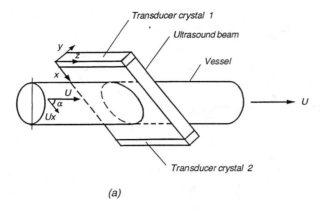

(a)

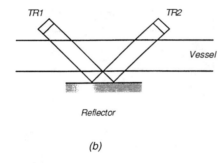

(b)

Figure 3.22 Transit-time flowmeter using a wide-beam **(a)** and an arrangement of crystals and reflector **(b)**.

different frequencies so that the signal having a frequency of $|f_1 - f_2|$ is generated, corresponding to each site of the crystal. While phase changes due to the fluid stream occur in the same direction, if the phase of one signal is inverted, the phase difference can be detected by a phase detector.

Figure 3.21(d) shows a method in which the phase difference can be obtained using single-frequency excitation.[72,73] Two crystals are driven by the same frequency, but the phases are opposite each other. If the driving signals are assumed to be $\pm V\cos\omega t$, the signal appearing at two crystals will be

$$e_1 = V\cos\omega t - kV\cos(\omega t + \sigma + \phi) \qquad (3.35)$$

$$e_2 = -V\cos\omega t + kV\cos(\omega t + \sigma - \phi) \qquad (3.36)$$

where k is the attenuation factor, σ is phase difference without flow, and ϕ is phase difference due to the flow. The sum of these two signals will be

$$e = e_1 + e_2 = 2kV\sin\phi \cdot \sin(\omega t + \sigma) \approx 2kV \cdot \phi \cdot \sin(\omega t + \sigma). \qquad (3.37)$$

This shows that the amplitude of the resultant signal is proportional to the phase difference due to the flow. Although it is difficult to balance the two driving signals so that the sum of these becomes exactly zero, intermittent excitation can be used, so that the received signal is selectively led to the adder by suppressing signal amplification during excitation. Rader et al.[73] used 5-MHz excitation for a 1-cm diameter tube, with a beam angle of 45°. They obtained about 130 mV for a flow velocity of 100 cm/s.

A modified transit-time flowmeter which measures volume flow rate directly has also been developed.[74,75] In this flowmeter, an ultrasound beam wider than the vessel diameter is used. If a wide transducer crystal generates a uniform ultrasound beam and is also received by a wide transducer crystal, as shown in Figure 3.22(a), the beam can be considered as a set of infinitesimal beams specified by the coordinates y and z on the transducer crystal. The transit time of an infinitesimal beam can be expressed as

$$T = \int_0^L \frac{dx}{c \pm U_x} \approx \frac{L}{c} \pm \frac{1}{c^2} \int_0^L U_x dx \qquad (3.38)$$

where L is the separation of two crystals, c is sound velocity, and U_x is the x-direction component of flow velocity at a coordinate x,y,z in the vessel and zero outside the vessel. Plus and minus signs correspond to the sound beam's travel in a downstream or upstream direction. For a single frequency excitation of $\sin\omega(t + L/c)$, the received signal will be the integral of all infinitesimal beam components represented as

$$A = G \iint \sin\omega(t \pm \delta T)dydz \qquad (3.39)$$

where δT is the transit time difference due to the flow in the vessel which is given by

$$\delta T = \frac{1}{c^2} \int_0^L U_x dx \qquad (3.40)$$

and G is the coupling constant of two transducer crystals. Then

$$A = G \iint (\sin\omega t \cdot \cos\omega\delta T \pm \cos\omega t \cdot \sin\omega\delta T)\, dydz. \qquad (3.41)$$

If $\omega \cdot \delta T \ll \pi$, then

$$A = G\left[\sin\omega t \iint dydz \pm \omega\cos\omega t \iint \delta T dydz \right]$$
$$= G\left[S \cdot \sin\omega t \pm \frac{\omega\cos\omega t}{c^2} \iiint U_x dxdydz \right] \qquad (3.42)$$

where S is the surface area of the crystal. The integral in the last term is equal to the x component of the total flow, Q, thus

$$A = G\left[S \cdot \sin\omega t \pm \frac{\omega Q \cos\alpha}{c^2} \cdot \cos\omega t \right] \qquad (3.43)$$

where α is the angle between U and U_x. This expression implies that the received signal contained a component with an amplitude proportional to the total volume flow. In the actual probe design, a reflected acoustic pathway, as shown in Figure 3.22(b), is employed. This configuration is advantageous for reducing the angular sensitivity for misalignment of the probe to the vessel axis.

A commercial flowmeter using this principle is available (Model T101, Transonic Systems, Inc.; Ithaca, NY). Probes of different sizes are supplied, in which vessel diameters ranging from 1 to 32 mm are covered by nine probe sizes. These probes can be used for both acute and chronic measurements.

According to this principle, volume flow rate can be measured regardless of the vessel diameter or flow velocity, as long as the whole cross-section of the vessel remains in the uniform ultrasound beam. Actually, an *in vitro* experiment showed that volume flow sensitivity changed by less than 10%, even though velocity for the given flow rate varied by a ratio of 9 to 1.[74] Also, an *in vivo* experiment in sheep showed that the blood flow rate of the carotid artery, measured by the chronically implanted ultrasonic

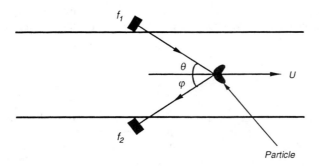

Figure 3.23 Measurement of Doppler shift.

probe, was highly linear with an absolute accuracy of 5.1% against that measured directly by draining from an implanted T-cannula. The zero offset was also negligible.[75] Another *in vivo* experiment, in which blood flow to the hindquarters of steers was measured by the ultrasonic probe and by the indicator dilution technique (see Section 3.2.3) using *p*-aminohippurate, showed that there was no difference in mean blood flow rates measured by these two techniques.[76]

3.2.2.3 Ultrasonic Doppler Flowmeters

When a flowing fluid contains particles by which ultrasound is scattered, the velocity of particles can be determined by the Doppler shift of the scattered ultrasound. The scattering in which particle size is less than the wavelength of the sound is called Rayleigh scattering, and in that case the intensity of the scattered sound wave is proportional to the fourth power of the sound frequency. That relation was confirmed experimentally for backscattering from the blood.[77] Thus, the scattered signal will be significantly stronger at higher sound frequencies. The sound intensity, however, decreases exponentially with increasing propagation distance, as seen in Equation (3.30), and the absorption coefficient, α, is almost proportional to the sound frequency. If the ratio of the absorption coefficient, α, and the frequency, f, is assumed to be a constant, k, the amplitude of the scattered sound wave after propagation over a distance, x, will be expressed as

$$I \propto f^4 e^{-2\alpha x} = f^4 e^{-2kfx}.$$ (3.44)

This function has a maximum at

$$f = \frac{2}{kx}$$ (3.45)

As an example, if $k = 2 \times 10^{-7}$ s·cm^{-1} and $x = 1$ cm, the scattered signal has a maximum at

$$f = 10^7 s^{-1} = 10 MHz .$$ (3.46)

When increasing the distance, x, the frequency which gives the maximum signal intensity is decreased inversely proportional to x.

The ultrasonic Doppler flowmeter is a measurement of Doppler shift in the scattered wave due to the red blood cells moving in the blood stream.[78,79] When an ultrasound transmitter and a receiver are arranged as shown in Figure 3.23, the Doppler shift caused by the scattering object with velocity, U, can be calculated in two steps. First, the frequency observed at the scattered object is calculated by a moving observer receiving a sound wave from a stationary source. At the moving observer, the observed frequency of a wave coming from a source having angle, θ, to the flow direction is given as

$$f_1 = \frac{c + U \cos \theta}{c} f_s$$ (3.47)

where f_s is the source frequency and c is sound velocity. Then, the frequency of the reflected wave observed at a stationary receiver is calculated from a stationary observer receiving a sound wave from a moving source. The received frequency of a wave going toward the observer having angle, φ, to the flow direction is given as

$$f_2 = \frac{c}{c - U\cos\varphi} f_1.$$

(3.48)

Rearranging the two equations above and assuming that $U\cos\varphi \ll c$, the Doppler shift, Δf, can be obtained by

$$\Delta f = f_2 - f_s = \frac{c + U\cos\theta}{c - U\cos\varphi} f_s - f_s \approx \frac{U(\cos\theta + \cos\varphi)}{c} f_s.$$

(3.49)

If $\theta = \varphi$,

$$\Delta f \approx \frac{2U\cos\theta}{c} f_s.$$

(3.50)

For example, if $U = 100$ cm/s, $\theta = \varphi = 45°$, $c = 1.5 \times 10^5$ cm/s, and $f = 5$ MHz, then Δf is about 4.7 kHz. This means that the observed frequency is 5 MHz + 4.7 kHz = 5.0047 MHz for a flow away from the transmitter and the receiver and 5 MHz –4.7 kHz = 4.9953 MHz for a flow towards them.

The Doppler shift is essentially a frequency modulation by the velocity of the scattering object. Thus, to extract the information of the flow velocity of the object fluid, a demodulation process has to be performed. Because the Doppler shift is very small compared to the source frequency, the demodulation process is usually performed in two steps. First, the detected signal is converted into a signal in audio frequency so as to obtain a higher resolution in determining the Doppler shift frequency. Then, flow information, such as the maximum velocity or the velocity distribution, is extracted by using appropriate signal processing techniques.

To convert the received signal into a lower frequency signal, different techniques have been used. Figure 3.24(a) is the simplest method in which transmitting and receiving signals, which are proportional to $\cos\omega t$ and $\cos(\omega - \Delta\omega)t$, are fed into a mixer so that the Doppler-shift component proportional to $\cos\Delta\omega t$ can be extracted as the beat of the transmitting and receiving signals.[79-82] In actual situations in the body, the receiving signal contains reflected components from stationary objects which have the same frequency as the transmitting wave. Besides this, the direct leakage of the transmitting wave, due to electric and ultrasonic coupling between transmitting and receiving elements, is contained in the receiving signal. Consequently, it is practically unnecessary to feed the oscillator output to the mixer. Instead of using the mixer, in which transmitting and receiving waves are superimposed and rectified, a phase-sensitive detector or a multiplier can be used to extract the Doppler-shifted signal, because the phase difference relative to the reference signal varies with the frequency difference between these signals. This method is called coherent demodulation, while the method using a mixer is called noncoherent demodulation.[63] In either method, the sign of the Doppler shift, $\Delta\omega$, which relates to the direction of the flow, cannot be discriminated.

To obtain information on flow direction, either heterodyne demodulation or quadrature demodulation has been used commonly. Figure 3.24(b) shows the principle of heterodyne demodulation. The received signal is superposed with a sinusoidal wave from a local oscillator that has a slightly different angular frequency, ω_2, from that of the transmitting wave, ω_1. Consequently, a beat frequency corresponding to the angular frequency difference $(\omega_1 - \omega_2)$ will appear when the Doppler shift is zero, and the beat frequency will increase or decrease corresponding to a positive or negative Doppler shift, $\Delta\omega$. Thus, flow direction can be discriminated by the difference in the beat frequency.[83-86] Because the frequency difference between two oscillators is usually one thousandth or less of the frequency of each oscillator, both oscillators must be stable enough to each other so as to obtain stable beat frequency. To make one oscillator stable relative to the other, a phase-lock loop is used which involves using a low-frequency oscillator corresponding to the frequency difference. The phase of the beat of two oscillator outputs is locked to that of the low-frequency oscillator output.[87]

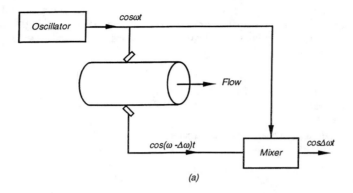

(a)

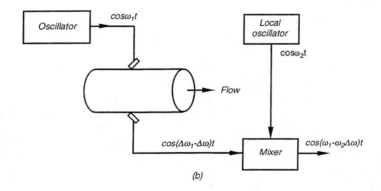

(b)

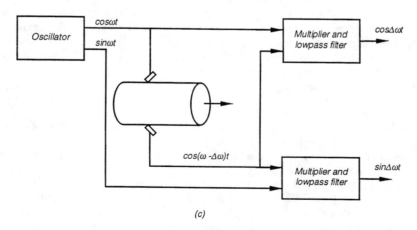

(c)

Figure 3.24 Techniques of extracting Doppler shift components: **(a)** simple noncoherent demodulation, **(b)** heterodyne demodulation, and **(c)** quadrature demodulation.

Figure 3.24(c) shows the configuration of so-called quadrature demodulation, which was also devised for directional Doppler measurement.[87,88] The upper half of this configuration is identical to the simple coherent demodulation, which provides Doppler frequency output proportional to $\cos\Delta\omega t$, preserving phase information. The lower part is also a coherent demodulation, but the phase of the reference wave is shifted 90° from that of the upper part. It also provides Doppler frequency output proportional to $\sin\Delta\omega t$, which has the same frequency as the first but the phase differs by 90°. Thus, the phase of the second output relative to that of the first one changes +90° to –90°, when the sign of $\Delta\omega$ changes. This means that flow direction can be discriminated by the relative phase relation between two coherent-demodulation outputs.

To extract velocity information from the signal containing Doppler shift components, either a zero-crossing counter or a spectrum analyzer is commonly used. The principle of the zero-crossing counter is based on the theory of random noise. Because the reflected signal involves a great number of waves scattered by independent red blood cells moving in the transmitting ultrasound beam, the signal can be regarded as a kind of the random noise. According to Rice,[89] the number of zero-crossings of the random noise signal in unit time is given as the root mean square of the frequency component, f, that is

$$N = 2 \left[\frac{\int_0^\infty f^2 P(f)df}{\int_0^\infty P(f)df} \right]^{1/2} \tag{3.51}$$

where $P(f)$ is the spectral power distribution function of the signal. Note that an alternating wave has two zero-crossings in one cycle, and if f is taken as the mean frequency $\overline{f}$ in the above equation, then $N = 2\,\overline{f}$. The spectral power distribution function, $P(f)$, is determined by the velocity distribution of the red blood cells, thus it depends on the velocity profile in the vessel. If the parabolic velocity profile is assumed, it is shown that the number of zero-crossings is expressed as

$$N = 1.03 \times \frac{2U_m \cos\theta}{c} f_s = 1.03 \times \frac{4\overline{U} \cos\theta}{c} f_s \tag{3.52}$$

where U_m and $\overline{U}$ are maximum and mean velocity, respectively.[90] Because a zero-crossing counter can easily be realized by ordinary electronic components, it has been used in simple flowmeters. However, the zero-crossing counter is influenced by velocity profile, and it does not give any information about velocity distribution. Besides that, the zero-crossing counter is sensitive to noise, so adjustment of the threshold level of the zero-crossing detection is critical.

The spectrum analysis is a technique that measures the power of each frequency component of the Doppler-shifted wave corresponding to each velocity component in the blood flow. Thus, spectrum analysis is a straightforward method for extracting velocity information from the received signal which contains reflected waves coming from many scattering objects moving at different velocities. Actually, this technique has been used in earlier studies of ultrasonic blood flow measurement.[78,91] In the artery, the blood flow is pulsatile, and the spectrum of the Doppler signal varies during the cardiac cycle. To observe the variation of each velocity component, spectrum analysis should be performed for every short time interval, and it is also desirable to have the obtained spectrum displayed without any time lag.

Spectrum analysis of the Doppler-shifted signal is usually performed either by bandpass filters or by the fast Fourier transform (FFT) analysis. Figure 3.25 shows a spectrum analyzer constructed of multi-channel bandpass filters which cover the whole frequency range of the Doppler shift. By rectifying the output of each bandpass filter, the power of that frequency component can be obtained in each channel. The obtained spectrum, as well as its time course, is usually visualized on a monitor screen and also recorded on a paper chart, as shown in Figure 3.26. Each line corresponds to a channel of a specific frequency, and the power level of each channel is represented as brightness on the monitor screen and darkness on the chart. The wave form of the maximum velocity can be recognized as the envelope of the time course of the spectral power density. The negative part corresponds to the reverse flow, which is detected by a directional Doppler system such as the heterodyne or quadrature demodulation. The flow signal corresponding to the frequency of near zero cannot be determined, because strong reflection from stationary objects always masks the reflected signal from stationary red blood cells.

The spectrum analysis can also be performed by numerical calculations of Fourier transforms, if the input wave is converted to digital quantities with a sampling rate at least twice as much as the highest frequency in the signal. By applying the fast Fourier transform, the amount of numerical calculation is greatly reduced. As an example, a calculation of 128 frequency components from 256 sample points required only 4.5 ms, and the spectrum could be displayed every 10 ms using a 16-bit microcomputer.[92]

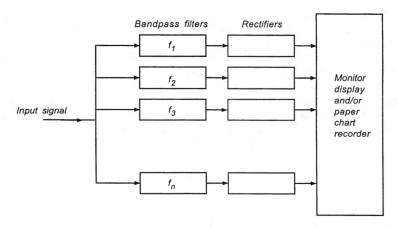

Figure 3.25   Spectrum analyzer constructed by bandpass filters and rectifiers.

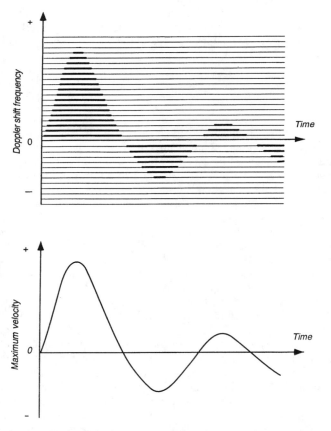

Figure 3.26   Visualized pulsatile flow pattern by spectral components (above) and the wave form of the maximum velocity obtained as the envelope of above spectral components (below).

The advantage of the spectrum analyzer is that it provides information on velocity distribution, whereas the zero-crossing counter provides only a mean velocity. However, measurement of velocity distribution is not so important in most clinical applications. When it is only necessary to measure the mean velocity, then either method is acceptable, as long as its accuracy is satisfactory. A study showed that the accuracy of a zero-crossing counter and that of an FFT spectrum analyzer were comparable for

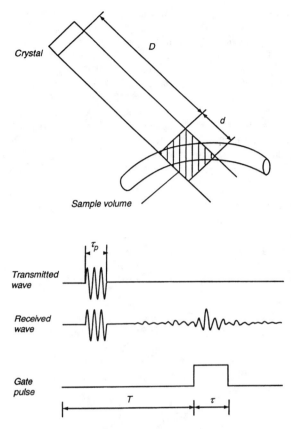

Figure 3.27 Principle of the pulse Doppler system. A short ultrasound pulse is transmitted, and the received wave is gated by an appropriate timing so that the signal from the sample volume is extracted.

measurement of the carotid arterial velocity waveform, while the zero-crossing counter was more accurate for measurement in the popliteal and radial arteries in which the signal-to-noise ratio was lower.[93]

3.2.2.4 Methods of Range Discrimination

The most significant advantage of the ultrasound Doppler blood flow measurement is the ability to observe blood flow in a vessel from a distant site. This feature allows the blood flow in deep arteries or veins to be assessed noninvasively from the body surface. However, if two or more large vessels exist in an ultrasound beam, Doppler shift signals from these vessels will also enter into the receiver. To observe the Doppler signal from the vessel in a specific site, the range-discriminating Doppler flowmeter has been developed.

Two different systems are commonly used for this purpose, i.e., the pulse Doppler system and the random-signal Doppler system. The principle of the pulse Doppler system is shown in Figure 3.27.[94,95] If a short ultrasound pulse is transmitted, then reflected signals from many reflecting objects in different ranges will be received, but each reflected component has different timing according to the distance from the transducer. By gating the received signal at a specific time interval, the signal reflected by the object in a specific range can be selectively obtained. The flow velocity information for this signal is obtained by a frequency spectrum analysis.

The distance from the transducer to the measurement region is determined by recording the time from the moment of emitting an ultrasound pulse to the onset of gate opening. The measurement range, d, is determined by the width of the gate pulse, τ, so that if the time length of the emitting ultrasound pulse, τ_p, is short enough compared to τ, then under such conditions d is nearly equal to $c\tau/2$. For example, if $\tau = 4$ μs and $c = 1.5 \times 10^3$ m/s, then $d \cong 3$ mm.

However, if the time length of the transmitted pulse is short, the phase shift due to the Doppler shift is very small, which is typically on the order of one thousandth of one cycle. When τ_p is a few microseconds and ultrasound frequency is several megahertz, the Doppler shift cannot be measured by the received signal for single pulse. In other words, a single transmitted pulse has a wide spectral width and the Doppler shift is too small to detect from the power spectrum of the single received signal.

This difficulty can be eliminated by the use of repetitive ultrasound pulses having a fixed phase relation, which is realized by gating the output of a master oscillator. Although the wave is not continuous but appears intermittently, a signal including many pulses can have a very sharp spectrum as long as the phase relation is maintained, and small Doppler shifts can be determined by a spectral analysis. Theoretically, there is no lower limit in Doppler shift frequency that can be detected by spectral analysis.

On the other hand, there is an upper limit for detectable flow velocity in the pulse Doppler system, due to the limit in the allowable repetition frequency of the ultrasound pulses. To avoid ambiguity among reflected waves of different pulses, a pulse should be emitted after all the returned echos of preceding pulses have been received. Thus, the pulse repetition frequency, f_p, is limited as

$$f_p = \frac{c}{2D_{max}} \tag{3.53}$$

where D_{max} is the maximum range of the measurement. Because the Doppler shift frequency is determined by sampled data corresponding to repetitive pulses, it is postulated by the Nyquist relation that the sampling frequency should be higher than twice that of the maximum signal frequency, f_{max}, i.e.,

$$f_{max} = \frac{f_p}{2}. \tag{3.54}$$

Applying an ordinary expression for the Doppler shift, the maximum velocity that can be measured by the pulse Doppler system is obtained by

$$U_{max} = \frac{cf_{max}}{2f_s} = \frac{c^2}{8f_s D_{max}}. \tag{3.55}$$

This expression shows that the maximum observable velocity is limited by the ultrasound frequency and the maximum range to be measured, although the maximum velocity and range can be increased by decreasing the ultrasound frequency, f_s, which will reduce the resolution of range discrimination.

The limitation of the maximum velocity and range in the pulse Doppler system can be eliminated by the random-signal Doppler system. The correlation technique has been effectively used to extract a specific signal among large unwanted waves, especially in radar and telecommunications. The principle is based on the fundamental property of the random noise. A random noise in a time interval can have a strong correlation only when the correlation is taken to the same wave in the same time interval. The correlation is practically zero for the wave in a different time interval, or any other signal, if the observed time interval is long enough.

In the random-signal Doppler system, the transmitted wave is modulated by a noise wave, and correlation is taken between the received wave and the delayed noise wave with a definite delay time, as shown in Figure 3.28.[96] Among the received waves, only the reflected wave components from an object in a definite range, at which the transit time is just the delay time, can have a strong correlation to the delayed noise wave, and the correlation to other reflected components from objects out of the range becomes small enough. Thus, the reflected wave component in any range can be extracted by adjusting the delay time of the noise wave.

Instead of a random noise, a pseudo-random signal can also be used for this purpose. The pseudo-random signal is a periodic function of time but has a strong autocorrelation just as the true random noise does, and the period is long enough. There are many kinds of pseudo-random signals.[97] Figure 3.29 shows an example which is called the shift-register-generated sequence. It takes only two values, +1 and −1; its time average is zero; and it can easily be generated by a shift register. By using an n-stage shift register, a pseudo-random sequence having a length of $(2^n - 1)$ times the shift pulse interval, t_p, can be generated. The range resolution of the system is determined by the time interval in which the modulation signal has a strong correlation, which is equal to t_p, thus the resolution, d, is given as

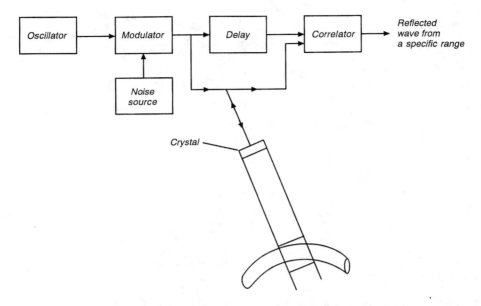

Figure 3.28 Principle of the random-signal Doppler system.

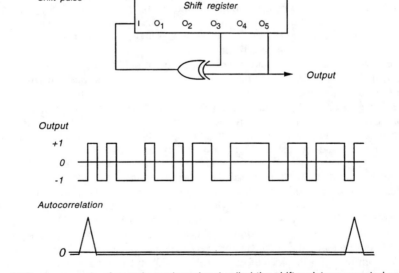

Figure 3.29 An example of pseudo-random signal called the shift-resister-generated sequence.

$$d = \frac{ct_p}{2}. \tag{3.56}$$

The random-signal Doppler system has at least two advantages. First, if a definite level of signal-to-noise ratio is postulated, the required peak amplitude of the transmitted wave of the random-signal Doppler system is less than that of the pulse Doppler system, because S/N corresponds to the averaged power of the transmitting wave. Thus, if the peak amplitude of the ultrasound wave emitted into the tissue is limited by safety considerations, higher S/N will be expected from the random signal Doppler system. Second, it has no range-velocity limitation, as seen in Equation (3.55), thus higher velocity in the deep vessel can be measured. However, it has been shown that the velocity and range resolution of the random-signal Doppler system is not superior to that of the pulse Doppler system.[63]

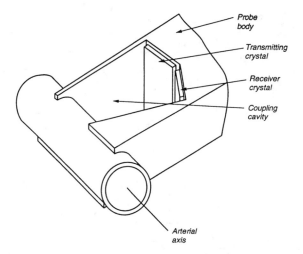

Probe body

Transmitting crystal

Receiver crystal

Coupling cavity

Arterial axis

Figure 3.30 A perivascular Doppler flow probe. (From Richardson, P. C. A. et al., *Med. Biol. Eng. Comput.*, 25, 661, 1987. With permission.)

The range-discriminating Doppler flowmeter can be used more effectively by combining it with a real-time, pulse-echo imaging system, because the measurement range or sample volume can be displayed on the screen together with the image of the target blood vessel, and the spectrum of the Doppler shift signal can also be monitored on the same screen.[98] This system is called a duplex scanner.

In duplex systems, the position of the sample volume can be determined by adjusting a mark of the sample volume to the image of the target blood vessel on the monitor screen. The beam axis is also displayed on the screen, and the angle between the beam axis and the axis of the vessel can be estimated. Besides that, the vessel diameter can be estimated from its image. From this information it is expected that the volume flow rate can be quantitatively estimated, but it must be pointed out that there are many sources of error, such as the effect of nonuniform ultrasound beams, high-pass filtering to eliminate the vessel wall thump, vessel pulsatility, ambiguities in the diameter measurement, and angle measurement.[99]

The flow imaging can be realized using the principle of the duplex system, if the location of the sample volume is scanned automatically over the whole field. Although it requires many spectrum analyses for each flow image, even real-time flow imaging systems have been developed using a parallel FFT processing scheme, and the instantaneous flow distribution is displayed in the color scale.[100]

3.2.2.5 Perivascular and Intravascular Doppler Probes

Although ultrasonic Doppler blood flowmeters have been developed mostly for transcutaneous measurements, there are some studies of Doppler flowmeters for perivascular and intravascular applications. Perivascular probes have been developed for implantable use in experimental animals and also for clinical use during operations. Di Pietro and Meindl[101] developed an implantable continuous-wave Doppler flowmeter for monitoring blood flow in the major artery in dogs. The probe had transmitting and receiving crystals of 4 × 4 mm on opposite sides of the vessel at a 60° angle to the vessel axis, and the volume was 3.8 ml. To monitor the flow, an FM radio telemetry system was used, and the power consumption of the implanted unit was 10 mW. Carter et al.[102] also developed a similar implantable, continuous wave Doppler flowmeter and applied it to monitor blood flow in the coronary, renal, and iliac arteries in sheep, during and after exercise.

An implantable pulsed Doppler flowmeter was also developed to provide information on the velocity profile in a vessel, as well as its diameter.[103] The transducer was a single 3 × 3-mm crystal, set at a 59° angle with respect to the vessel axis, and blood velocity was sampled at eight points evenly spaced across the vessel, up to 11 mm in diameter. They applied the probe to the abdominal aorta of dogs and obtained an accuracy of ± 8.7% in average volume flow rate compared with direct bleed-out measurements.

A perivascular Doppler flow probe for preoperative use has also been developed.[104,105] The probe has a transmitting and receiving crystal in a coupling cavity filled with a hydrophylic polymer (Figure 3.30) and can be used on vessels in the range 5 to 12 mm.

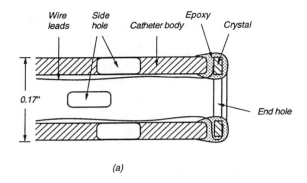

(a)

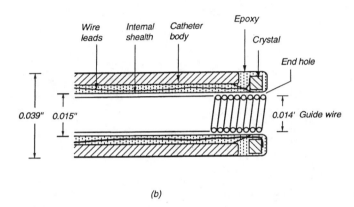

(b)

Figure 3.31 Pulse Doppler catheters of the original design **(a)** and having a guide wire **(b)**.

Intravascular Doppler flowmeters have been developed and applied for investigating the aortic flow velocity pattern. A continuous wave Doppler catheter, having two hemidisk-shaped crystals of 8 MHz and 1.5 mm in diameter, was developed and used for coronary arterial flow measurements in patients with cardiac arrhythmias.[106]

Pulse Doppler catheters have also been developed. The range gating capability allows the operator to measure flow velocity at a desired distance away from the tip of the catheter, thereby decreasing artifacts resulting from catheter-induced turbulence.[107] Figure 3.31(a) is a cross-sectional diagram of the tip of the catheter. It has a circular crystal 4.16 mm in outer diameter with a central hole operated at 20 MHz. The catheter has an inner lumen, which can be used for pressure measurement and injection of contrast media or other agents. Figure 3.31(b) shows a similar Doppler catheter, but one which has a flexible, steerable guide wire that allows precise adjustment and positioning of the Doppler crystal.[108] It also has a circular 20-MHz crystal, and the sample volume is adjustable from 1 to 12 mm distal to the catheter tip.

3.2.3 INDICATOR DILUTION METHOD

Flow measurement by indicator dilution is a method in which a definite amount of indicator is injected into the blood stream, and the average flow rate is estimated from the time course of the concentration of the indicator at the downstream. Dyes, radioisotopes, electrolytes, or heat can be used as indicators. Fick's method for cardiac output measurement is also a variation of the indicator dilution method in which oxygen, carbon dioxide, or other gases are used as the indicator.

3.2.3.1 Principle

Suppose a part of the circulatory system is simulated by a model, as shown in Figure 3.32. The indicator is injected at the upstream, mixed uniformly in the mixing chamber, and then moved to the downstream where the concentration of indicator is observed.

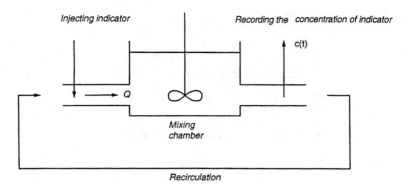

Figure 3.32 Principle of the indicator dilution method.

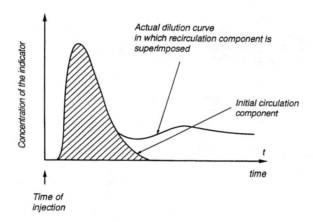

Figure 3.33 The dilution curve for a rapid injection of the indicator.

Two types of injection schemes are commonly used; one is rapid injection and the other is constant injection.[109] In the rapid injection methods, such as slug injection and bolus injection, a definite amount of indicator is injected in a short period of time. After a while, the concentration of indicator at the downstream will increase rapidly and then decrease until the indicator appears again by recirculation, as shown in Figure 3.33. The time course of the concentration of indicator is called the indicator dilution curve.

If the injected indicator does not leak out from the conduit, the whole amount of the injected indicator passes the observation point at the downstream. Thus, if the flow rate is Q and the concentration of indicator at time t is $c(t)$, then the amount of injected indicator, I, should be

$$I = \int_0^\infty Q \cdot c(t) dt \qquad (3.57)$$

as long as the indicator does not recirculate. If the flow rate is constant, it can be solved as

$$Q = I \bigg/ \int_0^\infty c(t) dt. \qquad (3.58)$$

This means that the flow rate can be determined from the amount of injected indicator and the integral of the indicator dilution curve without a recirculation component.

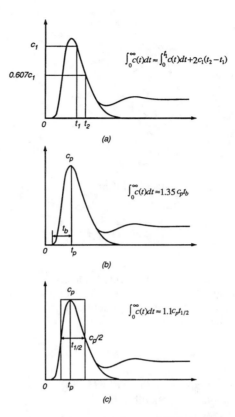

Figure 3.34 Empirical approximation methods of estimating the initial circulation component of the dilution curve.

In the actual circulatory system, however, the indicator will recirculate. To eliminate a recirculation component, some empirical methods were proposed for cardiac output measurement in which an indicator is injected into the vein or the pulmonary artery, and the dilution curve is observed at the artery.

Figure 3.34(a) shows a method of approximating the downslope of the indicator dilution curve by an exponential curve.[110] If two points (t_1,c_1) and (t_2,c_2) are taken so that

$$c_2 = c_1 e^{-1/2} = 0.607c_1 \tag{3.59}$$

then the integral of the dilution curve can be obtained as

$$\int_0^\infty c(t)dt \cong \int_0^{t_1} c(t)dt + 2c_1\left(t_2 - t_1\right). \tag{3.60}$$

Figure 3.34(b) shows a method called the forward triangle in which the integral of the dilution curve is approximated only by the buildup time, t_b, and the peak concentration, c_p, is used.[111] The approximation formula is

$$\int_0^{t_p} c(t)dt \cong 1.35c_p t_b \tag{3.61}$$

Figure 3.34(c) shows a method in which the half width, $t_{1/2}$, and the peak concentration, c_p, are used.[112,113] The approximation formula is

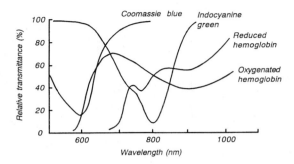

Figure 3.35 Relative spectral transmittance of dyes and reduced and oxygenated hemoglobin.

$$\int_0^{t_p} c(t)dt \cong 1.1 c_p t_{1/2}. \tag{3.62}$$

In the constant injection method, indicator is injected at a steady rate. The concentration of indicator at the downstream reaches an equilibrium after the indicator-free fluid in the system is washed out. At the equilibrium, the flow, Q, is determined by the amount of injection of the indicator per unit time, I, and the concentration of the indicator, C, as

$$Q = I/C. \tag{3.63}$$

In a closed circulation system, the concentration will be changed if the indicator is recirculated to the observation site. The concentration at the equilibrium should be determined before the fastest recirculation component reaches the observation site.

3.2.3.2 Dye Dilution Method

The dye dilution method is a kind of indicator dilution method in which a dye is used as the indicator and the concentration of dye is usually detected optically. The dye for blood flow measurement should be water soluble, nontoxic, and sterile and should permit precise determination of its concentration in whole blood or plasma. Besides this, it should not be lost or degenerated metabolically during initial circulation. It is preferable that the dye leaves the circulatory blood rapidly after initial circulation so that it does not disturb repeated measurements.[114]

In Figure 3.35, spectral transmission curves of typical dyes are shown. Coomassie blue has its absorption peak at about 600 nm. At this wavelength, reduced hemoglobin also has strong spectral absorption, and thus the dilution curve fluctuates due to the variation in oxygen saturation of the blood. To reduce the fluctuation, a high concentration of dye should be used. Evans blue has a similar spectral absorption; however, it remains in the circulatory blood for a longer time than does Coomassie blue. Indocyanine green has two absorption peaks near 800 nm. Around this wavelength, the absorption curves of oxygenated and reduced hemoglobin cross each other, thus the spectral absorption is not affected by oxygen saturation. It is also advantageous that it has no strong absorption in the visible range and does not cause staining of the tissues.

To record the dilution curve, the cuvette densitometer, the earpiece densitometer, or the fiberoptic catheter can be used. The cuvette densitometer is an instrument that measures dye concentration in the blood drained continuously via an intravascular catheter. Figure 3.36 shows the construction of a cuvette.[115] The blood flows through a gap between two glass plates, and the light transmitted to the blood layer is detected by a photodetector such as the photomultiplier tube.

When the concentration of dye is low enough and it is dispersed uniformly in the blood, the Lambert-Beer's law is valid for the optical density, I, and the concentration of dye, c, i.e.,

$$\varepsilon cd = \log\left(\frac{I_0}{I}\right) = \log I_0 - \log I \tag{3.64}$$

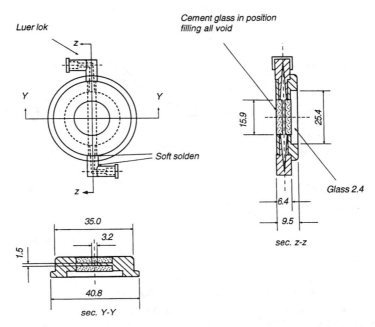

Figure 3.36 A densitometer cuvette. (From Gilford, S. R. et al., *Rev. Sci. Instrum.*, 24, 696, 1953. With permission.)

where ε is the absorption coefficient, d is the thickness of the blood layer, and I_0 is the optical density when $c = 0$. Although the optical density is a nonlinear function of the concentration of dye, linear output can be obtained owing to the logarithmic characteristic of the photomultiplier tube when the anode current is kept invariable.[115] The sensitivity of the densitometer can be absolutely calibrated by introducing blood with a known concentration of dye into the cuvette. The calibration factor may be changed by hematocrit values; however, the change can be compensated for by a dichromatic densitometer, in which two wavelengths are used so that one is sensitive and the other insensitive to the concentration of dye, but both are sensitive to hematocrit.

The earpiece densitometer is an instrument which measures the concentration of dye in the blood by light absorption in the ear lobe. Although the dye is not uniformly dispersed in the ear tissue, but remains only in the blood vessels, it is always assumed that Lambert-Beer's law is approximately valid.

Calibration of the earpiece densitometer is performed by taking a blood sample after the dye is uniformly mixed in the circulating blood, and the deflection at the terminal of the dilution curve corresponds to the concentration of dye in the blood sample. The concentration of dye at the terminal can be roughly estimated as the amount of injected dye divided by the circulating blood volume, which is about 7% of the body weight of a normal subject.

In order to compensate the variation in transmitted light, due to the change of the blood content in the tissue, dichromatic measurement is used. For example, 805 and 900 nm are used for Indocyanine green. To measure transmittance in the two wave lengths, the use of a single light beam divided into two photocells by a dichroic mirror is more stable than the use of two independent light beams.[116]

A fiberoptic catheter can be used for recording dye dilution curves *in vivo*. At the catheter tip, the blood is illuminated by a light beam transmitted through a bundle of optical fibers, and part of the backscattered light is transmitted to the detector through another bundle of optic fibers. Although this application differs from the cuvette or earpiece densitometer, in which a light beam passes through a layer of definite thickness, it was confirmed that a similar calibration curve could be obtained between the concentration of dye in the blood and the change in intensity of the backscattered light.[117,118] In a comparative study with a cuvette densitometer, it was also confirmed that cardiac output measurements obtained by both methods were highly correlated, having a correlation coefficient of about 0.9.[117]

An advantage of the fiberoptic catheter is that dilution curves in the large vessels can be recorded without distortion which may occur due to the velocity distribution in the catheter to the cuvette or the arterial branches to the ear tissue. The fiberoptic catheter can be made small enough for use in neonates.

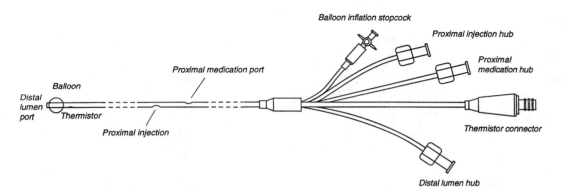

Figure 3.37 A thermodilution catheter. (Courtesy of Spectramed Co., Oxford, CA.)

Volz and Christensen[119] developed a fiberoptic catheter 0.8 mm in diameter which contained 14 fibers made from polymethyl methacrylate (PMMA).

3.2.3.3 Thermodilution Method

The thermodilution method is a kind of indicator dilution method in which a definite amount of heat is injected into the blood stream, and the corresponding temperature change is recorded downstream. In the thermodilution method, a cold fluid is often used as an indicator, because the cold fluid is less harmful to the blood and tissue than a hot one. For cardiac output measurement in a human adult, about 10 ml of cold saline or isotonic dextrose solution near 0°C is commonly used. The injected cold indicator is mixed and diluted in the warm blood stream and causes a slight temperature decrease in the blood at the downstream.

The thermodilution method has several advantages, i.e., the indicator has no toxic effect so that measurement can be performed repeatedly, the dilution curve can be easily recorded by a thermistor placed in the vessel, and the recirculation component is small enough so that integration of the dilution curve can be performed accurately. The fundamental assumption of the indicator dilution method — that the indicator should not leak out from the vascular system between the injection and detection sites — is not perfectly valid because heat can dissipate across the vessel wall. This effect is insignificant in the large vessels due to a small ratio of surface area to volume per unit length. Thus, the thermodilution method is more appropriate for flow measurements in large vessels than in smaller ones.[120]

Flow rate, Q, can be calculated from the blood temperature, T_B, as follows. If the volume and the temperature of the injected fluid are V_I and T_I and densities and specific heats of the blood and the injected fluid are ρ_B, ρ_I and C_B, C_I, respectively, then

$$Q = \frac{\rho_I C_I}{\rho_B C_B} \cdot \frac{V_I\left(T_B - T_I\right)}{\int_0^\infty \Delta T_B dt}$$

(3.65)

where ΔT_B is the change of blood temperature. The coefficient $\rho_B C_B/\rho_I C_I$ is about 0.93 for 5% dextrose solution and 0.91 for saline.[121,122]

For cardiac output measurements, a specially designed catheter, the so-called Swan-Ganz-thermodilution catheter, is widely used. Figure 3.37 shows the configuration of the catheter. It is 7 to 7.5 F (2.3 to 2.5 mm in outer diameter) and has a balloon, a thermistor having a time constant of 0.2 to 0.4 s attached near the tip, and an injection port located 25 to 30 cm from the tip.[121,123] Besides that, additional lumina are available for pressure measurement or medication in the commercial catheters.

Conventionally, the catheter is introduced from a peripheral vein into the pulmonary artery through the right ventricle, as shown to Figure 3.38. A bolus of cold saline or dextrose solution is injected into the right atrium, mixing occurs in the right atrium and the right ventricle, and the decrease in the resultant temperature is detected by the thermistor placed in the pulmonary artery.

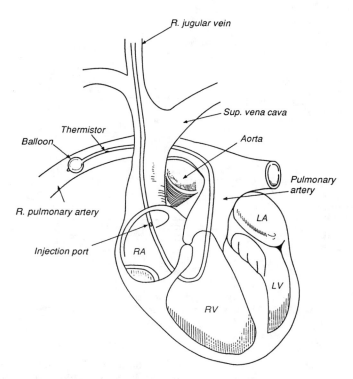

Figure 3.38   The thermodilution method for monitoring cardiac output. (From Philip, J. H. et al., *IEEE Trans. Biomed. Eng.,* BME-31, 393, 1984. With permission.)

There are many studies which have examined the validity of the thermodilution method in model experiments,[124,125] in animals,[124,126] and in human subjects.[121,127] An *in vitro* comparison with calibrated rotameter measurements showed the average error was 2.2% with a correlation coefficient of 0.993.[128] A clinical study[121] compared thermodilution with dye dilution measurements. The obtained cardiac outputs determined by both methods were close to each other in 63 measurements in 20 subjects and within a range of 2.9 to 8.0 l/min. From these studies, it has been concluded that the thermodilution method with rapid injection for measurement of cardiac output is acceptable and accurate for most clinical purposes.

The thermodilution method with continuous injection is used for local blood flow measurement, but it can also be applied to cardiac output measurement. Figure 3.39(a) shows an example of a simple thermodilution catheter.[124] It has an injection orifice directed upstream, and a change in blood temperature is detected by a thermistor downstream of the orifice. It is recommended that the Reynold's number at the injection of the fluid be at least 3000 in order to achieve good mixing. Because the injection rate should be small in continuous injection, injection fluid temperature will change during passage through the catheter. To measure the true injected fluid temperature, an additional thermistor is used, as shown in Figure 3.39(b).[122,129] It was shown that when the indicator was injected through an injection orifice 0.45 mm in diameter, oriented upstream 30 to 45° to the axis of the tube, an injection rate of 35 to 55 ml/min was required to obtain a uniform temperature 10 mm downstream of the orifice, for mixing at a blood flow rate of 300 to 500 ml/min.[129]

A different approach to the thermodilution method has been attempted in which intravascular heating is used instead of injecting a cold fluid.[130] To transfer heat into the blood stream, an electric heater can be used. In this study, heating wire is wound around a standard thermodilution catheter. When a sinusoidal thermal signal of 0.02 Hz with an average power of 4 W was applied in the right ventricle of the sheep and blood temperature change in the pulmonary artery was observed by performing synchronous detection, cardiac output could be measured over the range 1.8 to 9.5 l/min, and the obtained correlation coefficient was 0.977, compared to 0.993 for the standard rapid injection thermodilution measurements.

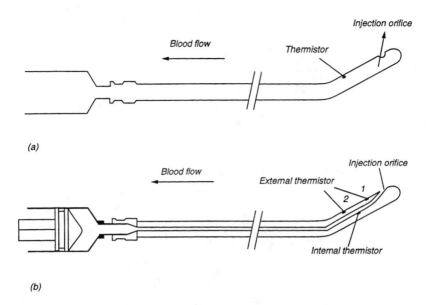

(a)

(b)

Figure 3.39 Thermodilution catheters for continuous injection.

3.2.3.4   Fick Method

Pulmonary blood flow can be determined by measuring the rate of gas intake to or extraction from the blood in the lung, and the change in the gas concentration in the blood stream through the lung. This technique is called the Fick method and has been used as a standard method of cardiac output measurement.[131]

If oxygen uptake, $\dot{V}O_2$, and oxygen concentration in the pulmonary arterial blood, C_aO_2, and the pulmonary venous blood, C_vO_2, are known, the blood flow rate through the lung, Q, can be obtained as

$$Q = \frac{\dot{V}_{O_2}}{C_{aO_2} - C_{vO_2}}. \tag{3.66}$$

This expression is equivalent to Equation (3.63) in which C is considered to be the increment of indicator concentration.

When a gas is used as an indicator, the amount of indicator is usually represented by volume, thus the concentration in Equation (3.66) should be represented by a volume fraction. The volume of oxygen in the blood can be calculated from oxygen saturation using the oxygen capacity of hemoglobin, i.e., 1 g hemoglobin can be combined with 1.38 ml oxygen in standard conditions.[132]

$\dot{V}O_2$ can be obtained by measuring tidal volume and oxygen content in the expiratory air (see Section 7.3.4). C_{aO_2} is given by the gas analysis of any arterial blood. C_{vO_2} is also given by the gas analysis of the venous blood. The determination should be performed on mixed venous blood, because the oxygen content of venous blood in different organs varies. Thus, the venous blood has to be sampled in the right ventricle or the pulmonary artery via cardiac catheterization.

Because blood samples cannot be taken so frequently, continuous information on cardiac output is difficult to obtain. Only in animal experiments has continuous monitoring of cardiac output been attempted, where arterial and venous blood were circulated through cuvettes and oxygen saturations were measured continuously.[133-135]

The Fick method is considered to be the most reliable method of cardiac output measurement as long as the mixed venous blood is sampled; however, a disadvantage is that the Fick method requires cardiac catheterization.

Another approach to cardiac output measurement is called the indirect Fick method. It employs the same principle as the direct Fick method but information on blood gas contents is obtained only by expiratory gas analysis, so that cardiac catheterization is unnecessary. In the indirect Fick method, carbon dioxide or other gases, such as acetylene or nitrous oxide, are used as an indicator instead of oxygen as in the direct Fick method.

When a carbon dioxide measurement is employed, the pulmonary blood flow is expressed as

$$Q = \frac{\dot{V}_{CO_2}}{C_{vCO_2} - C_{aCO_2}} \tag{3.67}$$

where $\dot{V}_{CO_2}$ is carbon dioxide output, and C_{vCO_2} and C_{aCO_2} are carbon dioxide contents in the mixed venous and the arterial blood. $\dot{V}_{CO_2}$ can be measured by expiratory gas analysis (see Section 7.3.4). C_{aCO_2} can be determined by the arterial carbon dioxide partial pressure, which is equal to the mean carbon dioxide partial pressure of the alveolar gas and is obtained as the carbon dioxide partial pressure of the expiration gas.[136] To convert partial pressure to volume fraction, the dissociation curve of carbon dioxide should be used, while the slope can be regarded as linear in the physiological range and is approximated as 3.46 vol%/kPa (0.46 vol%/mmHg).[136]

The most important part of the indirect Fick method is determination of the carbon dioxide content in the mixed venous blood, C_{vCO_2}, from respiratory gas analysis. To achieve this, the rebreathing method has been studied extensively,[137-140] while the breath-holding method has also been used.[136,141]

The principle of the rebreathing method is that when the subject breathes the air from an anesthetic bag for 15 to 30 s, the carbon dioxide partial pressure of the air in the bag is equilibrated to that of the mixed venous blood. By approximating the time course of the carbon dioxide content in the bag by an exponential function, the final asymptotic level can be predicted from the time course.[138] A fully automated system of this procedure has been developed.[142]

In the breath-holding method, the mixed venous carbon dioxide content is estimated from the time course of carbon dioxide partial pressure in the alveolar gas during breath-holding of about 20 s.[136] By inhaling air with different carbon dioxide concentrations from 5 to 7% before breath-holding and plotting the rate of change of carbon dioxide partial pressure in expired gas, the partial pressure of the mixed venous blood can be estimated as a condition where the rate of change is zero.[141]

A different method was proposed in which the information on gas contents in the mixed venous blood is not taken into account, but only carbon dioxide output and partial pressure in expired gas are measured.[143] The pulmonary blood flow is determined from the change in the carbon dioxide output corresponding to the change in the carbon dioxide partial pressure in the alveolar gas when the ventilation pattern is altered between hyper- and hypoventilation.

Gases other than oxygen and carbon dioxide can also be used as the indicator. If these gases do not exist in the blood initially, the concentration of the gas in the mixed venous blood is zero before appearing in recirculation, and pulmonary blood flow can only be determined by the concentration in the alveolar gas and the uptake of the gas. For the indicator gas, acetylene or nitrous oxide have been used.[144,145] By using a mass spectrometer, different gases can be measured simultaneously (see also Section 7.3.4).[146]

3.2.3.5 Other Dilution Methods

Besides the indicators described above, radioisotopes and solutions having different electric conductivities from that of blood have also been used as indicators for the dilution method. As radioactive indicators, substances labeled with ^{131}I, especially radio-iodinated serum albumin (RISA) have been used.[147-150] For cardiac output measurements, RISA of 0.37 to 3.7 MBq (mega-becquerei) or 10 to 100 μCi (micro-curies) is injected into the vein, and the concentration of RISA in the arterial blood is measured externally by a gamma-ray detector with a collimator directed to the left ventricle or the aortic arc. The sensitivity of the detector can be calibrated comparing the external gamma-ray count, 10 to 20 min after injection, with the direct count of the sampled blood at that moment.

The problem with the radio-iodine is its uptake by the thyroid gland. To minimize this, the patient is given potassium iodide before and after the examination to saturate the thyroid gland with nonradioactive iodide. To reduce radiation dosage further, ^{131}I-labeled Cholografin is recommended, because its half-life in the body is 5.34 hours, while that of RISA is 10.5 days.[151]

In the dilution method, because of the indicator having a different electric conductivity, a hypertonic saline is used. For cardiac output measurement, 1 to 10 ml of 2 to 8% saline is rapidly injected into the central vein, and the electric conductivity change is detected in the pulmonary artery[152-154] or in the aorta.[155] By simultaneous measurements with other methods, it was shown that cardiac outputs measured by saline dilution method were highly correlated to those measured by the electromagnetic flowmeter[154]

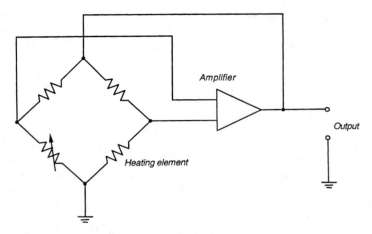

Figure 3.40 A bridge circuit with a feedback amplifier for maintaining the heating element at a constant temperature.

and the dye dilution method.[153] Detection of electric conductivity changes in the blood by perivascular electrodes[156] and the use of autogenous plasma as the indicator[157] were also attempted.

3.2.4 FLOW VELOCITY MEASUREMENTS BY HEAT DISSIPATION

The rate of heat dissipation from a heated element placed in the blood stream depends on the flow velocity. Although it also depends on many other factors — such as the temperature difference between the element and the surrounding fluid; the size and the shape of the element; viscosity, thermal conductivity, specific heat, and density of the fluid; and the state of the flow, i.e., if the flow is laminar or turbulent — flow velocity can still be determined from the rate of heat dissipation when changes in the above-mentioned parameters are insignificant. In such a situation, the rate of heat dissipation, H, can be approximated as

$$H = a + bU^m \tag{3.68}$$

where U is the flow velocity, and a, b, and m are constants. The constants a and b should be determined by calibration, while m is always taken as 0.5.[158]

To maintain the element at a constant temperature, electric heating is commonly used. In equilibrium the rate of heat dissipation is equal to heat production, i.e.,

$$H = RI^2 \tag{3.69}$$

where R is the resistance of the heating element and I is the applied current. If R has a temperature coefficient, the temperature of the heating element can be kept constant using a bridge with a feedback amplifier, as shown Figure 3.40.

3.2.4.1 Thermistor Velocity Probe

The thermistor is a convenient device for use in thermal velocity probes because it has a large temperature coefficient. Two thermistors are commonly used in the flow probe. One measures the fluid temperature, and the other is heated to a higher temperature. The temperature difference between the heated thermistor and the fluid is kept constant. Figure 3.41 shows two examples of such probes; (a) is a 6-F (2 mm o.d.) cardiac catheter having two thermistors near the tip. Because of a bend, the thermistors are placed near the center of the blood stream when the catheter is introduced into the vessel.[159] The response times were 0.2 s for an abrupt increase and 1.5 s for an abrupt decrease in temperature of the fluid, and it was used for venous flow measurements.

Figure 3.41(b) also shows a velocity probe with two thermistors.[160] This probe has the velocity sensing element at the center of five flexible springs, and the reference thermistor is placed at the tip. The sensing

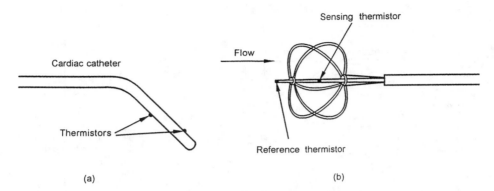

Figure 3.41 Thermistor velocity probes. (Part (a) modified from Katsura, S. et al., *IRE Trans. Med. Electronics,* ME-6, 283, 1959. Part (b) modified from Mellander, S. and Rushmer, R. F., *Acta Physiol. Scand.,* 48, 13, 1960.)

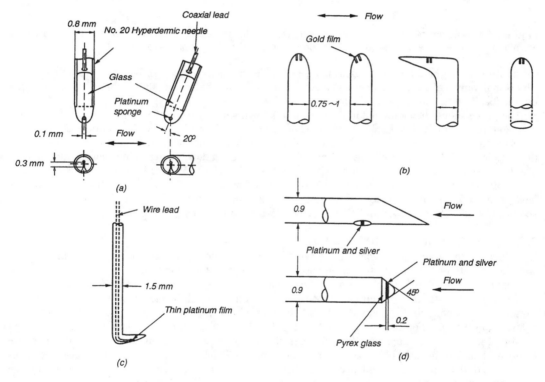

Figure 3.42 Hot-film velocity probes. (Part (a) from Ling, S. C. et al., *Circ Res.,* 23, 789, 1968. Part (b) from Seed, W. A. and Wood, N. B., *Cardiovasc. Res.,* 4, 253, 1970. Part (c) from Nerem, R. M. et al., *Circ. Res.,* 34, 193, 1974. With permission. Part (d) modified from Clark, C., *J. Phys. E Sci. Instrum.,* 7, 548, 1974.)

element consists of a tiny thermistor around which has been wound a fine wire for heating. Its frequency response was about 5 Hz, and it could be used for venous flow measurements.

3.2.4.2 Hot Film Velocity Probe

A thermal probe, consisting of a thin metal film used for local flow measurements, is called a hot film velocity probe or hot film anemometer. Due to the small heat capacity of the sensing element, it can have a wide frequency response of up to 10 kHz. Many probes of different shapes have been made, as shown in Figure 3.42. The sensing element consists of platinum, platinum-silver, or gold fused on a Pyrex glass or quartz rod. The size is 0.06 to 0.2 mm in width and 0.3 to 0.5 mm in length. The thickness

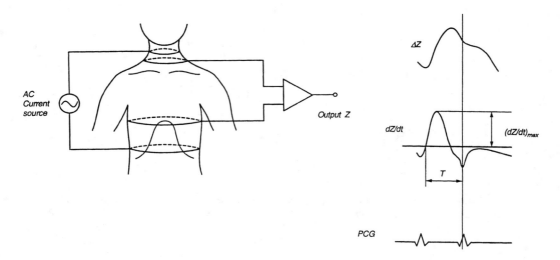

Figure 3.43 Diagram of impedance cardiography with the tetrapolar electrode arrangement (left) and typical waveforms of impedance change (ΔZ) and its derivative (dZ/dt); PCG: phonocardiogram.

of the film should be adjusted so as to realize suitable resistance, on the order of 0.01 µm. The metal film is maintained at a constant temperature at about 5°C above the flowing fluid temperature, employing the feedback bridge, as noted in Figure 3.40.

The probe should be calibrated in the stream of known velocity using the same fluid and at the same temperature as for the actual measurement, while the difference in the sensitivity (b in Equation (3.68)) is less than 10% between water at 20°C and blood at 37°C.[161] The dynamic response of a commercial probe (55M01, DISA Electronics; Allendale, NJ) was studied in detail using a sine wave excitation unit. The result showed that at laminar flow with a flow velocitiy above 50 cms^{-1}, the cutoff frequency was above 500 HZ. For flow velocities above 100 cms^{-1}, it was above 1.5 kHz.[162] Hot film velocity probes have been used in such cardiovascular studies as dynamic measurement of the velocity profile,[161,163] measurement of the velocity waveform,[162,164] detecting flow reversal,[165] or measurement of turbulence.[166,167]

3.2.5 IMPEDANCE CARDIOGRAPHY

Impedance cardiography is a technique in which stroke volume or cardiac output is estimated by the waveforms of transthoracic electric impedance. Since Kubicek et al.[168] introduced a simple equation for calculation of stroke volume, this technique has been studied extensively by many investigators.[169]

To record transthoracic electrical impedance, the tetrapolar electrode arrangement as shown in Figure 3.43 has commonly been used. An a.c. current within the range of 20 to 100 kHz at a current level within the range of 10 µA to 10 mA is supplied through current electrodes, which are placed at the top of the neck and at the end of the rib cage or distal to it. Voltage electrodes are placed at the base of the neck and at the level of the xiphisternal joint, and the induced voltage between voltage electrodes is measured. The thoracic impedance, Z, is then defined by the obtained voltage, V, divided by the supplied current, I.

Figure 3.43 shows a typical tracing of the change in thoracic impedance, ΔZ, its time derivative, dZ/dt, and the phonocardiogram. The decrease in thoracic impedance is customarily represented as an upward deflection, and the peak height in dZ/dt is represented as $(dZ/dt)_{max}$, which represents the maximum negative slope.

A steep change in the thoracic impedance is always observed at the early systole, and it is thought that this change is caused by the ejection of blood into the vessels in the thorax which creates an additional parallel conductor in the segment of thorax between voltage electrodes. Actual amplitude of the impedance waveform is about 0.1 to 0.2 Ω, while the average impedance Z_0 is about 20 to 30 Ω in adults.

Kubicek et al.[168] derived an equation for the stroke volume as

$$SV = \frac{\rho_b L^2}{Z_0^2} \Delta Z \tag{3.70}$$

Table 3.2 Evaluations of Impedance Cardiography Compared to Other Techniques of Cardiac Output Measurement

Subject	n	Compared Method	Regression Equation	Correlation Coefficient	Correction of Blood Resistivity	Ref.
Normal	8	Dye	0.90x + 0.62	0.94	No, $\rho = 150\ \Omega$	170
Patient	14	Fick	0.90x + 3.3	0.91	No, $\rho = 135\ \Omega$	171
Patient	20	Radioisotope dilution	0.72x + 0.47	0.88	Yes	172
				0.61	No, $\rho = 150\ \Omega$	
Patient	10	Dye	0.69x + 0.94	0.85	No, $\rho = 135\ \Omega$	173
Normal	10	Dye	0.99x − 0.25	0.90	Yes	174
Normal	6	CO_2 rebreathing technique	0.89x + 1.06	0.91	Yes	175
Normal	21	Dye	0.65x + 1.7	0.91	Yes	176
Normal	50	Thermal dilution	0.47x − 0.7	0.60	Yes	177

where ρ_b is blood resistivity, L is the distance between voltage electrodes, and Z_0 is the average thoracic impedance. This equation corresponds to a parallel conductor model (see Section 3.3.1.3). It is assumed that the blood is ejected uniformly at a constant rate, and the impedance change at the end of the systole is estimated as

$$\Delta Z = T(dZ/dt)_{max} \qquad (3.71)$$

where T is the ejection time. Studies evaluating cardiac output measurements by impedance have been carried out by comparing this technique with dye-dilution, radioisotope dilution, the Fick method, and the carbon dioxide rebreathing techniques, as listed in Table 3.2.[170-177] Although these studies reported fairly high correlation coefficients, there are also reports of correlation coefficients less than 0.6.[178-180] Even among reports which gave higher correlation coefficients, regression equations are different, and this implies a difficulty in quantitative estimation of cardiac output by applying a simple model, such as Equation (3.70).

A complexity arises from the fact that the thoracic impedance waveform can be affected by blood volume changes in various organs in the thorax during a cardiac cycle, and the contribution of each organ to the thoracic impedance may be different from subject to subject, due to the morphological difference. Quantitative estimation of the effect of each organ on the total impedance change has been investigated in animal experiments[181] and models.[182,183] These studies showed that the effect of the volume change in the aorta and large arteries is less than 30%, while that in the lung is about 60%.

A change in blood resistivity will affect the sensitivity of the impedance signal to the blood volume change in an organ, and as seen in Table 3.2 the correlation coefficient increased significantly when known blood resistivities were used in the calculation of cardiac outputs.[172] It was also mentioned that the blood resistivity changes 10 to 15% with the flow velocity,[184] and this effect will also influence the cardiac output estimation.

Consequently, cardiac output measurement by thoracic impedance cannot be accurate as long as a simple model of a homogeneous cylinder is assumed; however, accuracy will be improved significantly by taking account of many factors affecting thoracic impedance.[185]

Different approaches to measuring the stroke volume or cardiac output by impedance have also been proposed. Patterson[186] tried impedance measurement using an esophageal electrode, a bipolar pill electrode 2.1 cm long with thread-like wires. An a.c. current of 100 kHz and 4 mA was supplied to the ordinary band electrodes around the neck and the thorax. The subject swallowed the pill electrode, and its position was adjusted by the ECG recorded from that electrode. As the electrode moved near the atrium, the height and the shape of the P-wave changed sharply. Then the electrode was placed slightly below the estimated position of the left atrium for ventricular volume recording. The observed waveform showed an increasing impedance during systole, while ordinary thoracic impedance decreased. An increasing impedance can be explained by the decrease in ventricular blood volume.

An intraventricular impedance technique has also been attempted for measurement of the stroke volume of a ventricle.[187,188] In this technique, an electrode catheter was placed along the long axis of the left ventricle, as shown in Figure 3.44; a.c. current was supplied through two outer electrodes, and induced voltages between six adjacent pairs of inner electrodes were recorded. By assuming that the ventricular wall is insulated from the cavity blood, the stroke volume can be estimated as the sum of

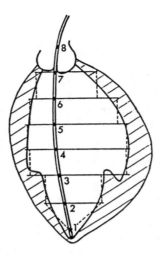

Figure 3.44 Schematic representation of the intraventricular impedance technique. (From Baan, J. et al., *Cardiovasc. Res.*, 15, 328, 1981. With permission.)

blood volume changes in five segments. An animal experiment showed that cardiac outputs measured by this technique were highly correlated with flows obtained by an electromagnetic flowmeter.[180] It was also shown that this technique can be applied to obtain an intraventricular pressure-volume diagram.[188]

3.2.6 BLOOD FLOW RECORDING IN SINGLE VESSELS BY LASER-DOPPLER FLOWMETRY

3.2.6.1 Introduction

The blood flow in single vessels of various sizes has been studied with the laser-Doppler technique. For a more detailed presentation of the theoretical background and the various applications in which the laser-Doppler method has been used, the reader is referred to Section 3.3.4. In principle, three different approaches have been used. First, airborne beams have been used to study the velocity of red cells in the retina of the eye. Low-powered laser beams are focused onto the vessels of the retina through the transparent parts of the eye. Second, fiberoptic catheters have been inserted into blood vessels both experimentally and in human patients. In these applications, measurements of flow along the vessel and perpendicular to the vessel have been performed. In the latter case, the fiber has been introduced through the vessel wall. Third, the laser-Doppler principle has been applied in microscope studies in which the incident laser beam and the light returning from the object pass the microscope. With this arrangement, extremely high resolution can be reached that allows detailed analysis of, for instance, the flow pattern in a vessel.

The measurement of the blood flow of whole blood using the laser-Doppler technique involves an optical problem that is extremely complicated. The mean free path of light in whole blood is approximately of the same order as the diameter of a red blood cell. Laser beams are likely to penetrate a few hundred microns into the blood, which means that a particular beam has been multiple-scattered (has hit many cells in motion) before it returns to the photodetector. Theoretically, this optical situation makes the analysis of the signal very complicated. Stern[189,190] has analyzed the basic theoretical problem of whole blood measurements using catheters.

3.2.6.2 Airborne Beams

Riva, Benedek, and colleagues[191,192] were the first to record blood flow from a single vessel using the laser-Doppler principle. They were interested in the blood flow of the retinal vessels, including both veins and arteries. This application required a reduction of the laser beam intensity to a few microwatts so as not to interfere with the biological structure of the retina. A spectral representation of the Doppler information gave the maximum Doppler frequency, which is related to the maximum velocity in the vessel. By assuming a parabolic velocity profile in the vessel, the mean velocity can be calculated.

Hill et al.[193] obtained high sensitivity for the retinal measurements by combining a polarizing monostatic optical system with photon correlation processing, this being the most sensitive signal processing method. A Zeiss fundus camera is modified to fit this application. Okamoto et al.[194] have described a system built around a fundus camera. This group has quantitatively analyzed the pulsatility of blood flow in retinal arteries and veins.

3.2.6.3 Microscope-Based Instrument

Mishina et al.[195,196] presented a laser-Doppler microscope for the study of very small circulatory volumes, such as a single capillary (10 μm in diameter). The microscope is of the penetrating type. Two parallel beams from a beamsplitter are combined by means of a focusing lens on a focal plane. Flow velocities in the range of 50 mm/s to 50 cm/s can be measured with this system. The applications for this instrument included local blood flow velocity mapping in smaller arteries and veins across the vessel, as well as capillary measurement in frogs exposed to different gas mixtures.

Later, the same group[197] published a report on a more developed microscope by which measurements of blood flow in the arterioles of the cat brain were recorded. Born et al.[198] modified the design by Mishina et al.[195,196] into a backward scatter system around an inverted microscope. The authors conclude that it is unlikely that laser-Doppler anemometry will be useful in *in vivo* applications because of the poor resolution obtained in whole blood. A similar system for small vessel studies was also suggested by Eniav et al.[199] and Cochrane et al.[200] Recently Okada et al.[201] have studied fluctuations of red cell velocity in microvessels with a microscopic technique.

Kilpatrick et al.[202,203] measured the coronary sinus blood flow of dogs using a specially designed laser-Doppler system, called the FOLDA technique. The optic probe inserted into the coronary sinus had an outer diameter of 0.3 mm and a core diameter of 0.05 mm. The fiber was positioned in the coronary sinus by using a polyethylene tube preventing the fiber from being directed towards the vascular wall. The position of the fiber was monitored by direct observation of the laser or by watching the spectra generated. A spectrum was generated using a fast Fourier transform analyzer. The maximum frequency in the recorded spectrum is proportional to the velocity of flow. When the optical fiber was directed in the same direction as the flow, a nonlinear response could be seen which probably can be explained by the disturbances in the flow pattern close to the fiber tip.

The laser-Doppler instrument was checked against an electromagnetic flowmeter. A good correlation was found between the two methods with a correlation coefficient of 0.97.[202,203] The laser-Doppler allows continuous recording of coronary sinus blood flow over long periods of time. One limitation was found, namely the inability to measure high velocities at the same direction of flow as discussed above. The version described in the above-mentioned papers cannot differentiate between forward and reverse flow. Technically, this problem can be solved. Another and maybe more serious limitation is that the exact position of the fiber tip is unknown, a problem when the flow velocity in most vessels varies with the position across the vessel. The authors concluded, though, that for many vessels the flow profile is flat enough to make the measurement worthwhile.

3.2.6.4 Catheter-Based Instruments

Tanaka and Benedek[204] were the first to use a multimode optical fiber to introduce laser light into a blood vessel. They measured the average blood flow in the rabbit femoral vein by taking autocorrelation functions of scattered light. However, their method was unable to detect instantaneous blood flow changes in pulsatile blood flow and to differentiate reverse flow.

Kajiya et al.[205] have used laser-Doppler flowmetry extensively for coronary artery and great cardiac vein blood flow studies in dogs. In a recent review article,[206] most of the work by this group is presented. The system generally used by this group is shown in Figure 3.45.[207] Three different access routes for the fibers into the blood vessels have been used. First the optical fiber has been introduced into large- and mid-sized epicardial coronary arteries and veins through the vessel wall in experiments on dogs. A cuff surrounding the vessel with a small hole in it supports the thin optical fiber (50 mm) which is introduced into the vessel at an angle of 60°. The fiber was moved stepwise, and the velocity profile throughout the lumen was obtained.[205] In a second approach, used for small epicardial arteries and veins thin enough to be transparent to laser light, the optical fiber was fixed to the outer side of the vessel wall by a drop of cyanoacrylate. Finally, in intramyocardial arteries and veins, the fiber was introduced into the vascular lumen from the portion penetrating the myocardium. With these methods, the Kajiya group was able to record unique intravascular flow patterns.[205]

Recently, Stern[190] designed an instrument for coronary blood flow measurements using an optical fiber (Figure 3.46). By separating the optical fibers of the catheters by 0.5 mm, he showed that it is possible to extend the sensitivity region for flow parallel to the fiber tip end surface. This result was obtained by using the Monte Carlo simulation method (Figure 3.47). The signal from the optical detector

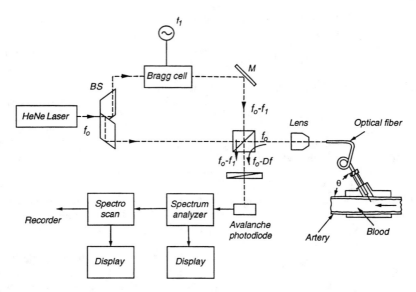

Figure 3.45 Diagram of a laser-Doppler flowmetry system for animal studies of coronary circulation. (From Nishihara, H. et al., *Appl. Optics,* 21, 1785, 1982. With permission.)

is analyzed in a specialized superheterodyne receiver that calculates two measures of the flow velocity, the algebraic mean Doppler shift, and the Doppler bandwidth. However, no biological measurements were reported in this paper.

3.2.7 CORRELATION METHODS FOR MICROVASCULAR RED BLOOD CELL VELOCITY MEASUREMENT

Blood flow velocity in a small vessel can be measured as the red blood cell velocity under microscopic observation. Figure 3.48(a) shows a method making use of a projection microscope and photodetectors.[98,208-212] Two photodetectors are placed on the image of a vessel with fixed separation along the vessel axis. The red blood cells intercept the light, and each photodetector provides output when a plasma spacing between red blood cells passes through the photodetector slit. The transit time of a plasma spacing between two slits can be determined simply by a time measurement from the moment of appearance of a plasma spacing at the upstream slit to that of the same plasma spacing at the downstream slit. To realize this measurement without ambiguity, the separation of two slits should be less than the size of the image of one red blood cell.

More accurate determination of the red blood cell velocity can be achieved by computing the correlation of outputs from two detectors. If the outputs from two detectors at the upstream slit and the downstream one are $f_1(t)$ and $f_2(t)$, then the cross-correlation function, $\phi(t)$:

$$\phi(t) = \int f_1(t) \cdot f_2(t + \tau)d\tau \tag{3.72}$$

will have a maximum at τ_m, so that the red blood cell velocity, U, is given by

$$U = \delta/\tau_m \tag{3.73}$$

where δ is the separation of two slits. As an example, Gaehtgens et al.[212] applied two phototransistors with a center-to-center spacing of 1 mm on a projection screen (magnification of 300×). Using the cross-correlation technique, they could measure red blood cell velocities in the capillaries, arterioles, and venules up to 60 μm in diameter in the mesenteric microcirculation of the cat.

Instead of using photodetectors on the projection microscopic screen, videosystems are now commonly used. In the videosystem, brightness signals at any point on the image of the vessel can be extracted by sampling the videosignal at a time window corresponding to the observation point.

OPTICAL CONFIGURATION

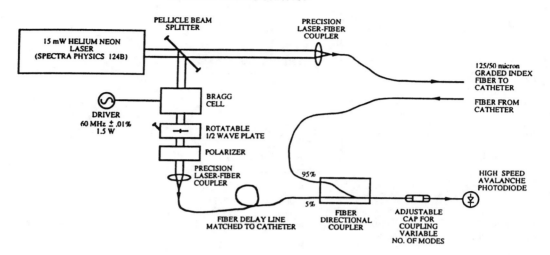

ELECTRONIC CONFIGURATION

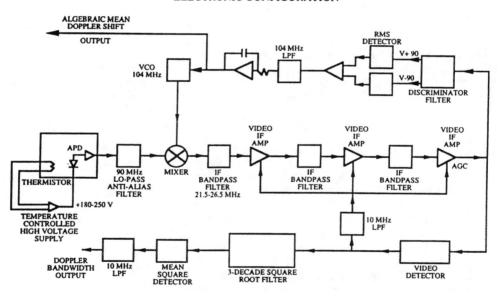

Figure 3.46 Diagram of a dual-fiber laser-Doppler velocimeter for coronary blood flow measurement. (From Stern, M. D., *SPIE Optic. Fibers Med. II,* 713, 132, 1986. With permission.)

In the first of two methods employing videosystems, the averaged transit time for a red blood cell to cross a fixed distance is obtained by seeking the maximum cross-correlation function, as shown in Figure 3.48(b).[213,214] The principle of this method is the same as that of the two-slit method. In the second method, the time interval is fixed, and the transit distance in this time interval is determined by searching maximum spatial cross-correlation, as shown in Figure 3.48(c).[215,216] In this method, the spatial cross-correlation function, $\phi(\delta)$, is expressed as

$$\phi(\delta) = \int g_1(x) \cdot g_2(x+\delta)dx \qquad (3.74)$$

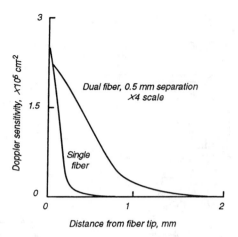

Figure 3.47 Doppler sensitivity profiles for single fiber and dual fiber for plane flows parallel to the surface in which the fibers are terminated at 45°.

where $g_1(x)$ and $g_2(x)$ are the light intensity signals along the vessel at time t and $t + \tau$. If $\phi(\delta)$ has a maximum at δ_m, then the red blood cell velocity, U, is obtained as

$$U = \delta_m/\tau. \tag{3.75}$$

The second method is advantageous for measurements of higher red blood cell velocities, because, in the first method, time resolution in the determination of τ_m is limited by the framing rate of the videosystem, while the second method is not restricted by the framing rate.

For real-time measurement, however, the time of computing cross-correlations limits the response to the change in the red blood cell velocity. To achieve a faster response, a method of using a grating was proposed.[217] The principle is shown in Figure 3.49(a). A grating is placed on the image of a microvessel on the projection screen, and the light passed through the grating is collected by a photosensor. When a red blood cell moves across the grating with a velocity, U, light signals having a frequency component of

$$f = U/d \tag{3.76}$$

will be generated, where d is the grating constant. The frequency component can be extracted from the signal by the zero-crossing count (see Section 3.2.2.3). Figure 3.49(b) shows a modified method in which a prism grating and three photodetectors are used so as to discriminate flow direction.[217] It was shown that by using a prism grating with a grating constant of 0.3 μm red blood cell velocity could be measured in a range of −15 to +15 mm/s, and it could respond to pulsatile flow at heart rates up to 300 beats/min.[217]

3.2.8 MISCELLANEOUS MECHANICAL FLOWMETERS

The flowing fluid has kinetic energy, and when the flow is disturbed by a mechanical element, it may cause a force or displacement in the sensing element, corresponding to the flow rate or flow velocity. Actually, many transducers in industrial use are based on such mechanical principles. Although most commercial flowmeters cannot be used for blood flow measurements, there are still several mechanical flowmeters, as shown in Figure 3.47, which have been designed and used for physiological studies.

Figure 3.50(a) shows a method in which the flow rate is measured from the pressure difference between two points along the axis of the vessel.[218,219] For a steady flow, the pressure difference can be determined as the frictional pressure drop. However, when the flow rate is changed, additional pressure will be developed due to the inertia of the fluid, which is proportional to the time derivative of the flow velocity. If the velocity is $U(t)$, the pressure difference, ΔP, is given as

$$\Delta P = \rho\delta\frac{dU(t)}{dt} + \alpha U(t) \tag{3.77}$$

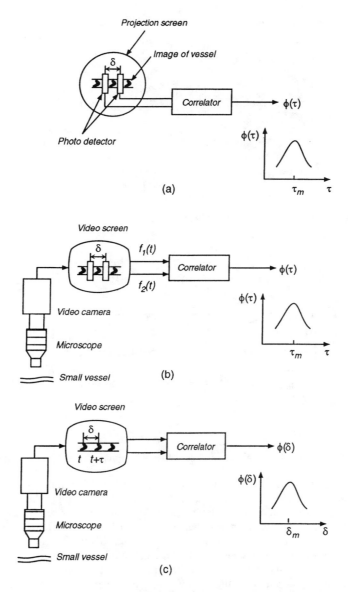

Figure 3.48 Correlation methods for red cell velocity measurement: **(a)** using a projection microscope and photodetectors, the cross-correlation of signals from two different points on a microvessel is obtained; **(b)** using the same principle as above but with the introduction of a video system; **(c)** special cross-correlation between two images at fixed time intervals.

where ρ is the density of the fluid, δ is the distance between the two points, and α is the proportionality constant between the velocity and the frictional pressure drop between the two points.[218] As long as α is constant, differential Equation (3.77) is linear, and the instantaneous velocity $U(t)$ can be obtained as the solution of the differential equation. Fry[219] reported that this method could be applied in animals and human subjects, and pulsatile blood velocity patterns in the ascending aorta could be obtained by using a specially designed double-lumen catheter.

Figure 3.50(b) shows a flowmeter head designed for use in the vena cava of the animal.[220] The kinetic energy of the blood flow causes a pressure difference between the upstream and downstream orifices, and the flow rate can be determined from the differential pressure output while it is nonlinear to the flow rate.

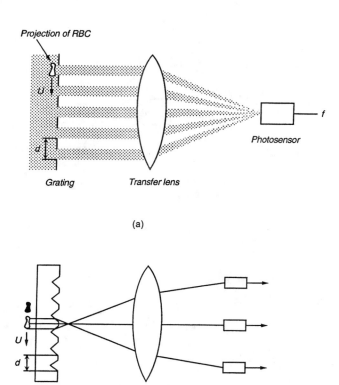

Projection of RBC

Grating   Transfer lens

Photosensor

f

(a)

Prism
grating

Transfer lens   Photosensor

(b)

Figure 3.49 Red blood cell velocity measurements with single grating **(a)** and three-stage prism grating **(b)**.

Figure 3.50(c) also shows a flowmeter head for detecting the differential pressure.[221] The pressure difference is measured between the tube that has a funnel-shaped opening to the upstream flow and the adjacent tube which has an opening to the downstream flow. The catheter head is held in midstream by five stainless-steel springs. The catheter also has an additional lumen, which can be used for arterial pressure measurement when the catheter is introduced into the superior vena cava.

Figure 3.50(d) shows a flowmeter called the Potter electroturbinometer.[222,223] It has an armature with blades and an inside magnet, and the rotation of the armature is detected by a pickup coil. This type of flowmeter is relatively insensitive to change in viscosity. However, it has disadvantages: i.e., it causes a significant pressure drop and it cannot measure flow below a certain critical level due to the friction between the armature shaft and the bearing surface.

Figure 3.50(e) shows an apparatus called a rotameter.[224] The flow causes elevation of the float, and the float position is detected by a differential transformer (see Section 4.2.1). The flowmeter is also relatively insensitive to changes in viscosity, and it can be designed so that the float elevates in proportion to the flow rate. The pressure drop is not proportional to the flow rate, but remains in a narrow range. As an example, it was shown that, in a rotameter, the pressure drop remained between 0.2 kPa (2 cmH$_2$O) and 0.35 kPa (3.5 cmH$_2$O) when the flow varied from a very low rate up to 3 l/min.

Figure 3.50(f) shows a technique called the bristle flowmeter.[225-227] It measured the deviation of a fine bristle caused by a stream. If the deviation of the bristle is proportional to the drag force due to the kinetic energy of the fluid, it will be proportional to the square of the flow velocity. A transducer vacuum tube was used for detecting the deviation of the bristle, in which the deflection of a plate pin could be converted to the plate current.

Figure 3.50(g) shows a technique of mass flow measurement in which the Coriolis force induced by a rotational motion of the flowing fluid is detected.[228] An elastic tube is supported at three points, and

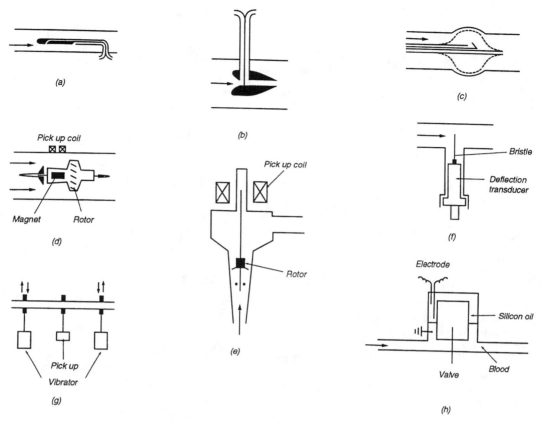

Figure 3.50 Mechanical flowmeters: **(a)** pressure gradient technique, **(b)** and **(c)** methods of detecting kinetic energy, **(d)** the turbinometer, **(e)** the rotameter, **(f)** the bristle flowmeter, **(g)** the vibration flowmeter, and **(h)** the density flowmeter.

both ends are driven sinusoidally in opposite directions so that a rotational vibration of the segment is generated. Then, the induced force, F, at the center of the segment is given as

$$F = 2QL\frac{d\theta}{dt} \tag{3.78}$$

where Q is the mass flow rate, L is the length of the segment, and θ is the angular deflection of the segment from the resting position. When a silastic tube of 3-mm i.d. and 10 cm long was driven at 640 Hz, the force was detected at the center and synchronous rectification was performed with the driving signal. Linear output was obtained in a range of 0 to 4 g/s.

Figure 3.50(h) shows a technique for direct measurement of volume displacement.[229] When the flow is obstructed at the valve, the fluid flows into the bypass where silicon oil is added, and two electrodes detect the moment of arrival of the boundary between conductive and nonconductive fluid. Then, the valve is opened and the oil moves to the initial position because of the difference in density between the two fluids. Results showed that when the volume of fluid contained between the two electrodes was 1.5 ml, flow ranging from 2 to 50 ml/min could be measured with an accuracy of ±4%.

For continuous measurement of a low flow rate, such as blood perfusion to a small organ or urinary and lymphatic flow measurement, the drop counter can be used. To detect drops, various techniques have been attempted, as shown in Figure 3.51, such as the (a) optical method,[230] (b) use of mechanical pickup,[231] (c) method of detecting electrostatic change,[232] and (d) method of electric conductivity.[233] Although the size of a drop depends on the diameter of the dropping tube and the density and surface tension of the fluid, flow rates ranging from 0 to 20 ml/min can be measured by such drop counters. At

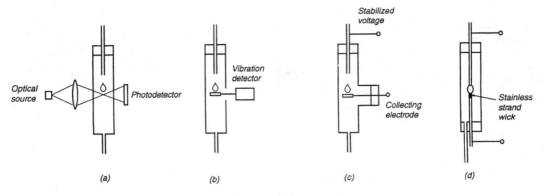

Figure 3.51   Drop counters.

a higher flow rate, the size of a drop is reduced, and thus the number of drops becomes nonlinear to the flow rate.

3.3 TISSUE BLOOD FLOW MEASUREMENT

The amount of blood flow perfused into a definite region in the tissue can be measured by many different techniques, such as volume plethysmographies, impedance plethysmographies, clearance methods, measurement techniques by heat transport, laser-Doppler techniques, and measurement techniques by nuclear magnetic resonance.

Depending upon requirements of the measurement — such as the object site, spatial resolution, response time, or absolute or relative accuracy — the appropriate method should be selected. For the average tissue blood flow measurement in the limb, plethysmography combined with the venous occlusion technique can be successfully applied. The laser-Doppler technique provides a measurement of a very local superficial tissue blood flow. The clearance method using radioisotopes and measurement by nuclear magnetic resonance provides information about the deep tissue blood flow, and by employing the technique of computed tomography the blood flow distribution over a cross-section of the body can be obtained.

3.3.1   VENOUS OCCLUSION PLETHYSMOGRAPHY

Plethysmography is a technique for recording the volume change in a tissue. When the volume change occurs only through a change of the blood volume in the tissue, information about the tissue blood flow can be obtained by the plethysmographic observation. A small pulsatile change of the volume is generally observed in almost any tissue due to the arterial pulsation. However, quantitative measurement of the tissue blood flow can only be realized by employing the venous occlusion method, and it has been used for several decades as a standard technique of tissue blood flow measurement.[234]

3.3.1.1   Venous Occlusion Method

The venous occlusion method is a fundamental technique of quantitative determination of the limb blood flow from the plethysmographic measurement. Figure 3.52 illustrates the principle of the venous occlusion method applied to a segment of the lower leg. Two cuffs are attached to the proximal and distal sites of the segment, and a pressure higher than the maximum arterial pressure is applied to the distal cuff so that all arteries and veins under the cuff are occluded. Then a pressure slightly lower than the minimal arterial pressure is applied instantaneously to the proximal cuff so that only the veins are occluded at this level, but arterial blood flow is not obstructed. Then the volume change of the segment is recorded; consequently, the blood flow, Q, is determined by the rate of increase of the volume, dV/dt. This technique can also be applied to the whole limb or the extremity of the limb and has been widely used in physiological studies.[235-239]

For normal subjects, the proximal cuff pressure for venous occlusion is chosen to be around 6.7 kPa (50 mmHg), and the distal cuff pressure is around 20 kPa (150 mmHg). To perform venous occlusion quickly, the cuff pressure for venous occlusion should be applied in a short time. Barendsen et al.[240] employed a system in which the maximum cuff pressures reached were within 0.1 s.

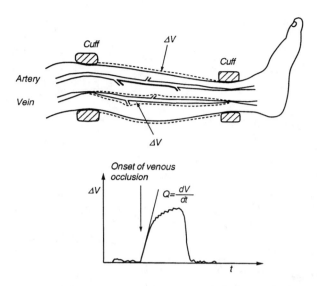

Figure 3.52 Principle of the venous occlusion method.

The validity of the venous occlusion method has been examined by many researchers, and it was shown that as long as an appropriate cuff pressure is applied, the venous flow can be completely obstructed while the arterial pressure distal from the venous occlusion cuff remains unchanged.[237,238] However, it was pointed out that there are still many causes of error due to local tissue deformation at the cuff inflation and to the technique of volume recording.[237]

3.3.1.2 Displacement Plethysmography

The change of volume in the limb can be detected at the skin surface as a displacement. The term "displacement plethysmography" has not been well defined; however, it is used here for the sake of convenience to distinguish a technique of detecting displacement of the skin surface of the limb from impedance plethysmography described in the next section. Included in the category of displacement plethysmography are water-filled or air-filled plethysmographies (both are called fluid displacement plethysmography), mercury strain-gauge plethysmography, and capacitance plethysmography. Constructions of these techniques are shown in Figure 3.53.

Water-filled plethysmography consists of a rigid chamber filled with water which can be attached to the segment of the limb, as shown in Figure 3.53(a), and the limb volume change is detected by a change of the level of the open water surface. The water chamber should fit tightly to the limb. For this purpose, a soft rubber diaphragm or an iris diaphragm surrounding the limb is used at each end of the chamber, and a thin rubber or polythene sleeve is used as a water seal.[241-243] The area of the water surface where the level change is detected should be sufficiently large so that the influence of the change of the chamber pressure due to the level change at the water surface will be small enough. Dahn[243] recommended an area of at least 50 cm^2 for measurements at the lower leg, although a smaller area has been used by some researchers. The water level can be obtained from a chamber pressure, while other methods such as a water conductance level recorder[244] or a photoelectric transducer[245] have also been used.

When the chamber does not fit the limb satisfactorily, the volume change of the limb cannot be reflected in the water level. To confirm this, Dahn[243] conducted an experiment in which both proximal and distal cuffs were inflated to above arterial pressure, and a saline was injected into the vein in the limb segment. It was found that when the fitting was poor, the transmittance of the volume change in the limb to the water chamber was reduced to 50% or less.

Air-filled plethysmography consists of an air-filled chamber (Figure 3.53(b)), with a transducer that detects the air flow between the chamber and the outer air. The pneumotachograph (see Section 3.4.1.2) has been commonly used as an air flow transducer in which the air flow rate was detected by the pressure difference across a small air resistance. The volume change can thus be obtained by integrating the air flow. To keep the pressure change in the chamber small enough, the resistance at the flow transducer should be small. In one example, the pressure drop for a maximal air flow of 7 ml/s was 2.6 Pa (0.027 cmH$_2$O), and it had a linear frequency response of up to 25 Hz.[246]

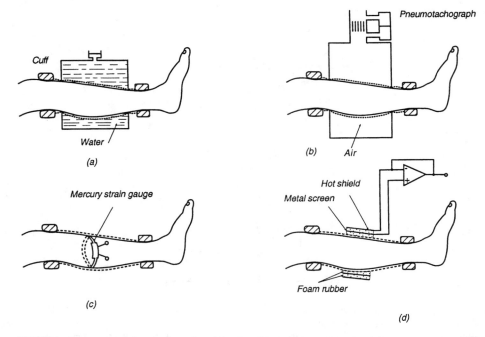

Figure 3.53 Displacement plethysmographies: **(a)** water-filled plethysmography, **(b)** air-filled plethysmography, **(c)** mercury-strain-gauge plethysmography, and **(d)** capacitance plethysmography.

A simplified method of air-filled plethysmography was also proposed in which a cuff was used instead of a rigid air chamber.[247,248] The cuff was approximately 4 to 8 cm in width, and a pressure of approximately 1.3 to 2.7 kPa (10 to 20 mmHg) was maintained in it. The pressure difference between the cuff and a chamber maintained at a constant pressure of the averaged cuff pressure was detected by a sensitive diaphragm. In comparison with the water-filled plethysmograph in the contralateral arm, it was shown that both measurements correlated fairly well.

Figure 3.53(c) illustrates a technique known as mercury strain-gauge plethysmography, in which changes in the circumferential length of the limb are detected by the electrical resistance change of the mercury conductor in a fine distensible tube. This technique has been investigated extensively.[249-254] In order to estimate the volume change of the limb segment correctly from the change of circumferential length, it is postulated that the tissue does not deform in the direction of the limb axis, and the cross-sectional area is uniform along the limb segment. Although these assumptions are not fully valid in actual human limbs, it was pointed out that a fairly accurate estimation of the volume change can be possible.[250]

The actual mercury strain-gauges are usually composed of fine and highly distensible tubes. For example, a latex tube with an internal diameter of 0.5 mm and a wall thickness of 0.8 mm was used, and the developed tension for 1% elongation from the natural length was 2.5 gf.[250] The resistance is on the order of 1 Ω.

The sensitivity of the mercury strain-gauge is given as

$$\frac{dr_m}{r_m} = 2\frac{dL}{L} \tag{3.79}$$

where r_m is gauge resistance, and L is the length of the gauge. Because the resistance of the gauge is small, the resistance of the lead wire cannot be neglected. If the lead wire has a resistance, r_w, relative change in the total resistance, r, is

$$\frac{dr}{r} = 2\frac{dL}{L}\frac{1}{1+r_w/r_m}. \tag{3.80}$$

On the other hand, if the gauge does not cover the entire circumference of the limb, but a gap of length g exists, the circumferential length is $c = L + g$, and the resulting strain is given as

$$\frac{dL}{L} = \frac{dc}{c}\left(1 + \frac{g}{L}\right). \tag{3.81}$$

If the gauge resistance is written as $r_m = Lk$, then

$$\frac{dr}{r} = 2\frac{dc}{c} \cdot \frac{1 + g/L}{1 + r_m/kL}. \tag{3.82}$$

Thus, if the wire resistance is chosen as $r_m = k \cdot g$, the effects of the gap and the wire resistance can compensate each other.[251]

Although the mercury strain-gauge plethysmograph measures the circumferential length change only at one or a few sections in the limb segment, it was shown that plethysmographs could be obtained that are practically equal to those obtained by the water plethysmograph, as long as the gauge was placed near the center of the segment.[254]

Figure 3.53(d) shows a technique called capacitance plethysmography. A cuff-type electrode is fastened around the limb with a small space to the skin surface beneath it, and the change in capacitance between the electrode and the limb is detected.[255,256] In the actual cuff, the electrode was made of a metal screen, and polyurethane foam was used to keep the spacing. The outside of the electrode was shielded by another electrode that was driven at the same potential as the inner electrode, realizing a so-called hot, or active, shield.

If the limb is assumed to be a uniform circular cylinder of diameter, D, and spacing, s, the capacitance, C, is given by

$$C = \frac{\varepsilon L \pi D}{s} \tag{3.83}$$

where ε is the dielectric constant of the medium between the electrode and the limb, and L is the length of the electrode. The volume, V, of the segment of the cylinder beneath the electrode is given by

$$V = \frac{\pi L (D - 2s)^2}{4}. \tag{3.84}$$

If $D \gg 2s$, then the sensitivity of the capacitance change to the volume change is given by

$$\frac{dC}{dV} = \frac{dC}{ds} \Big/ \frac{dV}{ds} = \frac{\varepsilon}{s^2} \cdot \frac{D}{D - 2s} \cong \frac{\varepsilon}{s^2} \tag{3.85}$$

One result showed that when an electrode length was 38 mm, the mean diameter 64 mm, and the spacing 6 mm, the capacitance change was linear for up to a 2% volume change.[257]

3.3.1.3 Impedance Plethysmography

Impedance plethysmography is a technique in which volume change in the tissue is measured by the change of electrical impedance. Combined with the venous occlusion method, blood flow in a limb or digit can be measured by this technique.

Figure 3.54 shows the method of blood flow measurement in a segment of a limb. Four band electrodes are commonly used. A constant a.c. current of 0.1 to 10 mA and 20 to 200 kHz is supplied to the outer electrodes, and the voltage is measured at the inner electrodes. The impedance, Z, of the limb segment is defined as the ratio of the voltage across the segment and the supplied current. The reciprocal of Z is the admittance, Y.

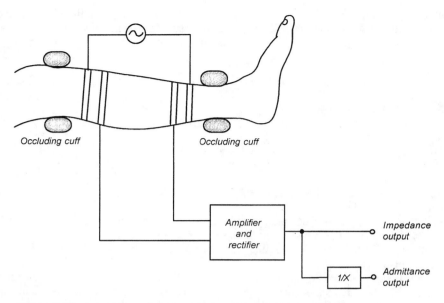

Figure 3.54   Impedance plethysmography system for limb blood flow measurement.

The measurement of blood volume change in the limb by electrical impedance is based on the parallel conductor model, illustrated in Figure 3.55. Consider a limb segment of length L, volume V_0, and tissue resistivity ρ_0. The cross-sectional area, A_0, is assumed to be constant along the segment axis, so that $A_0 = V_0/L$. When additional blood of volume V_b with resistivity ρ_b is added to this segment, the parallel conductor model postulates that the blood is distributed uniformly in the segment, forming a parallel conductor having length L, cross-sectional area $A_b = V_b/L$, and resistivity ρ_b. The impedance of each conductor is given as

$$Z_0 = \frac{\rho_0 L}{A_0} = \frac{\rho_0 L^2}{V_0} \tag{3.86}$$

$$Z_b = \frac{\rho_b L}{A_b} = \frac{\rho_b L^2}{V_b} \tag{3.87}$$

and the impedance, Z, composed of parallel conductors is expressed by

$$Z = \frac{Z_0 Z_b}{Z_0 + Z_b}. \tag{3.88}$$

If $\Delta Z = Z - Z_0 \ll Z_0$, then V_b is shown in Equation (3.89):[257]

$$V_b = -\frac{\rho_b L^2}{Z_0^2} \Delta Z. \tag{3.89}$$

This equation implies that the blood volume added to the segment can be estimated by the impedance change due to the increase of the blood volume, as long as the blood resistivity is known.

A similar equation can be derived from the admittance change, ΔY, as

$$V_b = \rho_b L^2 \Delta Y. \tag{3.90}$$

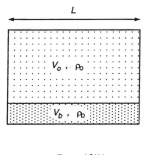

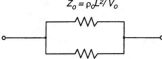

Figure 3.55 The parallel conductor model for a limb segment of length, L, (above) and equivalent circuit (below). V_o and V_b are volumes of the limb segment and increment of the blood, respectively, and ρ_o and ρ_b are resistivities of the tissue and the blood, respectively.

In the actual human limb, validity of the parallel conductor model is confirmed when it is applied to the venous occlusion method. A comparison with water plethysmography showed that the ratio of limb blood flow measured by impedance and that by water plethysmography was 1.01 ± 0.091 in the forearm and 1.01 ± 0.092 in the calf, and the correlation coefficients were 0.987 and 0.989, respectively.[258]

While blood resistivity is different in each subject, it can be estimated from the hematocrit value, *Hct*, by empirical formulas. An example is[259]

$$\rho_b = 53.2 \exp(0.022 \, Hct) \, \Omega \cdot cm. \tag{3.91}$$

Impedance plethysmography was also applied to digital measurement of blood flow.[260,261]

Automated systems have been developed for routine clinical use.[262,263] In those systems, all processes, including venous occlusion, detection of the initial slope of the impedance change, and calculation of blood flow, were performed automatically.

3.3.2 CLEARANCE TECHNIQUE
3.3.2.1 Principle
When a tissue containing some indicator substance is perfused by either indicator-free blood or blood of lower indicator concentration, the indicator in the tissue will be washed out into the venous blood. Applying a simple model to this process, perfused blood flow rate to the tissue can be estimated from the time course of the indicator concentrations in the arterial and venous blood. When the arterial blood is indicator free, the remaining indicator in the tissue decreases exponentially, and the perfusion rate can be estimated only by the rate of decrease of the indicator in the tissue. Thus, measurement of the absolute amount of indicator is unnecessary.

In the theory of clearance technique, it is usually assumed that the indicator diffuses freely from the tissue to the blood and vice versa. Actually, inert gases, hydrogen gas, or nitrous oxide (N_2O) almost satisfy this assumption and can be used for the clearance technique.

When a region in the tissue is perfused uniformly with a blood flow, Q, the change in the amount of the indicator in this region is given as

$$dI = Q(c_a - c_v)dt \tag{3.92}$$

where c_a and c_v are concentrations of the indicator in arterial and venous blood. If the amount of indicator and the blood flow in a unit weight of tissue are denoted by c and q, then

$$dc = q(c_a - c_v)dt. \tag{3.93}$$

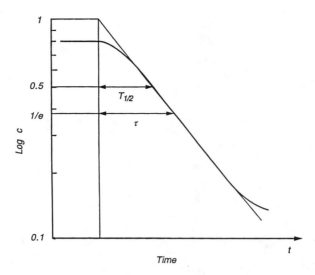

Figure 3.56 Typical clearance curve of an indicator plotted on a logarithmic scale of the indicator concentration, c.

If the indicator can diffuse quickly enough so that an equilibrium between concentrations of the indicator in the tissue and in the capillary blood is established in every moment, then

$$c = \lambda c_v \tag{3.94}$$

where λ is the partition coefficient, and c_v is the indicator concentration in the tissue. Conventionally, c is expressed as the amount of indicator in unit weight of tissue, while c_v is expressed as that in unit volume of blood; thus, λ has a dimension of volume per weight.

If the indicator is completely removed from blood at the lung, the arterial concentration, c_a, will be zero after the lung is ventilated by indicator-free air. From Equation (3.93), then,

$$dc = -(qc/\lambda)dt. \tag{3.95}$$

Integrating this equation, one can obtain

$$c = c_0 \exp(-qt/\lambda) = c_0 \exp(-t/\tau) \tag{3.96}$$

where $\tau = \lambda/q$. The perfusion blood flow rate per unit weight of the tissue, q, is then obtained as

$$q = \lambda/\tau. \tag{3.97}$$

This result implies that the tissue blood flow can be determined only by the time constant of the washout curve, as long as the partition coefficient is known. Figure 3.56 shows a typical clearance curve plotted on a logarithmic scale of the indicator concentration. The time constant, τ, can then be obtained by fitting a line at the downslope.

The perfusion blood flow rate can also be determined from the process of increasing the concentration of the indicator in the tissue, when the indicator is introduced continuously into the arterial blood. This method was first applied for cerebral blood flow measurement by Kety and Schmidt [257] using N_2O and is known as the Kety-Schmidt method. As shown in Figure 3.57, part of the indicator in the arterial blood diffuses into the tissue, until the concentration in the tissue is equilibrated to that of the arterial blood. As long as the flow rate, q, does not change throughout the process, Equation (3.93) can be integrated as

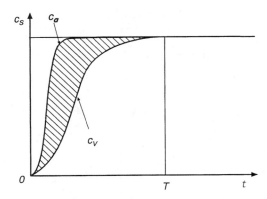

Figure 3.57 Principle of the Kety-Schmidt method.

$$\int_0^T dc = q \int_0^T \left(c_a - c_v \right) dt \tag{3.98}$$

where T is the time of attaining saturation. Then, q, is solved as

$$q = c_s \Bigg/ \int_0^T \left(c_a - c_v \right) dt = \lambda c_{vs} \Bigg/ \int_0^T \left(c_a - c_v \right) dt \tag{3.99}$$

where c_s and c_{vs} are c and c_v at saturation. In this method, c_a and c_v should be continuously measured by continuous blood sampling from the intra-arterial and intravenous catheter. One advantage of this method is that complete removal of the indicator from the lung is not required.

3.3.2.2 Use of Radioactive Indicators

Radioactive nuclei which radiate gamma rays can be convenient indicators for the clearance method, because the quantity of the indicator in a definite region in the tissue can be measured externally by a scintillation counter with a collimator, gamma camera, or emission-computed tomography system. Radioisotopes of inert gases such as ^{85}Kr, ^{79}Kr, and ^{133}Xe have been used as such indicators. These nuclei are gamma radiators with half-lives of 10.8 years, 34 hours, and 5.27 days, respectively. They can diffuse in the tissue and are removed completely from the lung.

A clearance method using these radioactive gases has been applied most widely to cerebral blood flow measurement. Both Kr and Xe pass through the blood brain barrier and rapidly diffuse into the tissue, so that the equilibrium between the concentration in the tissue and that in the blood is established. The partition coefficient of Kr between the blood and the cortex is 0.92 ml/g, and the white matter 1.26 ml/g;[265,266] while that of Xe between the blood and the cortex is 0.8 ml/g, and the white matter 1.5 ml/g.[267]

To introduce such an indicator into the tissue, either intra-arterial injection of indicator-dissolved saline or inhalation of indicator-contained air can be used. For the cerebral blood flow measurement by arterial injection, a solution containing indicator is injected into the carotid artery. If the indicator is injected at a constant rate, it requires 10 to 15 min to achieve a saturation for which the indicator concentration in the brain-tissue reaches an equilibrium equal to that of the arterial blood. The clearance curve is then recorded after stopping the injection for the following 10 to 20 min. Injection time can be shortened by a stepwise procedure in which indicator is injected rapidly in the initial 20 to 40 s. This is followed by a slower injection for about 3 min at a rate of about 1/5.[266]

In the inhalation method, the mixture of oxygen and indicator gas is inhaled for 5 to 15 min, and then the clearance curve is recorded. The advantage of the inhalation method is that it does not require arterial catheterization. However, in order to obtain the same indicator concentration in the tissue by inhalation, a larger amount of indicator should be taken into the body than for the arterial injection

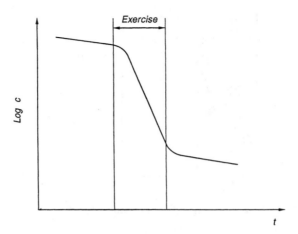

Figure 3.58 The clearance curve in the muscle before, during, and after excercise plotted on a logarithmic scale.

method. Typically, the necessary amount of indicator is about 10 times that of the carotid arterial injection. A higher indicator concentration in the venous blood will also increase the error caused by the recirculation of indicator due to the pulmonary shunt flow component and the residual air in the lung, while the amount of arterial recirculation is empirically estimated, and the perfusion flow rate can be corrected.[267]

The Kety-Schmidt method can also be performed by inhaling indicator gas and sampling blood from the artery and the vein. Because indicator concentration in the artery is measured, arterial recirculation does not cause error. The Kety-Schmidt method with arterial and venous blood sampling is valuable when oxygen consumption is estimated simultaneously from the arteriovenous oxygen saturation difference and the blood flow rate.[268]

The principle of the clearance method is used extensively to observe tissue blood flow distribution for the brain, in particular, using emission computed tomography (ECT).[269] From sequential ECT images, clearance curves of each pixel can be obtained, and the tissue blood flow is then determined at each site. Both single photon ECT (SPECT) and positron emission tornography (PET) have been used for this purpose. Single photon emitters such as ^{85}Kr and ^{133}Xe are detected by SPECT, which provides a spatial resolution of about 1 cm. Positron emitters such as ^{15}O and ^{13}N are detected by positron emission CT and provide a spatial resolution of about 3 to 4 mm or less.

The blood flow in the skeletal muscle at rest and during work is also measured by the clearance method. For example, a small amount of saline containing about 1.85 MBq (50 µCi) of ^{133}Xe is injected into the muscle with a hypodermic needle, and then the clearance curve is recorded externally by a scintillation counter. The muscular blood flow increases during work 10 to 20 times that at rest. The slope of the clearance curve in the logarithmic scale changes significantly, as shown in Figure 3.58, and the blood flow in each phase can be determined from the slope.[270-272]

Blood flow of the skin can be also measured using the clearance method. A radioactive indicator such as ^{133}Xe is either injected into the skin tissue with a hypodermic needle or transported transcutaneously by diffusion. Figure 3.59 shows a method of epicutaneous application of the indicator, in which ^{133}Xe gas is added to a small chamber adhered to the surface of the skin for about 3 min. The chamber is then removed so that the skin surface is exposed to the air. It was shown that the retrograde diffusion of ^{133}Xe from the skin to the environment is negligible compared to transport by blood flow, thus the skin blood flow is determined from the slope of the clearance curve.[273,274]

3.3.2.3   Hydrogen Gas Clearance Method

Hydrogen gas (H$_2$) can be a convenient indicator for tissue blood flow measurement by the clearance method. Because it diffuses freely into the tissue, the dissolved hydrogen gas in the blood is rapidly removed from the lung and its concentration can be detected by a fine platinum electrode. The partition coefficient of hydrogen gas is close to 1 ml/g in most tissue, including the brain.[275]

To introduce hydrogen gas into the tissue, methods such as inhalation, injection of hydrogen-dissolved saline into the artery, or generation of hydrogen gas by electrolysis can be used. The clearance curve is recorded as the change of hydrogen partial pressure in the tissue which can be measured polarographically.

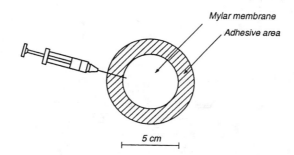

Figure 3.59 A method of skin blood flow measurement by epicutaneous application of ^{133}Xe. (From Sejrsen, P., *Circ. Res.*, 25, 215, 1969. With permission.)

When a platinum electrode and a reference electrode are placed on or inserted into the tissue, and a positive potential is given to the platinum electrode relative to the reference electrode, the oxidation reaction

$$H_2 \rightarrow 2H^+ + 2e^- \tag{3.100}$$

occurs on the platinum electrode surface. Although H_2 is consumed at the electrode surface by this reaction, it is supplied from the surrounding media by diffusion, and thus the reaction will continue. Consequently, a constant current will be obtained at an equilibrium, which is proportional to the H_2 flux, corresponding to the partial pressure of hydrogen gas in the surrounding media.

Many different types of electrode have been used in the hydrogen gas clearance method. Figure 3.60 shows some examples: (a) is a tissue electrode, (b) is an intravascular electrode,[275] and (c) is a contact electrode for gastric mucosal measurement.[276] A light plating of the bare platinum tip by cathodizing in platinum chloride solution is recommended to obtain a sensitive, rapid, and stable electrode.[275] The hydrogen clearance method has been applied to tissue blood flow measurements, mostly in animal studies in organs such as the myocardium, kidney, skeletal muscle,[275] brain,[277,278] intestine,[279] or stomach.[276]

A probe in which hydrogen gas is generated electrochemically has also been used.[280] Figure 3.61 shows a surface probe of this type. It consists of two platinum wires; one is used to generate and the other to measure hydrogen. A reference electrode is situated about 2 to 3 cm from this probe. To generate hydrogen, a constant current of 0.3 to 1.0 µA is applied for about 1 s between the reference electrode and the larger platinum wire, 200 µm in diameter, so that the reference electrode is an anode and the platinum wire is a cathode. Then, hydrogen partial pressure at the surface of the probe is measured polarographically by the smaller platinum wire. It was shown that it measures the local blood flow within a tissue volume of about 2 mm^3, and measurements can be repeated at intervals of about 4 min.[280]

Figure 3.62 shows a hybrid probe for simultaneous hydrogen clearance and laser-Doppler velocimetry.[281] It has two platinum wire tips — one is for generating hydrogen and the other is for measuring hydrogen partial pressure — and two optical fibers for laser-Doppler velocimetry. The hybrid probe can have unique advantages for both methods. The hydrogen clearance method yields an absolute quantity of blood flow rate in the tissue, while the measurement is discontinuous. Conversely, laser-Doppler velocimetry is difficult to calibrate in an absolute quantity, while it yields a continuous measurement. Furthermore, laser-Doppler velocimetry yields a reliable zero velocity, while this is difficult to obtain by hydrogen clearance due to the nonconvective loss of hydrogen gas.

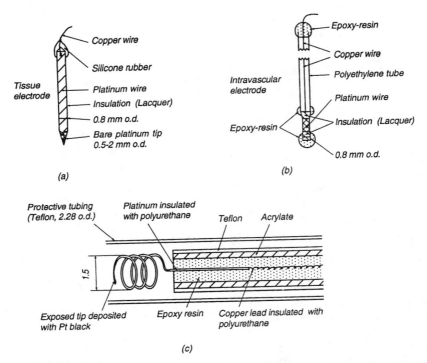

Figure 3.60 Electrodes for hydrogen clearance method: **(a)** tissue electrode, **(b)** intravascular electrode, and **(c)** contact electrode. (Modified from Aukland, K. et al., *Circ. Res.*, 14, 164, 1964; Murakami, M. et al., *Gastroenterology*, 82, 457, 1982.)

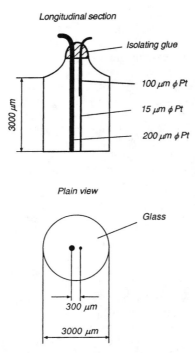

Figure 3.61 A surface probe for hydrogen clearance in which hydrogen gas is generated electrochemically and measured polarographically. (Modified from Stosseck, K. et al., *Pflügers Arch.*, 348, 225, 1974.)

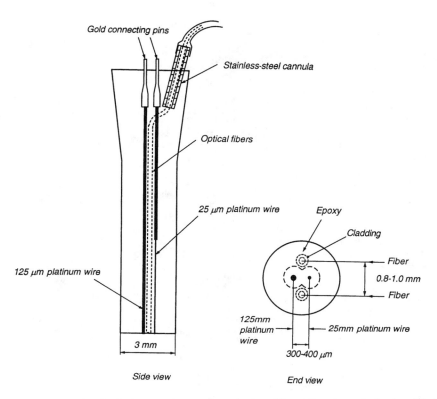

Figure 3.62 A hybrid probe for hydrogen clearance method and laser-Doppler velocimetry. (Modified from DiResta, G. R. et al., *Am. J. Physiol.*, 253, G573, 1987.)

3.3.3 TISSUE BLOOD FLOW MEASUREMENT BY HEAT TRANSPORT

When a region of tissue is heated or cooled, continuously or stepwise, heat transport occurs between the tissue and the perfusing blood, and the tissue blood flow can be estimated from the amount of heat transfer. For the perfusing blood, the arterioles, capillaries, and venules act as perfect heat exchangers in which the blood quickly reaches the tissue temperature.[282]

The rate of heat transfer from the tissue of temperature T_s and the arterial blood of temperature T_a is expressed as

$$H = \rho c q \left(T_s - T_a\right) \tag{3.101}$$

where ρ and c are density and specific heat of the blood, respectively, and q is the tissue blood flow rate. Thus, tissue blood flow can be determined from the rate of heat transfer when a constant temperature difference is maintained between a region of the tissue and the arterial blood. In a steady state, when the tissue is heated by an electric heater in a probe, the rate of heat transfer from the tissue to the blood is equal to the heat applied to the tissue in unit time, which is the power consumption of the heater in the probe.

Many different probes working on this principle have been tried.[283-286] For blood flow measurements in a small region of the tissue, needle type probes are inserted into the region. Figure 3.63 shows examples of probes of this type.[286] On the other hand, contact probes are used for skin blood flow measurements, some of which are shown in Figure 3.64.[287-291] Experimental evaluations showed that the outputs of these thermal probes are linearly related to tissue blood flows measured by such other methods as venous occlusion,[287] isotope clearance,[102] or direct measurement of venous flow in isolated perfused organs.[292]

Transient operation of thermal probes has been studied mainly for apparent thermal conductivity measurement in living tissues, but this technique also provides an estimation of tissue blood flow, because the apparent thermal conductivity of a tissue varies with the tissue blood flow rate. A small thermistor bead is commonly used for this purpose. A thermistor bead generates heat when an electric current is

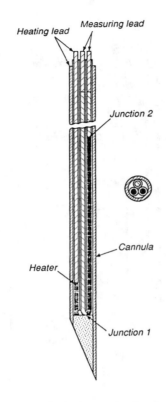

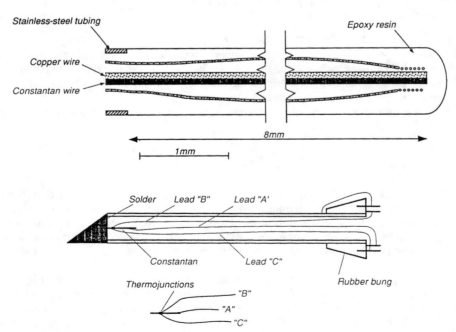

Figure 3.63 Needle-type thermal probes for tissue blood flow measurement. (Modified from References 283, 284, and 285.)

applied, and the temperature of the bead can be measured by its resistance. For the measurement of apparent thermal conductivity of the tissue, different modes of applying electric power to the thermistor bead (Figure 3.40) are used.

3.3.4   TISSUE BLOOD FLOW MEASUREMENT BY LASER-DOPPLER FLOWMETERS

3.3.4.1 Introduction

Laser-Doppler flowmetry (LDF) is a method for continuous and noninvasive measuring of tissue blood flow, utilizing the Doppler shift of laser light as the information carrier. The method has already proved its potential usefulness in the clinical assessment of blood flow within such disciplines as dermatology, plastic surgery, and gastrointestinal surgery. In experimental medicine, laser-Doppler flowmetry has been used in the study of spontaneous rhythmical variations, as well as in the study of spatial and temporal fluctuations in human skin blood flow. The method has facilitated further investigations of the nature of blood flow in a variety of tissue types. The method is rapidly spreading to many new applications in microvascular blood flow measurements. Several commercial laser-Doppler instruments are available. This section is a presentation of the theory and practice of laser-Doppler flowmetry.

3.3.4.2   Interaction between Light and Tissue

3.3.4.2.1 *Light Scattering in Tissue.*   Tissue is a scattering and absorbing medium which has a higher refractive index than air. A laser beam that is brought to impinge on the skin surface will be partly reflected back (only 4 to 5% of the incident energy) owing to specular surface reflections. The remaining light beam penetrates the tissue and will undergo scattering and absorption processes.

Scattering of electromagnetic waves in any medium is related to the heterogeneity of the medium.[293] Tissue is, in its finest parts, composed of molecules containing discrete electrical charges, electrons, and protons.[294] If a molecule or a cluster of molecules in a cell membrane is illuminated by an electromagnetic wave, the electrical charges in the molecules are set into oscillatory motions by the electric field of the incident wave. Moving or vibrating charges radiate electromagnetic energy in all directions; this is the secondary, re-radiated light that is called scattered radiation.

Scattering occurs whenever there are differences in refractive indices between different parts of a medium. The directions of scattering are very much dependent on the size and shape of the scattering particles.[295,296] When particles are small in comparison to the wavelength of incident light, we have what is generally called Rayleigh scattering. The phase differences between the secondary waves leaving the particles are small. In this case, the result will be a homogeneous zone of light surrounding the particle. If the particle dimensions are large in comparison to the wavelength, the resulting radiation is more or less concentrated in the forward direction, i.e., moving in the same direction as the exciting or incident beam. This scattering geometry is generally called Mie scattering. In time, the scattered radiation from one molecule will, of course, interact with new molecules, and a very complicated multifaceted problem occurs in which closely located structures interact with one another electromagnetically. Biological tissue is an extremely heterogeneous matter in which structures of various sizes and orientations occur. As a consequence, the scattered field of radiation is very complicated.

One usually assumes that the incident light is scattered and thereby forms spherical zones of diffuse light in which the incident direction of light is lost; the scattered and multiple scattered light will hit a specific structure from randomly distributed directions. At a specific point in a tissue volume, the resultant radiation can be obtained by superimposing the scattered waves from surrounding particles or structures.

3.3.4.2.2 *Doppler Shift.*   The frequency shift resulting from relative motion of the source and observer of a propagating wave is known as the Doppler effect. The magnitude of the frequency shift is given by

$$\Delta f = \frac{v}{c} V_o \qquad (3.102)$$

where v is the velocity of the source with respect to the observer, c is the velocity of the carrier wave, and V_o is the unshifted frequency. The velocity of the source can be determined if c and V_o are known. Laser-Doppler flowmetry has excellent spatial resolution. Generally, the volume resolution is on the order of 1 mm^3. In some types of tissue (for instance, highly absorbing liver tissue), this resolvable volume becomes less. On the other hand, neural tissues, which have excellent scattering properties, can modify the spatial resolution to values much higher than 1 mm^3. In capillary blood flow, the velocity is around 10^{-3} m/s and the Doppler shift of laser light becomes quite small. Such small frequency shifts are hard to measure with an absolute technique. Instead, the frequency-shifted light is mixed in a nonlinear detector with unshifted light to extract the information of interest, i.e., the frequency difference of the Doppler frequencies. In blood flow studies in microcirculation, the frequency difference is on the order of 10^1 to 10^4 Hz.

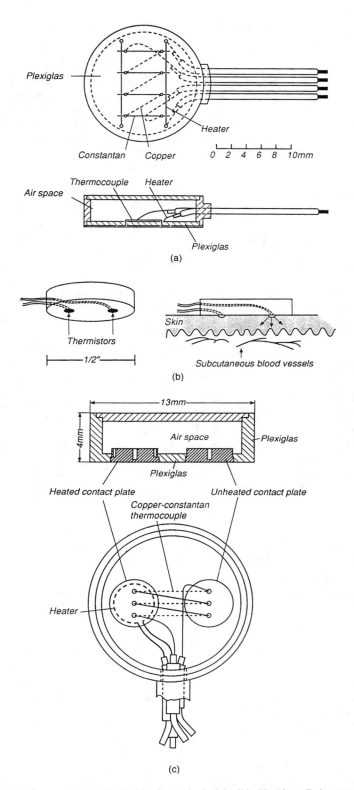

Figure 3.64 Skin blood flow probes working on the thermal principle. (Modified from References 287 through 291.)

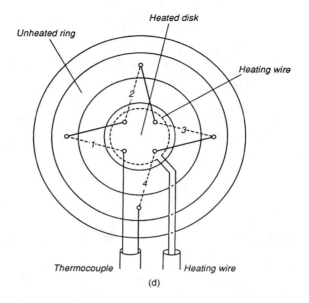

(d)

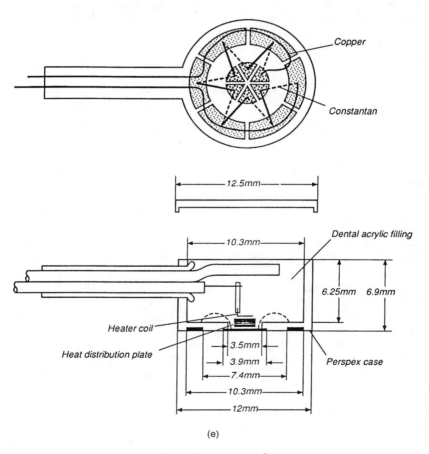

(e)

Figure 3.64 (continued)

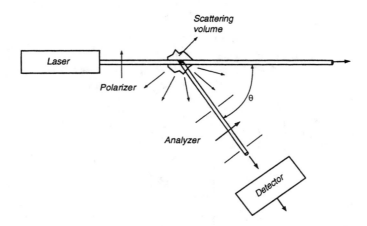

Figure 3.65   Representation of the light-scattering measurement.

3.3.4.2.3 Instrumentation.   In a typical flow measurement experiment, light from a laser passes through a polarizing element. The incident beam then impinges on the medium in which flow is to be determined. (Figure 3.65), The position of the detector defines the scattering angle, θ. The scattered light then passes through an analyzer which selects the polarization angle. The light finally reaches a detector.

The angle between the incident light and the scattered light to the detector is called the scattering angle, θ. There are two Doppler shifts to consider: (1) motion with respect to the source changes the frequency of the scattered light, and (2) motion relative to the detector shifts the frequency observed. It can be shown that the total Doppler shifts are

$$\Delta f = \left(\mathbf{K_S} - \mathbf{K_o}\right) \cdot \mathbf{V}/2\pi = \mathbf{K} \cdot \mathbf{V}/2\pi \qquad (3.103)$$

where $\mathbf{V}$ is the velocity vector, and the scattered and incident wave vectors are $\mathbf{K_s}$ and $\mathbf{K_o}$, respectively:

$$|\mathbf{K}| = \left|\mathbf{K_s} - \mathbf{K_o}\right| = \frac{4\pi}{\lambda}\sin\frac{\theta}{2}. \qquad (3.104)$$

The unshifted light is usually brought to the detector in the form of reflected light from stationary surfaces. In other applications, though, a portion of the incident light must be directed towards the photodetector where it is mixed with the portion of the light that has passed the scattering volume under study.

The signal from the photodetector can be processed as discrete photopulses (photon-counting) or as analog signals. Signal processing gives either the autocorrelation function or the frequency spectrum of the signal. The power spectral density can be obtained by a Fourier transformation of the autocorrelation function. The choice of autocorrelation function vs. power spectral density for presentation of the measurement is dependent upon the application and the traditions within different fields in which the LD method is applied. As a rule, correlation is used for higher frequency shifts, whereas the power spectral density representation is more appropriate for low frequency applications. Figure 3.66 shows the typical shape of correlation curves as well as spectral density curves for two types of LD experiments.

In instruments for microvascular blood flow measurements, the laser light is launched and the scattered light picked up by using optical fibers, the properties of which are described below. More detailed presentations of the physics of LD flowmetry (light scattering in tissue, photon diffusion, and models for the interaction between light and tissue, etc.) can be found in numerous books[295,297-299] and articles.[189,300-310]

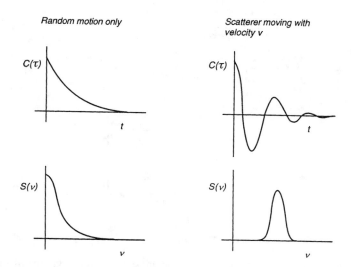

Figure 3.66 Illustration of anticipated data forms in a laser-Doppler velocimetry experiment. In the case of random motion only, the autocorrelation function $C(\tau)$ will be an exponential decay, and the spectrum $S(v)$ will be a Lorentzian distribution about zero frequency. In the presence of a directed velocity, v, $C(\tau)$ is a damped cosine and $S(v)$ is a Doppler-shifted peak.

3.3.4.3 Instrument Design Principles

3.3.4.3.1 *Light Sources.*
For most LDF instruments designed during the last 15 years the gas laser has been the light source of choice. This type of laser is inexpensive, easy to use, and reliable and has reasonable stability. In several recent constructions, the semiconductor laser diode has been used as an interesting alternative light source.[311-314]

The helium-neon laser has been used extensively in LD flowmetry. It is beyond the scope of this chapter to discuss basic laser physics but readers interested in this topic are referred to Shimada,[315] Young,[316] Yariv,[317] and Chu[295] or many other specialized textbooks.

Semiconductor laser diodes have been used successfully in LD flowmetry by several groups.[311-314] Several commercially available instruments use laser diodes at wavelengths around 800 nm. The main advantage of the laser diode is related to its physical size and energy supply. The diode is very small in comparison with gas lasers and can be housed in small probes near the skin. In some applications, fiberoptics can be entirely eliminated.[314] High voltage power supplies necessary for gas lasers can be eliminated because a laser diode can be fed from a low voltage source. The laser diodes designed for use in compact disc players and for communication purposes are low priced compared to gas lasers. Disadvantages of the diode laser for LD work are the limited coherence length as well as the divergence of the emitted light beam. The need for careful temperature stabilization to avoid instability is another. Temperature variations will induce a wavelength drift at a rate of 0.23 nm per °C. Variations in temperature will result in mode jumps, i.e., jumps in frequency, a condition devastating to any type of laser-Doppler application.

3.3.4.3.2 *Light Detectors.*
The photomultiplier is a photodetector with inherent high current gain and low noise. It can be used to detect power levels as low as 10^{-19} W.[318] Photomultipliers have been used extensively in the analysis of scattered laser light. Chu[295] has conducted a survey of the technical factors that must be taken into account when choosing and operating photomultipliers.

Semiconductor diodes are the most common type of photodetector for LD flowmeter applications. The p-n and p-i-n diodes basically utilize the same physical mechanism as that of the p-n junction. An incoming photon is absorbed and generates a hole and a free electron, both of which will drift across the p-n junction under the influence of an externally applied electric field.

The p-i-n diode has an intrinsic (pure) high resistivity layer which is sandwiched between the p and n regions. The intrinsic layer decreases the diffusion time and thus the response time of the diode. Examples of the p-i-n diodes most popular in LD work include the UDT 450 (UDT Sensors; Hawthorne, CA), the HP 2-4203 (Hewlett Packard; San Jose, CA), and the BPX 40 diode (Phillips; Eindhoven, The Netherlands). An alternative to the UDT 450 is the HUV 1000B from Hamamatsu Corp. (Japan). The

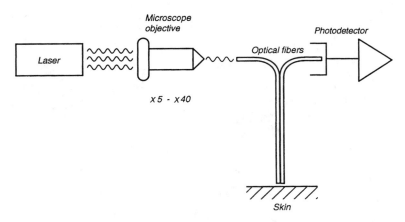

Figure 3.67   An arrangement of optical fibers in laser-Doppler flowmetry.

diodes have been used in the conventional way following the recommendations of the manufacturers. Kajiya et al.[319] have used an avalanche diode (PF-1005, Mitsubishi; Tokyo, Japan) in their system for blood flow measurements in the coronary and femoral arteries. Their motivation for using an avalanche diode is the high signal-to-noise ratio that is obtained at low signal levels (~1 μW)

3.3.4.4   Fiberoptic Arrangements
The way optical fibers are arranged in laser-Doppler flowmetry is illustrated in Figure 3.67. The laser light is coupled to the optical fiber by means of a lens (microscope objective, ×5 to ×40; N.A., numerical aperture, 0.1 to 0.6). This fiber leads the light to the tissue under study. Another fiber, or a set of fibers, picks up the light and guides the information-carrying light from the skin to the photodetector(s). A variety of optical fiber types (plastic as well as silica fibers) with diameters ranging from 50 μm to 2000 μm have been used.

Nilsson et al.[309] suggested the use of a dual channel detector to improve the signal-to-noise ratio. One of their fiber arrangements is presented in Figure 3.68. The pick-up fibers later were arranged differently. Bonner et al.[320] used 125-μm, graded index glass fibers (10200, Corning Glass; Corning, NY). In their design, the fibers were separated by 0.5 mm, which was regarded to be an optimum.

The way optical fibers are arranged, relative to the tissue volume under study, is of great importance to the outcome of the measurement. One example is given by Tenland et al.[321] who rotated the fiberoptic probe in 90° steps and found significant changes in skin blood flow. The explanation for this result is that an asymmetrical arrangement of the optical fibers probes different microvascular beds for every new angular orientation. This phenomenon is related to the spatial variability of skin blood supply.

In certain applications of LD flowmetry, the standard fiberoptical setup is not particularly well-suited for blood flow measurements. The optical properties of tissue, the choice of measurement site, or the theoretical requirements for heterodyne detection may require a redesign of the fiberoptic system. As long as the basic requirements for a certain detection scheme are fulfilled, there is a great deal of freedom in the design of new fiberoptic systems to fit certain applications. However, redesign of the fiberoptic system in such a way that detection requirements are fulfilled does not necessarily mean the results obtained with this new design can always be compared with results obtained with earlier probes. A couple of examples of innovative fiberoptic systems for special applications are given below.

Damber et al.[322] studied the acute effects of catecholamines on testicular blood flow. These authors found that the total intensity of backscattered light was too low to ensure heterodyne mixing requirements. The problem was solved by leading unshifted light directly from the laser to the photodetector via an optical fiber (Figure 3.69). The portion of light that is brought to the photodetectors in this way could be adjusted and was checked for every experimental situation. This way of arranging the fiberoptic probe can be of value in all applications in which tissue absorption is high.

Salerud and Öberg[323] suggested the use of a single optical fiber technique for deep tissue perfusion measurement. By using small fiber diameters and by introducing the fiber into the tissue volume under study in a non-traumatic way, blood flow from deeply located tissue volumes can be recorded.

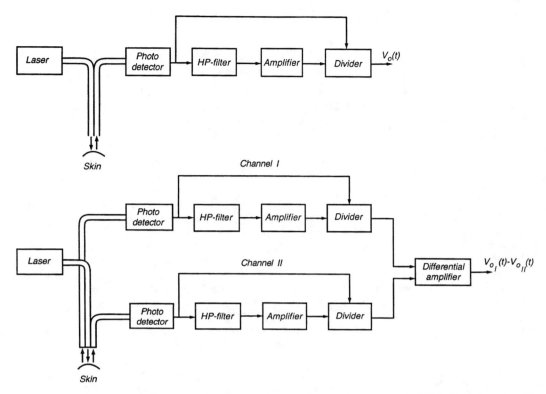

Figure 3.68 Block diagrams of the single channel and dual channel detector principle in laser-Doppler skin blood flowmetry.

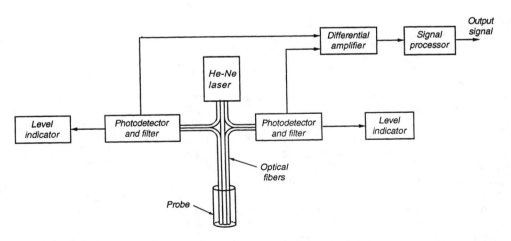

Figure 3.69 A laser-Doppler measurement system in which unshifted light is directly lead from the laser to each photodetector via an optical fiber.

3.3.4.5 Signal Processing Principles

During the development of laser-Doppler flowmetry, several approaches to signal processing have been presented. Of course, the ideal algorithm for flow calculation is one which is linear for all magnitudes of flow, tissue types, and hematocrits and also takes into account multiple scattering effects. So far, no such algorithm exists. The various approaches that have been suggested follow.

Stern et al.[324] used a "heuristic flow parameter":

$$F = \sqrt{\int\left(\omega^2 P(\omega) d\omega\right)} \qquad (3.105)$$

where $P(\omega)$ is the power spectral density of the Doppler signal. This algorithm is based on empirical findings, but Stern et al.[324] have shown that in a number of applications it was found to give an accurate estimate of flow. With this formulation of the flow parameter, F could easily be calculated by means of analog electronic circuitry. In practice, the flow value can be calculated as

$$F = \frac{\sqrt{R^2 - SI}}{I} \qquad (3.106)$$

where R is the output of a root-mean-square (RMS) detector that receives the signal, I is the mean photocurrent, and S is a gain constant. Normalization with respect to the photocurrent serves to make the flow parameter independent of laser intensity and total reflectivity of the tissue.

Bonner and Nossal[305] developed a mathematical model for LD measurement of blood flow in tissue. Their model is based on the following assumptions. The scattering cross-section for particles of the size of a blood cell is sharply peaked in the forward direction (Mie scattering). The radiation in a tissue matrix of this kind mostly contains photons which have been scattered at least once by surrounding tissue elements. A small percentage of the back-scattered light comes from photons scattered only by moving blood cells. Such a complex tissue matrix generally cannot be characterized in detail. As long as the tissue scattering cross-section is much greater than that of the moving cells, a model can be designed that can be of guidance in instrument construction.

In a tissue matrix, photons generally suffer several collisions before they interact with a blood cell, which means that the direction of light will be randomized before the collision between a photon and a moving blood cell occurs. The same authors[305] showed that the normalized first moment of the spectrum, $\langle\omega\rangle$, is proportional to the RMS speed of moving particles, $\langle V^2 \rangle^{1/2}$, as

$$\langle\omega\rangle = \frac{\langle V^2 \rangle^{1/2}}{(12\xi)^{1/2} \cdot a} \beta f(\overline{m}) \qquad (3.107)$$

where a is the radius of an average spherical scatterer; ξ is an empirical factor related to the shape of the cell; β is an instrumental factor, namely the optical coherence of the signal at the detector surface; and $\overline{m}$ is the average number of collisions which a detected photon makes with a moving cell. $\overline{m}$ is a function of the hematocrit of tissue and is linear if $\overline{m} \ll 1$ (low hematocrits). In experiments with a flow model system built of hollow silicon fibers surrounded by a scattering medium, these authors could demonstrate that the Doppler shift varied linearly with the mean velocity of the flowing cells. Nilsson et al.[325] independently came to a similar conclusion concerning the algorithm using a more experimental approach.

So, for low and moderate red cell volume fractions, the first moment of the power spectral density is approximately proportional to the flux of red blood cells. However, for higher hematocrits this algorithm underestimates the flux. Ahn et al.[326] have evaluated the first moment algorithm in experiments on an isolated part of the feline small intestine which was perfused by a controlled blood supply from the superior mesenteric artery. Their conclusion was that the first moment-type of signal processor (not linearized) seriously underestimated the flow at higher flow rates (>100 ml $\cdot$ min^{-1} $\cdot$ 100 g tissue).

3.3.4.6   Calibration and Standardization of Laser-Doppler Flowmeters
3.3.4.6.1   *Calibration.*   Most measurement methods must be calibrated at regular intervals by comparing them with a standard to maintain accuracy over longer periods of time. The standard usually has a well-specified accuracy and long-term stability. An instrument under calibration is adjusted to give readings coinciding with those of the standard. An alternative way is to make up a calibration curve, i.e., a diagram describing the difference between the "true" value (standard) and the reading of the

instrument under calibration. The calibration curve is then used for correcting the actual measurements obtained with the instrument.

Such calibrations are not possible to perform with LD instruments, simply because there is no "gold standard" for measuring blood flow in tissue. Nevertheless, the LD method has been compared with a variety of other methods. Such comparisons generally cannot be regarded as "calibrations" as the other flow measurements have unspecified accuracies or even measure a different physiological variable related to the flux of red blood cells, but comparisons of this type can provide useful information in a specific experiment or preparation.

Some examples of such comparisons are given below:

- Comparisons with the xenon isotope technique have been performed by Holloway,[327] Kastrup et al.,[328] Neufeld et al.,[329] Engelhart and Kristensen,[330] Nicholson et al.,[331] and Engelhart et al.[332]
- Radioactive microspheres for blood flow measurements have been studied in rabbits by Eyre et al.[333]
- Matsen et al.[334] used an ischemic human model to compare transcutaneous pO_2 with laser-Doppler.
- Kvietys et al.[335] have compared LD, hydrogen clearance, and microspheres in cat models.
- Sundberg and Castrén[336] used drug- and temperature-induced changes in peripheral circulation to compare LD flowmetry and digital-pulse plethysmography.

These types of evaluations and comparisons give, in general, a more or less linear relation between the LD method and other methods if the flow and local hematocrit do not reach extreme values. Each relation depends on the specific organs and sites used. The correlation coefficient can vary between different measurement sites in one and the same organ.

3.3.4.6.2 Standardization.

To standardize an instrument or a method is to make sure it maintains stability, linearity, and reproducibility. However, such standardization does not mean that an absolute value of flow is achievable. Several methods have been suggested for the standardization of LD instruments. Several authors[314,325,337-339] have designed rotating discs for the standardization of LD instruments.

The Brownian motion of particles in a suspension can be used to standardize LD instruments. For short-term use ($\approx$ 1 to 2 hours), ordinary milk can be used. For standardization during longer periods ($\approx$ 2 to 3 months), it is important to use a suspension that is not affected by sedimentation or particle aggregation.

A variety of mechanical flow models has been described in the LD literature.[325,339-341] Although most of them were originally used for the study of the principles of LD flowmetry, they can all be of great value for standardizing or calibrating LD flowmeters. They all seem to give similar results when effects of variations of hematocrit, average velocity, and volume flow are studied.

The laser-Doppler method has been used in a great variety of applications in medicine and biology. It is beyond the scope of this chapter to present the way in which the LD principle has been applied to various problems. The reader is referred to Shepherd and Öberg[342] or Öberg[343] for a review. The bibliography presented by Perimed[344] also gives an extensive presentation of the various applications.

3.3.4.7 Laser-Doppler Perfusion Imaging

Laser-Doppler flowmetry is usually applied to single point measurements in a tissue volume. The volume monitored is small (on the order of 1 mm³). The spatial variability[321,339] for microvascular blood perfusion can be very high in some tissue types. The combination of small monitoring volumes and high spatial heterogeneity of flow can lead to serious misinterpretation of the actual flow. In some applications, a variant of the LD method that can give spatial information about flow is to be preferred.

Most LD flowmeters have so far utilized optical fibers to lead light to and from the skin. In instruments based on optical fibers the probe must be attached to the tissue during the recording, which will introduce a risk for infection or even pain in some applications. Movements of the fiber can also introduce noise in the recording. This is the reason for the development of a number of perfusion imaging devices.

Fujii and collaborators[345] have used a line-scanning technique based on laser speckle image sensing. This group has described both one- and two-dimensional systems.[345] In the one-dimensional system, a laser beam impinges on the skin. Light returning from the skin is collected and brought to a charge-coupled device (CCD) sensor which is scanned frequently. Differences among the data of successive scans are calculated and integrated. The two-dimensional system uses a continuous line scanner device. The He-Ne laser is linearly expanded by a rod lens. The image is reflected in a rotating mirror which illuminates the skin surface (Figure 3.70). The scattered signal is collected by a lens (L2) which is focused on a CCD image sensor. With this device and adequate signal processing, two-dimensional plots

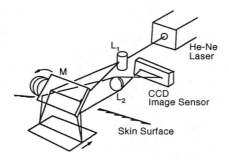

Figure 3.70 A scanning system for laser speckle image sensing. (Modified from Fujii, H. et al., *Appl. Optics*, 26, 5321, 1987.)

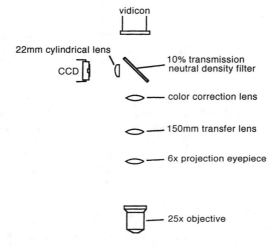

Figure 3.71 The optical arrangement for measurement of red cell velocity in microvessels. (From Goodman, A. H., *J. Biomed. Eng.*, 8, 32, 1986. With permission.)

of skin blood flow can be produced. The devices described by the Fujii group are based on speckle analysis. This principle was first suggested by Fercher and Briers in 1981.[346] Notice that speckle analysis does not utilize the Doppler shift of laser light but rather the magnitude of the speckle intensity.

Goodman[347] describes a CCD line-scan image sensor for measurement of red cell velocity in microvessels. The setup is outlined in Figure 3.71. The single microvessel is transilluminated. A CCD scan device is sampling the optical data along the axis of a microvessel to obtain a voltage pattern which is a representation of the pattern of bright and dark spots along the sample line. An analogue signal is digitized and stored in the memory of the unit. This process is repeated with short time intervals. Two subsequent traces are compared from which the direction and distance of the motion can be assessed. The blood velocity can be computed from the distance of motion and from the time interval. This device can be calibrated by means of a transilluminated, slowly rotating Perspex plate which carries a dried smear of red blood cells on its upper surface. The output voltage is a linear function of the tangential velocity and the optical magnification.

Essex and Byrne[348] have described a scanning laser-Doppler-based device for imaging of blood flow of the skin (Figure 3.72). A 2-mW He-Ne laser is mounted behind the array of photodiodes and lenses. The beam is passed coaxially through the detector arrangement to the center of a mirror which can be moved in two directions. Scattered light from the skin is collected by the same mirror and focused via lenses to the four photodiodes. The scanning mirror is driven by a d.c. motor. This mirror is moving continuously at a constant angular velocity during each scan. Mirror position information is derived from optical encoders on the gear box shaft but also by counting pulses sent to the stepper motor. The imager designed by Essex and Byrne[348] scans an area of 50×70 cm with a distance from the scanner

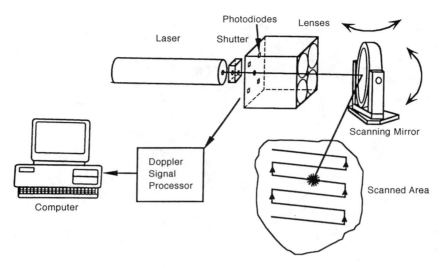

Figure 3.72 A scanning laser-Doppler system for imaging of blood flow of the skin. (From Essex, T. J. H. and Byrne, P. O., *J. Biomed. Eng.,* 13, 189, 1991. With permission.)

to the object of about 1.6 m. The scanning time is 6 min for a complete image. In the device the laser beam is moving continuously over the tissue. Samples of the reflected light intensities are taken at discrete time intervals. The continuous scanning motion of the laser beam across the skin generates a movement artifact that can be eliminated through the use of a differential input to the Doppler processor and a second-order high pass filter with a cutoff frequency of 250 Hz. The scanner has been used in clinical work for burn-depth assessment.[349] In this application, the advantage of a noncontact scan image of the skin is utilized.

Nilsson and collaborators[350] have presented an imaging device that is based on LD perfusion measurements. The instrument has been described in detail by Wårdell et al.[351-353] This perfusion imager analyzes coherent light scattered in the skin. Color-coded images of tissue perfusion are presented. The optical scanning device uses two mirrors, the positions of which are controlled by two separate stepping motors (Figure 3.73). A low-powered He-Ne laser beam is directed towards a mirror which deviates the laser beam towards the tissue under study. A full format image takes 4.5 min to generate and covers an area 12 × 12 cm (4096 pixels) at a distance of 20 cm between the scanning head and the tissue. The advantage of this scanner compared to the one of Essex and Byrne is that no scanning artifact is generated because the mirrors are stopped before the Doppler signal is sampled. This scanner has been utilized in a number of experimental[354-356] and clinical[357-360] studies. The scanners by Essex and Byrne and by Nilsson and collaborators are available commercially through Moore Instruments (Axminster, U.K.) and Lisca Developments (Linköping, Sweden).

3.3.5 TISSUE BLOOD FLOW MEASUREMENTS BY NUCLEAR MAGNETIC RESONANCE

Nuclear magnetic resonance is the phenomenon of absorbing or emitting energy when a nucleus having magnetic moment is placed in a static magnetic field with a superimposed excitation field of a specific frequency. Under appropriate static and excitation magnetic fields, the amplitude and the phase of the resonant signal depend on the velocity component of the nuclei. Hence, the measurement of nuclear magnetic resonance can be used for flow detection.

In blood flow measurement, the magnetic resonance of the hydrogen nucleus, which is ^1H, in the blood (mostly in water molecules) is observed. The frequency at which specific absorption or emission of magnetic energy occurs is called resonant frequency and is determined by the applied static field strength and the type of the nucleus. In a static field, H, the resonant frequency, ν, is given as

$$\nu = \gamma H / 2\pi \qquad (3.108)$$

where γ is called the gyromagnetic ratio and is dependent on the nucleus. For ^1H, $\nu = 42.5$ MHz when $H = 1T$.

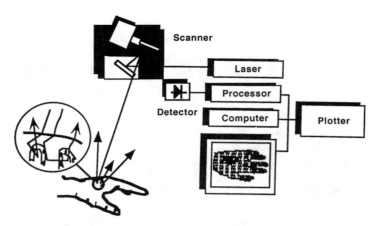

Figure 3.73 A blood flow imaging system consists of a scanning system using two mirrors. (From Wårdell, K. et al., *IEEE Trans. Biomed. Eng.*, 40, 309, 1983. With permission.)

In the absence of the excitation field, each nucleus having a magnetic moment aligns along the direction of the static magnetic field. As a consequence, all nuclei produce a magnetization, *M,* which is proportional to the magnetic field strength, *H,* so that

$$M = \chi_m H \qquad\qquad (3.109)$$

where χ_m is magnetic susceptibility. χ_m of ^{1}H in the blood at 37°C is 3.229 × 10^{-9}.

When an excitation field at resonance frequency is applied perpendicularly to the direction of the static field, as shown in Figure 3.74, precessional motion of the magnetic moment of the nucleus is induced, and this motion produces an oscillating magnetic field which can be detected by a pickup coil (for example, in the *x* direction in this figure). The precession of the nuclear magnetic moment does not start instantly, but it builds gradually. The time constant of this process is called the longitudinal relaxation time or T_1. Many factors such as static magnetic field strength, molecular structure, elemental composition, and temperature affect T_1. For the whole blood at 25°C, T_1 is about 0.1 s when the resonant frequency is 20 kHz, and about 1 s at 50 kHz.[361]

Many different techniques of flow measurement using nuclear magnetic resonance have also been proposed, and they are classified into two groups, i.e., continuous wave excitation and pulsed wave excitation. Methods using continuous wave excitation provide flow-dependent signal amplitude. If the fluid passes through the resonance region in a short time compared to T_1, resonance does not fully build up, and the signal amplitude decreases with increasing flow rate. However, if the excitation field is strong enough, the saturation effect occurs so that the signal amplitude tends to be zero, because populations of nuclear magnetic moments parallel and antiparallel to the excitation field tend to be equalized; hence, the net magnetization is reduced to zero. Under such conditions, a flesh medium flowing into the resonance region limits the saturation effect, and the signal amplitude increases with increasing flow rate.[362]

Techniques using continuous excitation have been applied to human limb blood flow measurements.[363-367] In one study, a permanent magnet of 0.075 T was used, and the forearm was placed in a coil with an internal diameter of 12.5 cm. The study with a model arm showed that signal amplitude proportional to the blood flow rate could be obtained, and it was applied to the measurement of total arm blood flow in patients with atrio-venous fistulas.[368]

In the flow measurement using pulsed wave excitation, a pulse or a sequence of pulses is applied so as to induce appropriate rotational motion of nuclear magnetic moments, and then the signal induced by the excited nuclei is detected. During this process, if part of the nuclei in the resonance region is transported to another region where the static field strength is different, amplitude, frequency, or phase of the signal will be changed. This change in the signal can be used for flow detection. While there are different modalities in the sequence of pulse wave excitation, most of them employ a principle called the spin echo method.

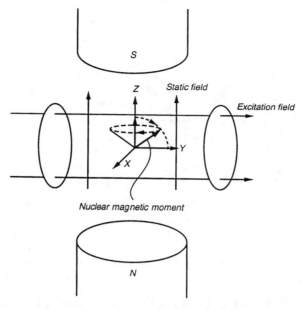

Figure 3.74 Principle of nuclear magnetic resonance. Precessional motion of the nuclear magnetic moment placed in a static field is induced by an excitation field at resonant frequency.

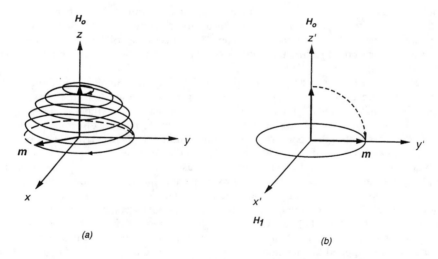

Figure 3.75 The locus of the net magnetization vector, **m**, when a 90° pulse is applied: **(a)** observed on the laboratory frame of reference, and **(b)** observed on the rotating frame of reference.

The principle of the spin echo method can be visualized by considering the motion of the net magnetization vector **m**, which is equivalent to the sum of all magnetic moments in the resonance region.[361] Before excitation, **m** remains in equilibrium and is oriented toward the static field vector, $\mathbf{H}_0$. After the excitation field $\mathbf{H}_1$ is applied, **m** begins to move, and the angle between **m** and $\mathbf{H}_0$ increases gradually, as illustrated in Figure 3.75(a). Introducing a rotational frame of reference, the motion of **m** can be illustrated as in Figure 3.75(b), where **m** moves about $\mathbf{H}_1$. The angle of rotation produced by the excitation pulse depends on the strength of excitation field $\mathbf{H}_1$ and the duration of the excitation. If an appropriate strength and duration of excitation are applied, the angle of rotation about $\mathbf{H}_1$ can be 90°. The excitation pulse which realizes this condition is called the 90° pulse.

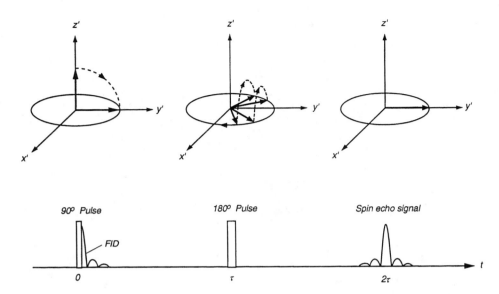

Figure 3.76 Principle of the spin echo method. By applying a 90° pulse along *x′*, the net magnetization vector, **m**, rotates 90° around *x′*. Spin groups having slightly different rates of precession will fan out in the *x′-y′* plane. By applying a 180° pulse along with *y′* at time τ, these spins rotate 180° around *y′* and then rephase at time 2τ so that a signal appears again.

Immediately after applying a 90° pulse, **m** has the same magnitude as that in equilibrium but is diminished due to the effect on different phase angles by groups of nuclear spins having slightly different rates of movement so as to fan out in the *x′-y′* plane. This process occurs through inhomogeneity of the static field strength. As a consequence, the signal decays quickly as illustrated in Figure 3.76. This signal is called fast induction decay, or FID.

After an appropriate time, τ, has passed, a 180° pulse is applied along the *y′* direction (as in the figure), which causes reversal of the magnetization vector of each group about the *y′* axis. Then, at the time 2τ, all spin groups are rephased, and the signal appears again. This signal is called the spin echo signal. The amplitude of the spin echo signal is proportional to the number of excited nuclei remaining in the region at the moment of the appearance of the spin echo signal; hence, it decreases with increasing flow by which part of the excited spins are removed from the detecting region.

If the static field has a gradient in one direction, it causes a phase difference of precession between nuclei at different locations. Under such conditions, a nucleus moving toward the field gradient gains phase in proportion to the distance of motion which corresponds to the velocity component along the field gradient times 2τ. Consequently, nuclei having different velocity components have different phases, and the amplitude of the spin echo signal is affected by the velocity distribution. It is shown that the velocity distribution can be calculated from amplitudes of spin echo signals obtained with many different τ. Grover and Singer[369] applied this method to examine blood flow in the human fingers and obtained a plot of the density of the nuclei vs. velocity component along the finger axis.

Spacial distribution of the blood flow in a tissue can be measured using nuclear magnetic resonance imagers developed mostly for noninvasive anatomical observation as well as for chemical analysis using magnetic resonance spectroscopy. The magnetic resonance imager provides resonance signals from local volume elements in the body. If the signal from each volume element contains information on the local blood flow, the spacial distribution of the blood flow can be visualized. Versatile techniques of nuclear magnetic imagings are described in detail in many textbooks.[370]

The algorithm of ordinary imaging technique is explained simply as follows. First, the excitation pulse is applied with a linear gradient field along one direction, say the *z* axis, so that only nuclear spins in a narrow range of the *z* axis satisfy the resonance condition and excitation occurs only in a thin region, which is usually called a slice. Then, signals corresponding to two other axes, say the *x* and *y* axes, are discriminated by the technique in which field gradients and the Fourier transformation are combined.

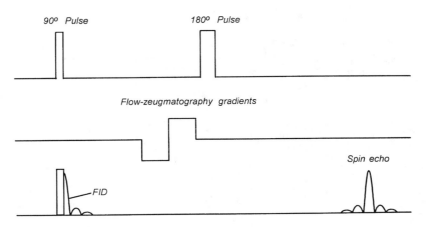

Figure 3.77 Time sequence of the biphasic gradient modulation for selective detection of flow components along a specific direction.

Either the reconstruction method or the two-dimensional Fourier transformation method is used for this purpose.

In the reconstruction method, a gradient field in one direction in the slice, say the x direction, is applied when the signal is detected. Because the signal frequency depends on the field strength, signals coming from each x have different frequencies corresponding to the value of x, thus, by applying the Fourier transformation to the obtained signal, a frequency spectrum can be calculated which corresponds to the distribution of the signal amplitude along the x direction. By repeating this procedure many times for slightly different directions, signal distribution over the slice can be calculated by the ordinary reconstruction method of X-ray computed tomography.

The two-dimensional Fourier transformation method employs a technique called phase encoding to obtain information along a direction. In this technique, a linear gradient field in one direction, say the y direction, is given for a definite period of time so as to produce different phase shifts in excited spin groups at different y values. Then, after the application of a linear gradient field in the x direction, signal components which are in phase to the excitation wave are detected selectively by a phase-sensitive detector. The obtained amplitude corresponds to the population of excited nuclei in specific y ranges where phase shifts are multiples of 360°. The above procedure is repeated many times, varying the magnitude of the gradient field in the y direction. Consequently, information in the y direction is obtained as a spacial frequency spectrum, and distribution of spins in the y direction can be calculated using the inverse Fourier transformation. The distribution in the x direction can be calculated by the Fourier transformation of the signal, as for the reconstruction method.

The blood flow distribution can be obtained, in principle, by a decrease in the amplitude of the spin echo signal corresponding to each volume element. However, if the change in signal amplitude due to the blood flow is small, compared with other factors such as anatomical differences, flow measurement will be difficult. To extract the effect of flow selectively, a 90° pulse with a field gradient is first applied so as to distribute spins randomly in the plane perpendicular to the static field, and after an appropriate time has passed, ordinary spin echo measurement is performed. The spin echo signal amplitude increases in proportion to the amount of fresh blood entering the randomized region during that period. This procedure of flow detection is called the transit time method. By this method, Singer and Crooks[371] demonstrated blood flow imaging in the human brain.

The velocity component in the arbitrary direction can be detected selectively by a method in which a biphasic gradient modulation, called the flow-zeugmatography gradient, is applied in the image sequence as illustrated in Figure 3.77.[372] The gradient causes no phase shift for spins having no velocity components in the direction of the gradient. But if spins have velocity components in that direction, effects of positive and negative parts of the biphasic gradient field do not cancel out each other, and the velocity component can be detected by a phase shift of the spin echo signal. To detect phase shifts, quadrature detection is needed in which both the in-phase component and the component having a 90° phase difference to the excitation wave are detected. Using this method, Moran et al.[373] demonstrated flow imaging in flow phantoms and in living human hearts.

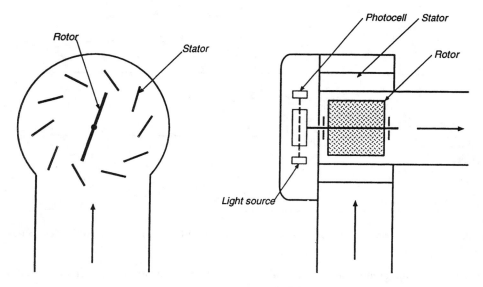

Figure 3.78   Wright respirometer.

3.4 RESPIRATORY GAS FLOW MEASUREMENTS

3.4.1 GAS FLOW SENSORS

Many different types of gas flow sensors have been proposed and used for respiratory gas flow measurement. According to the measurement principle, available gas flowmeters may be tentatively classified as rotameters, pneumotachographs, hot-wire anemometers, time-of-flight flowmeters, ultrasonic flowmeters, and vortex flowmeters.

3.4.1.1 Rotameter

A rotameter has a rotor which responds to the air flow, and the flow rate is measured by the revolutions of the rotor. Figure 3.78 shows a rotameter called the Wright Respirometer© (Warren E. Collins, Inc.; Braintree, MA). A rotor of this type responds only to air flow in one direction. Rotation of the rotor is detected optically. This type has been widely used for respiration monitoring and spirometry.

According to the specifications of a commercial model, the measurement range is 2 to 300 l/min, the rotor rotates 150 turns per 1 liter of air flow, resolution is 0.01 liters, and the pressure drop is about 0.2 kPa (2 cmH$_2$O) at 100 l/min. An evaluation shows that the measurement error was within ±10% in the flow range of 10 to 60 l/min.[374] The measurement error increases significantly at lower flow rates less than 5 l/min, and the rotor does not respond to a flow less than 2.5 l/min. The effect of respiration frequency is insignificant under physiological respiratory conditions. According to the nominal specifications, the error remains within 5% when the respiration frequency is in a range of 10 to 40 per minute.

The effect of gas composition is insignificant under clinical situations. Although the sensitivity is increased with the density of the gas, the response to air with 50% N$_2$O is almost the same as that of air. However, because the rotameter has a moving part with mechanical inertia, it cannot respond to a rapid change of air flow. Actually, an evaluation of spirometers shows that a spirometer using a rotameter type of air flow sensor causes errors of up to 10% in the measurement of forced vital capacity (FVC) and forced expiratory volume in one second (FEV$_1$) and does not satisfy the recommendations of the American Thoracic Society.[3]

Sensitivity in this kind of mechanical flow sensor is always stable so that recalibration is practically unnecessary, but differences in sensitivity between sensors may, to some extent, be unavoidable. An evaluation of commercial rotameters shows that the difference in sensitivity of individual sensors was up to ±15%.[375]

Figure 3.79 shows a flow sensor with a rotor that rotates in both directions. A spirometer using this type of sensor was supplied commercially. The flow-sensing part has a small rotor placed at the center of a transparent plastic tube 124 mm long with a diameter of 22 mm. The revolution of the rotor is detected optically through the tube. Despite its unique design and the availability of a disposable sensor,

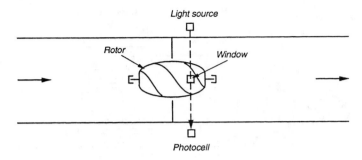

Figure 3.79 Flowmeter head of the Spirostat®.

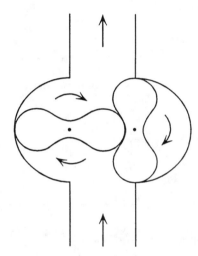

Figure 3.80 Dräger volumeter.

it has not been used widely. The flow resistance is about twice that of the Wright Respirometer, and the errors in the measurement of FVC and FEV_1 have been up to ±15%.[376]

Figure 3.80 shows a gas flowmeter called the Dräger Volumeter.[377] It has two lozenge-shaped rotors which respond to flow in both directions. This type of flowmeter provides accurate volume flow rates for dry gases. However, rotor friction is greatly increased by the condensation of water, and insertion in the inspiratory pathway is recommended when it is mounted on a ventilator or anesthetic machine.

3.4.1.2 Pneumotachograph

The term pneumotachograph usually means a device in which flow rate is detected by a differential pressure across a small flow resistance. The flow resistance consists of either bundles of parallel lumens placed longitudinally in the gas pathway or a wire screen. The flow rate is proportional to the pressure drop across these resistances in a certain flow range. A flowmeter with a chamber in which turbulent flow is induced may also be considered to be a device of similar type, although the flow rate is not proportional to the pressure drop, but rather to the square root of it. A flowmeter that consists of a venturi tube is based on measurement of a change in dynamic pressure at the narrow region of the tube. In this flowmeter, the flow rate also is proportional to the square root of the pressure difference.

Figure 3.81 shows the Fleisch pneumotachograph head.[378] It has bundles of lumens consisting of metal foils, so as to maintain a laminar flow over a wide range of flow rates. The maximum pressure drop is limited to about 0.07 kPa (7 mmH$_2$O), while four sizes are required to cover the flow range from 0–60 to 0–1000 l/min. To avoid water condensation in the flow resistance, electrical heating is applied. A flowmeter head with a wire screen as a flow resistance has also been used, especially in flowmeters for infants.[379] A 400-mesh stainless-steel gauze is a typical material used in flow resistance. In this type

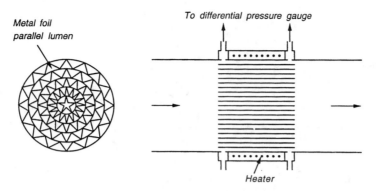

Figure 3.81   Fleisch pneumotachograph head.

Table 3.3   Physical Properties of Gases

Gas	Density (g/l)	Specific Heat at Constant Pressure (J/g·K)	Thermal Conductivity (mW/cm·K)	Viscosity (μcP)	Sound Velocity (m/s)
N_2	1.251	1.034 (10°C)	0.2374	178.1 (10.9°C)	334 (0°C)
O_2	1.429	0.922 (10°C)	0.2424	189	316
Air	1.293	1.006 (20°C)	0.1619	170.8	331.45
CO_2	1.977	0.837 (16°C)	37.61	139.0	259
Ar	1.783	0.523 (15°C)	41.33	209.6	319
He	0.178	5.208 (25°C)	1.411	186.0	965
N_2O	1.977	0.892 (26°C)	39.30	135	324 (10°C)
Water vapor	—	2.051 (100°C)	0.248 (107°C)	125.5	494 (134°C)

of flowmeter, nonlinearity larger than that of the Fleisch pneumotachography is recognized. A study shows that the flow range in which nonlinearity remains ±1% is up to 12 l/s in a Fleisch pneumotachograph, while it is up to 2.5 l/s in a flowmeter of the mesh type having almost the same flow resistance.[380] It is also pointed out that linearity depends upon the geometry upstream from the flowmeter head.[381]

The performance of a pneumotachograph depends on the viscosity of the gas. In a Fleisch pneumotachograph, it was shown that the pressure drop across the flow resistance is directly proportional to gas viscosity.[382] As seen in Table 3.3,[383] there are significant differences in viscosities among gases appearing in clinical situations. As an example, a large difference in the gas viscosity is expected between pure oxygen and a mixture of oxygen and nitrous oxide (N_2O). The viscosity of N_2O is about 70% that of O_2, while the concentration of N_2O sometimes increases by up to 90%. Actually, it was shown that the pressure difference at the Fleisch pneumotachograph head depended linearly on the O_2 concentration in N_2O.[384] While these studies were mostly carried out with Fleisch pneumotachographs, the performance of the screen type pneumotachograph may be similar.

There are gas flowmeters which sense pressure differences but are based on different principles. One of them is shown in Figure 3.82 and is called the turbulent air flowmeter.[385] It has a cylindrical chamber with two flow-through tubes offset from one another by a difference in size of slightly more than their internal diameter. The actual size of the reported chamber is 2.54 cm in diameter with a 1.9-cm separation between the end walls and flow-through tubes 9.5 mm in internal diameter. This geometry produces turbulent flow over a wide range of flow rates, and it causes a pressure drop proportional to the square of the flow rate between the upstream and downstream tubes. It was confirmed that this square law is valid over a flow range from 3.5 to 270 l/min for the chamber of the above-mentioned size.[385]

The advantage of the turbulent flowmeter is the fact that water condensation does not affect the measurement as much as it does the Fleisch or screen-type pneumotachograph. However, due to the square relation between the pressure drop and flow rate, resolution decreases in lower flow, and the pressure drop increases seriously in higher flow.

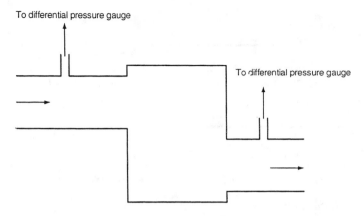

Figure 3.82 Turbulent airflow meter.

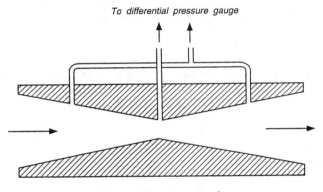

Figure 3.83 Ventigrator®.

A venturi tube, as shown in Figure 3.83, is also a flowmeter detecting flow rate by a pressure difference. The principle of this type of flowmeter is based on the measurement of the change in dynamic pressure. According to Bernouli's theorem,

$$\frac{1}{2}\rho U^2 + P = const. \tag{3.110}$$

where ρ is density, U is velocity, and P is static pressure. When the cross-sectional area of the tube is reduced from A_1 at the entrance to A_2 at the center, the pressure difference will be expressed as

$$\Delta P = \frac{1}{2}\rho\left(1 - \frac{A_2}{A_1}\right)U^2. \tag{3.111}$$

The pressure difference is proportional to the square of the flow velocity.

A commercial model (Ventigrator®, Ohio Medical Co.; Madison, WI) consists of a venturi tube with a throat diameter of 7.5 mm and produces a differential pressure of 0.06 kPa (6 mmH$_2$O) at 10 l/min oxygen flow. As predicted from Equation (3.111), the developed pressure is proportional to the density of a gas. Actually, the ratio of sensitivities observed in oxygen and nitrous oxide was roughly in accordance with the ratio of densities of these gases.[374]

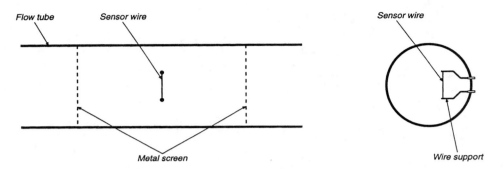

Figure 3.84   A hot-wire anemometer for respiratory measurement.

3.4.1.3   Hot-Wire Anemometer

The hot-wire anemometer consists of a fine metal wire heated by an electric current. The gas flow velocity is estimated by the amount of heat transfer from the wire to the gas and the temperature difference between the wire and gas temperatures. As shown in Figure 3.84, the hot-wire anemometer for respiratory measurement has a platinum or tungsten wire about 10 μm in diameter placed in a flow tube, and the wire is heated to 30 ~ 400°C above flowing gas temperature. To obtain a definite flow profile and thus a unique relationship between gas flow rate and flow velocity at the wire, fine metal meshes are placed at both ends of the tube.

Heat transfer from the wire to the gas is determined by the wire geometry, temperature difference between the wire and the gas, and the thermal conductivity of the gas. For a diameter and length of the wire, d and l, respectively, and the wire and gas temperature, T_w and T_g, respectively, the heat transfer, H, is expressed as

$$H = \alpha \pi dl \left(T_w - T_g \right) \tag{3.112}$$

where α is the heat transfer coefficient, which is expressed as

$$\alpha = N_u k / d \tag{3.113}$$

where N_u is a nondimensional quantity called the Nusselt number, and k is the thermal conductivity of the gas. According to Collis and Williams,[386] this gives an experimental formula

$$N_u = \left(\frac{T_w - T_g}{2T_g} \right)^{0.17} \left(A + BU^n \right). \tag{3.114}$$

In the Reynold's number range 0.02 to 44, $n = 0.45$, $A = 0.24$, and $B = 0.56$. The Reynold's number, Re, is defined as

$$\text{Re} = Ud\rho / \mu \tag{3.115}$$

where ρ and μ are the density and viscosity, respectively, of the gas. For a tube 3 cm in diameter with a wire 10 mm in diameter (and if the gas is air), the above Reynold's number range corresponds to a flow range from about 0.02 to 40 l/s. This range fully covers the requirements for respiratory measurements.

If the wire temperature is not very high, Equation (3.112) can be approximated by

$$H = \left(a + bU^{0.45} \right) \left(T_w - T_g \right) \tag{3.116}$$

where a and b are constants depending on the properties of the gas and the wire geometry.

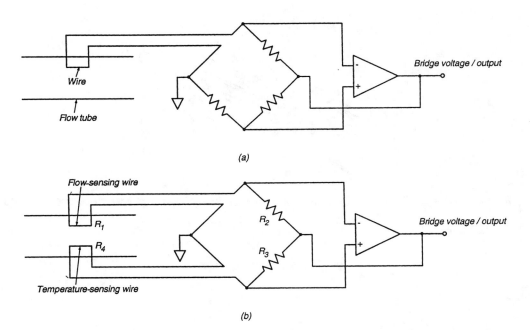

Figure 3.85 Servo-control circuits for hot-wire anemometer of a constant wire temperature **(a)** and that of a constant temperature difference between the wire and the flowing gas **(b)**.

According to King's law,[158] the heat dissipation, H, is written as

$$H = \left(a + bU^{0.5}\right)\left(T_w - T_g\right) \tag{3.117}$$

which can be considered as an approximation of the above equation.

The heat dissipation, H, can be obtained by the current, I, and wire resistance, R, as

$$H = RI^2. \tag{3.118}$$

The wire resistance in the operating condition can be determined by the ratio of the voltage across the wire and the current through it. Because the wire resistance depends on its temperature, the wire temperature, T_w, can be determined by the wire resistance. Consequently, as long as $(T_w - T_g)$ is constant, U can be calculated from the wire current using the above formulae.

To obtain a constant wire temperature, a feedback circuit, as shown in Figure 3.85(a), is commonly employed. The unbalanced voltage in a bridge containing the wire as one arm is amplified and fed back to the bridge voltage. Due to the temperature dependency of the wire resistance, a change in the wire current will change the heat generation in the wire, which causes a change in its temperature and thus a change in its resistance so as to compensate for the unbalanced voltage. If the feedback gain is high enough, the unbalanced voltage is reduced to practically zero, which means the resistance and thus the temperature of the wire are kept constant. In equilibrium, the total resistance of the bridge is constant, hence the bridge voltage is proportional to the wire current.

Even if the wire temperature is unchanged, $(T_w - T_g)$ will change if the gas temperature, T_g, fluctuates. To eliminate this effect, T_w should be kept high enough above T_g if it is constant.

To compensate for fluctuation of the flowing gas temperature, a servo-controlled compensation method, as shown in Figure 3.85(b), was proposed.[387] If temperature coefficients of flow-sensing wire resistance R_1 and temperature-sensing wire resistance R_4 are equal, then

$$\frac{1}{R_1}\frac{dR_1}{dT_1} = \frac{1}{R_4}\frac{dR_4}{dT_2}. \tag{3.119}$$

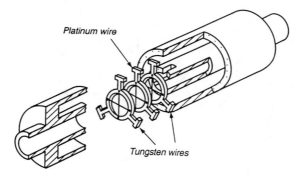

Figure 3.86   A bi-directional hot-wire anemometer. (Modified from Yoshiya, L. et al., *J. Appl. Physiol.*, 38, 360, 1975.)

When the bridge is balanced,

$$R_1 R_3 = R_2 R_4 \qquad\qquad (3.120)$$

thus,

$$R_3 dR_1 = R_2 dR_4, \qquad\qquad (3.121)$$

and

$$dT_1 = dT_2. \qquad\qquad (3.122)$$

This implies that a change in gas temperature causes the same temperature change in the flow-sensing wire temperature, and thus the temperature difference between the gas and the flow-sensing wire is kept constant. However, to avoid the effect of self-heating in the gas temperature sensing wire, its resistance should be small compared to the flow sensing wire resistance, i.e.,

$$R_4 \ll R_1. \qquad\qquad (3.123)$$

One problem with a hot-wire anemometer used for respiratory measurements is the lack of capability for detecting flow direction. When a bi-directional flow measurement is required, an additional flow direction sensor should be used. An example of a simple flow detection sensor consists of two electrically heated wires mounted close to each other, so that a gas heated by the upstream wire passes the downstream wire.[388] The flow direction can be detected by a deviation from the equilibrium of wire resistances. Using two wires about 25 μm in diameter and separated by about 0.5 mm from each other, the sensor could detect gas flow as low as 1 m/min and had a response time of 2 ms.

A bi-directional hot-wire anemometer using three wires has also been developed.[389] It has a platinum wire for flow detection and two tungsten wires for detecting flow direction, which are assembled in a probe as shown in Figure 3.86. The tungsten wires are at a distance of 1.6 mm from the platinum wire of about 375°C, and the heated gas passes a tungsten wire downstream, causing a change in its temperature, which provides a signal corresponding to the flow direction.

Generally, each hot-wire anemometer probe requires calibration, because the heat dissipation from the flow-sensing wire strongly depends on its physical arrangement. Although the nonlinearity of the calibration curve is an inconvenience, the validity of $U^{0.45}$ dependency in Equation (3.117) is experimentally tested, and thus linearization can be achieved by appropriate analog or digital processing.

The sensitivity of a hot-wire anemometer is affected by physical properties of the gas such as density, specific heat, and thermal conductivity, but these effects are complicated and not fully understood theoretically. Calibration for each gas is thus required in quantitative measurements. In respiratory flow measurements, gas composition may change, and different calibration curves should be applied for

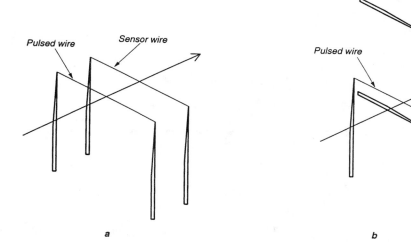

Figure 3.87 Parallel **(a)** and right-angle **(b)** arrangement of pulsed and sensor wires in the time-of-flight flowmeter.

different compositions. However, it was shown that the change in heat dissipation from a wire in a mixture of two gases is approximately linear to the molar fraction of a gas.[389,390]

The response time of a hot-wire anemometer depends on the time constant of the wire and is proportional to the square of the wire diameter, $U^{-0.5}$, and the overheating ratio, which is defined as the ratio of the overheating temperature and the air temperature. For a platinum wire 2.5 μm in diameter, flow velocity of 1 m/s, and an overheating ratio of 0.4, the time constant was about 0.1 ms.[391] Even for a wire diameter of about 10 μm, the time constant is on the order of milliseconds, except in the case of very low flow velocity, and the response is always quick enough for respiratory measurements.

The accuracy of an actual hot-wire flowmeter depends not only on the physical characteristics of the wire, but also on the performance of the electronics, including the linearizer. An evaluation of two commercial hot-wire anemometers showed that, in 67 subjects, the variation range was greater than $\pm 11\%$ of water-shield spirometer measurements for spirometric parameters such as the forced vital capacity (FVC), the forced expiratory volume in one second (FEV$_1$), the FEV$_1$/FVC%, the mean forced expiratory flow between 200 and 1200 ml of the FVC (FEF$_{200-1200}$), the mean forced expiratory flow during the middle half of the FVC (FEF$_{25-75\%}$), the mean forced expiratory flow between 75 and 85% of FVC (FEF$_{75-85\%}$), and the maximum voluntary ventilation (MVV).[392]

3.4.1.4 Time-of-Flight Flowmeter

The time-of-flight flowmeter is an instrument that measures the flow velocity by introducing a tracer into the upstream and detecting it at the downstream. The most convenient tracer is a heated gas bolus. As long as the separation between the introducing and detecting sites is known, flow velocity can be determined by the time of flight of the tracer.

A small bolus of heated gas can be produced by applying pulse current to a fine wire placed in the stream, and the heated gas bolus can be detected by another wire. This method is known as the pulsed-wire technique. The parallel wire arrangement, shown in Figure 3.87(a), is commonly used, but the right angular arrangement shown in Figure 3.87(b) is also used when the flow is highly turbulent.

Similar to the hot-wire anemometer, fine wires of 5 to 10 μm in diameter are used in this technique. Both platinum and nickel are suitable materials for pulsed wires, because they are capable of operating at high temperatures, even in excess of 600°C, without oxidizing. Tungsten wire is not suitable, because it oxidizes at a lower temperature of about 300°C, but it is suitable for a sensor wire.[393]

Time courses for temperatures of the pulsed wire, the gas, and the sensor wire, when a short pulse current is applied to the wire, are shown in Figure 3.88. The pulsed wire temperature elevates quickly, and then decreases gradually due to heat dissipation from the wire to the gas. The heated gas moves downstream and reaches the sensor wire. During this period, the heated gas spreads transversely and longitudinally to some extent by a diffusion. Then the sensor wire temperature increases but is delayed from the gas temperature due to the thermal inertia of the wire. These processes are well analyzed theoretically and experimentally.[393]

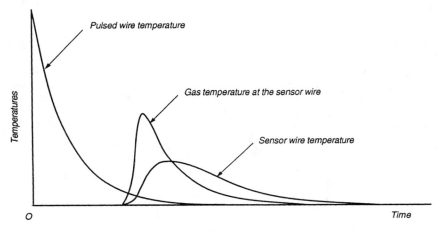

Figure 3.88 Typical time course of temperatures of the pulsed wire, the gas, and the sensor wire. The scale for pulsed wire temperature is reduced.

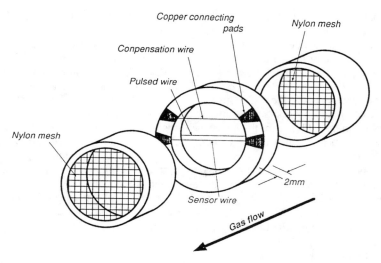

Figure 3.89 Schematic diagram of a time-of-flight flowmeter. (From Mosse, C. A. and Roberts, S. P., *Med. Biol. Eng. Comput.*, 25, 34, 1987. With permission.)

Figure 3.89 shows an example of the time-of-flight gas flowmeter designed for respiratory flow measurement particularly in neonates.[394] It has three wires in a flow tube; a pulsed wire, a sensor wire, and a compensation wire — for detecting ambient temperature fluctuation. A nylon mesh is attached to either side of the tube to protect wires and obtain definite velocity profile. A pulse current of 2 A and 30 μs is applied to a wire 10 μm in diameter, and the heat pulse is detected by the sensor wire separated by 2 mm from the pulsed wire. The result showed that the reciprocal time of flight was almost proportional to the flow ranging from 0.2 to 24 l/min.

A time-of-flight flowmeter using the "sing-around" technique provides a frequency output proportional to the flow velocity.[395] In this system, a signal detected by the sensor wire triggers the next pulse applied to the pulsed wire. Once an initial pulse is applied to the pulsed wire, an oscillation is sustained as long as the gas flow is continued. In an actual system, delayed triggering was employed in which the triggering time is delayed just the same as the time of flight so that the repetition period is twice as much as the time of flight. By using 5-μm wires with a separation of 12 mm assembled in a flow tube 35 mm in diameter, a frequency output proportional to the flow rate was obtained in a flow range from 0.05 to 12 l/min, and the calibration curves for O_2, CO_2, and N_2 did not deviate more than 5% from that for the air.

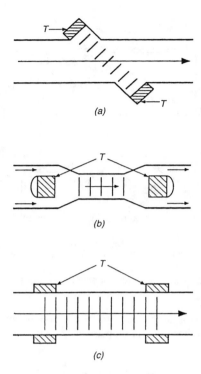

Figure 3.90 Ultrasonic respiratory flowmeters: **(a)** diagonal beam configuration, **(b)** coaxial configuration, **(c)** configuration using cylindrical shell transducers; T: ultrasonic transducer.

3.4.1.5 Ultrasonic Gas Flowmeter

An ultrasonic technique has been applied to respiratory gas flow measurement similar to that in the ultrasonic blood flowmeters described in Section 3.2.2. The gas, however, does not contain any particle which scatters ultrasound effectively, and hence only the transit time or phase shift measurement can be applied, whereas the Doppler shift measurement for scattered ultrasound cannot be applied.

Different configurations, as shown in Figure 3.90, have been used in ultrasonic respiratory flowmeters: (1) the diagonal-beam configuration in which a transmitting and a receiving crystal are arranged at an angle to the flow, (2) the coaxial configuration,[396] and (3) a configuration using cylindrical shell transducers operating at hoop mode.[397]

In the diagonal-beam configuration, transit times of downstream and upstream sound waves, T_1 and T_2, are given as

$$T_1 = \frac{D}{c + U\cos\theta}, \quad T_2 = \frac{D}{c - U\cos\theta} \tag{3.124}$$

where c is sound velocity, D is the distance between two crystals, and θ is the beam angle with respect to the flow tube. In the coaxial or the cylindrical configurations, $\theta = 0$. Then, if $U \ll c$, flow velocity, U, can be determined from the transit time difference, ΔT, or the phase difference, $\Delta\phi$, of downstream and upstream sound waves as

$$U = \frac{c^2 \Delta T}{2D\cos\theta} = \frac{c^2 \Delta\phi}{2\omega D\cos\theta}. \tag{3.125}$$

In this expression, sound velocity, c, appears explicitly, and it varies according to gas composition and water vapor pressure. An estimate shows that a difference of about 8.9% is expected between determinations of gas flow velocities in air and in pure oxygen, and about 5.3% between dry air at 30°C

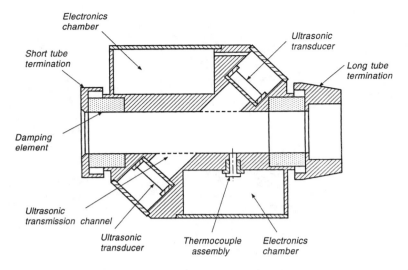

Figure 3.91   A diagonal beam ultrasonic flowmeter. (Modified from Bues, C. et al., *IEEE Trans. Biomed. Eng.*, BME-33, 768, 1986.)

and saturated air at 40°C.[396] The error due to the difference in sound velocity can be eliminated by taking the difference of reciprocals of downstream and upstream transit times, T_1 and T_2, as

$$\frac{1}{T_1} - \frac{1}{T_2} = \frac{2U\cos\theta}{D}. \tag{3.126}$$

Figure 3.91 shows the cross-section of a diagonal-beam ultrasonic flowmeter head.[398,399] This flowmeter head contains a respiratory tube of length 90 mm and diameter 20 mm, and a sound transmission channel with two capacitive ultrasound transducers. Short ultrasonic pulse trains are transmitted downstream and upstream simultaneously at a 500-Hz rate. Its measurement range was 0 to 9 l/s, it had a flat frequency response of up to 70 Hz, and volume accuracy with room air was ±0.7%. Figure 3.92 shows the exploded view of a flowmeter head using the cylindrical shell transducers.[400] The cylindrical shell transducer, made of lead zirconate titanate (PZT), has three resonant modes of vibration: hoop or radial mode, length mode, and wall-thickness mode. Using the hoop mode, ultrasound transmission parallel to the flow axis can be generated effectively. The operating frequency was about 47 kHz. In this configuration, the inner surface is flush with the wall of the flow tube, and flow disruption and fluid accumulation are prevented.

3.4.1.6   Vortex Flowmeter

The vortex flowmeter is a kind of flowmeter which utilizes the dependency of the flow rate in a channel upon the frequency of vortex shedding in the stream. Two different types, the Kármán vortex flowmeter and the swirlmeter, are used in respiratory flow measurements.

The Kármán vortex flowmeter is based on the principle that vortex streets are generated behind a rod placed in the stream perpendicular to the flow, as shown in Figure 3.93, and the number of vortices generated in unit time is proportional to the flow velocity. The number of vortices generated in unit time is expressed as

$$f = S_t U/d \tag{3.127}$$

where S_t is a coefficient called the Strouhal number, U is the flow velocity, and d is the diameter of the tube. The Strouhal number is a dimensionless quantity and is about 0.2 when the Reynold's number remains in a range from roughly 10^3 to 10^5.[401]

The respiratory flowmeter shown in Figure 3.94 consists of two rods having a triangular cross-section with a vortex angle of 40° and a base length of 2 mm installed in a pipe 12 mm in diameter. The Kármán

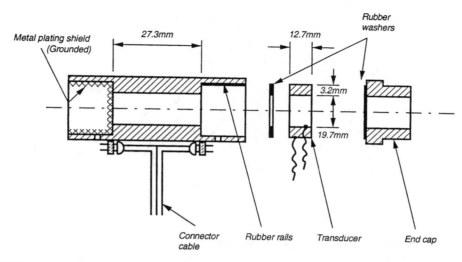

Figure 3.92 A flowmeter using cylindrical shell transducers (From Plaut, D. and Webster, J. G., *IEEE Trans. Biomed. Eng.,* BME-27, 590, 1980. With permission.)

Figure 3.93 Kármán vortex street generated behind a rod in the stream.

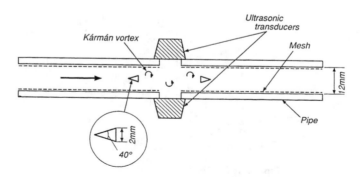

Figure 3.94 A Kármán vortex respiratory flowmeter.

vortices are detected by an ultrasonic transducer. It was shown that the flowmeter was almost insensitive to gas composition and the temperature and humidity of the gas.[402] A simple electronic spirometer with the Kármán vortex flowmeter is available commercially (Nihon Kohden Co. Ltd.; Tokyo, Japan).

The swirlmeter consists of swirl blades installed in a pipe (Figure 3.95). When the gas passes through the blades, it spins and forms vortices. The vortices can be detected by a thermal sensor, and the gas flow rate can be determined by the number of vortices passing the sensor during a unit time interval. A simple spirometer using a swirlmeter is available (Chest Co.; Tokyo, Japan).

3.4.2 VOLUME-MEASURING SPIROMETERS

To examine the ventilatory function of the lung, different parameters of expiratory air flow have to be measured. Many parameters in standard pulmonary function tests have a dimension of volume such as

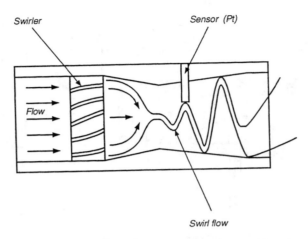

Figure 3.95   The swirlmeter. (Courtesy of Chest Co.; Tokyo, Japan).

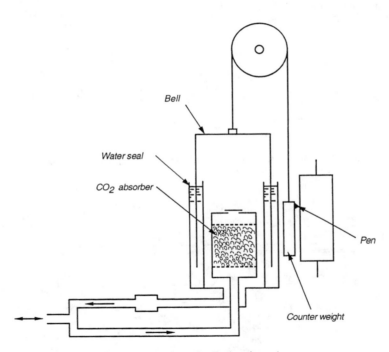

Figure 3.96   Benedict-Roth spirometer.

the vital capacity (VC), timed forced expiratory volumes (FEV_t), and maximal voluntary ventilation (MVV), and thus direct measurements of the air volumes have been performed using volume-measuring spirometers.

A typical volume-measuring spirometer is the water-seal spirometer, as shown in Figure 3.96. It has a cylinder with a dome at the top called a bell which is movable at the water-seal without friction. When the expired air is led into the bell, its volume can be measured by the elevation of the bell. To keep the pressure inside the bell at atmospheric pressure, the weight of the bell is balanced by a counter-weight. The original type of water-seal spirometer was developed in the 1920s and is called the Benedict-Roth spirometer.[403]

Using a carbon dioxide absorber in the closed respiratory circuit, respiratory gas volumes of several breaths can be measured. If the bell is initially filled with pure oxygen, its volume decreases according to the amount of oxygen consumed, and thus the rate of oxygen uptake can be estimated by the slope of the volume record.

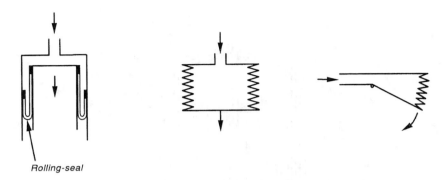

Rolling-seal

Figure 3.97 Various bellows for dry spirometers.

Instantaneous flow rate can be obtained from volume measurement by differentiating instantaneous volume signal or simply measuring the slope of the recorded volume on a chart. However, when the air flow changes greatly, measurement accuracy depends on the dynamic response of the volume measurement, which is mainly determined by the inertia of the moving parts in the spirometer. Thus, to improve dynamic response, the weight of the bell and other moving parts has to be reduced. Actually, by using a lightweight plastic bell weighing 175 g and attaching a pen to a plastic block projecting from the edge of the dome, a resonance frequency of 52 to 48 Hz and an overshoot of 2% at 4 Hz have been attained.[404,405] It was shown that this type of spirometer satisfies the recommendations of the American College of Chest Physicians Committee for frequency response, which requires that the measurement error in volume should be less than 5% at 6 Hz.[406]

So-called dry spirometers which do not employ water seals have also been used. In a dry spirometer, bellows are used (Figure 3.97) of the wedge type (a), straight bellows type (b), or rolling-seal type (c). Dry spirometers are preferred as lightweight portable instruments. Many comparative studies showed that the performances of well-designed dry spirometers were comparable to those of the water-seal spirometer.[407-411] According to the results of an evaluation of commercial spirometers, most volume-measuring spirometers, including both water-seal and dry, met ATS recommendations (see Section 3.1.2.3), whereas most flow-measuring spirometers experienced difficulties.[3]

In modern spirometers, parameters of the lung function are computed automatically and displayed immediately. To realize this, the displacement of the bell or bellows has to be detected electrically. Figure 3.98 shows an example of the spirometer in which the displacement of the bell is detected by a linear potentiometer.

3.4.3 LUNG PLETHYSMOGRAPHY

Lung plethysmography is a measurement technique to assess respiration by the volume change in the lung. The volume change is measured mechanically in body plethysmography and inductance plethysmography, and it is measured electrically in impedance pneumography.

3.4.3.1 Body Plethysmography

Body plethysmography is a technique of measuring volume change of the body by placing the subject in an airtight chamber. When the subject is allowed to breathe air from the outside of the chamber through a mouthpiece, the volume change of gas in the lung can be measured by the volume change of the body. The volume change of the body is measured either by a change of the air pressure in the chamber or by a change of the whole volume when the pressure in the chamber is maintained as constant.

As shown in Figure 3.99, different techniques have been used for body plethysmographic measurement. Figure 3.99(a) is a method in which pressure change due to the change of lung volume is detected by a sensitive pressure gauge.[412-415] When the volume of the air in the chamber is 600 l, the pressure change per 1-liter change in volume of the body is about 0.16 kPa (1.7 cmH$_2$O). Pressure changes of this order can be measured accurately enough by a capacitive-type pressure transducer. Compared with the method of volume measurement described later, the method of pressure measurement has advantages: it is easy to attain a fast response and simple to check the zero level by exposing the pressure transducer to the atmosphere. On the other hand, the change in chamber pressure causes force to be exerted on the

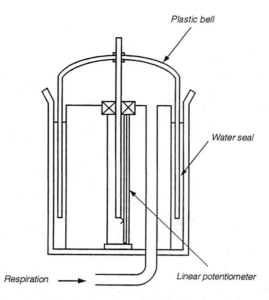

Figure 3.98 A spirometer in which displacement of the bell is detected by a linear potentiometer.

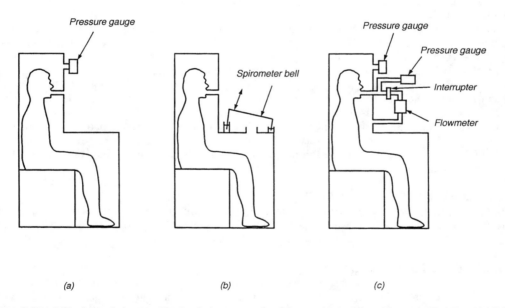

(a)   (b)   (c)

Figure 3.99 Different techniques of body plethysmography: **(a)** a method of detecting pressure change, **(b)** a method of detecting volume change, **(c)** a closed-circuit system measuring various parameters such as lung volume, alveolar pressure, lung compliance, and airway resistance.

wall of the chamber, and hence the chamber should be built rigidly enough so that it is not deformed by the pressure change in the chamber.

Figure 3.99(b) shows the method of detecting volume change without allowing pressure change in the chamber. Because the pressure in the chamber remains at atmospheric pressure, the requirement for mechanical strength in the chamber is not so great as for that of the pressure-measurement type. The volume of the chamber does not affect the characteristics of the measurement as it does in the pressure measurement type, and the volume of the chamber can be minimized. Additionally, no thermodynamic change of the state of the gas in the chamber accompanied by compression and decompression occurs in this method of volume measurement type.

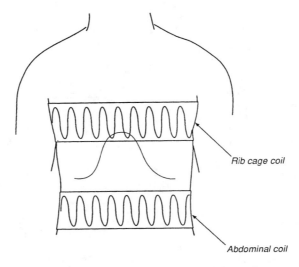

Figure 3.100 Inductance plethysmograhy. Two elastic bands containing coils are used by which changes in cross-sectional areas at the rib cage and the abdomen are measured.

Mead[415] employed a water-seal spirometer for volume measurement. Using a wedge-shaped bell of 36 × 46 cm, a resonance frequency of 16.5 Hz, damping constant of 0.52, and an almost flat response up to 12 Hz were attained. When a flow transducer was connected at the mouthpiece and measured the flow simultaneously, the delay of the volume measurement from the integration of the flow was less than 12 ms.

In order to measure various parameters of pulmonary function such as lung volume, alveolar pressure, lung compliance, and airway resistance, a body plethysmography as shown in Figure 3.99(c) can be used, in which the subject breathes through a mouthpiece with a flowmeter and an interrupter. When the subject is asked to make a respiratory effort against a closed airway, a pressure change at the mouth, ΔP, which corresponds to the alveolar pressure, and volume change of the lung, ΔV, can be measured. Then, according to Boyle's law, gas volume in the lung, V_L, can be obtained as

$$V_L = \frac{\Delta V}{\Delta P} P_0 \tag{3.128}$$

where P_0 should be the atmospheric pressure minus saturated water vapor pressure at body temperature, because water vapor is always saturated in the alveoli so that the water vapor pressure in the lung remains constant, regardless of compression or decompression of the air in the lung. Once the gas volume in the lung is determined at a moment, its change can be obtained from the air flow at the mouthpiece even when the airway is open, and then, applying Boyle's law, the alveolar pressure change can be determined. Consequently, lung compliance and airway resistance can be determined.[414]

3.4.3.2 Inductance Plethysmography

For continuous monitoring of ventilation, a vest which measures the changes in the thoracic and abdominal cross-sectional area (as shown in Figure 3.100) has been used.[416-418] The vest consists of two separate elastic bands, which are positioned at the rib cage and abdomen. Each band contains a coil in zig-zag fashion, and its inductance changes according to changes in the cross-sectional area.

It is suspected that the changes in these two inductances correspond to the volume changes of the rib cage and abdominal compartments, and the lung volume change, ΔV, can be represented as a sum of these two components by

$$\Delta V = K_1 \Delta R + K_2 \Delta A \tag{3.129}$$

where ΔR and ΔA are changes in outputs from the rib cage and abdominal coils, and K_1 and K_2 are coefficients representing sensitivities of the rib cage and abdominal outputs on the lung volume. These

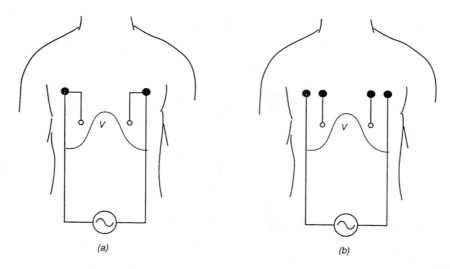

Figure 3.101 Typical electrode positions in impedance pneumography: **(a)** two-electrode system and **(b)** four-electrode system.

coefficients can be determined statistically when compared to simultaneously performed spirometric measurements. Relative contributions of rib cage and abdominal components on the lung volume can also be checked when the subject voluntarily shifts air volume from the rib cage to abdomen and vice versa with the mouth occluded. A study of 20 normal subjects showed that 93% of obtained values remain within ±10% of spirometric measurement values in supine and standing positions.[419]

To detect the change in a cross-sectional area at rib cage and abdomen, a long spring coil can also be used instead of the zig-zag fashioned coil.[420] In this method, inductance of the coil decreases when the cross-sectional area is increased, and a larger inductance change than that in the zig-zag fashioned coil can be obtained.

3.4.3.3 Impedance Pneumography

Impedance plethysmography has been used for measuring various physiological volume changes. Impedance pneumography is the term commonly used to describe impedance plethysmography for respiration measurement.

To obtain impedance changes due to respiration, electrodes are attached to the thorax. Usually, a.c. current with a frequency range of between 20 and 100 kHz and a current level of between 25 and 500 μA are applied to a pair of electrodes, and the potential is detected from that electrode pair or other electrodes, as shown in Figure 3.101. Different kinds of electrodes, such as the conventional ECG-monitoring electrodes and strip-shaped tape electrodes, have been used. Figure 3.102 shows an electrode structure for impedance pneumography, which has a strip-shaped electrode for applying current and a disc-shaped electrode for detecting signal.[421]

While the electric impedance across the thorax increases with increasing lung volume, the relation between impedance change, ΔZ, and lung volume change, ΔV, depends on many factors such as electrode position, size and shape of the body, and body fluid distribution in the thorax. Baker and Geddes[422] investigated $\Delta Z/\Delta V$ in humans and animals at different electrode positions for different sizes and shapes of the bodies. In the study, the two-electrode method was used by placing a pair of electrodes bilaterally along the midaxillary lines at different thoracic levels. As shown in Figure 3.103, the maximum $\Delta Z/\Delta V$ was obtained at the xiphoid level in ectomorphic human subjects and dogs, whereas the thoracic level for maximum $\Delta Z/\Delta V$ was not well defined in mesomorphic and endomorphic humans. There is a distinctive relation between the ratio $\Delta Z/\Delta V$ and the body size. According to Valentinuzzi et al.,[423] for the body weight, W, in kilograms, $\Delta Z/\Delta V$ in ohms per liter is given as

$$\Delta Z/\Delta V = 453.23 \; W^{-1.084}$$

$$(3.130)$$

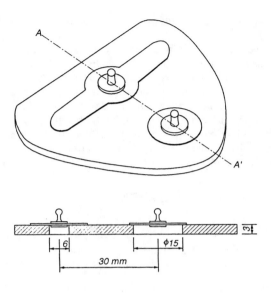

Figure 3.102 An example of the electrode for impedance pneumography. (From Itoh, A. et al., *Med. Biol. Eng. Comput.*, 20, 613, 1982. With permission.)

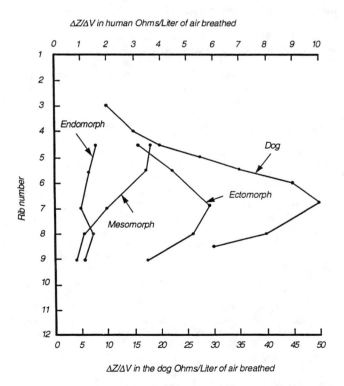

Figure 3.103 The ratio of impedance change, ΔZ, and lung volume change, ΔV, at different thoracic levels. (From Baker, L. E. and Geddes, L. A., *Ann. N. Y. Acad. Sci.*, 170, 667, 1970. With permission.)

and it was confirmed that this relation is valid in a wide range of body weight from mice of about 30 g to horses of about 500 kg.

The reliability of impedance pneumography has been confirmed by various researchers.[421,424-429] While the relation between impedance and lung volume is usually regarded as linear in normal respiratory conditions, it was pointed out that nonlinearity should be considered in accurate measurements. Baker et al.[427] introduced a fourth degree polynomial to describe the nonlinear relation. Using that approximation, lung volumes could be determined with a standard deviation of 0.0873 l in seven subjects, when the electrode pair was placed about 6 cm above the xiphoid level. Itoh et al.[430] developed an instrument in which all procedures, including calibration by a spirometer and computation of the lung volumes using the calibration curve, were automatized and showed that breath-by-breath tidal volumes could be monitored with an accuracy of ±10%.

REFERENCES

1. Kamiya, A. and Togawa, T., Adaptive regulation of wall shear stress to flow change in the canine carotid artery, *Am. J. Physiol*, 239, H14, 1980.
2. Folkow, B. and Neil, E., *Circulation,* Oxford University Press, London, 1971, 3.
3. Gardner, R. M., Hankinson, J. L., and West, B. J., Evaluating commercially available spirometers, *Am. Rev. Respirat. Dis.,* 121, 73, 1980.
4. Weast, R.C., Ed., *CRC Handbook of Chemistry and Physics,* 1st student ed., 1988, F31.
5. Shercliff, J. A., Relation between the velocity profile and the sensitivity of electromagnetic flowmeters, *J. Appl. Physiol.,* 25, 817, 1954.
6. Shercliff, J. A., Experiments on the dependence of sensitivity on velocity profile in electromagnetic flowmeters, *J. Sci. Instrum.,* 32, 441, 1955.
7. Goldman, S. C., Marple, N. B., and Scolnik, W. L., Effects of flow profile on electromagnetic flowmeter accuracy, *J. Appl. Physiol.,* 18, 652, 1963.
8. Shercliff, J. A., The effect of nonuniform magnetic fields and variations of the velocity distribution on electromagnetic flowmeters, in *New Findings in Blood Flowmetry,* Cappelen, C. H. R., Ed., Universitetsforlaget, Oslo, 1968, 45.
9. Clark, D. M. and Wyatt, D. G., The effect of magnetic field inhomogeneity on flowmeter sensitivity, in *New Findings in Blood Flowmetry,* Cappelen, C. H. R., Ed., Universitetsforlaget, Oslo, 1968, 49.
10. Gessner, U., Effects of the vessel wall on electromagnetic flow measurement, *Biophys. J.,* 1, 627, 1961.
11. Ferguson, D. J. and Landahl, H. D., Magnetic meters: effects of electrical resistance in tissue on flow measurements, and an improved calibration for square-wave circuits, *Circ. Res.,* 14, 917, 1966.
12. Edgerton, R. H., The effect of arterial wall thickness and conductivity on electromagnetic flowmeter readings, *Med. Biol. Eng.,* 6, 627, 1968.
13. Wyatt, D. G., Dependence of electromagnetic flowmeter sensitivity upon encircled media, *Phys. Med. Biol.,* 13, 529, 1968.
14. Wyatt, D. G., The electromagnetic blood flowmeter, *J. Sci. Instr.,* 1, 1146, 1968.
15. Wyatt, D. G., Blood flow and blood velocity measurement *in vivo* by electromagnetic induction., *Med. Biol. Eng. Comput.,* 22, 193, 1984.
16. Kolin, A., Circulatory system: methods, blood flow determination by electromagnetic method, in *Medical Physica,* Vol. 3, Glasser, O., Ed., Year Book Publlishing, Chicago, 1960, 141.
17. Westersten, A., Herrold, G., Abbott, E., and Assali, N. S., Gated sine-wave electromagnetic flowmeter, *IRE Trans. Med. Electron.,* 6, 213, 1959.
18. Westersten, A., Herrold, G., and Assali, N. S., A gated sine wave flowmeter, *J. Appl. Physiol.,* 15, 533, 1960.
19. Denison, Jr., A. B., Spencer, M. P., and Green, H. D., A square wave electromagnetic flowmeter for application to intact blood vessels, *Circ. Res.,* 3, 39, 1955.
20. Shirer, H. W., Shackelford, R. B., and Jochim, K. E., A magnetic flowmeter for recording cardiac output, *IRE Trans. Med. Electron.,* 6, 232, 1959.
21. Spencer, M. P. and Denison, Jr., A. B., The square-wave electromagnetic flowmeter: theory of operation and design of magnetic probes for clinical and experimental applications, *IRE Trans. Med. Electron.,* 6, 220, 1959.
22. Goodman, A. H., A transistorized squarewave electromagnetic flowmeter. I. The amplifier system, *Med. Biol. Eng.,* 7, 115, 1969.
23. Goodman, A. H., A transistorized squarewave electromagnetic flowmeter. II. The flow transducer, *Med. Biol. Eng.,* 7, 133, 1969.
24. Yanof, H. M., A trapezoidal-wave electromagnetic blood flowmeter, *J. Appl. Physiol.,* 16, 566, 1961.
25. Yanof, H. M., Salz, P., and Rosen, A. L., Improvements in trapezoidal-wave electromagnetic flowmeter, *J. Appl. Phys.,* 18, 230, 1963.
26. Dension, Jr., A. B. and Spencer, M. P., Circulatory system: methods, magnetic flowmeters, in *Medical Physica,* Vol. 3, Glasser, O., Ed., Year Book Publishing, Chicago, 1960, 178.

27. Wyatt, D. G., The design of electromagnetic flowmeter heads, in *New Findings in Blood Flowmetry*, Cappelen, C. H. R., Ed., Universitetsforlaget, Oslo, 1968, 69.

28. Cox, P., Arora, H., and Kolin, A., Electromagnetic determination of carotid blood flow in the anesthetized rat, *IRE Trans. Bio-Med. Electron.*, 10, 171, 1963.

29. Kolin, A. and Vanyo, J., New design of miniature electromagnetic blood flow transducers suitable for semi-automatic fabrication, *Cardiovasc. Res.*, 1, 274, 1967.

30. Wyatt, D. G., Baseline errors in cuff electromagnetic flowmeters, *Med. Biol. Eng.*, 4, 17, 1966.

31. Wyatt, D. G., Noise in electromagnetic flowmeters, *Med. Biol. Eng.*, 4, 333, 1966.

32. Hognestad, H., Some problems in square-wave electromagnetic flowmeter system design, in *New Findings in Blood Flowmetry*, Cappelen, C. H. R., Ed., Universitetsforlaget, Oslo, 1968, 55.

33. Woodcock, J.P., *Theory and Practice of Blood Flow Measurement*, Butterworths, London, 1975.

34. Dobson, A., Sellers, A. F., and McLeod, F. D., Performance of a cuff-type blood flowmeter *in vivo*, *J. Appl. Physiol.*, 21, 1642, 1966.

35. Folts, J. D. and Rowe, G. G., A nonerosive electromagnetic flowmeter probe for chronic aortic implantation, *J. Appl. Physiol.*, 31, 782, 1971.

36. Folts, J. D. and Rowe, G. G., Silicone rubber encapsulated flow probes for chronic implantation on the ascending aorta, in *Chronically Implanted Cardiovascular Instrumentation*, McCutcheon, E. P., Ed., Academic Press, New York, 1973, 35.

37. Williams, B. T., Barefoot, C., and Schenk, Jr., W. G., A removable electromagnetic flow probe: preliminary report, *Rev. Surg.*, 26, 227, 1969.

38. Williams, B. T., Sancho-Fornos, S., Clarke, D. B., Abrams, L. D., and Schenk, Jr., W. G., Continuous, long-term measurement of cardiac output after open-heart surgery, *Ann. Surg.*, 174, 357, 1971.

39. Mills, C. J., A catheter tip electromagnetic velocity probe, *Phys. Med. Biol.*, 11, 323, 1966.

40. Mills, C. J. and Shillingford, J. P., A catheter tip electromagnetic velocity probe and its evaluation, *Cardiovasc. Res.*, 1, 263, 1967.

41. Bevir, M. K., Sensitivity of electromagnetic velocity probes, *Phys. Med. Biol.*, 16, 229, 1971.

42. Uther, J. B., Peterson, K. L., Shabetai, R., and Braunwald, E., Measurement of ascending aortic flow patterns in man, *J. Appl. Physiol.*, 34, 513, 1973.

43. Gabe, J. T., Gault, J. H., Ross, Jr., J., Mason, D. T., Mills, C. J., Schillingford, J. P., and Braunwald, E., Measurement of instantaneous blood flow velocity and pressure in conscious man with a catheter-tip velocity probe, *Circulation*, 40, 603, 1969.

44. Jones, M. A. S. and Wyatt, D. G., The surface temperature of electromagnetic velocity probes, *Cardiovasc. Res.*, 4, 388, 1970.

45. Buchanan, Jr., J. W. and Shabetai, R., True power dissipation of catheter tip velocity probes, *Cardiovasc. Res.*, 6, 211, 1972.

46. Warbasse, J. R., Hellman, B. H., Gillilan, R. E., Hawley, R. R., and Babitt, H. I., Physiologic evaluation of a catheter tip electromagnetic velocity probe. A new instrument, *Am. J. Cardiol.*, 23, 424, 1969.

47. Stein, P. D. and Schuette, W. H., New catheter-tip flowmeter with velocity flow and volume flow capabilities, *J. Appl. Physiol.*, 26, 851, 1969.

48. Kolin, A., Archer, J. D., and Ross, G., An electromagnetic catheter-flowmeter, *Circ. Res.*, 21, 889, 1967.

49. Kolin, A., Ross, G., Grollman, Jr., J. H., and Archer, J., An electromagnetic catheter flowmeter for determination of blood flow in major arteries, *Proc. Nat. Acad. Sci.*, 59, 808, 1968.

50. Kolin, A., A radial field electromagnetic intravascular flow sensor, *IEEE Trans. Biomed. Eng.*, BME-16, 220, 1969.

51. Kolin, A., An electromagnetic catheter blood flowmeter of minimal lateral dimensions, *Proc. Nat. Acad. Sci.*, 66, 53, 1970.

52. Biscar, J. P., Three-electrode probe for catheter-type blood flowmeters, *IEEE Trans. Biomed. Eng.*, BME-20, 62, 1973.

53. Togawa, T., Okai, O., and Oshima, M., Observation of blood flow EMF in externally applied strong magnetic field by surface electrodes, *Med. Biol. Eng.*, 5, 169, 1967.

54. Okai, O., Togawa, T., and Oshima, M., Magnetorheography: observation of blood flow EMF in static magnetic field by surface electrodes, in *Dig. 7th Int. Conf. Med. Biol. Eng. Stockholm*, 1967, 212.

55. Okai, O., Togawa, T., and Oshima, M., Magnetorheography: nonbleeding measurement of blood flow in man, *Jpn. Heart J.*, 15, 469, 1974.

56. Okai, O., Togawa, T., and Oshima, M., Magnetorheography: nonbleeding measurement of blood flow, *J. Appl. Physiol.*, 30, 564, 1971.

57. Lee, B. Y., Trainor, F. S., Kavner, D., and Madden, J. L., A clinical evaluation of a noninvasive electromagnetic flowmeter, *Angiology*, 26, 317, 1973.

58. Boccalon, H., Candelon, B., Tillie, I. J., Graulle, A., Doll, H.G., Puel, P., and Enjalbert, A., New noninvasive device for pulsatile blood flow measurement, in *Dig. 11th Int. Conf. Med. Biol. Eng. Ottawa*, 1976, 428.

59. Boccalon, H., Candelon, B., Doll, H. G., Puel, P. F., and Enjalbert, A. P., Noninvasive electromagnetic pulsatile blood flow: experimental study and clinical applications, *Cardiovasc. Res.*, 12, 66, 1978.

60. Boccalon, H., Lozes, A., Newman, W., and Doll, H. G., Noninvasive electromagnetic blood flowmeter: theoretical aspects and technical evaluation, *Med. Biol. Eng. Comput.*, 20, 671, 1982.

61. Salles-Cunha, S. X., Battocletti, J. H., and Sances, Jr., A., Steady magnetic fields in noninvasive electromagnetic flowmetry, *Proc. IEEE,* 68, 149, 1980.
62. Kanai, H., Yamano, E., Nakayama, K., Kawamura, N., and Furuhata, H., Transcutaneous blood flow measurement by electromagnetic induction, *IEEE Trans. Biomed. Eng.,* BME-21, 144, 1974.
63. Atkinson, P. and Woodcock, J. P., *Doppler Ultrasound and its Use in Clinical Measurement,* Academic Press, London, 1982, 22.
64. Goldstein, A., Ultrasonic imaging, in *Encyclopedia of Medical Devices and Instrumentations,* Webster J. G., Ed., John Wiley & Sons, New York, 1988, 2803.
65. Henning, E. M., Piezoelectric sensors, in *Encyclopedia of Medical Devices and Instrumentations,* Webster, J. G., Ed., John Wiley & Sons, New York, 1988, 2310.
66. Gessner, U., The performance of the ultrasonic flowmeter in complex velocity profiles, *IEEE Trans. Biomed. Eng.,* BME-16, 139, 1969.
67. Franklin, D. L., Baker, D. W., Ellis, R. M., and Rushmer, R. F., A pulsed ultrasonic flowmeter, *IRE Trans. Med. Electron.,* 6, 204, 1959.
68. Franklin, D. L., Baker, D. W., and Rushmer, R. F., Pulsed ultrasonic transit time flowmeter, *IRE Trans. Biomed Eng.,* BME-9, 44, 1962.
69. Franklin, D. L., Techniques for measurement of blood flow through intact vessels, *Med. Electron. Biol. Eng.,* 3, 27, 1965.
70. Zarnstorff, W. C., Castillo, C. A, and Crumpton, C. W., A phase-shift ultrasonic flowmeter, *IRE Trans. Biomed. Eng.,* BME-9, 199, 1962.
71. Noble, F. W., Dual frequency ultrasonic fluid flowmeter, *Rev. Sci. Instr.,* 39, 1327, 1968.
72. Plass, K. G., A new ultrasonic flowmeter for intravascular application, *IEEE Trans. Biomed. Eng.,* BME-11, 154, 1964.
73. Rader, R. D., Meehan, J. P., and Henriksen, J. K. C., An implantable blood pressure and flow transmitter, *IEEE Trans. Biomed. Eng.,* BME-20, 37, 1973.
74. Drost, C. J., Vessel diameter-independent volume flow measurements using ultrasound, in *Proc. San Diego Biomed. Symp.,* 17, San Diego, CA, 1978, 299.
75. Burton, R. G. and Gorewit, R. C., Ultrasonic flowmeter uses wide-beam transit-time technique, *Med. Electron.,* 15, 68, 1984.
76. Eisenmann, J. H., Huntington, G. B., and Ferrell, C. L., Blood flow to hindquarters of steers measured by transit time ultrasound and indicator dilution, *J. Dairy Sci.,* 70, 1385, 1987.
77. Shung, K.K., Sigelmann, R. A., and Reid, J. M., Scattering of ultrasound by blood, *IEEE Trans. Biomed. Eng.,* BME-23, 460, 1976.
78. Satomura, S., Study of the flow patterns in peripheral arteries by ultrasonics, *J. Acoust. Soc. Jpn.,* 15, 151, 1959.
79. Franklin, D. L., Schlegel, W., and Rushmer, R. F., Blood flow measured by Doppler frequency shift of back-scattered ultrasound, *Science,* 134, 564, 1961.
80. Kato, K., Motomiya, M., Izumi, T., Kaneko, Z., Shiraishi, J., Omizo, H., and Nakao, S., Linearity of readings on ultrasonic flowmeter, in *Dig. 6th Int. Conf. Med. Biol. Eng. Tokyo,* 1965, 284.
81. Rushmer, R. F., Baker, D. W., and Stegall, H. F., Transcutaneous Doppler flow detection as a nondestructive technique, *J. Appl. Physiol.,* 21, 554, 1966.
82. Stegall, H. F., Rushner, R. F., and Baker, D. W., A transcutaneous ultrasonic blood-velocity meter, *J. Appl. Physiol.,* 21, 707, 1966.
83. Yao, S. T., Needham, T. N., and Ashton, J. P., Transcutaneous measurement of blood flow by ultrasound, *Biomed. Eng.,* 5, 230, 1970
84. Gross, G. and Light, L. H., Direction-resolving Doppler instrument with improved rejection of tissue artifacts for transcutaneous aortovelography, *J. Physiol.,* 217, 5P, 1971.
85. Light, L. H., Direction-resolving Doppler system and real-time analogue spectral recorder for transcutaneous aortovelography, in *Dig. 9th Int. Conf. Med. Biol. Eng. Melbourne,* 1971, 228.
86. Hatterland, K. and Eriksen, M., A heterodyne ultrasound blood velocity meter, *Med. Biol. Eng. Comput.,* 19, 91, 1981.
87. Mackay, R. S., Noninvasive cardiac output measurement, *Microvasc. Res.,* 4, 438, 1972.
88. Arts, M. G. J. and Roevros, J. M. J. G., On the instantaneous measurement of bloodflow by ultrasonic means, *Med. Biol. Eng.,* 10, 23, 1972.
89. Rice, S. O., Mathematical analysis of random noise, *Bell System Tech. J.,* 23/24, 1, 1944–45.
90. Flax, S. W., Webster, J. G., and Updike, S. J., Statistical evaluation of the Doppler ultrasonic blood flowmeter, *Biomed. Sci. Instr.,* 7, 201, 1970.
91. Light, L. H., Non-injurious ultrasonic technique for observing flow in the human aorta, *Nature,* 224, 1119, 1969.
92. Pedersen, J. E., Fast dedicated microprocessor for real-time frequency analysis of ultrasonic blood-velocity measurements, *Med. Biol. Eng. Comput.,* 20, 681, 1982.
93. Cote, G. L. and Fox, M. D., Comparison of zero-crossing counter to FFT spectrum of ultrasound Doppler, *IEEE Trans. Biomed. Eng.,* 35, 498, 1988.
94. Wells, P. N. T., A range-gated ultrasonic Doppler system, *Med. Biol. Eng.,* 7, 641, 1969.
95. Baker, D. W., Pulsed ultrasonic Doppler blood-flow sensing, *IEEE Trans. Sonics Ultrasonics,* SU-17, 170, 1970.
96. Jethwa, C. P., Kaveh, M., Cooper, G. R., and Saggio, F., Blood flow measurements using ultrasonic pulsed random signal Doppler system, *IEEE Trans. Sonics Ultrasonics,* SU-22, 1, 1975.

97. Kedem, B., Spectral analysis and discrimination by zero-crossings, *Proc. IEEE,* 74, 1477, 1986

98. Baker, M. and Wayland, H., One-line volume flow rate and velocity profile measurement for blood in microvessels, *Microvasc. Res.,* 7, 131, 1974.

99. Evans, D. H., Can ultrasonic duplex scanners really measure volumetric blood flow?, in *Physics in Medical Ultrasound,* Evans, J. A., Ed., The Institute of Physical Sciences in Medicine, London, 1986, 145.

100. Namekawa, K., Kasai, C., Tsukamoto, M., and Koyano, A., Imaging of blood flow using auto-correlation, *Ultrasound Med. Biol.,* 8, 138, 1982.

101. Di Pietro, D. M., and Meindl, J. D., Optimal system design for an implantable CW Doppler ultrasonic flowmeter, *IEEE Trans. Biomed. Eng.,* BME-25, 255, 1978.

102. Carter, J., Reynoldson, J. A., Thorburn, G. D., and Bates, W. A., Blood flow measurement during exercise in sheep using Doppler ultrasonic method, *Med. Biol. Eng. Comput.,* 19, 373, 1981.

103. Allen, H. V., Anderson, M. F., and Meindl, J. D., Direct calibration of a totally implantable pulsed Doppler ultrasonic blood flowmeter, *Am. J. Physiol.,* 232, H537, 1977.

104. Richardson, P. C. A., Stevens, A. L., Cowan, D., Calil S., and Roberts, V. C., Design of a continuous-wave Doppler ultrasonic flowmeter for perivascular application. Part 1. Probe design, *Med. Biol. Eng. Comput.,* 25, 661, 1987.

105. Cowan, D., Stevens, A. L., and Roberts, V. C., Design of a continuous-wave Doppler ultrasonic flowmeter for perivascular application. Part 2. Signal processing system, *Med. Biol. Eng. Comput.,* 26, 153, 1988.

106. Benchimol, A., Stegall, H. F., Maroko, P. R., Gartlan, J. L., and Brener, L., Aortic flow velocity in man during cardiac arrhythmias measured with the Doppler catheter-flowmeter system, *Am. Heart J.,* 78, 649, 1969.

107. Cole, J. S. and Hartley, C. J., The pulsed Doppler coronary artery catheter. Preliminary report of a new technique for measuring rapid changes in coronary artery flow velocity in man, *Circulation,* 56, 18, 1977.

108. Sibley, D. H., Millar, H. D., Hartley, C. J., and Whitlow, P. L., Subselective measurement of coronary blood flow velocity using a steerable Doppler catheter, *J. Am. Coll. Cardiol.,* 8, 1332, 1986.

109. Meier, P. and Zierler, K. L., On the theory of the indicator-dilution method for measurement of blood flow and volume, *J. Appl. Physiol.,* 6, 731, 1954.

110. Hamilton, W. F., Moore, J. W., Kinsman, J. M., and Spurling, R. G., Studies on the circulation. IV. Further analysis of the injection method, and of changes in hemodynamics under physiological and pathological conditions, *Am. J. Physiol.,* 99, 534, 1932.

111. Hetzel, P. S., Swan, H. J. S., Ramirez de Arellano, A. A., and Wood, E. H., Estimation of cardiac output from first part of arterial dye-dilution curves, *J. Appl. Physiol.,* 13, 92, 1958.

112. Kamiya, A., Togawa, T., and Oshima, M., Simple formulae for calculation of cardiac output, *Jpn. J. MEBE,* 6, 224, 1968.

113. Matsunaga, K., Imamura, N., and Ueda, M., Cardiac output by dye dilution using a left carotid loop in conscious rats, *J. Pharmacol. Meth.,* 4, 1, 1980.

114. Fox, I. J., Indicators and detectors for circulatory dilution studies and their application to organ or regional blood-flow determination, *Circ. Res.,* 10, 447, 1962.

115. Gilford, S. R., Gregg, D. E., Shadle, O. W., Ferguson, T. B., and Marzetta, L. A., An improved cuvette densitometer for cardiac output determination by the dye-dilution method, *Rev. Sci. Instrum.,* 24, 696, 1953.

116. Reed, J. H., Jr., and Wood, E. H., Use of dichromatic earpiece densitometry for determination of cardiac output., *J. Appl. Physiol.,* 23, 373, 1967.

117. McCarthy, B., Hood, Jr., W. B., and Lown, B., Fiberoptic monitoring of cardiac output and hepatic dye clearance in dogs, *J. Appl. Physiol.,* 23, 641, 1967.

118. Hugenholtz, P. G., Wanger, H. R., Gamble, W. J., and Polonyi, M. L., Direct read-out of cardiac output by means of the fiberoptic indicator dilution method, *Am. Heart J.,* 77, 178, 1969.

119. Volz, R. J. and Christensen, D. A., A neonatal fiberoptic probe for oximetry and dye curves, *IEEE Trans. Biomed. Eng.,* BME-26, 416, 1979.

120. Hosie, K. F., Thermal-dilution technics, *Circ. Res.,* 10, 491, 1962.

121. Ganz, W., Donoso, R., Marcus, H. S., Forrester, J. S., and Swan, H. J. C., A new technique for measurement of cardiac output by thermodilution in man, *Am. J. Cardiol.,* 27, 392, 1971.

122. Ganz, W. and Swan, H. J. C., Measurement of blood flow by thermodilution, *Am. J. Cardiol.,* 29, 241, 1972.

123. Swan, H. J. C., Ganz, W., Forrester, J. S., Marcus, H., Diamond, G., and Chonette, D., Catheterization of the heart in man with use of a flow-directed balloon-tipped catheter, *New Engl. J. Med.,* 283, 447, 1970.

124. Fronek, A. and Ganz, V., Measurement of flow in single blood vessels including cardiac output by local thermodilution, *Circ. Res.,* 8, 175, 1960.

125. Clark, C., A local thermal dilution flowmeter for the measurement of venous blood flow in man, *Med. Biol. Eng.,* 6, 133, 1968.

126. Goodyear, A. V. N., Huvos, A., Eckhardt, W. F., and Ostberg, R. H., Thermal dilution curves in the intact animal, *Circ. Res.,* 7, 432, 1959.

127. Branthwaite, M. A. and Bradley, R. D., Measurement of cardiac output by thermal dilution in man, *J. Appl. Physiol.,* 24, 434, 1968.

128. Forrester, J. S., Ganz, W., Diamand, G., McHugh, T., Chonette, D. W., and Swan, H. J. C., Thermodilution cardiac output determination with a single flow-directed catheter, *Am. Heart J.,* 83, 306, 1972.

129. Ganz, W., Tamura, K., Marcus, H. S., Donoso, R., Yoshida, S., and Swan, H. J. C., Measurement of coronary sinus blood flow by continuous thermodilution in man, *Circulation,* 44, 181, 1971.

130. Philip, J. H., Long, M. C., Quinn, M. D., and Newbower, R. S., Continuous thermal measurement of cardiac output, *IEEE Trans. Biomed. Eng.,* BME-31, 393, 1984.

131. Guyton, A. C., Measurement of cardiac output by the direct Fick method, in *Circulatory Physiology: Cardiac Output and its Regulation,* W.B. Sanders, Philadelphia, 1963, 21.

132. Roughton, F. J. W., Transport of oxygen and carbon dioxide, in *Handbook of Physiology,* Vol. 1, Fenn, W. O. and Rahn, H., Eds., American Physiology Society, Washington, D.C., 1964, 767.

133. Guyton, A. C., A continuous cardiac output recorder employing the Fick principle, *Circ. Res.,* 7, 661, 1959.

134. Guyton, A. C., and Farish, C. A., A rapidly responding continuous oxygen consumption recorder, *J. Appl. Physiol.,* 14, 143, 1959.

135. Guyton, A. C., Farish, C. A., and Williams, J. W., An improved arteriovenous oxygen difference recorder, *J. Appl. Physiol.,* 14, 145, 1959.

136. Dubois, A. B., Britt, A. B., and Fenn, W. O., Alvelar CO_2 during the respiratory cycle, *J. Appl. Physiol.,* 4, 535, 1952.

137. Collier, C. R., Determination of mixed venous CO_2 tensions by rebreathing, *J. Appl. Physiol.,* 9, 25, 1956.

138. Defares, J. G., Determination of $P\bar{V}CO_2$ from the exponential CO_2 rise during rebreathing, *J. Appl. Physiol.,* 13, 159, 1958.

139. Campbell, E. J. M. and Howell,J. B. L., Rebreathing method for measurement of mixed venous PCO_2, *Br. Med. J.,* Sept. 8, 630, 1962.

140. Farhi, L. E., Nesarajah, M. S., Olszowka, A. J., Metildi, L. A., and Ellis, A. K., Cardiac output determination by simple one-step rebreathing technique, *Resp. Physiol.,* 28, 141, 1976.

141. Knowles, J. H., Newman, W., and Fenn, W. O., Determination of oxygenated, mixed venous blood CO_2 tension by a breath-holding method, *J. Appl. Physiol.,* 15, 225, 1960.

142. Leavell, K., Finkelstein, S. M., Warwick, W. J., and Budd, J. R., Automated noninvasive determination of mixed venous PCO_2, *Med. Instrum.,* 20, 248, 1986.

143. Gedeon, G., Forslund, L., Hedenstierna, G., and Romano, E., A new method for noninvasive bedside determination of pulmonary blood flow, *Med. Biol. Eng. Comput.,* 18, 411, 1980.

144. Asmussen, E. and Nielsen. M., The cardiac output in rest and work determined simultaneously by the acetylene and the dye injection methods, *Acta Physiol. Scand.,* 27, 217, 1952.

145. Becklake, M. R., Varvis, C. J., Pengelly, L. D., Kenning, S., McGregor, M., and Bates, D. V., Measurement of pulmonary blood flow during exercise using nitrous oxide, *J. Appl. Physiol.,* 17, 579, 1962.

146. Sackner, M. A., Greeneltch, D., Heiman, M. S., Epstein, S., and Atkins, N., Diffusing capacity, membrane diffusing capacity, capillary blood volume, pulmonary tissue volume, and cardiac output measured by a rebreathing technique, *Am. Rev. Resp. Dis.,* 111, 157, 1975.

147. Huff, R. L., Feller, D. D., Judo, O. J., and Bogardus, G. M., Cardiac output of men and dogs measured by *in vivo* analysis of iodinated (I^{131}) human serum albumin, *Circ. Res.,* 3, 564, 1955.

148. Seldon, W. A., Hickie, J. B., and George, E. P., Measurement of cardiac output using a radioisotope and a scintillation counter, *Br. Heart J.,* 21, 401, 1958.

149. Schreiner, Jr., B. F., Lovejoy, Jr., F. W., and Yu, P. N., Estimation of cardiac output from precordial dilution curves in patients with cardiopulmonary disease, *Circ. Res.,* 7, 595, 1959.

150. Gorten, R. J. and Stauffer, J. C., A study of the techniques and sources of error in the clinical application of the external counting method of estimating cardiac output, *Am. J. Med. Sci.,* 238, 274, 1959.

151. Austin, W. H., Poppell,J. W., and Baliff, R. J., Cardiac output measurement by external counting of a rapidly excreted indicator, *Am. J. Med. Sci.,* 242, 457, 1961.

152. Bourdillon, P. J., Becket, J. M., and Duffin, P., Saline conductivity method for measuring cardiac output simplified, *Med. Biol. Eng. Comput.,* 17, 323, 1979.

153. Trautman, E. D. and Newbower, R. S., The development of indicator-dilution techniques, *IEEE Trans. Biomed. Eng.,* BME-31, 800, 1984.

154. Voorhees, III, W. D., Bourland, J. D., Lamp, M. L., Mullikin, J. C., and Geddes, L. A., Validation of the saline-dilution method for measuring cardiac output by simultaneous measurement with a perivascular electromagnetic flowprobe, *Med. Instrum.,* 19, 34, 1985.

155. Grubbs, D. S., Worley, D. S., and Geddes, L. A., A new technique for obtaining values of cardiac output in rapid succession, *IEEE Trans. Biomed. Eng.,* BME-29, 769, 1982.

156. Smith, M., Geddes, L. A., and Hoff, H. E., Cardiac output determined by the saline conductivity method using an extraarterial conductivity cell, *Cardiovasc. Res. Cent. Bull.,* 5, 123, 1967.

157. Goodwin, R. S. and Sapirstein, L. A., Measurement of the cardiac output in dogs by a conductivity method after single intravenous injections of autogenous plasma, *Circ. Res.,* 5, 531, 1957.

158. King., L. V., On the convection of heat from small cylinders in a stream of fluid: determination of the convection constants of small platinum wire with application to hot-wire anemometry, *Phil. Trans.,* A.214, 373, 1914.

159. Katsura, S., Weiss, R., Baker, D., and Rushmer, R. F., Isothermal blood flow velocity probe, *IRE Trans. Med. Electronics,* 6, 283, 1959.

160. Mellander, S. and Rushmer, R. F., Venous blood flow recorded with an isothermal flowmeter, *Acta Physiol. Scand.,* 48, 13, 1960.

161. Clark, C., Thin film gauges for fluctuating velocity measurements in blood, *J. Phys. E. Sci. Instrum.*, 7, 548, 1974.
162. Paulsen, P. K., Hasenkam, J. M., Nygaard, H., and Gormsen, J., Analysis of the dynamic properties of a hot-film anemometer system for blood velocity measurements in humans, *Med. Biol. Eng. Comput.*, 25, 195, 1987.
163. Ling, S. C., Atobek, H. B., Fry, D. L., Patel, D. J., and Jamicki, J. S., Application of heated-film velocity and shear probes to hemodynamic studies, *Circ. Res.*, 23, 789, 1968.
164. Nerem, R. M., Rumberger, Jr., J. A., Gross, D. R., Hamlin, R. L., and Geiger, G. L., Hot-film anemometer velocity measurements of arterial blood flow in horses, *Circ. Res.*, 34, 193, 1974.
165. Seed, W. A. and Wood, N. B., Development and evaluation of a hot-film velocity probe for cardiovascular studies, *Cardiovasc. Res.*, 4, 253, 1970.
166. Stein, P. D. and Sabbah, H. N., Turbulent blood flow in the ascending aorta of humans with normal and diseased aortic valves, *Circ. Res.*, 39, 58, 1976.
167. Yamaguchi, T., Kikkawa, S., Yoshikawa, T., Tanishita, K., and Sugawara, M., Measurement of turbulence intensity in the center of the canine ascending aorta with a hot-film anemometer, *J. Biomed. Eng.*, 105, 177, 1983.
168. Kubicek, W.G., Karnegis, J. N., Patterson, R., Witsoe, D. A., and Mattson, R. H., Development and evaluation of an impedance cardiac output system, *Aerospace Med.*, 37, 1208, 1966.
169. Mohapatra, S. N., Impedance cardiography, in *Encyclopedia of Medical Devices and Instrumentations,* Webster, J. G., Ed., John Wiley & Sons, New York, 1988, 1622.
170. Smith, I. J., Bush, J. E., Wiedmeier, V. T., and Tristani, F. E., Application of impedance cardiography to study of postural stress, *J. Appl. Physiol.*, 29, 133, 1970.
171. Nagger, C. Z., Dobnik, D. B., Hessas, A. P., Kripke, B. J., and Ryan, T. J., Accuracy of the stroke index as determined by the transthoracic electrical impedance method, *Anaesthesiology*, 42, 201, 1975.
172. Hill, D. W. and Thompson, F. D., The importance of blood resistivity in the measurement of cardiac output by the throracic impedance method, *Med. Biol. Eng.*, 13, 187, 1975.
173. Gabriel, S., Atterhög, J. H., Orö, L., and Ekelund, L.-G., Measurement of cardiac output by impedance cardiology in patients with myocardial infarction, *Scand. J. Clin. Lab. Invest.*, 36, 29, 1976.
174. Denniston, J. C., Maher, J. T., Reeves, J. T., Cruz, J. C., Cymerman, A., Grover, R.F., Measurement of cardiac output by electrical impedance at rest and during exercise, *J. Appl. Physiol.*, 40, 91, 1976.
175. Miyamoto, Y., Takahashi, M., Tamura, T., Nakakmura, T., Hiura, T., and Mikami, M., Continuous determination of cardiac output during exercise by the use of impedance plethysmography, *Med. Biol. Eng. Comput.*, 19, 638, 1981.
176. Takada, K., Fujinami, T., Okuda, N., Hokimoto, S., Ouhashi, N., Nakayama, K., Okamoto, M., and Okutani, H., Reliability and usefulness of impedance cardiography to measure cardiac response during exercise, in *Proc. 5th Int. Conf. Electrical Bio-Impedance*, Japanese Society of Medical, Electrical, and Biological Engineers, Tokyo, 1981, 157.
177. Perrino, Jr., A. C., Lippman, A., Ariyan, C., O'Connor, T. Z., and Luther, M., Intraoperative cardiac output monitoring: comparison of impedance cardiography and thermodilution, *J. Cardiothorac. Vasc. Anesthesiol.* 8, 24, 1994.
178. Judy, W. V., Largely, F. M., McCowen, K. D., Stinnett, D. M., Baker, L. E., and Lohnson, P. C., Comparative evaluation of the thoracic impedance and isotope dilution methods for measuring cardiac output, *Aerospace Med.*, 40, 532, 1969.
179. Keim, H. J., Wallace, J. M., Thurston, H., Case, D. B., Drayer, J. I. M., and Laragh, J. H., Impedance cardiography for determination of stroke index, *J. Appl. Physiol.*, 41, 797, 1976.
180. Donovan, K. D., Dobb, G. J., Woods, W. P., and Hockings, B. E., Comparison of transthoracic electrical impedance and thermodilution methods for measuring cardiac output, *Crit. Care Med.*, 14, 1038, 1986.
181. Patterson, R. P., Kubicek, W. G., Witsoe, D. A., and From, A. H. L., Studies on the effect of controlled volume change on the thoracic electrical impedance, *Med. Biol. Eng. Comput.*, 16, 531, 1978.
182. Sakamoto, K., Muto, K., Kanai, H., and Iizuka, M., Problems of impedance cardiography, *Med. Biol. Eng. Comput.*, 17, 697, 1979.
183. Patterson, R. P., Sources of the thoracic cardiogenic electrical impedance signal as determined by a model, *Med. Biol. Eng. Comput.*, 23, 411, 1985.
184. Sakamoto, K. and Kanai, H., Electrical characteristics of flowing blood, *IEEE Trans. Bio. Eng.*, BME-26, 686, 1979.
185. Patterson, R. P., Fundamentals of impedance cardiography, *IEEE Eng. Med. Biol. Mag.*, 8-1, 35, 1989.
186. Patterson, R. P., Possible technique to measure ventricular volume using electrical impedance measurements with an esophageal electrode, *Med. Biol. Eng. Comput.*, 25, 677, 1987.
187. Baan, J., Aouw Jong, T. T., Kerkhof, P. L. M., Moene, R. J., van Dijk, A. D., van der Velde, E. T., and Koops, J., Continuous stroke volume and cardiac output from intra-ventricular dimensions obtained with impedance catheter, *Cardiovasc. Res.*, 15, 328, 1981.
188. Valentinuzzi, M. E. and Spinelli, J. C., Intracardiac measurements with the impedance technique, *IEEE Eng. Med. Biol. Mag.*, 8-1, 27, 1989.
189. Stern, M. D., Laser Doppler velocimetry in blood and multiple scattering fluids: theory, *Appl. Optics*, 24, 1968, 1985.
190. Stern, M. D., Laser Doppler measurement of coronary blood flow velocity, *SPIE Optical Fibers in Med. II*, 713, 132, 1986.
191. Riva, C., Ross, B., and Benedek, G. B., Laser Doppler measurements of blood flow in capillary tubes and retinal arteries, *Invest. Ophthalmol.*, 11, 936, 1972.
192. Tanaka, T., Riva, C., and Ben-Sira, I., Velocity measurements in human retinal vessels, *Science*, 186, 830, 1974.

193. Hill, D. W., Young, S., Parker, P., and Pike, E. R., Photon correlation velocimetry of blood flow in the retina, *J. Quantum Electron.*, 13, 85D, 1977.

194. Okamoto, S., Shimizu, H., and Ozawa, T., Measurement of blood flow in the retina by differential laser Doppler method, *Jpn. J. Ophthalmol.*, 24, 128, 1980.

195. Mishina, H., Koyama, T., and Asakura, T., Velocity measurements of blood flow in the capillary and vein using a laser Doppler microscope, *Appl. Optics*, 14, 2326, 1975.

196. Mishina, M. and Asakura, T., Measurement of velocity fluctuations in laser Doppler microscope by the new system employing the time-to-pulse height converter, *Appl. Phys.*, 5, 351, 1975.

197. Koyama, T., Horimoto, M., Mishina, H., and Asakura, T., Measurement of blood flow velocity by means of a laser Doppler microscope, *Optic*, 61, 411, 1982.

198. Born, G. V. R., Melling, A., and Whitelaw, J. W., Laser Doppler microscope for blood velocity measurements, *Biorheology*, 15, 163, 1987.

199. Eniav, S., Berman, H. J., Fuhro, R. L., DiGiovanni, P. R., Fridman, J. D., and Fine, S., Measurement of blood flow *in vivo* by laser Doppler anemometry through a microscope, *Biorheology*, 12, 203, 1975.

200. Cochrane, T., Earnshaw, J. C., and Love A. H., Laser Doppler measurements of blood velocity in microvessels, *Med. Biol. Eng. Comput.*, 19, 589, 1981.

201. Okada, E., Fukuoka, Y., Umetani, J., Sekizuka, E., Oshio, V. C., and Minamitani, H., Spectrum analysis of fluctuations of RBC velocity in microvessels by using microscopic laser Doppler velocimeter, *IEEE Engineering in Medicine and Biology Society, 11th Int. Meeting*, Seattle, WA, 1989, 74.

202. Kilpatrick, D., Thomas Linderer, T., Sievers R. E., and Tyberg, J. A., Measurement of coronary sinus blood flow by fiberoptic laser Doppler anemometry, *Am. J. Physiol.*, 242, H1111, 1982.

203. Kilpatrick, D., Tyberg, J. V., and Parmley, W. W., Blood velocity measurement by fiberoptic laser Doppler anemometry, *IEEE Trans.*, BME-29, 142, 1982.

204. Tanaka, T. and Benedek, G. B., Measurement of the velocity of blood flow (*in vivo*) using a fiberoptic catheter and optical mixing spectroscopy, *Appl. Optics*, 14, 189, 1975.

205. Kajiya, F., Tomogawa, G., Tsujioka, K., Ogasawara, Y., and Nishihara. H., Evaluation of local blood flow velocity in proximal and distal coronary arteries by laser Doppler method, *Trans. ASME J. Biomech. Eng.*, 107, 10, 1985.

206. Kajiya, F., Progress and recent topics in blood flow measurements, *Frontiers Med. Biol. Eng.*, 1, 271, 1989.

207. Nishihara, H., Koyama, J., Hoki, N., Kajiya, F., Hironaga, M., and Kano, M., Optical-fiber laser Doppler velocimeter for high-resolution measurement of pulsatile blood flows, *Appl. Optics*, 21, 1785, 1982.

208. Asano, M., Yoshida, K., and Tatai, K., Observation of the behavior of microcirculation by rabbit ear chamber technique. I. On a modification of rabbit ear chamber, development of microphotoelectric plethysmography, and rhythmic fluctuation of microcirculation, *Bull. Inst. Publ. Health*, 12, 34, 1963.

209. Asano, M., Yoshida, K., and Tatai, K., Blood flow rate in the microcirculation as measured by photoelectric microscopy, *Bull. Inst. Publ. Health*, 13, 201, 1964.

210. Wayland, H. and Johnson, P. C., Erythrocyte velocity measurement in microvessels by a two-slit photometric method, *J. Appl. Physiol.*, 22, 333, 1967.

211. Intaglietta, M., Tompkins, W. R., and Richardson, D. R., Velocity measurements in the microvasculature of the cat omentum by one-line method, *Microvasc. Res.*, 2, 462, 1970.

212. Gaehtgens, P., Meiselman, H. J., and Wayland, H., Erythrocyte flow velocities in mesenteric microvessels of the cat, *Microvasc. Res.*, 2, 151, 1970.

213. Goodman, A. H., Guyton, A. C., Drake, R., and Loflin, J. H., A television method for measuring capillary red cell velocities, *J. Appl. Physiol.*, 37, 126, 1974.

214. Intaglietta, M., Silverman, N. R., and Tompkins, W. R., Capillary flow velocity measurements *in vivo* and *in situ* by television methods, *Microvasc. Res.*, 10, 165, 1975.

215. Fu, S. E. and Lee, J. S., A video system for measuring the blood flow velocity in microvessels, *IEEE Trans. Biomed. Eng.*, BME-25, 295, 1978.

216. Tyml, K. and Sherebrin, M. H., A method for one-line measurements of red cell velocity in microvessels using computerized frame-by-frame analysis of television images, *Microvasc. Res.*, 20, 1, 1980.

217. Slaaf, D. W., Rood, J. P. S. M., Tangleder, G. J., Jeurens, T. J. M., Alewijnse, R., Reneman, R. S., and Arts, T., A bidirectional optical (BDO) three-stage prism grating system for one-line measurement of red blood cell velocity in microvessels, *Microvasc. Res.*, 22, 110, 1981.

218. Fry, D. L., Mallos, A. J., and Casper, A. G. T., A catheter tip method for measurement of the instantaneous aortic blood velocity, *Circ. Res.*, 4, 627, 1956.

219. Fry, D. L., The measurement of pulsatile blood flow by the computed pressure gradient technique, *IRE Trans. Med. Electron.*, 6, 259, 1959.

220. Mixter, Jr., G., Respiratory augmentation of inferior vena caval flow demonstrated by a low-resistance phasic flowmeter, *Am. J. Physiol.*, 172, 446, 1953.

221. Brecher, G. A., Venous return during intermittent positive-negative pressure respiration studied with a new catheter flowmeter, *Am. J. Physiol.*, 174, 299, 1953.

222. Sarnoff, S. J., Berglund, E., and Wathe, P. E., The measurement of systemic blood flow, *Proc. Soc. Exp. Biol. Med.*, 79, 414, 1952.

223. Sarnoff, S. J., and Berglund, E., The Potter electroturbinometer. An instrument for recording total systematic blood flow in the dog, *Circ. Res.*, 1, 331, 1953.

224. Shipley, R. E. and Wilson, C., An improved recording rotameter, *Proc. Soc. Exp. Biol. Med.*, 78, 724, 1951.

225. Brecher, G. A. and Praglin, J., A modified bristle flowmeter for measuring phasic blood flow, *Proc. Soc. Exp. Biol. Med.*, 83, 155, 1953.

226. Brecher, G. A., Cardiac variations in venous return studied with a new bristle flowmeter, *Am. J. Physiol.*, 176, 423, 1954.

227. Brecher, G. A., Critical review of bristle flowmeter techniques, *IRE Trans. Med. Electron.*, 6, 294, 1959.

228. Togawa, T., and Suma, K., A study on vibrational flowmeter, in *Dig. 6th Int. Conf. Med. Elect. and Biol. Eng.*, Tokyo, 1965, 48.

229. Dawes, G. S., Mott, J. C., and Vane, J. R., The density flowmeter, a direct method for the measurement of the rate of blood flow, *J. Physiol.*, 121, 72, 1953.

230. Hilton, S. M. and Lywood, D. W., A photoelectric drop-counter, *J. Physiol.*, 123, 64P, 1954.

231. Geddes, L. A., Moore, A. G., Bourland, J., Vasku, J., and Contrell, G., An efficient indirect drop transducer, *Med. Res. Eng.*, 8, 27, 1969.

232. Constantinou, C. E. and Briggs, E. M., Precision electrostatic urine flowmeter, *J. Appl. Physiol.*, 29, 396, 1970.

233. Togawa, T., Kamiya, A., Yamazaki, Z., and Fujimori, Y., A drop transducer for the simultaneous recording of flow rate and electrolyte concentration of urine, *Jpn. J. Med. Elecron. Biol. Eng.*, 12, 16, 1974.

234. Hyman, C. and Winsor, T., History of plethysmography, *J. Cardiovasc. Surg.*, 2, 506, 1961.

235. Brodie, T. G., and Russell, A. E., On the determination of the rate of blood-flow through an organ, *J. Physiol. Lond.*, 32, 47P, 1905.

236. Hewlett, A. W. and van Zwaluwenburg, J. G., The rate of blood flow in the arm, *Heart*, 1, 87, 1909.

237. Landowne, M. and Katz, L. N., A critique of the plethysmographic method of measuring blood flow in the extremities of man, *Am. Heart J.*, 23, 644, 1942.

238. Wilkins, R. W. and Bradley, S. E., Changes in arterial and venous blood pressure and flow distal to a cuff inflated on the human arm, *Am. J. Physiol.*, 147, 260, 1946.

239. Formel, P. F. and Doyle, J. T., Rationale of venous occlusion plethysmography, *Circ. Res.*, 5, 354, 1957.

240. Barendsen, G. J., Venema, H., and van den Berg, Jw., Semicontinuous blood flow measurement by triggered venous occlusion plethysmography, *J. Appl. Physiol.*, 31, 288, 1971.

241. Greenfield, A. D. M., A simple water-filled plethysmograph for the hand or forearm with temperature control, *J. Physiol. Lond.*, 123, 62P, 1959.

242. Greenfield, A. D. M., Whitney, R. J., and Mowbray, J. F., Methods for the investigation of peripheral blood flow, *Br. Med. Bull.*, 19, 101, 1963.

243. Dahn, I., On the calibration and accuracy of segmental calf plethysmography with a description of a new expansion chamber and a new sleeve, *Scand. J. Clin. Lab. Invest.*, 16, 347, 1964.

244. Follett, D. H. and Preece, A. W., A new principle for level recording in venous occlusion plethysmography, *Med. Biol. Eng.*, 7, 217, 1969.

245. Stolinski, C., Sirs, J. A., Aroill, B. L., and Fentem, P. H., A photoelectric volume transducer for use with a water-filled plethysmograph, *J. Appl. Physiol.*, 22, 1161, 1967.

246. Dahn, I., Jonson, B., and Nilsèn, R., A plethysmographic method for determination of flow and volume pulsations in a limb, *J. Appl. Physiol.*, 28, 333, 1970.

247. Winsor, T., The segmental plethysmograph: a description of the instrument, *Angiology*, 8, 87, 1957.

248. Hyman, C. and Winsor, T., The application of the segmental plethysmograph to the measurement of blood flow through the limbs of human beings, *Am. J. Cardiol.*, 6, 667, 1960.

249. Whitney, R. J., The measurement of changes in human limb-volume by means of a mercury-in-rubber strain gauge, *J. Physiol. Lond.*, 109, 5P, 1949.

250. Whitney, R. J., The measurement of volume changes in human limbs, *J. Physiol. Lond.*, 121, 1, 1953.

251. Brakkee, A. J. M. and, Vendrik, A. J. H., Strain-gauge plethysmography: theoretical and practical notes on a new design, *J. Appl. Physiol.*, 21, 701, 1966.

252. Sigdell, J. E., A critical review of the theory of the mercury strain-gauge plethysmograph, *Med. Biol. Eng.*, 7, 365, 1969.

253. Hallböök, T., Månsson, B., and Nilsèn, R., A strain gauge plethysmograph with electrical calibration, *Scand. J. Clin. Lab. Invest.*, 25, 413, 1970.

254. Dahn, I. and Hallböök, T., Simultaneous blood flow measurement by water and strain gauge plethysmography, *Scand. J. Clin. Lab. Invest.*, 25, 419, 1970.

255. Hyman, C., Burnap, D. and Figar, S., Bilateral differences in forearm blood flow as measured with capacitance plethysmograph, *J. Appl. Physiol.*, 18, 997, 1963.

256. Wood, J. R. and Hyman, C., A direct reading capacitance plethysmograph, *Med. Biol. Eng.*, 8, 59, 1970.

257. Nyboer, J., Electrical impedance plethysmography. A physical and physiologic approach to peripheral vascular study, *Circulation*, 2, 811, 1950.

258. Yamakoshi, K., Shimazu, H., Togawa, T., and Ito, H., Admittance plethysmography for accurate measurement of human limb blood flow, *Am. J. Physiol.*, 235, H821, 1978.

259. Geddes, L. A. and Sadler, C., The specific resistance of blood at body temperature, *Med. Biol. Eng.*, 11, 336, 1973.

260. van den Berg, J. and Alberts, A. J., Limitations of electric impedance plethysmography, *Circ. Res.*, 2, 333, 1954.
261. Bashour, F. A. and Jones, R. E., Digital blood flow. I. Correlative study of the electrical impedance and the venous occlusive plethysmographs, *Dis. Chest*, 47, 465, 1965.
262. Yamakoshi, K., Shimazu, H., Bukhari, A. R. S., Togawa, T., and Ito, H., Clinical evaluation of an electrical admittance blood flow monitor, *J. Clin. Eng.*, 4, 341, 1979.
263. Sinton, A. M., Seagar, A. D., and Davis, F. M., Automated venous occlusion plethysmograph, *Med. Biol. Eng. Comput.*, 26, 295, 1988.
264. Kety, S. S. and Schmidt, C. F., The nitrous oxide method for the quantitative determination of cerebral blood flow in man: theory procedure and normal values, *J. Clin. Invest.*, 27, 476, 1948.
265. Lassen, N. A. and Ingvar, D. H., The blood flow of the cerebral cortex determined by radioactive krypton[85], *Experientia*, 17, 42, 1961.
266. Ingvar, D. H. and Lassen, N. A., Regional blood flow of the cerebral cortex determined by krypton[85], *Acta Physiol. Scan.*, 54, 325, 1962.
267. Veall, N. and Mallett, B. L., Regional cerebral blood flow determination by ^{133}Xe inhalation and external recording: the effect of arterial recirculation, *Clin. Sci.*, 30, 353, 1966.
268. Lassen, N. A., Brain, in *Peripheral Circulation,* Johnson, P. C., Ed., John Wiley & Sons, New York, 1978, 337.
269. Gur, D., Good, W. F., Wolfson, Jr., S. K., Yonas, H., and Shabason, L., *In vivo* mapping of local cerebral blood flow by xenon-enhanced computed tomography, *Science,* 215, 1267, 1982.
270. Lassen, N. A., Lindbjerg, J., and Munck, O., Measurement of blood flow through skeletal muscle by intramuscular injection of xenon-133, *Lancet,* 1, 686, 1964.
271. Tønnesen, K. H., Blood-flow through muscle during rhythmic contraction measured by [133]xenon, *Scand. J. Clin. Lab. Invest.*, 16, 646, 1964.
272. Tønnesen, K. H., The blood-flow through the calf muscle during rhythmic contraction and in rest in patients with occlusive arterial disease measured by [133]xenon, *Scand. J. Clin. Lab. Invest.*, 17, 433, 1965.
273. Sejrsen, P., Epidermal diffusion barrier to ^{133}Xe in man and studies of clearance of ^{133}Xe by sweat, *J. Appl. Physiol.*, 24, 211, 1968.
274. Sejrsen, P., Blood flow in cutaneous tissue in man studied by washout of radioactive xenon, *Circ. Res.*, 25, 215, 1969.
275. Aukland, K., Bower, B. F., and Berliner, R. W., Measurement of local blood flow with hydrogen gas, *Circ. Res.*, 14, 164, 1964.
276. Murakami, M., Moriga, M., Miyake, T., and Uchino, H., Contact electrode method in hydrogen gas clearance technique: A new method for determination of regional gastric mucosal blood flow in animals and humans, *Gastroenterology,* 82, 457, 1982.
277. Haining, J. L., Turner, M. D., and Pantall, R. M., Measurement of local cerebral blood flow in the unanesthetized rat using a hydrogen clearance method, *Circ. Res.*, 23, 313, 1968.
278. Shinohara, Y., Meyer, J. S., Kitamura, A., Toyoda, M., and Ryu, T., Measurement of cerebral hemispheric blood flow by intracarotid injection of hydrogen gas. Validation of the method in the monkey, *Circ. Res.*, 25, 735, 1969.
279. Mishima, Y., Shigematsu, H., Horie, Y., and Satoh, M., Measurement of local blood flow of the intestine by hydrogen clearance method; experimental study, *Jpn. J. Surgery*, 9, 63, 1979.
280. Stosseck, K., Lübbers, D. W., and Cottin, N., Determination of local blood flow (microflow) by electrochemically generated hydrogen. Construction and application of the measuring probe, *Pflügers Arch.*, 348, 225, 1974.
281. DiResta, G. R., Kiel, J. W., Riedel, G. L., Kaplan, P., and Shepherd, A. P., Hybrid blood flow probe for simultaneous H_2 clearance and laser-Doppler velocimetry, *Am. J. Physiol.*, 253, G573, 1987.
282. Chato, J. C., Heat transfer to blood vessels, *Trans. ASME*, 102, 110, 1980.
283. Hensel, H., Ruef, J., Fortlaufende Registrierung der Muskeldurchblutung am Menschen mit einer Calorimetersonde, *Pflügers Arch.*, 259, 267, 1954.
284. Mowbray, J. F., Measurement of tissue blood flow using small heated thermocouple, *J. Appl. Physiol.*, 14, 647, 1959.
285. Levy, L., Grainchen, H., Stolwijk, J. A. J., and Calabresi, M., Evaluation of local tissue blood flow by continuous direct measurement of thermal conductivity, *J. Apply. Physiol.*, 22, 1026, 1967.
286. Adams, T., Heisey, S. R., Smith, M. C., Steinmetz, M. A., Hartman, J. C., and Fly, H. K., Thermodynamic technique for the quantification of regional blood flow, *Am. J. Physiol.*, 238, (*Heart Circ. Physiol.,* 7), H682, 1980.
287. Hensel, H. and Bender, F., Fortlaufende Bestimmung der Hautdurchblutung am Menschen mit einem Elektrischen Wärmeleitmesser, *Pflügers Arch.*, 263, 603, 1956.
288. Hensel, H., Messkopf zur Durchblutungsregistrierung an Oberflächen, *Pflügers Arch.*, 268, 604, 1959
289. Harding, D. C., Rushmer, R. F., Baker, D. W., Thermal transcutaneous flowmeter, *Med. Biol. Eng.*, 5, 623, 1967.
290. van de Staak, W. J. B. M., Brakkee, A. J. M., de Rijke-Herweijer, H. E., Measurements of the thermal conductivity of the skin as an indication of skin blood flow, *J. Invest. Dermatol.*, 51, 149, 1968.
291. Holti, G. and Mitchell, K. W., Estimation of the nutrient skin blood flow using a noninvasive segmented thermal clearance probe, in *Noninvasive Physiological Measurements*, Vol. 1, Rolfe, P., Ed., Academic Press, London, 1979, 113.
292. Grayson, J., Internal calorimetry in the determination of thermal conductivity and blood flow, *J. Physiol.*, 118, 54, 1952.
293. Ischimaru, S., Theory and application of wave propagation and scattering in random media, *Proc IEEE,* 65, 1030, 1977.
294. Anderson, R. R. and Parrish, J. A., The optics of human skin, *J. Invest. Dermatol.*, 77, 13, 1981.

295. Chu, B., *Laser Light Scattering,* Academic Press, New York, 1974.
296. Bohren, C. F., and Hoffman, D. R., *Absorption and Scattering of Light by Small Particles,* John Wiley & Sons, New York, 1983.
297. Berne, B. J. and Pecora, R., *Dynamic Light Scattering,* John Wiley & Sons, New York, 1979.
298. Drain, L. E., *The Laser Doppler Technique,* John Wiley & Sons, New York, 1980.
299. Cummins, H. Z. and Swinney, H. L., Light beating spectroscopy, in *Progress in Optics,* Vol. 8, Wolf, E., Ed., North-Holland, Amsterdam, 1970, 133.
300. Duteil, L., Bernengo, J. C., and Schalla, W., A double wavelength laser Doppler system to investigate skin microcirculation, *IEEE Trans. Biomed. Eng.,* BME-32, 439 1985.
301. Nossal, R., Laser light scattering, *Meth. Explor. Phys.,* 20, 299, 1982.
302. Johnson, C., Optical diffusion in blood, *IEEE Trans. Biomed. Eng.,* BME-17, 129, 1970.
303. Yeh, Y. and Cummins, H. Z., Localized fluid flow measurements with a HeNe laser spectrometer, *J. Appl. Physiol.,* 4, 171, 1964.
304. Khan, A., Schall, L. M., Tur, E., Maibach, H. I., and Guy, R. H., Blood flow in psoriatic skin lesions: the effect of treatment, *Br. J. Dermatol.,* 117, 193, 1987.
305. Bonner, R. and Nossal, R., Model for laser Doppler measurements of blood flow in tissue, *Appl. Optics,* 20, 2097, 1981.
306. Bonner, R. F., Nossal, R., Havlin, S., and Weiss, G. H., Model for photon migration in turbid biological media, *J. Opt. Soc. Am.,* A4, 423, 1987.
307. Maret, G. and Wolf, P. E., Multiple light scattering from disordered media. The effect of Brownian motion of scatterers, *Z. Phys.,* B65, 409, 1987.
308. Steinke, J. M. and Shepherd, A. P., Diffusion model of the optical absorbance of whole blood, *J. Opt. Soc. Am.,* 5, 813, 1988.
309. Nilsson, G. E., Tenland, T., and Öberg, P. Å., A new instrument for continuous measurement of tissue blood flow by light beating spectroscopy, *IEEE Trans. Biomed. Eng.,* BME-27, 12, 1980a.
310. Twersky, V., Absorption and multiple scattering by biological suspensions, *J. Opt. Soc. Am.,* 60, 1084, 1970.
311. Haumschild, D. J., Microvascular blood flow measurement by laser Doppler flowmetry, in *TSI Application Note,* TSI, St. Paul, MN, 1650, 1986.
312. Pettersson, H., Öberg, P. Å., Rohman, H., Gazelius, B., and Olgart, L., Vitality assessment in human teeth by laser Doppler flowmetry, in *Images of the Twenty-First Century,* Vol. 11, Kim. Y. and Spelman, F. A., Eds., Proc. Ann. Int. Conf. of the IEEE Engineering in Medicine and Biology Society, Seattle, WA, 1989.
313. Kolari, P. J., Optoelectronic Doppler velocimetry based on semi-conductor laser diode for measurements of cutaneous blood flow, *Int. J. Microcirc.,* 3, 476, 1984.
314. de Mul, F. F. M., van Spijker, J., van der Plas, D., Greve, J., Aarnoudse, J. G., and Smits, T. M., Mini laser-Doppler (blood) flow monitor with diode laser source and detection integrated in the probe, *Appl. Opt.,* 23, 2970, 1984.
315. Shimada, K., *Introduction to Laser Physics,* Springer-Verlag, Berlin, 1984.
316. Young, M., *Optics and Lasers,* 2nd ed., Springer-Verlag, Berlin, 1984.
317. Yariv, A., *Optical Electronics,* 3rd ed., Holt, Rinehart, and Winston, New York, 1985.
318. Engström, R. W., Multiplier phototube characteristics: application to low light levels, *J. Opt. Soc.,* 37, 420, 1947.
319. Kajiya, F., Hoki, N., Tomonaga, G., and Nishihara, H., A laser-Doppler-velocimeter using an optical fiber and its application to local velocity measurement in the coronary artery, *Experientia,* 37, 1171, 1981.
320. Bonner, R. F., Clem, T. R., Bowen, P. D., and Bowman, R. L., Laser-Doppler continuous real-time monitor of pulsatile and mean blood flow in tissue microcirculation, in *Scattering Techniques Applied to Supra-Molecular and Nonequilibrium Systems,* Chen, S. H., Chu, B., and Nossal, R., Eds., Plenum, New York, 1981, 685.
321. Tenland, T., Salerud, E. G., Nilsson, G. E., and Öberg, P. Å., Spatial and temporal variations in human skin blood flow, *Int. J. Microcirc. Clin. Exp.,* 2, 81, 1983.
322. Damber, J. E., Lindahl, O., Selstam, G., and Tenland, T., Testicular blood flow measured with a laser Doppler flowmeter: acute effects of cathecolamines, *Acta Physiol. Scand.,* 115, 209, 1982.
323. Salerud, E. G. and Öberg, P. Å., Single-fiber laser Doppler flowmetry. A method for deep tissue perfusion measurements, *Med. Biol. Eng. Comput.,* 25, 329, 1987.
324. Stern, M. D., Lappe, D. L., Bowen, P. D., Chimosky, J. E., Holloway, G. A., Keiser, H. R., and Bowman, R. L., Continuous measurement of tissue blood flow by laser-Doppler spectroscopy, *Am. J. Physiol.,* 232, H441, 1977.
325. Nilsson, G. E, Tenland, T., and Öberg, P. Å., Evaluation of laser Doppler flowmeter for measurement of tissue blood flow, *IEEE Trans. Biomed. Eng.,* BME-27, 597, 1980b.
326. Ahn, H., Johansson, K., Lundgren, O., and Nilsson, G. E., *In vivo* evaluation of signal processors for laser Doppler tissue flowmeters, *Med. Biol. Eng. Comput.,* 25, 207, 1987.
327. Holloway, G. A., Cutaneous blood flow responses to injection trauma measured by laser Doppler velocimetry, *J. Invest. Dermatol.,* 74, 1, 1980.
328. Kastrup, J., Bülow, J., and Lassen, N. A., A comparison between [133]Xenon washout technique and laser Doppler flowmetry in the measurement of local vasoconstrictor effects on the microcirculation in subcutaneous tissue and skin, *Clin. Physiol.,* 7, 403, 1987.
329. Neufeld, G. R., Galante, S. R., Whang, J. M., deVries, D., Baumgardner, J. E., Graves, D. J., and Quinn, J. A., Skin blood flow from gas transport: helium xenon and laser Doppler compared, *Microvasc. Res.,* 35, 143, 1988.

330. Engelhart, M. and Kristensen, J. K., Evaluation of cutaneous blood flow responses by [133]Xenon washout and a laser-Doppler flowmeter, *J. Invest. Dermatol.,* 80, 12, 1983.

331. Nicholson, C. D., Schmitt, R. M., and Wilke, R., The effect of acute and chronic femoral artery ligation on the blood flow through the gastrocnemius muscle of the rat examined using laser Doppler flowmetry and xenon-133 clearance, *Int. J. Microcirc. Clin. Exp.,* 4, 57, 1985.

332. Engelhart, M., Petersen, L. J., and Kristensen, J. K., The local regulation of blood flow evaluated simultaneously by 133-xenon washout and laser Doppler flowmetry, *J. Invest. Dermatol.,* 91, 451, 1988.

333. Eyre, J. A., Essex, T. J. H., Flecknell, P. A., Bartholomew, P. H., and Sinclair, J. I., A comparison of measurements of cerebral blood flow in the rabbit using laser Doppler spectroscopy and radionuclide labelled microspheres, *Clin. Phys. Physiol. Meas.,* 9, 65, 1988.

334. Matsen, F. A., Wyss, C. R., Robertson, C. L., Öberg, P. Å., and Holloway, G. A., The relationship of transcutaneous pO_2 and laser Doppler measurement in a human model of local arterial insufficiency, *Surg., Gynecol. Obstetr.,* 159, 418, 1984.

335. Kvietys, P. R., Shepherd, A. P., and Granger, D. N., Laser-Doppler H_2 clearance, and microsphere estimates of mucosal blood flow, *Am. J. Physiol.,* 249, G221, 1985.

336. Sundberg, S. and Castrén, M., Drug- and temperature-induced changes in peripheral circulation measured by laser-Doppler flowmetry and digital-pulse plethysmography, *Scand. J. Clin. Lab. Invest.,* 46, 359, 1986.

337. Öberg, P. Å., A method for frequency stabilization of multimode He-Ne lasers in laser-Doppler flowmetry, in *Proc. World Congr. on Med. Phys. and Biomed. Eng.,* BE18-E.1, San Antonio, TX, 1988, 376.

338. Jentink, H. W., Hermsen, R. G. A. M., de Mul, F. F. M., Suichies, H. E., Aarnoudse, J. G., and Greve, J., Tissue perfusion measurements using a mini diode laser Doppler perfusion sensor, in *Microsensors and Catheter-Based Imaging Technology,* Vol. 904, SPIE, Los Angeles, 1988.

339. Shepherd, A. P., Riedel, G. L., Kiel, J. W., Haumschild, D. J., and Maxwell, L. C., Evaluation of an infrared laser Doppler blood flowmeter, *Am. J. Physiol.,* 252, G832, 1987.

340. Roman, R. J. and Smits, C., Laser-Doppler determination of papillary blood flow in young and adult rats, *Am. J. Physiol.,* 251, F115, 1986.

341. Smits, G. J., Roman, R. J., and Lombard, J. H., Evaluation of laser-Doppler flowmetry as a measure of tissue blood flow, *J. Appl. Physiol.,* 61, 666, 1986.

342. Shepherd, A. P. and Öberg, P. Å., *Laser Doppler Flowmetry,* Kluwer Academic, London, 1990.

343. Öberg, P. Å., Laser Doppler flowmetry, *CRC Crit. Rev. Bioeng.,* 18(2), 125, 1990.

344. *Perimed Literature Reference List,* No. 9 (Sept.), Perimed AB, Stockholm, 1988.

345. Fujii, H., Nohira, K., Yamamoto, Y., Ikawa, H., and Ohura, T., Evaluation of blood flow by laser speckle image sensing. Part 1. *Appl. Opt.,* 26, 5321, 1987.

346. Fercher, A. F. and Briers, J. D., Flow visualization by means of single-exposure speckle photography, *Opt. Comm.,* 37, 326, 1981.

347. Goodman, A. H., CCD line-scan image sensor for the measurement of red cell velocity in microvessels, *J. Biomed. Eng.,* 8, 32, 1986.

348. Essex, T. J. H. and Byrne, P. O., A laser Doppler scanner for imaging blood flow in skin, *J. Biomed. Eng.,* 13, 189, 1991.

349. Niazi, Z. B. M., Essex, T. J. H., Papini, R., Scott, D., MacLean, N. R., and Black, M. J., New laser doppler scanner, a valuable adjunct in burn depth assessment, *Burns,* 19, 485, 1993.

350. Nilsson, G. E., Jakobsson, A., and Wårdell, K., Tissue perfusion monitoring and imaging by coherent light scattering, in *Biooptics in Biomedicine and Environmental Science,* Vol. 1524, SPIE, Los Angeles, 1991, 90.

351. Wårdell, K., Laser Doppler Perfusion Imaging. Linköping Studies in Science and Technology Dissertations No. 329, Linköping University, Sweden, 1994.

352. Wårdell, K., Jakobsson, A., and Nilsson, G. E., Laser Doppler perfusion imaging by dynamic light scattering, *IEEE Trans. Biomed. Eng.,* 40, 309, 1993.

353. Wårdell, K. and Nilsson, G.E., Laser Doppler Imaging of skin, in *Handbook of Noninvasive Methods and the Skin,* Serup, J., Jemec, B. E., Eds., CRC Press, Boca Raton, FL, 1995, 421.

354. Uhl, E., Sirsjö, A., Nilsson, G., Nylander, G., Influence of Ketamine and Pentobarbital on microvascular perfusion in normal skin and skin flaps, *Int. J. Microcirc.,* 14, 308, 1994.

355. Arnold, F., He, C. F., Jia, C. Y., and Cherry, G. W., Perfusion imaging of skin island flap blood flow by a scanning laser-Doppler technique, *Br. J. Plast. Surg.,* 48, 280, 1995.

356. Uhl, E., Sirsjö, A., Haapaniemi, T., Nilsson, G., Nylander, G., Hyperbaric oxygen improves wound healing in normal and ischemic skin tissue, *Plast. Reconstr. Surg.,* 93, 835, 1994.

357. Ljung, P., Bornmyr, S., Svensson, M., Wound healing after total elbow replacement in rheumatoid arthritis. Wound complications in 50 cases and laser-Doppler imaging of skin microcirculations, *Acta Orthop. Scand.,* 66, 59, 1995.

358. Algotsson, A., Vascular reactivity and autonomic functions in Alzheimer's disease, thesis, Karolinska Institutet, Huddinge University Hospital, Huddinge, Sweden, 1995.

359. Bornmyr, S., Arner, M., Svensson, H., Laser Doppler imaging of finger skin blood flow in patients after microvascular repair of the ulnar artery at the wrist, *J. Hand Surg.,* 19B, 295, 1994.

360. Andersson, T., Cutneous microdialysis, Linköping Medical Dissertation No. 456, Linköping University, Sweden, 1995.

361. Brooks, R. A., Battocletti, J. R., Sances, Jr., A., Larson, S. J., Bowman, R. L., and Kudravcev, V., Nuclear magnetic relaxation in blood, *IEEE Trans. Biomed. Eng.,* BME-22, 12, 1975.

362. Bowman, R. L, and Kudravcev, V., Blood flowmeter utilizing nuclear magnetic resonance, *IRE Trans. Med. Electron.*, 6, 267, 1959.
363. Battocletti, J. H., Halbach, R. E., Sances, Jr., A. Larson, S. L., Bowman, R. L., and Kudravcev,V., A flat crossed coil detector for blood flow measurement using nuclear magnetic resonance, *Med. Biol. Eng. Comput.*, 17, 183, 1979.
364. Battocletti, J. H., Halbach, R. E., Salles-Cunha, S. X., Sances, Jr., A., Towne, J. B., Hebert, L. A., and Kauffman, H. M., Clinical applications of the nuclear magnetic resonance limb blood flowmeter, *Proc. IEEE*, 67, 1359, 1979.
365. Halbach, R. E., Battocletti, J. H., Sances, Jr., A., Bowman, R. L., and Kudravcev, V., Cylindrical crossed coil NMR limb blood flowmeter, *Rev. Sci. Instrum.*, 50, 428, 1979.
366. Salles-Cunha, S. X., Halbach, R. E., Battocletti, J. H., and Sances, Jr., A., Nuclear magnetic resonance (NMR) cylindrical blood flowmeter: *in vitro* evaluation, *J. Clin. Eng.*, 5, 205, 1980.
367. Battocletti, J. H., Blood flow measurement by NMR, *CRC Crit. Rev.*, 13, 311, 1986.
368. Singer, J. R., NMR diffusion and flow measurements and an introduction to spin phase graphing, *J. Phys. E*, 11, 281, 1978.
369. Grover, T. and Singer, J. R., NMR spin-echo flow measurements, *J. Appl. Physiol.*, 42, 938, 1971.
370. Kaufman, L., Crocks, L., Sheldon, P., Hricak, H., Herfkens, R., and Bank, W., The potential impact of nuclear magnetic resonance imaging on cardiovascular diagnosis, *Circulation*, 67, 251, 1983.
371. Singer, J. R. and Crooks, L. E., Nuclear magnetic resonance blood flow measurements in the human brain, *Science*, 221, 654, 1983.
372. Moran, P. R., A flow velocity zeugmatographic interface for NMR imaging in humans, *Magn. Reson. Imaging*, 1, 197, 1982.
373. Moran, P. R., Moran, R. A., and Karstaedt, N., Verification and evaluation of internal flow and motion, *Radiology*, 154, 433, 1985.
374. Nunn, J. F. and Ezi Ashi, T. I., The accuracy of the respirometer and venti-grator, *Br. J. Anaesth.*, 34, 422, 1962.
375. Lunn, S. N. and Hillard, E. K., The effect of repairs of the Wright respirometer, *Br. J. Anaesth.*, 42, 1127, 1970.
376. FitzGerald, M. X., Smith, A. A. and Gaensler, E. A., Evaluation of "electronic" spirometers, *N. Engl. J. Med.*, 289, 1283, 1973.
377. Mushin, W. W., Rendell-Baker, L., Thompson, P. W., Mapleson, W. W., *Automatic Ventilation of the Lung*, 2nd ed., Blackwell Scientific, Oxford, 1969, 44.
378. Fleisch, A., Le pneumotachographe, *Helv. Physiol. Pharmacol. Acta*, 14, 363, 1956.
379. Gregory, G. A. and Kitterman, J. A., Pneumotachograph for use with infants during spontaneous or assisted ventilation, *J. Appl. Physiol.*, 31, 766, 1971.
380. Fry, D. L., Hyatt, R. E., McCall, C. B., Mallos, A. J., Evaluation of three types of respiration flowmeter, *J. Appl. Physiol.*, 10, 210, 1957.
381. Finucane, K. E., Egan, B. A., and Dawson, S. V., Linearity and frequency response of pneumotachographs, *J. Appl. Physiol.*, 32, 121, 1972.
382. Johns, D. P., Pretto, J. J., and Streeton, J. A., Measurement of gas viscosity with a Fleisch pneumotachograph, *J. Appl. Physiol. Resp., Environ. Exercise Physiol.*, 53, 290, 1982.
383. Weast, R. C., *Handbook of Chemistry and Physics*, First student edition, CRC Press, Boca Raton, FL, 1987, F-8.
384. Hobbes, A. F. T., A comparison of methods of calibrating the pneumo-tachograph, *Br. J. Anaesthesiol.*, 39, 899, 1967.
385. Elliott, S. E., Shore, J. H., Barnes, C. W., Lindauer, J., and Osborn, J. J., Turbulent air flow meter for long-term monitoring in patient-ventilator circuits, *J. Appl. Physiol. Resp. Environ. Exercise Physiol.*, 42, 456, 1977.
386. Collins, D. C. and Williams, M. J., Two-dimensional convection from heated wire at low Reynolds numbers, *J. Fluid Mech.*, 6, 357, 1959.
387. Lundsgaard, J. S., Grønlund, J., and Einer-Jensen, N., Evaluation of a constant-temperature hot-wire anemometer for respiratory gas flow measurements, *Med. Biol. Eng. Comput.*, 17, 211, 1979.
388. Micco, A. J., A sensitive flow direction sensor, *J. Appl. Physiol.*, 35, 420, 1973.
389. Yoshiya, I., Nakajima, T., Nagai, L., and Jitsukawa, S., A bidirectional respiratory flowmeter using the hot-wire principle, *J. Appl. Physiol.*, 38, 360, 1975.
390. McQuaid, J. and Wright, W., The response of a hot-wire anemometer in flow of gas mixtures, *Int. J. Heat Mass Transfer.*, 16, 819, 1973.
391. Comte-Bellot, G., Hot-wire anemometry, *Ann. Rev. Fluid Mech.*, 8, 209, 1976.
392. Shanks, D. E. and Morris, J. F., Clinical comparison of two electronic spirometers with a water sealed spirometer, *Chest*, 69, 461, 1976.
393. Bradbury, L. J. S. and Castro, L. P., A pulsed wire technique for velocity measurements in high turbulent flows, *J. Fluid Mech.*, 49, 657, 1971.
394. Mosse, C. A. and Roberts, S. P., Microprocessor-based time-of-flight respirometer, *Med. Biol. Eng. Comput.*, 25, 34, 1987.
395. Nemoto, T. and Togawa, T., A pulsed-wire spirometer using sing-around method, *Trans. SICE Jpn.*, 19, 314, 1983.
396. Blumenfeld, W., Wilson, P. D., and Turney, S., A mathematical model for the ultrasonic measurement of respiratory flow, *Med. Biol. Eng.*, 12, 621, 1974.
397. Plaut, D. and Webster, J. G., Ultrasonic measurement of respiratory flow, *IEEE Trans. Biomed. Eng.*, BME-27, 549, 1980.
398. Buess, C., Pietsch, P., Guggenbühl, W., and Koller, E. A., Design and construction of a pulsed ultrasonic air flowmeter, *IEEE Trans. Biomed. Eng.*, BME-33, 768, 1986.

399. Buess, C., Fietsch, P., Guggenbühl, W., and Koller, E. A., A pulsed diagonal-beam ultrasonic airflow meter, *J. Appl. Physiol.*, 61, 1195, 1986.
400. Plaut, D. and Webster, J. G., Design and construction of an ultrasonic pneumotachometer, *IEEE Trans. Biomed. Eng.*, BME-27, 590, 1980.
401. Tsuhchiya, K., Ogata, S., and Ueta, N., Kármán Vortex flow meter, *Bull. JSME*, 13, 573, 1970.
402. Utsunomiya, H. and Kyozuka, S., Development of an ultrasonic ventilation monitor by counting vortexes, *JJME*, 19 (Suppl.), 138, 1981.
403. Roth, P., Modifications of apparatus and improved technique adaptable to the Benedict type of respiration apparatus, *Boston Med. Surg. J.*, 186, 457, 491, 1922.
404. Stead, W. W., Wells, H. S., Gault, N. L., and Ognanovich, J., Inaccuracy of the conventional water-filled spirometer for recording rapid breathing, *J. Appl. Physiol.*, 14, 448, 1959.
405. Wells, H. S., Stead, W. W., Rossing, T. D., and Ognanovich, J., Accuracy of an improved spirometer for recording of fast breathing, *J. Appl. Physiol.*, 14, 451, 1959.
406. Glindmeyer, H. W., Anderson, S. T., Diem, J. E., and Weill, H., A comparison of the Jones and Stead-Wells spirometer, *Chest*, 73, 596, 1978.
407. Collins, M. M., McDermott, M., McDermott, T. J., Bellows spirometer and transistor timer for the measurement of forced expiratory volume and vital capacity, *J. Physiol. Lond*, 172, 39P, 1964.
408. Ledwith, J. W., Comparative spirometry measurements using a bellows-type and water-sealed spirometer, *Rev. Dis. Chest*, 95, 512, 1967.
409. Drew, C. D. M. and Hughes, D. T. D., Characteristics of the Vitalograph spirometer, *Thorax*, 24, 701, 1969.
410. Levison, H., Kamel, M., Weng, T. R., Kruger, K., Expiratory flow rates determined in children and young adults, *Acta. Ped. Scand.*, 59, 648, 525, 1970.
411. Wever, A. M. J, Britton, M. G., and Hughes, D. D. T., Evaluation of two spirometers: a comparative study of the Stead-Wells and the Vitalograph spirometers, *Chest*, 70, 244, 1976
412. Dubois, A. B., Botelho, S. Y., Bedell, G. N., Marshall, R., Comroe, J. H., A rapid plethysmographic method for measuring thoracic gas volume: a comparison with a nitrogen washout method for measuring functional residual capacity in normal subjects, *J. Clin. Invest.*, 35, 322, 1956.
413. Dubois, A. B., Botelho, S. Y., Comroe, J. H., A new method for measuring airway resistance in man using a body plethysmograph: value in normal subjects and in patients with respiratory disease, *J. Clin. Invest.*, 35, 327, 1956.
414. Comroe, Jr., J. H., Botelho, S. Y., Dubois, A. B., Design of a body plethysmograph for studying cardiopulmonary physiology, *J. Appl. Physiol.*, 14, 439, 1959.
415. Mead, J., Volume displacement body plethysmograph for respiratory measurements in human subjects, *J. Appl. Physiol.*, 15, 736, 1960.
416. Milledge, J. S. and Stott, F. D., Inductive plethysmography — a new respiratory transducer, *J. Physiol.*, 267, 4P, 1977.
417. Sackner, J. D., Nixon, A. J., Davis, B., Atkins, N., and Sackner, M. A., Noninvasive measurement of ventilation during exercise using a respiratory inductive plethysmograph, 1, *Am. Rev. Respir. Dis.*, 122, 867, 1980.
418. Hill, S. L., Blackburn, J. P., and Williams, T. R., Measurement of respiratory flow by inductance pneumography, *Med. Biol. Eng. Comput.*, 20, 517, 1982.
419. Chadha, T. S., Watson, H., Birch, S., Jenouri, G. A., Schneider, A. W., Cohn, M. A., and Sackner, M. A., Validation of respiratory inductive plethysmography using different calibration procedures, *Am. Rev. Respir. Dis.*, 125, 644, 1982.
420. Imai, S., Noshiro, M., Ishida, A., Prototype of a respiratory monitor using inductive transducer, *Jpn. J. Med. Electron. Biol. Eng.*, 23, 46, 1985.
421. Itoh, A., Ishida, A., Kikuchi, N., Okazaki, N., Ishihara, T., Kira, S., Noninvasive ventilatory volume monitor, *Med. Biol. Eng. Comput.*, 20, 613, 1982.
422. Baker, L. E. and Geddes, L. A., The measurement of respiratory volumes in animals and man with use of electrical impedance, *Ann. N.Y. Acad. Sci.*, 170, 667, 1970.
423. Valentinuzzi, M. E., Geddes, L. A., and Baker, L. E., The law of impedance pneumography, *Med. Biol. Eng. Comput.*, 9, 157, 1971.
424. Allison, R. D., Holmes, E. L., and Nyboer, J., Volumetric dynamics of respiration as measured by electrical impedance plethysmography, *J. Appl. Physiol.*, 19, 166, 1964.
425. Kubicek, W. G., Kinnen, E., and Edin, A., Calibration of an impedance pneumograph, *J. Appl. Physiol.*, 19, 557, 1964.
426. Pallett, J. E., and Scopes, J. W., Recording respirations in newborn babies by measuring impedance of the chest, *Med. Electron. Biol. Eng.*, 3, 161, 1965.
427. Baker, L. E., Geddes, L. A., and Hoff, H. E., A comparison of linear and nonlinear characterizations of impedance spirometry data, *Med. Biol. Eng.*, 4, 371, 1966.
428. Logic, J. L., Maksud, M. G., and Hamilton, L. H., Factors affecting transthoracic impedance signals used to measure breathing, *J. Appl. Physiol.*, 22, 251, 1967.
429. Matsumura, N., Nishijima, H., Kojima, S., and Yasuda, H., Electrical impedance pneumography for the respiratory monitoring during exercise in cardiac patients, in *Proc. 5th Int. Conf. Electr. Bio. Impedance*, Tokyo, 1981, 355.
430. Itoh, A., Kikuchi, N., Ishida, A., Kuratomi, Y., Ishihara, T., Okazaki, N., Arai, T., and Kira, S., Noninvasive respiratory function monitoring system, in *Proc. 5th Int. Conf. Electr. Bio. Impedance*, Japanese Society of Medical, Electrical, and Biological Engineers, Tokyo, 1981, 341.

Chapter 4

Motion and Force Measurement

4.1 OBJECTS OF MEASUREMENT

4.1.1 UNITS OF QUANTITIES

The fundamental quantities that represent a motion are time and length. In the SI unit system, the unit of time is a second (s), and the unit of length is a meter (m). The units of velocity and acceleration are m/s (meter per second) and m/s^2 (meter per second squared), respectively. The standard gravitational acceleration is 9.806 65 m/s^2. To express rotation, the angle, angular velocity, and angular acceleration are used, and their units in the SI system are rad (radian), rad/s (radian per second), and rad s^2 (radian per second squared), respectively. Degree, minute, and second have been used conventionally and they are denoted by the symbols °, ′, and ″, respectively; $1° = \pi/180$ rad, $1′ = (1/60)°$, and $1″ = (1/60)′$. The unit of frequency is Hz (hertz), and 1 Hz = 1 s^{-1}.

The mass is a quantity of matter, and the unit of mass in the SI system is a kilogram (kg). The force is defined by mass and acceleration so that when a unit force is applied to a unit mass in free space, unit acceleration is generated. The unit of force in the SI system is N (newton), defined by a relation, 1 N = 1 kg m/s^2.

The force acting on a mass under gravity is the weight. The gravity field is not uniform in strength over the earth. Influences of centrifugal force and atmospheric buoyancy exist, but the weight of a unit mass can still be a convenient unit of force. Thus, kgf (kilogram-force) has been widely used, which is the weight of 1 kg mass under standard gravity, and

$$1 \text{ kgf} = 9.80665 \text{ N (exactly)}.$$

The following approximate conversions are sometimes helpful:

$$1 \text{ kgf} \approx 10 \text{ N}$$

$$1 \text{ gf} \approx 10^{-2} \text{ N}$$

$$1 \text{ mgf} \approx 10^{-5} \text{ N}$$

$$1 \text{ μgf} \approx 10^{-8} \text{ N}$$

In the SI system, the unit of momentum is kgm/s (kilogram meter per second), and the unit of the moment of force, or torque, is Nm (newton meter). The unit of angular momentum is kgm^2/s (kilogram square meter per second). The unit of moment of inertia is kgm^2 (kilogram square meter). The unit of work is J (joule) defined by a relation, 1 J = 1 Nm. The unit of power is W (watt) defined by a relation, 1 W = 1 J/s.

Table 4.1   Characteristics of Muscle Contraction

	Contraction Speed (l_0/s)	Change of Length (%)	Tension (kgf/cm²)
Smooth muscle	4 ~ 24	−40 ~ +80	0.5 ~ 5
Cardiac muscle	1 ~ 2	0 ~ 50	0.4 ~ 1
Smooth muscle	0.1 ~ 3	−60 ~ +80	0.4 ~ 2

Note: l_0 = muscle length at rest.

Table 4.2   Physiological Dimension Based on Body Length L

Quantities	Physiological Dimensions
Length	L
Mass, body weight	L^3
Time	L
Cross-sectional area	L^2
Surface area	L^2
Volume	L^3
Velocity	L
Frequency	L^{-1}
Acceleration	L^{-1}
Force	L^2
Energy	L^3
Power	L^2

4.1.2   OBJECTS OF MEASUREMENTS

Many kinds of body motions generated by muscular activities are objects of motion measurements, while passive motions due to externally applied forces are also sometimes of interest. Three types of muscle exist: skeletal, cardiac, and smooth muscle. The characteristics of the mechanical activities of these muscles are different from each other, as shown in Table 4.1. The skeletal muscle contracts quickly and develops large tensions. The cardiac muscle contracts more slowly, and the relative change in its length and developed tension are less. The smooth muscle contracts slowly, but the developed tension is sometimes comparable with that of the skeletal muscle.

The body motion generated by muscles depends both on the characteristics of the muscle as an actuator and on the mechanical characteristics of the body as the load to this actuator. A large muscle can generate a large tension, but it cannot contract very quickly because the mechanical load is large. Actually, the mechanical properties of muscles are related to their size, even if their configurations are geometrically similar to each other. The relations between the body size and various mechanical quantities are listed in Table 4.2.[1] For example, if the linear dimension of a body is doubled, the body mass increases eight times while the force and power are increased four times, velocity is unchanged, and acceleration and frequency are decreased to one half. This knowledge is useful when one tries to prepare instruments for measurements in animals or humans of different sizes.

4.1.3   COORDINATE SYSTEM

When a body motion is studied, a coordinate system, to which the observer refers, is always chosen to describe the motion. While the motion can be described by using any coordinate system in principle, the correct choice of the coordinate system simplifies the theory and data analyses. For example, the body motion of a subject can be studied by using videocameras placed on the ground. The recordings obtained provide data which represent motion referring to the coordinate system fixed to the ground. However, if one tries to extract rotation of a joint from the videorecording, fairly complicated procedures will be required, even though the motion is simply a rotation around a joint axis. Such a motion can be recorded much more easily and more accurately using a goniometer, which is a transducer attached directly to the joint and which can provide output corresponding to the rotation angle for each part of the body that moves freely in the space. A measurement performed by a transducer, such as a goniometer attached to the body, provides output corresponding to variables referring to the moving coordinate system, whereas a measurement performed by a transducer placed on the ground, such as a videocamera, provides output corresponding to variables referring to the fixed coordinate system. The moving coordinate system is convenient when the purpose of the measurement is to observe a specific part of the body relative to the resting part.

When a force measurement is performed on a moving body, however, inertial forces may appear, unless the body motion is linear and the velocity is constant. If the origin of a moving coordinate system has an acceleration, A_0, and rotates with an angular velocity, ω, and if a mass, m, at position vector, $\mathbf{r}'$, and velocity, u', refer to the moving coordinate system, then the inertial force observed in reference to the moving coordinate system is represented as

$$mA' = -mA_0 + 2mu' \times \omega + m\omega \times (\mathbf{r}' \times \omega) + m\mathbf{r}' \times d\omega/dt \qquad (4.1)$$

where letters with a prime (') indicate variables referring to the moving coordinate system. The right side of Equation (4.1) has four terms corresponding to the linear inertial force, the Coriolis force, the centrifugal force, and the apparent force due to the angular acceleration. As seen in this expression, great complexity arises when a moving coordinate system having acceleration and rotation is introduced. When force measurement is performed with a transducer fixed on the body having acceleration and rotation, as mentioned above, inertial forces appear in the measurement object.

On the other hand, the above relation shows the possibility of determining body motion through acceleration measurements. If several accelerometers with masses m are attached at different sites on the body, each accelerometer provides an output corresponding to mA' with a different position vector $\mathbf{r}'$ according to the attached site; $u' = 0$ if the accelerometer does not move relative to the body. Then, the motion of the body, which is expressed by A_0 and ω, can be calculated. In principle, these two vectors consist of six independent values. A measurement with six adequately arranged accelerometers will provide a complete set of equations. Even when the sites for attaching accelerometers on the body surface are limited, it was shown that the motion can be determined by at least nine accelerometers.[2,3] This means that the motion of a rigid body can be determined (except for the constant linear velocity component) by accelerometers attached to the body and without using any transducers on the fixed coordinate system.

The actual human body is not rigid but is composed of soft tissue and the skin. Thus, it is always difficult to select reference points by which the coordinate system can be defined when referring to the body. In actual measurement situations, adequate techniques have to be considered. For example, a bite bar which is attached to the upper jaw is sometimes used to determine the coordinate system fixed to the head, and a bite bar is also used to attach transducers.

4.2 MOTION MEASUREMENTS

The linear motion of an object can be measured as its displacement, velocity, or acceleration. The rotation can be measured by its rotating angle, angular velocity, or angular acceleration. While a linear motion can be fully determined only by its displacement so that its velocity and acceleration can be obtained by the first and second derivatives of the displacement, there are many situations where direct measurements of velocity and acceleration are much easier and more convenient than computing derivatives of displacement data. For example, a Doppler velocimeter provides direct velocity output. Angular velocities can also be easily measured by using convenient transducers, such as a tachometer which generates an electromotive force proportional to the rotating speed.

Vibration or sound can also be measured by displacement of the vibrating part or the medium; however, transducers which detect velocity or acceleration are sometimes more convenient. In a measurement of body motion outside the body, the limitations of transducers are not so severe, so that even general purpose transducers can be used; however, when a measurement has to be performed inside the body many restrictions arise, and only specially designed transducers can be acceptable.

4.2.1 DISPLACEMENT AND ROTATION MEASUREMENTS BY CONTACT TRANSDUCERS
4.2.1.1 Displacement and Rotation Measurements of the Body and in Extracted Tissue
Displacement and rotation of a part of the body can be measured using various types of transducers, such as resistive potentiometers, photo-encoders, and capacitive and magnetic transducers. The resistive

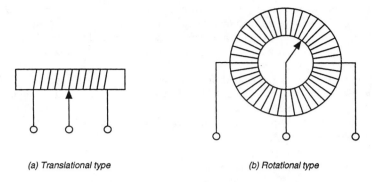

(a) Translational type   (b) Rotational type

Figure 4.1   Potentiometers: **(a)** translational type, **(b)** rotational type.

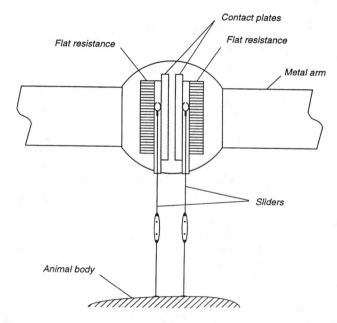

Figure 4.2   A transducer for measuring tissue displacement during blast experiment. (Modified from Clemedson, C.-J. and Jönsson, A., *J. Appl. Physiol.*, 24, 430, 1968.)

potentiometer consists of a resistant element and a movable contact or slider. Translational or rotational potentiometers, as shown in Figure 4.1, are commonly used. Many kinds of potentiometers having different measurement ranges and precision are commercially available. Typical models of the translational type have stroke lengths from 10 to 250 mm with a resolution of about 0.1 mm, nonlinearity of about 0.1%, and friction of about 10 gf. Those of the rotational type have a resolution of about 0.01%, a nonlinearity of about 0.1%, and a torque of about 3 gcm.

The translational potentiometer is used when the motion of the object is linear. Figure 4.2 shows an example of a transient motion measurement in a shockwave experiment.[4,5] An originally designed translational potentiometer having a stroke length of 60 mm was used for recording displacement of the chest and abdominal walls and other parts of the body of laboratory animals. The weight of the slider system attached to the object was about 1 g, the force necessary to overcome the friction of the slider was between 20 and 50 gf, and resolution of the displacement was about 0.1 mm.

The rotational potentiometer is used to measure rotational motions of the body such as joint motion. The goniometer is an instrument which is attached to the body and measures angular displacements of a joint. A simple goniometer consists of a rotational potentiometer as shown in Figure 4.3.[6] However, an actual joint motion is not a simple rotation around one fixed axis, but has a higher degree of freedom.

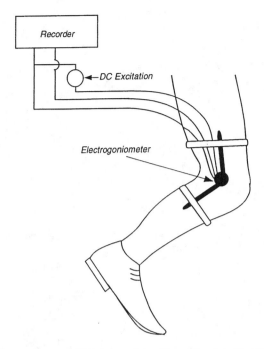

Recorder

←—DC Excitation

Electrogoniometer

Figure 4.3 An example of an electrogoniometer system. (From Contini, R. and Drills, R., in *Advances in Bioengineering and Instrumentation,* Alt, F., Ed., Plenum Press, New York, 1966, 3. With permission.)

For precise measurement of a joint motion, the goniometer used has three rotational potentiometers so that it measures rotations in sagittal, coronal, and transverse planes separately. Such goniometers were used in motion studies of the knee joint[7] and hip joint.[8]

A simple goniometer, in which a pendulum weight is attached to an arm, can be used when only the angle between one body segment and the vertical direction is studied.[9] In this type of goniometer, accurate alignment with a center of joint rotation is unnecessary.

The rotation of the head is sometimes measured by a rotational potentiometer; a flexible rod is attached to a helmet, and its end is connected to a rotational potentiometer.[10]

The friction of a potentiometer may cause measurement errors when the force is small. Friction-free measurement is possible by using optical, capacitive, or inductive types of transducers.

The photopotentiometer consists of a strip of photoconductive material placed between a resistor and conductor elements (Figure 4.4). A constant current is applied to the resistor element. When a light spot falls on the photoconductive strip, a potential output can be obtained between the conductor element and one end of the resistor element which is proportional to the distance between the light spot and the end of the resistor. Figure 4.4(a) shows a photopotentiometer which consists of the conductor and the resistor deposited on a CdS photoconductive layer. Figure 4.4(b) shows a photopotentiometer which consists of a thin layer of n-type silicon formed on a p-type silicon substrate. The p-n junction is nonconductive when a negative bias potential is applied to the p-type substrate, but the junction becomes conductive when a light spot falls on it.

The photopotentiometer has been used in an experimental apparatus that measures smooth muscle contraction by using a CdS-type photopotentiometer.[11] In these experiments, the authors reported that the contraction could be measured in a range from 0.127 to 0.635 cm, and the sensitivity of the output was 2.95 V/mm. A photopotentiometer which measures the skeletal muscle contraction was also used in the experimental apparatus.[12] This photopotentiometer had a measurement range of 10 cm, and its nonlinearity was within 2% in a range of 5 cm near the center.

The one-dimensional image sensor can be used to determine the position of a light spot. Two types of image sensors have been used. As shown in Figure 4.5(a), the metal-oxide semiconductor (MOS)-type image sensor has a photodiode array. Each photodiode is connected to the output port through a switching transistor driven sequentially by a scanner. The charge-coupled device (CCD) image sensor also has a photodiode array as shown in Figure 4.5(b), but the outputs of all photodiodes are transferred

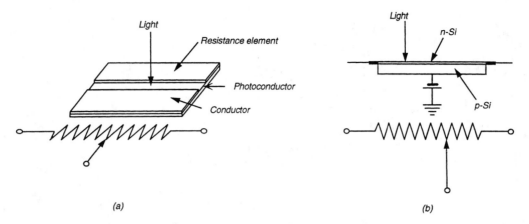

Figure 4.4 Photopotentiometers: **(a)** conductor and resistance element deposited on CdS, **(b)** thin layer of n-type silicon formed on a p-type silicon substrate.

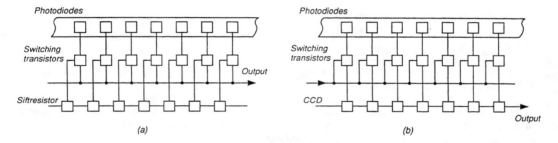

Figure 4.5 Two types of image sensors: **(a)** an image sensor in which outputs of photodiodes are scanned by switching transistors, **(b)** another type of image sensor in which a CCD transfers the outputs from the photodiodes sequentially to the common output port.

simultaneously to the CCD, which then transfers the outputs from the photodiodes sequentially to the common output port.

The photoencoder is a translational or rotational transducer in which the displacement is converted into a pulse sequence, or a series of coded signals, by interrupting light beams at a plate with a slit pattern. Figure 4.6 shows a simple rotary encoder in which the light beam is interrupted by a plate with equally separated slits so that a pulse train is generated when it rotates. A typical commercial rotary encoder provides 100 to 2000 pulses per turn. Also available are rotary and linear photoencoders with coded output; the displacement is converted to a series of digital signals which can be transferred to a computer.

Various types of capacitive displacement transducers have been studied, although most of them are designed for industrial use. Fairly accurate transducers of this type are available. For example, the cylindrical type capacitive transducer shown in Figure 4.7 has a measurement range of 250 mm with a resolution of 0.5 μm.[13]

The differential transformer consists of a core attached to the moving element and primary and secondary coils (Figure 4.8). While higher linearity and sensitivity can be attained, this transformer has the disadvantage that the inertia of the moving element is increased by the core attached to it.

The variable reluctance (or variable inductance) pickup is a displacement transducer which consists of a coil and a magnetic circuit having separate cores, as shown in Figure 4.9. The change of the gap between cores causes a change in magnetic reluctance and results in a change in the inductance of the coil.

Figure 4.10 shows an example of the variable reluctance pickup designed to measure isotonic muscle contraction.[14] A coil with a ferrite sleeve is fitted into the bottom of a flask. A ferrite disk is placed in the cup at the center of the flask and the muscle is attached to the disk, so that the weight of the disk

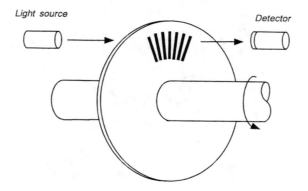

Figure 4.6 A rotary encorder.

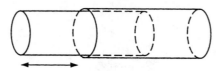

Figure 4.7 A cylindrical type of capacitive transducer for measurement of large displacements.

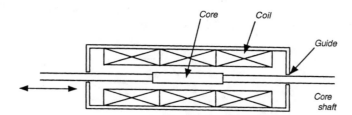

Figure 4.8 The differential transformer.

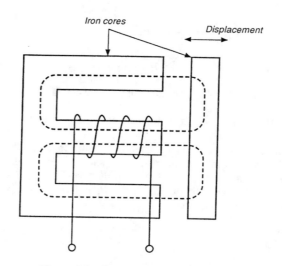

Figure 4.9 A variable reluctance pickup.

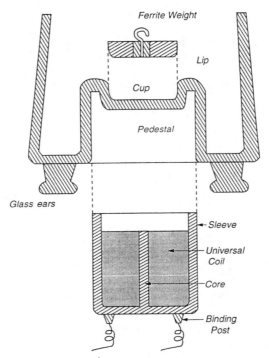

Figure 4.10 A variable reluctance pickup for isotonic muscle contraction measurements. (From Lentini, E. A. and Guyton, W., *J. Appl. Physiol.*, 18, 636, 1963. With permission.)

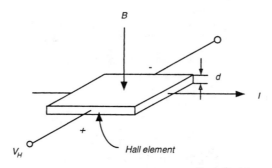

Figure 4.11 The principle of the Hall-effect transducer.

produces isotonic load to the muscle. When the muscle shortens, the disk is shifted vertically and the magnetic reluctance is increased.

The Hall effect transducer is also a kind of magnetic transducer in which displacement is detected by a Hall element placed in an inhomogeneous magnetic field. The Hall element generates a potential along the direction perpendicular to the applied current and magnetic field, as shown in Figure 4.11. When the applied current is I, the magnetic flux density is B, and the thickness of the element is d, then the generated electromotive force, V_H, is given as

$$V_H = R_H IB/d \qquad (4.2)$$

where R_H is the Hall coefficient. The materials InSb and InAs are commonly used in practical Hall elements.

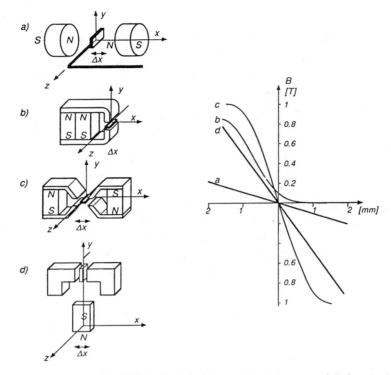

Figure 4.12 Different configurations of Hall-effect displacement transducers and their output characteristics. (From Nalecz, M. et al., *J. Biomed. Eng.,* 4, 313, 1982. With permission.)

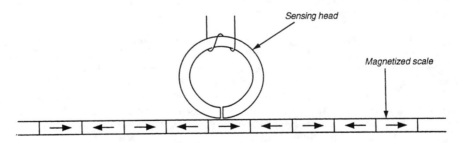

Figure 4.13 The magnetic scale.

Figure 4.12 shows different configurations of the Hall effect displacement transducers and their output characteristics.[15] While (a) and (d) provide linear output, (c) provides the highest sensitivity so that it detects the displacement to the order of 10 Å.[16] The Hall effect displacement transducers have been used widely for bladder motility detection,[17] apex cardiography,[15] mandibular position measurement,[18,19] and measurement of diaphragmatic motion.[20]

The magnetic scale is a displacement transducer in which a magnetic head detects the magnetization pattern on the moving element (just like a tape-recorder head detects magnetization on a tape) as shown in Figure 4.13. Typically, the resolution is about 0.2 mm, and measurement ranges from 0.2 to 3 m are available commercially (i.e., Sony Magnescale Co.; Tokyo, Japan).

The angle of inclination can be measured as the angle to the direction of the gravity, and such an instrument is called a clinometer or inclinometer. Simple clinometers can be made from a potentiometer with a pendulum, but the swing of the pendulum has to be damped when it is used to measure an angle by attaching it on a moving body. By using silicone oil of suitable viscosity, appropriate damping can be obtained so that it responds to frequencies below a few hertz in order to reduce extraneous oscillations.[21]

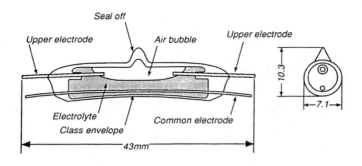

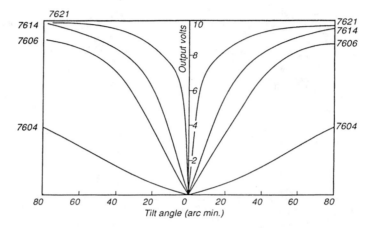

Figure 4.14 A simple clinometer: **(a)** an electrolyte solution is encapsulated in a glass tube with an air space, and **(b)** output vs. tilt angle characteristics of clinometers of different configurations.

A clinometer which measures the displacement of a liquid surface is also available. In the commercial clinometer, AccuStar® (Lucas Control Systems Products; Hampton, VA), displacement of a liquid surface is detected by capacitance change, and it attains a measurement range of ±60°, a resolution of 0.001°, a repeatability of 0.05°, and a time constant of 0.3 s.

A simple clinometer shown in Figure 4.14(a) is also available. In this device, an electrolyte solution is encapsulated in a glass tube with an air space, and the displacement of the electrolyte is detected by the changes in electric impedance between two electrode pairs. Different characteristics, as shown in Figure 4.14(b), can be realized by different geometrical designs.

4.2.1.2 Displacement Measurements *In Vivo*

Deformation of the organs and tissue in the living body can be observed only in the physiological environment, and thus *in vivo* measurement is required. In *in vivo* measurements, commercial transducers are hardly applicable, and appropriate transducer design is postulated according to the measurement situation. Actually, many kinds of laboratory-made transducers have been used in physiological studies.

Simple displacement transducers can be made by using strain-gauges. Metal wire, metal foil, or semiconductor strain-gauges can be attached to a beam so that the deformation of the beam is detected by the change in the length of the gauge. The wire and foil gauges can have a resolution of 10^{-6} in relative changes in their length. For example, a gauge 1 cm long can discriminate 0.01 μm.[22]

Figure 4.15 shows an example of a displacement transducer for a study of myocardial contraction.[23] A strain-gauge sensing element is attached to a c-shaped, highly compliant beam with two prongs that are inserted into the ventricular wall. Segmental length changes of the myocardium can be recorded with this transducer. One example of a transducer of this type weighed about 0.25 g, had a frequency response of up to 30 Hz, and could be used for up to 6 hours; the maximum load on the ventricular muscle was less than 10 gf.[24]

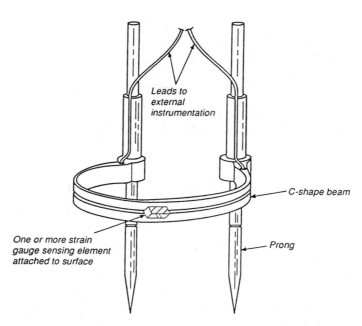

Figure 4.15 A displacement transducer for the study of myocardial contraction. (From Feldstein, C. F. et al., *Med. Instrum.,* 14, 277, 1980. With permission.)

Intravascular measurement of the arterial diameter change is sometimes required, and special transducers were devised for this purpose. Figure 4.16(a) shows an example of a simple inductive transducer which can be inserted into the artery through a catheter.[25] The sensing part consists of a spring steel wire loop and two copper wire loops. All wire loops are cemented together so as to form a single unit. When an a.c. current is applied to the primary coil, the electromagnetic force induced in the secondary coil is approximately proportional to the logarithm of the coil diameter, which is the arterial internal diameter as long as the coil stays in contact with the inside wall by means of the spring force. It was shown that variations as small as 0.01 mm in a 10-mm diameter lumen could be recorded.

Figure 4.16(b) also shows an arterial diameter transducer that uses a miniature transformer.[26] It consists of primary and secondary coils wound around a catheter and an elliptical-shaped spring steel wire to which a core rod is attached at the distal end so that the coupling between two coils depends on the position of the core. The resolution was 0.005 mm, the error of the mean diameter measurement *in vivo* was 2%, and the maximum force exerted to the arterial wall was 4.5×10^{-2} N. While the transducer has a limited range of linearity, transducers of four different sizes could cover the range from 3.5 to 10 mm of arterial diameter. A combined transducer which measures vessel diameter and flow velocity by using miniature transformers was also reported earlier.[27]

In order to measure cardiac ventricular diameter, the sonomicrometer has been used. It consists of two ultrasound transducer crystals and the electronic circuit as shown in Figure 4.17. The ultrasonic pulse transmitted from a transmitter crystal is received by another crystal. The distance between two crystals can be estimated by the transit time of the ultrasonic pulse, if the sound velocity in the medium between the crystals is known. Figure 4.18 shows an example of the construction of the transducer.[28] A plastic lens is glued to the crystal face to diffuse the ultrasonic beam so that the beam reaches the receiver crystal even when the alignment of these two crystals is inaccurate.

A study showed that the error of a sonomicrometer was less than 2% when the transducer crystals were placed 50 mm apart in water.[29] Another study showed that a resolution of 0.15 mm was obtained by using a 10 MHz clock for time measurement, and the resolution reached 0.03 mm when a 50-MHz clock was used.[30] The sonomicrometer has also been used in chronic experiments. In one study, it was in operation for more than 5 months.[31]

Displacement of a surface that reflects ultrasound can be tracked by the pulse-echo technique. This technique has been widely used in clinical examinations, especially for investigating the motion of the cardiac valve. Figure 4.19 shows a method of observing motion of the mitral valve. A transducer is placed on the chest wall. It generates ultrasound pulses, and reflected ultrasounds are received by the same transducer. The

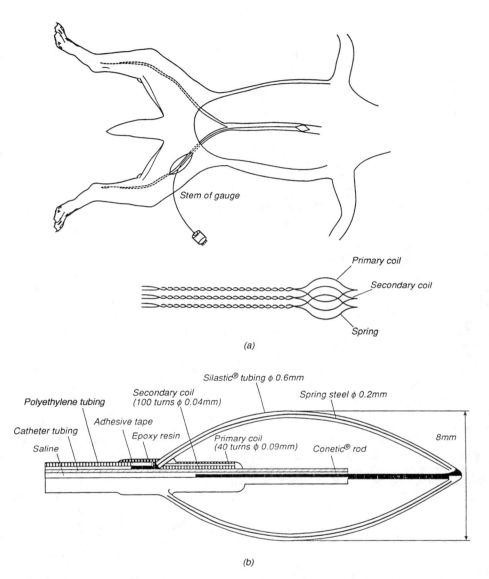

Figure 4.16 Inductive transducers for vascular diameter measurements: **(a)** an intraarterial induction coil, **(b)** a catheter-tip transducer using a miniature transformer. (Part (a) modified from Kolin, A. and Culp, G. W., *IEEE Trans. Bio-Med. Eng.,* BME-18, 110, 1971. Part (b) from van der Schee, E. J. et al., *Med. Biol. Eng. Comput.,* 19, 218, 1981. With permission.)

motion of the reflecting surface can be traced by the change in transit time. It can be visualized by the M-mode display, as shown on the right side of the figure. In the vertical scan on the CRT display, the brightness of the spot is modulated by the amplitude of echo, so that strong echoes produce bright spots on the vertical line. Then, the vertical line display is slowly swept horizontally, and the change of the depth of the reflecting surface corresponding to each specified echo is displayed as a trace of bright spots.

The spatial resolution along the ultrasound beam axis is on the order of the wavelength of the ultrasound when the beam is narrow and the reflecting surface is perpendicular to the beam. The effect of the beam width on the spatial resolution is more pronounced if the reflecting surface is not perpendicular to the beam. To reduce the beam width, the spherically curved transducer (as shown in Figure 4.20) can be used.[32] The beam width has a minimum at the focal distance, f, and the beam width — defined by the range in which the intensity of the sound is above half of that on the axis, which is called the full width at half maximum (FWHM) — is given by

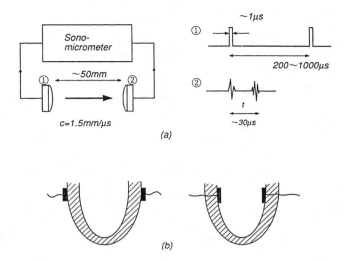

Figure 4.17 **(a)** A sonomicrometer using two ultrasound transducer crystals, and **(b)** transducer crystal arrangements in the measurement of cardiac ventricular diameter.

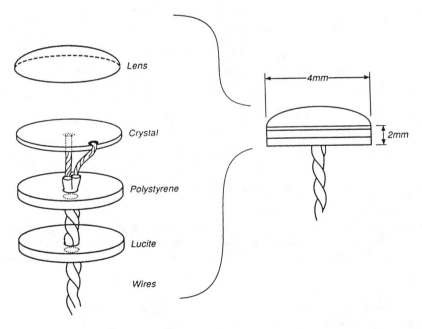

Figure 4.18 An example of construction of the transducer crystal in the sonomicrometer shown in Figure 4.17. (From Horwitz, L. D. et al., *J. Appl. Physiol.*, 24, 738, 1968. With permission.)

$$\text{FWHM} = 1.41 f \lambda / 2a \qquad (4.3)$$

where f is focal distance, λ is the average wavelength, and $2a$ is the diameter of the transducer crystal.

When the echo is received by the same transducer, the echo near the axis is stronger than that from off the axis, and lateral resolution becomes about $f\lambda/2a$, as long as the reflecting surface is perpendicular to the beam. For 5 MHz, λ is about 0.3 mm, and if $2a = 13$ mm and $f = 70$ mm, then the lateral resolution is about 1.6 mm, while FWHM is about 2.28 mm. However, if the reflecting surface is inclined from the plane perpendicular to the axis, components of the beam cause differences in the transit time of echoes; consequently, axial resolution is reduced.[33]

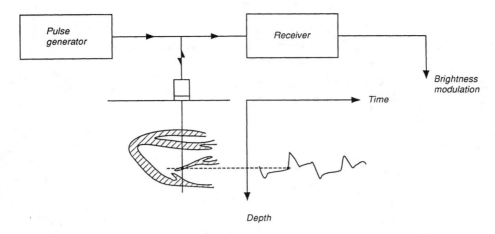

Figure 4.19 A method of observing motion of the mitral valve using M-mode display.

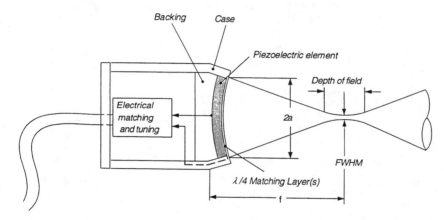

Figure 4.20 A focused-beam ultrasound transducer. (From Hunt, J. W. et al., *IEEE Trans. Biomed. Eng.*, BME-30, 453, 1983. With permission.)

The axial resolution for a small displacement of the reflecting surface can be improved by the phase-locked echo tracking system. A study showed that vascular diameter changes as small as 2 μm could be resolved using 7.8 MHz pulses.[34] However, absolute accuracy of a measurement of vascular diameter is limited by the wavelength of the ultrasound.

4.2.2 NONCONTACT MEASUREMENT OF DISPLACEMENT AND ROTATION
4.2.2.1 Optical Methods
In order to investigate body motion optically, motion picture and other photographic methods have been used widely.[6] Automated motion measurement systems are also available and are more convenient for data analysis. In these systems, the spacial coordinates of specific positions on the body are measured and stored in a computer.

Figure 4.21 shows an example of the motion measurement system for gait analysis.[35] The subject traverses the gait path, and the motion is observed by using a videocamera. Passive retro-reflecting markers are attached to some specific part of the body, such as the neck, shoulder, elbow, wrist, waist, knee, ankle, heel, or toe. Markers are illuminated by near-infrared light so that the measurement can be performed in a room illuminated by fluorescent light. The position of each marker can be determined by an intelligent tracking algorithm. It was shown that the measurement error was less than 0.1%. Figure 4.22 shows an example of the tracker motion data from a normal adult.[35] Three-dimensional analysis can be performed using two or three cameras.

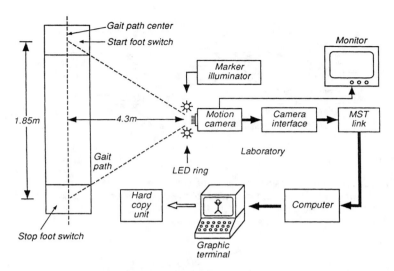

Figure 4.21 An example of the motion measurement system for gait analysis. (From Tayler, K. D. et al., *J. Biomech.*, 15, 505, 1982. With permission.)

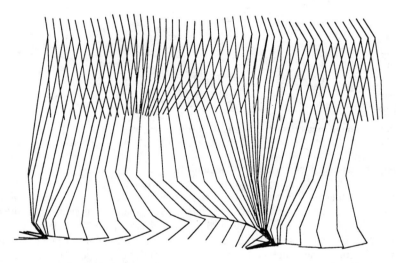

Figure 4.22 An example of the tracker motion data. (From Tayler, K. D. et al., *J. Biomech.*, 15, 505, 1982. With permission.)

A light-emitting diode (LED) can also be used as the marker. If many LEDs are driven sequentially, the spacial coordinate of each LED can be determined more easily than in a situation where many reflecting markers appear simultaneously, because in the sequential operation only one light spot is seen at one time.

The position of a light spot in a two-dimensional space can be directly determined by a two-dimensional position-sensitive detector. The principle is similar to that of the photopotentiometer described in the preceding section. Figure 4.23 shows two different configurations of position-sensitive detectors:[36] (a) the tetralateral position-sensitive detector in which the photoconductive layer is formed at one side of the device, and (b) the duolateral position-sensitive detector which has two photoconductive layers at both sides of the device. While the distortion of the image is small in the duolateral type, the tetralateral type is advantageous because of less dark current. Position-sensitive detectors of both types are available commercially (Hamamatsu Photonics Co.; Hamamatsu, Japan).

The position of a light marker in a three-dimensional space can be determined by using three one-dimensional position sensors. As shown in Figure 4.24, if a target light spot is projected by a cylindrical

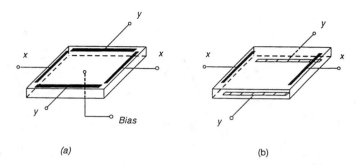

Figure 4.23 Two configurations of position-sensitive detectors: **(a)** tetralateral configuration in which the photoconductive layer is formed at one side of the device; **(b)** the duolateral configuration which has two photoconductive layers at both sides of the device.

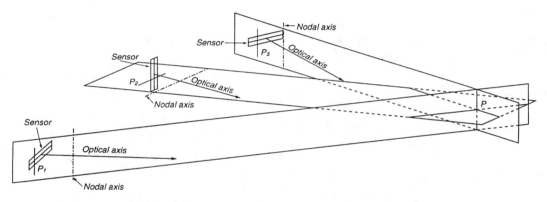

Figure 4.24 Measurement method of the position of a light marker using three one-dimensional position sensors. (From Macellari, V., *Med. Biol. Eng. Comput.*, 21, 311, 1983. With permission.)

lens onto a one-dimensional position sensor placed at the focal plane of the lens so that the sensor axis is orthogonal to the nodal axis of the lens, a plane is determined which includes the target light spot and the nodal axis of the lens. By using three sensors with cylindrical lenses arranged in different directions, the position of the target can be computed as the intercept of these three planes.[37] It was reported that by using CCD-type, one-dimensional position sensors with 2048 elements, a resolution of 1/4000 was attained in a measurement field 2.2 × 0.6 × 2.2 m at a viewing distance of 5 m.

Instead of positioning many cameras or image devices, multiple perspective views can be obtained by using mirrors. While two perspective views are sufficient for space localization of an object, redundant data obtained by three or more perspective views will improve the accuracy. A study showed that a three-mirror configuration provides numerical data of motions of body segments, and the body motion can be reconstructed on a display.[38]

In order to detect the rotational angle without having contact, polarizing filters can be used. If a light beam passes through two polarizing filters, the transmitted light intensity depends on the angle between the planes of polarization of two filters. Higher resolution and accuracy are achieved by using rotating polarizers.[39] The intensity of a light beam passed through a rotating polarizer and a stationary one varies sinusoidally. Also, the phase angle of the sinusoidal intensity variation changes when the orientation of the polarization of the stationary polarizer is changed, as shown in Figure 4.25. In a study, this principle was applied to joint angle measurement, and it was reported that by using a polarizer 6 in. in diameter and rotating at 15,000 rpm, an accuracy of 0.1° and a nonlinearity of ±1.5° were attained over a measurement angle of 90°.[39] While the misorientation of two polarizers in respect to the optical axis causes error, it was reported that the measurement error was less than 1° if the relative angle of two polarizers remains within 15°.

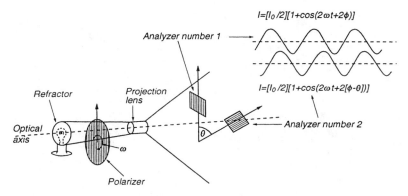

Figure 4.25 A method of rotational angle measurement using rotating and stationary polarizers. (From Reed, D. J. and Reynolds, P. J., *J. Appl. Physiol.*, 27, 745, 1969. With permission.)

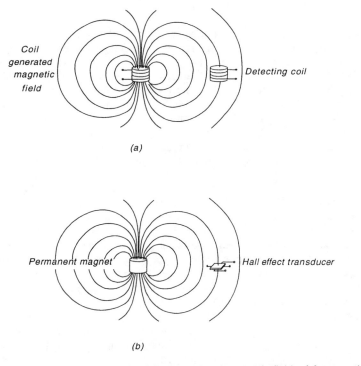

Figure 4.26 Displacement measurements using inhomogeneous magnetic fields: **(a)** magnetic field is generated by a coil and detected by another coil; **(b)** a permanent magnet and a Hall effect element are combined.

4.2.2.2 Magnetic Methods

If a magnetic pole is attached to an object, an inhomogeneous magnetic field is generated in the surrounding space. The position of the magnetic pole can be determined by measurement of the magnetic field distribution. Figures 4.26(a) and (b) show simple examples. In configuration (a), a magnetic field is generated by a coil and detected by another coil. In configuration (b), a permanent magnet and a Hall element are combined. Both methods involve the use of magnetometers. While the magnetic flux density is inversely proportional to the cube of distance from a magnetic dipole to the observation point, quantitative measurement of distance is possible over a limited range. Simple magnetometers are used for respiratory monitoring in which the respiratory movements of the chest wall or the abdomen are detected.[40-45]

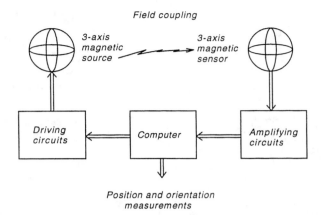

Figure 4.27 A method of determination of relative position and orientation in three-dimensional space using a three-axis magnetic dipole source and a three-axis magnetic sensor. (From Raab, F. H. et al., *IEEE Trans. Aerospace Electron. Syst.*, AES15, 709, 1979. With permission.)

In order to measure relative position and orientation in three-dimensional space, a system consisting of both a three-axis magnetic dipole source and a three-axis magnetic sensor can be used. In a system developed by Raab et al.,[46] both the source and the sensor consist of three orthogonally arranged coils wound around a ferrite core. Three coils in the source are excited sequentially by means of a controlling computer, and signals from three coils in the sensor are led to the computer as shown in Figure 4.27. Then, the position and orientation of the sensor relative to the source, which consists of 6° of freedom, can be computed. Such a system is available commercially (3 Space, Polhemus Inc.; Colchester, VT). A typical application of this system is to attach the sensor on a helmet and track the wearer's line-of-sight.[46] The system has also been used in measurements of various kinds of body movement, such as wrist motion analysis,[47] and lumbar spine mobility measurement,[48] and twisting mobility of the human back.[49]

4.2.3   LINEAR AND ANGULAR VELOCITY MEASUREMENTS

When a displacement of an object is measured continuously and precisely, the instantaneous velocity of the object can be obtained as the time derivative of the displacement. However, the noise contained in the displacement measurement may cause significant errors in determining velocity by differentiation of the output. Thus, direct measurement of velocity by a velocity transducer is sometimes advantageous.

The mean velocity in a definite time interval can be determined simply as the displacement occurring divided by the time interval. If the position of a marker is measured repetitively, the mean velocity in each time interval can be determined. Such a measurement is commonly performed in body motion analysis.

By integrating the acceleration of an object, its velocity can be determined. The advantage of acceleration measurement is that the measurement can be attained by an accelerometer attached to the moving object, and it does not require any observation from the stationary coordinate. On the other hand, determination of velocity from acceleration requires the initial velocity, and systematic error in the acceleration output is accumulated in the integration process and may cause a drift in the data. While velocity transducers have not been widely used in the biomedical field, many linear and angular velocity transducers of different principles exist, and they are advantageous in some measurement situations.

4.2.3.1   Electromagnetic Velocity Transducers

When the magnetic flux across a coil is varied, an electromotive force proportional to the time derivative of the magnetic flux is induced. In addition, when a conductor is moved in a magnetic field an electromotive force proportional to the velocity of the motion is induced in a direction perpendicular to directions of the motion and magnetic field. Both phenomena are employed in velocity transducers.

Figure 4.28 shows two types of velocity transducers that are applicable to measurements in a limited range of motion. Figure 4.28(a) shows a transducer that has a moving magnet in a solenoid coil. By using a coil consisting of two symmetrical parts in which wires are wound in opposite directions, the linear measurement range is extended. Commercial transducers of this type have a measurement range

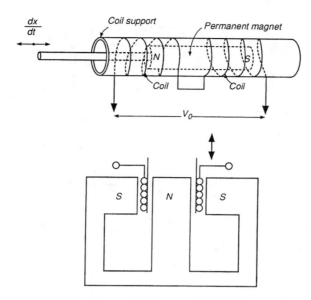

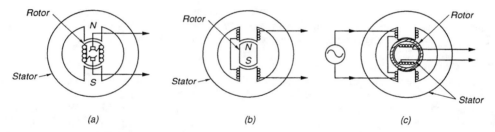

Figure 4.28 Two types of electromagnetic velocity transducers.

Figure 4.29 Angular velocity transducers: **(a)** the d.c. tachometer, **(b)** the a.c. tachometer, and **(c)** the drag-cup tachometer.

of 12.7 to 22.8 mm and a sensitivity of 4.7 to 5.6 mV/mm/s, and the weight of the moving parts is in the range of 3.5 to 66 g (Shinko Co.; Kobe, Japan). Figure 4.28(b) shows a transducer in which a moving coil is placed in a stationary magnetic field, and it provides an electromotive force proportional to the velocity.

The tachometer is essentially an angular velocity transducer that generates an electromotive force roughly proportional to the rotational speed. As shown in Figure 4.29, there are different types of tachometers, all based on electromagnetic induction. Figure 4.29(a) shows the configuration of the d.c. tachometer generator. The stator consists of a magnet, and the rotor has a rotating coil in which an electromotive force is generated when it rotates. While the polarity of generated potential in the coil alternates, d.c. output is provided when the output is derived through a commutator. The polarity of the output is inverted when the direction of rotation is inverted.

Figure 4.29(b) shows the configuration of the a.c. tachometer generator. The rotor is a permanent magnet, and the output is derived from the coil at the stator. It generates a.c. output proportional to the speed of rotation.

Figure 4.29(c) shows the configuration of the drag-cup tachometer. It has two stators with coils that are perpendicular to each other; a.c. excitation is applied to a coil, and the output is derived from another coil. A cup-shaped rotor made from a conductor is placed in between two stators. During rotation, eddy currents are induced in the cup, the current produces a magnetic field, and the a.c. output of the excitation frequency is induced in the second coil. The output amplitude is proportional to the rotation speed. The phase is inverted when the direction of rotation of the cup is inverted.

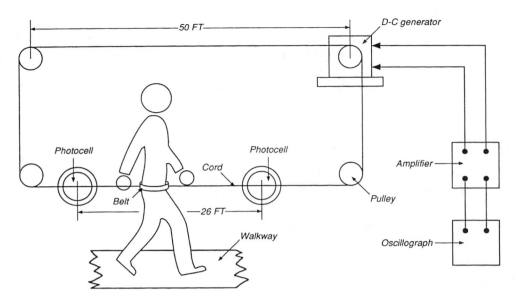

Figure 4.30 Measurement system for level walking using a tachometer. (From Contini, R. and Drills, R., in *Advances in Bioengineering and Instrumentation,* Alt, F., Ed., Plenum Press, New York, 1966, 3. With permission.)

Using a pulley, a tachometer can be used to measure translational motion. Figure 4.30 shows an example in which the instantaneous horizontal velocity of the subject's trunk during level walking is recorded by a tachometer activated by a cable cord passing over a pulley.[6]

4.2.3.2 Doppler Methods

When a moving object transmits a sound or an electromagnetic wave of a definite frequency, the observed frequency of the wave received by a stationary observer differs from that of the transmitter. The difference in frequency is the Doppler shift and is proportional to the velocity of the transmitter. When a transmitter approaches the receiver with a velocity, U, emitting a wave of frequency, f_s, then the Doppler shift is approximately Uf_s/c, where c is the propagation velocity of the wave, if $U \gg c$. When a stationary observer transmits a wave and receives a wave reflected by a moving object, the Doppler shift is approximately $2Uf_s/c$, as described in Section 3.2.2.

While the ultrasonic Doppler technique has been used widely in blood velocity measurements, it has also been used for detecting motions in the body, such as the fetal heartbeat and breathing in the uterus. The velocity of the cardiac ventricular wall is smaller than the blood velocity in large vessels, and the chest wall movement accompanying fetal breathing is much slower, about 1 to 3 cm/s at maximum.[50] Besides the lower object velocity, fetal measurements have to be performed through the thick abdominal and uterine wall, and thus ultrasound of relatively lower frequency has to be used because the absorption coefficient is almost proportional to the frequency. In practice, the frequency of the ultrasound has to be 2 MHz or less, and thus the Doppler shift for a motion of 3 cm/s is about 80 Hz. Consequently, Doppler shifts in the lower frequency range have to be detected in ultrasonic Doppler measurement of fetal motion.

If many moving objects exist in the measurement space, each object cannot be specified by a continuous wave Doppler measurement. In order to detect the Doppler shift in a specific range selectively, the pulsed Doppler measurement (as described in Section 3.2.2.4) has to be employed. An example showed that the motion of the anterior wall of the heart down to 10 cm/s could be recorded satisfactorily in a range of 2 cm at a distance of 8 cm from the transducer using 2-MHz ultrasound pulses with a repetition frequency of 1 kHz.[51]

4.2.3.3 Angular Velocity Transducer

Angular velocity can be measured by using a gyroscope which consists of a wheel rotating at high speed. A rotating wheel has a tendency to keep the axis of rotation unchanged. When the axis is forced to change its direction, a torque arises that is proportional to the angular velocity of the axis inclination,

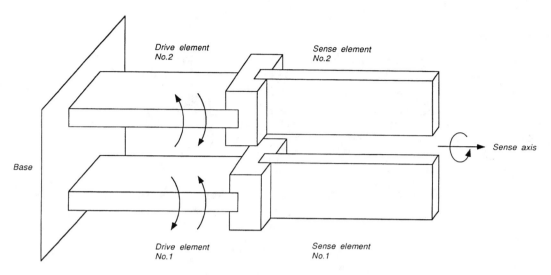

Figure 4.31 Angular velocity transducer called the tuning fork.

and hence it can be used to detect angular velocity. This type of transducer is known as the rate gyro, and a resolution on the order of 0.005°/s is attained in the model having a measurement range of ±10°/s.[52]

Angular velocity can also be detected directly using a transducer called the tuning fork. Figure 4.31 shows its configuration. It consists of four piezoelectric elements (bimorphs). Two are for driving and the other two are for sensing. The driving element is attached to the base and is connected to the sensing element so that they are perpendicular to each other. The driving element bends resonantly through electric excitation, and hence the sensing element swings. Under zero angular velocity, the sensing element does not produce a signal. But, when it rotates, a bending motion is induced due to the Coriolis force, and consequently a signal proportional to the angular velocity is induced in the sensing element. While a translational acceleration, having a component in the direction normal to the surface of the sensing element, may also induce a signal, such a component can be compensated for by employing differential operation of two pairs of elements. A commercial model (Angular Rate Sensor, Watson Industries Inc.; Eau Clair, MA) has the following specifications: a measurement range of ±30°/s, ±100°/s, or ±300°/s; a resolution of either 0.04°/s or 0.1% of the maximum angular velocity; a precision of 2%; a nonlinearity of 0.1% of a maximum angular velocity; a frequency response of 70 Hz; a power consumption of 300 mW; and a weight of 110 g. A model consisiting of two or three axes is also available.

4.2.4 TRANSLATIONAL AND ANGULAR ACCELERATION MEASUREMENTS

While the acceleration of an object can be derived from the velocity as its first derivative or from the displacement as its second derivative, differentiation of the signal usually causes noise. Direct measurement of the acceleration is often more convenient and easier. The acceleration of a linear motion can be measured simply as the force acting on a mass, and the angular acceleration on an axis can be measured as the torque appearing in a body having a moment of inertia at the axis of rotation. However, when both translational and rotational motions exist, the situation is more complicated so that apparent forces, such as centrifugal and Coriolis forces, may appear (as described in Section 4.1.3).

Many kinds of accelerometers of different specifications are commercially available. In acceleration measurement of human body motions, the maximum acceleration appearing due to voluntary activities is no more than a few *g* and remain in relatively higher sensitivity ranges among commercial accelerometers. By using silicon micromachining technologies, very small but sensitive accelerometers have been produced.

4.2.4.1 Translational Accelerometers

Most accelerometers having a sensitivity in the acceleration range of body motion are of the beam type. An elastic beam is fixed to the base at one end and a mass, called the seismic mass, is attached to the other end as shown in Figure 4.32(a). When the seismic mass is accelerated, a force proportional to the mass times the acceleration appears, and the beam bends elastically in proportion to the force. To avoid

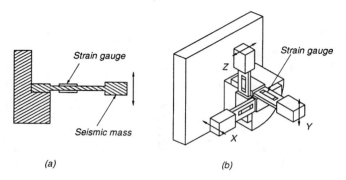

Figure 4.32 Configurations of (a) one-axis and (b) tri-axis accelerometers.

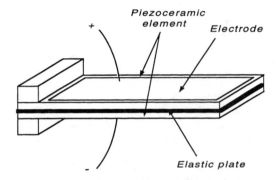

Figure 4.33 The bimorph consists of a beam with two piezoelectric elements of different polarities.

resonant oscillation of the seismic mass after a transient input, the mechanical system should be designed so as to have an adequate damping coefficient. Instead of a beam, a diaphragm, spring, or any other elastic material can be used in the accelerometer.

In order to determine the amplitude and the direction of acceleration in a three-dimensional space, either three translational accelerometers or a three-axis accelerometer must be used. Figure 4.32(b) shows an example of the three-axis accelerometer in which three beams with seismic masses are assembled on a base so that the sensitive directions of these beams are arranged perpendicularly to each other.

The displacement of the seismic mass can be detected by different sensing principles such as piezoresistive, piezoelectric, or capacitive. The semiconductor strain-gauge type of accelerometer has been used widely in measurements of motions in humans and animals. For example, a small but sensitive uniaxial accelerometer is available which has a measurement range of ± 5 g, a dimension of $3.56 \times 3.56 \times 6.86$ mm, and a weight of 0.5 g (Entran Devices, Inc.; Little Falls, NJ).

The piezoelectric type of accelerometer has been used when only an a.c. component of acceleration needs to be measured. In the piezoelectric material, a polarization voltage appears which is proportional to the deformation. The polarity of the polarization voltage depends on the molecular structure of the material. Figure 4.33 shows an example of a beam with two piezoelectric elements of different polarities in order to produce double or differential output. This configuration is called the bimorph.

The amount of the developed charge can be measured by the terminal voltage, because the piezoelectric element has a capacitance, and the terminal voltage is proportional to the stored charge. More accurately, the amount of the generated charge is measured by a charge amplifier as shown in Figure 4.34. When the input capacitance, which includes the capacitance of the piezoelectric element, C, and the stray capacity is C_i, then the following relations will be valid:

$$Q = C_i V_i + C(V_i - V) \tag{4.4}$$

$$V = -A V_i \tag{4.5}$$

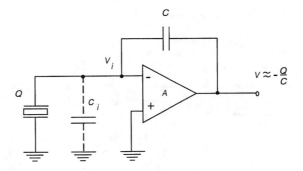

Figure 4.34 A charge amplifier.

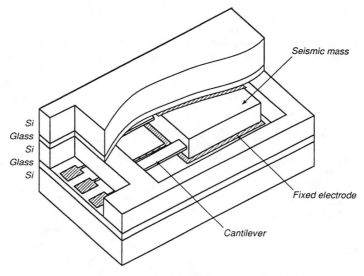

Figure 4.35 A capacitive-type accelerometer produced by silicon fabrication technology. (From Rudolf, F. et al., *Sensors Actuators,* A21-23, 297, 1990. With permission.)

where Q is the generated charge; V_i, V, and A are the input and output voltages and the gain of the operational amplifier, respectively. If $A \gg 1$, and $AC \gg C_i$, then

$$V = -Q/C \qquad (4.6)$$

This means that the output voltage is proportional to the generated charge regardless of the input capacitance. While the actual circuit has a finite time constant mainly determined by the leakage current of the feedback capacitor, very long time constants can be achieved by using a high quality capacitor. For example, a commercial charge amplifier has a time constant on the order of 10^5 s (Kistler 5007; Winterthur, Switzerland). Another advantage of using a charge amplifier is that the sensitivity is not affected by the stray capacitance of the connecting cable and the capacitance of the piezoelectric element.

Simple acceleration pickups making use of the piezoelectric principle have been used. For example, an acceleration pickup to be attached to the wrist was used for monitoring body motion during sleep.[53,54] Making use of the micromachining technology, very small but highly sensitive accelerometers have been developed. Figure 4.35 shows a typical configuration of an accelerometer with capacitive detection. A seismic mass supported by a beam is formed by silicon fabrication technology, and the displacement of the seismic mass is detected by the change of capacitance between the electrode on the mass and the fixed electrode. For example, very high sensitivity, such as a working range of ±0.1 g, can be attained in a device which has a seismic mass of 14.7 mg, an electrode gap of 7 μm, chip size of 8.3 × 5.9 × 1.9 mm, and a resonance frequency of 126 Hz.[55]

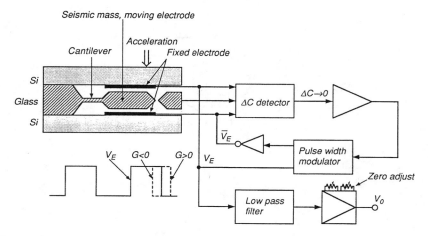

Figure 4.36 An accelerometer in which a servo-control operation is introduced so that the force appearing at the seismic mass by acceleration is balanced by an electrostatic force. (From Suzuki, S. et al., *Sensors Actuators, A21-23*, 316, 1990. With permission.)

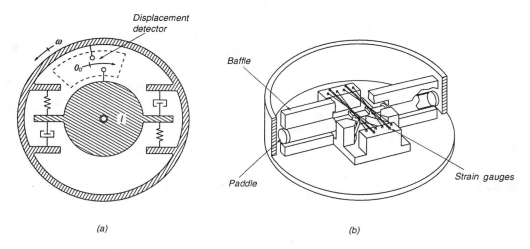

Figure 4.37 Angular accelerometers: **(a)** the moment appearing at a rigid body is measured by the displacement of a spring; **(b)** the fluid flow generated by angular acceleration is detected. (Part (a) from Doebelin, E. O., *Measurement System Application and Design*, 4th ed., McGraw-Hill, New York, 1990. With permission. Part (b) modified from Rosa, G. N., Instrument notes, 26, Stratham Instruments, Inc., Oxnard, CA, 1954.)

By introducing an electrostatic servo-control operation (Figure 4.36)[56] the force appearing in the seismic mass caused by acceleration can be balanced by an electrostatic force so the seismic mass stays at the equilibrium point. This method is advantageous in that nonlinearity and fracture, due to a large displacement of the seismic mass, can be avoided and adequate damping can be applied by adjusting feedback circuit parameters.

4.2.4.2 Angular Accelerometers
Direct measurement of angular acceleration is attained by measuring a torque appearing on a body which has a moment of inertia around the axis of rotation. If this torque is T, and the moment of inertia is I, then the angular acceleration is given by T/I, as long as the axis of rotation is fixed.

Figure 4.37(a) shows an example of an angular accelerometer in which the momentum appearing at a rigid body supported by a shaft is measured by the displacement of the spring connected to it.[52] Figure 4.37(b) shows a different type of angular accelerometer in which a liquid is used instead of a rigid body, and the flow generated by angular acceleration is detected by the force exerted on a paddle.[57]

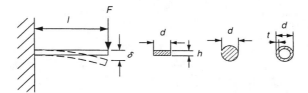

Figure 4.38 Cantilever beams for muscle contraction measurements.

4.3 FORCE MEASUREMENTS

Muscles produce forces and may cause body movements which in turn cause a reactive force from the environment, such as the reaction from a floor when a subject walks on it. Many forces can be applied to the body passively from the environment due to natural phenomena or artificial origins. Force measurements are required when these forces are studied quantitatively.

4.3.1 MUSCLE CONTRACTION MEASUREMENTS

In physiological studies of the muscles, the contractile force of the isolated muscle is measured. During the measurement, one end of the muscle is connected to a force transducer, and the other end is fixed. In isometric contraction studies, the muscle length is kept constant, and displacement of the moving part of the force transducer will thus be kept to a minimum. In such measurements, the beam-type transducer is commonly used, in which the force is converted into a small displacement of the beam. The sensitivity and displacement of the transducer depend on the design of the beam.

4.3.1.1 Design of the Elastic Beam for Muscle Contraction Measurements

In muscle contraction measurements, the cantilever beam as shown in Figure 4.38 is commonly used. The relation between the applied force and its displacement depends on the shape and the dimensions of the beam, as well as the mechanical properties of the beam material. When a force, F, is applied to the free end of a rectangular cross-sectional beam of length L, width b, thickness h, and Young's module E, perpendicular to the beam axis, the displacement, δ, at the end of the beam is given as

$$\delta = 4L^3 F/Ebh^3. \tag{4.7}$$

For a beam with a circular cross-section of diameter d, the displacement is given as

$$\delta = 64L^3 F/3\pi Ed^4. \tag{4.8}$$

For a beam made from thin wall pipe with a thickness, t, the displacement is given as

$$\delta = 4L^3 F/3\pi Ed^3 t. \tag{4.9}$$

As seen in these expressions, the displacement of a beam is generally proportional to higher powers of geometrical parameters. This means that the sensitivity of the beam-type force transducer is largely affected by its geometry, and accurate design is required to realize the desired sensitivity. On the other hand, transducers having different sensitivities can be made by a slight change in the geometrical parameters. The details of cantilever beam design for force transducers are described by McLaughlin.[58] Instead of a cantilever, any kind of elastic component can be used for converting force to small displacement or strain. Sometimes helical or spiral springs are also used for this purpose.

4.3.1.2 Force Measurements in Isolated Muscles

In physiological studies of the muscle, the contraction is measured by using a force transducer. Because of the difficulty in preparing a large specimen, small isolated muscle strips only a few millimeters or less in length are commonly used. A small specimen is advantageous as it is maintaining its activity only by a diffusive oxygen supply from the surrounding solution. A force transducer for such a measurement has to be sensitive and of a low compliance and have low drift and a fast response.

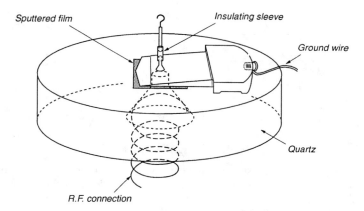

Figure 4.39 A cantilever-type force transducer with capacitive detection. (From Hamrell, B. B. et al., *J. Appl. Physiol.*, 38, 190, 1975. With permission.)

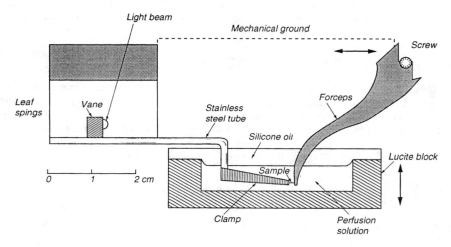

Figure 4.40 An isometric force recording system. (Modified from Hellam, D. C. and Podolsky, R. J., *J. Physiol. (Lond.)*, 200, 807, 1969.)

For example, Chapman[59] reported a force transducer with a Perspex cantilever 1.5 × 3.0 × 26.0 mm in size. The strain of the cantilever was detected by a piezoelectric strain-gauge, and it gave a linear output from 10^{-7} to 10^{-1} N with a resonant frequency of 1.1 kHz and a compliance of 2 mm/N. The same cantilever with a piezoresistive strain-gauge was also examined and provided a resonance frequency of 460 Hz and a compliance of 8 mm/N, and the noise and drift level was less than 10^{-7} N.

A cantilever-type force transducer with capacitive detection as shown in Figure 4.39 was reported by Hamrell et al.[60] A cantilever made by Invar was cemented to a quartz disk at one end and had at the other end an air gap of 0.025 mm between the cantilever and the vacuum-sputtered film on the quartz disk. The estimated movement of the cantilever was less than 0.25 mm, the resonance frequency was 600 Hz, and the signal-to-noise ratio was 100 at a force level above 10^{-2} N.

In order to detect a small displacement of a cantilever or spring, optical methods have also been used. Figure 4.40 shows an isometric force recording system.[61] A muscle preparation is held by forceps at one end, and the other end is held by a tube with a clamp that is coupled to a pair of phosphor-bronze leaf springs. The tube carries a vane that blocks the light beam partially. The characteristics of the system at a moving part at weight of 300 mg included the measurement range of 5×10^{-6} to 9×10^{-3} N, compliance of 18 mm/N, and time constant of 0.02 s.

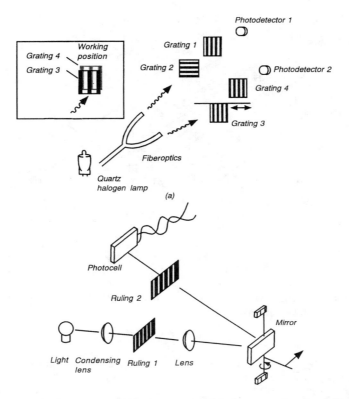

Figure 4.41 Methods of detecting small displacements using light beams and gratings. (Part (a) modified from Canaday, P. G. and Fay, F. S., *J. Appl. Physiol.,* 40, 243, 1976. Part (b) modified from Minns, H. G. and Franz, G. N., *J. Appl. Physiol.,* 33, 529, 1972.)

To detect small displacements using a light beam, parallel gratings with equally spaced, alternating clear and opaque bands were also used. Figure 4.41 shows an example.[62] It has two beams for differential operation so as to compensate for the fluctuation of the light source. One beam passes through two fixed gratings arranged at a right angle, and another beam passes through two parallell gratings in which one of them moves in respect to the other. The gratings used had 250 lines per inch, and the force ranged from 10^{-7} to 2×10^{-5} N with a compliance of 160 mm/N and a resonant frequency of 105 Hz. Minns and Franz[63] also reported a force transducer using gratings, which had a measurement range of up to 3×10^{-4} N with a compliance of 33 mm/N and a resonance frequency of 400 Hz.

4.3.1.3 *In Vivo* Measurements of Muscle Contraction

In order to measure the contracting force of a muscle *in vivo,* special transducers that can be attached to the muscle are required. Sometimes, chronic observations over several days or weeks are required, using implantable transducers in experimental animals.

Figure 4.42 shows an implantable myocardial force transducer which is sutured to the ventricular wall of a heart.[64] A bonded strain-gauge element is cemented on a spring bronze plate and glued with epoxy resin. The measurement range was 0 to 2 N, and the resonance frequency was 450 Hz.

Implantable force transducers have also been used for recording activities of the gastrointestinal tract. Jacoby et al.[65] reported a transducer (Figure 4.43) that consisted of a Be-Cu sheet, 3×9 mm, onto which a foil strain-gauge was bonded. The transducer was molded with silicone rubber, and Dacron mesh was embedded in it for suturing to the intestine at a right angle to it. The average lifetime of the transducer was 24 days. Similar transducers for recording gastrointestinal activity were also reported by Bass et al.[66] and Lambert et al.[67]

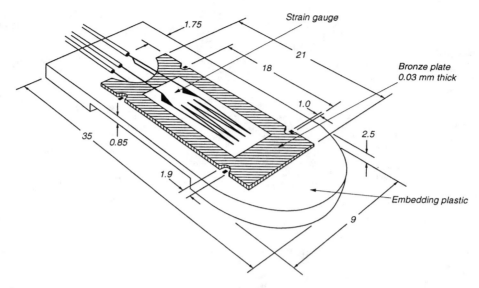

Figure 4.42 An implantable myocardial force transducer. (From Sutfin, D. C. and Lefer, A. M., *Med. Electron. Biol. Eng.,* 1, 371, 1963. With permission.)

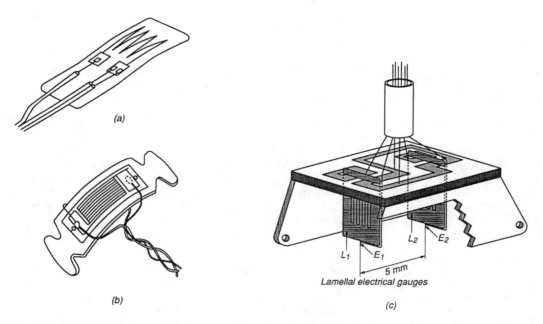

Figure 4.43 Implantable force transducers for recording of gastrointestinal tract activities. (Part (a) modified from Jacoby, H. I et al., *J. Appl. Physiol.,* 18, 648, 1963. Part (b) modified from Bass, P. and Wiley, J. N., *J. Appl. Physiol.,* 32, 567, 1972. Part (c) modified from Lambert, A. et al., *J. Appl. Physiol.,* 41, 942, 1976.)

4.3.2 MEASUREMENTS OF STRESSES IN THE BONE

Direct measurement of internal stress distribution in a bone is always difficult, and thus it has to be estimated indirectly. If the stress-strain relation of a bone is known, the stress can be estimated by measuring the strain. In animal experiments, the strain of a bone can be obtained by attaching strain-gauges onto it. Measurement of strain can be performed *in vivo* as long as the strain-gauges can be attached to the bone surgically. The stress-strain relation of a bone can be determined if the bone is extracted after the measurement without removing the strain-gauges. By applying known forces and recording the strain-gauge outputs, using the same instrument used in the experiment, the calibration curve between the applied strain and the strain-gauge output can be obtained.[68-75]

Cochran[71] described the procedure of implantation of strain-gauges onto the bone for chronic animal experiments. A temperature-compensated, foil-type gauge sealed with a polyimide matrix was used. After connecting insulated lead wires, exposed metal surfaces were covered with a polysulfide/epoxy waterproofing compound and heat cured. Then the gauge was sandwiched between a piece of polyimide film and a tetrafluorethylene film with a bondable surface. As the adhesive for the bone, methyl or isobutyl 2-cyanoacrylate was used. The bone surface was scraped, cauterized, degreased, and dried so that a smooth, clean dry surface was obtained. The adhesive was applied to both surfaces, and the gauge was bonded for 1 min under finger pressure or by a padded implement preshaped to conform to the bone contour. The gauge could be used in live animals for at least 3 weeks.

Baggott and Lanyon[73] confirmed the reliability of the bone strain measurement *in vivo* by comparing it with a direct strain measurement using an optical lever in dissected bone specimens after chronic studies. The result showed that the difference in strains between the optical and the gauge measurements was only 0.8% 3 weeks after implantation.

In order to measure strain components in different directions at a point of a bone, a strain-gauge rosette can be used in which two or more gauges are stacked or formed in planar technique on a base sheet. When using a strain-gauge rosette, the torsional component of the strain at a long bone section can be measured. To measure a bending component, two or more axial gauges bonded at different sites have to be used. In an animal study of locomotion, two axial strain-gauges and one strain-gauge rosette were bonded at three circumferential locations in the midshaft of a dog.[75]

4.3.3 GROUND FORCE MEASUREMENTS

A human body standing on the ground exerts a gravitational force to the ground. The reaction force is exerted from the ground to the body. When the body moves, the force exerted to the ground or to the body varies in its strength and direction due to the acceleration or deceleration of the body. Thus, the force exerted to the ground, which is called the ground force, provides information about the body motion. Ground force measurements have been performed clinically for gait analysis, in stabilometry, and for the evaluation of athletic capacity in sports medicine.

The ground force can be measured by a transducer placed on the ground or attached to the foot. While forces measured by these two methods are the same in their strength and the directions are opposite to each other, the obtained data are apparently different, because (as mentioned in Section 4.1.3) the transducer on the ground measures the force with reference to the fixed coordinate system whereas the transducer attached to the foot measures the force referring to the moving coordinate system.

4.3.3.1 Force Plates

The instrument installed on the ground so as to measure the ground force is called the force plate or force platform. As shown in Figure 4.44, the force plate has a plate supported by force transducers. The simple force plate has transducers which measure only vertical forces and provide outputs of the vertical component and the point of application of the ground force. The force plates used most commonly have transducers that measure one vertical and two horizontal (shear) components of the force, so that the strength, direction, point of application, and torque of the ground force can be obtained.

Each component of the ground force is determined from the transducer outputs as follows. If the positions of four transducers in the horizontal x,y plane are (x_1, y_1), (x_2, y_2), (x_3, y_3), and (x_4, y_4) and the transducer outputs, which are two horizontal and one vertical component, are (Fx_1, Fy_1, Fz_1), (Fx_2, Fy_2, Fz_2), (Fx_3, Fy_3, Fz_3) and (Fx_4, Fy_4, Fz_4), then the x, y, and z components of the ground force are given as

$$Fx = \sum_{i=1}^{4} Fx_i, \quad Fy = \sum_{i=1}^{4} Fy_i, \quad \text{and} \quad Fz = \sum_{i=1}^{4} Fz_i. \tag{4.10}$$

The point of application of the ground force, $(\bar{x}, \bar{y})$, is given as

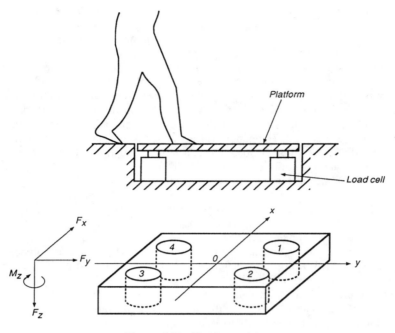

Figure 4.44   The force plate.

$$\bar{x} = \frac{\sum\limits_{i=1}^{4} Fz_i x_i}{Fz} + \frac{Fx}{Fz}\Delta z, \quad \bar{y} = \frac{\sum\limits_{i=1}^{4} Fz_i y_i}{Fz} + \frac{Fy}{Fz}\Delta z \qquad (4.11)$$

where Δz is the depth of the transducer elements measured from the ground surface level.

The torque referring to the vertical axis at the point of application is given as

$$Mz = \sum_{i=1}^{4} Fy_i(x_i - \bar{x}) - \sum_{i=1}^{4} Fx_i(y_i - \bar{y}). \qquad (4.12)$$

Force transducers in a force plate are of the strain-gauge type or piezoelectric type. Figure 4.45 shows an example of the strain-gauge type of force plate.[76] The platform is supported by four metal pipe pylons on which strain gauges are attached so that the compression, bend, and twist of the pylon can be detected.

Figure 4.46 shows a piezoelectric-type force transducer developed by Kistler Instrument AG (Winterthur, Switzerland). It consists of three pairs of quartz disks sandwiched between the steel base and the top of the transducer and provides outputs corresponding to the three orthogonal components of the applied force. While static force cannot be measured by a piezoelectric transducer, a fairly long time constant can be realized using charge amplifiers, as mentioned in Section 4.2.4.1. Actually, in Kistler's force plate, a time constant of up to 1000 s is attained. Other typical performances of the Kistler force plate are sensitivies of 5 mN for horizontal and 10 mN for vertical force components, a measurement range from −10 to 10 kN, and a resonance frequency of 800 Hz (for the Kistler, Type 9281 B).

The length and width of a force plate are typically 600 and 400 mm, respectively. To measure the ground forces of many steps in gait analysis, a walkway is required along which many force plates are installed. For example, eight force plates arranged in line were used in a study to estimate mechanical work during walking and running.[77]

4.3.3.2   Stabilometers

Clinical examination of the stability of posture is called stabilometry, and the force plate designed for stabilometry is called the stabilometer. Simple stabilometers only measure the locus of the point of

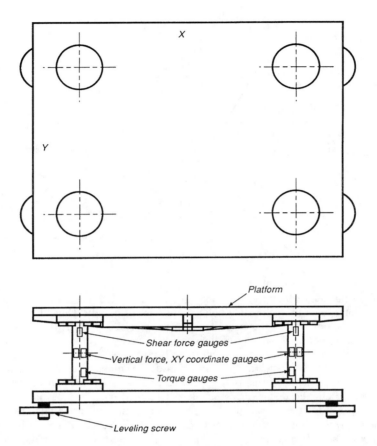

Figure 4.45 A strain-gauge force plate. (Modified from Cunningham, D. M. and Brown, G. W., *Proc. Soc. Exper. Stress Anal.,* 9, 75, 1951.)

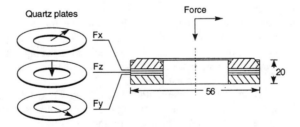

Figure 4.46 A piezoelectric force transducer for a force plate.

application of the ground force using their triangularly arranged vertical force transducers, as shown in Figure 4.47. If the coordinates of these transducers in the x,y plane are $(0,a)$, $(-b,-c)$, and $(b,-c)$ and their vertical force components are F_1, F_2, and F_3, then the x- and y-coordinates of the application point of the ground force, $(\bar{x}, \bar{y})$, are given as

$$\bar{x} = \frac{b(F_3 - F_2)}{F}, \quad \bar{y} = \frac{aF_1 - c(F_2 + F_3)}{F} \tag{4.13}$$

where $F = F_1 + F_2 + F_3$.

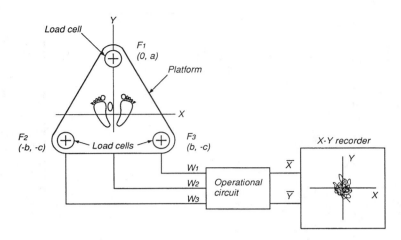

Figure 4.47   The stabilometer.

When the subject is standing still, the point of application of the ground force stays right below the center of gravity of the body. However, when the body is accelerated in a horizontal direction, the point of application deviates from the point right below the center of gravity. Thus, the locus of the point of application does not simply reflect the excursion of the center of gravity. If the center of gravity has a horizontal acceleration, a, and the height of the center of gravity from the ground level is h, then the deviation of the application point of the ground force from the point right below the center of gravity is ah/g. In an extreme situation (such as to jump forward) deviations of around 30 cm can occur.[78]

In order to study the characteristics of the posture control system in humans and in animals, the response to externally applied disturbances has been measured by using the moving force plate. For example, a motor-driven force plate was used to apply pseudo-random acceleration to a standing human subject.[79] For studying the control system of posture in small quadruped animals, a force plate was built which can be moved laterally and vertically.[80] This platform consists of four limb-support pads, each of which can be moved vertically by a linear hydraulic actuator. Two additional actuators drive the platform in horizontal directions. To monitor the forces exerted on each limb-support pad, a specially designed, small triaxial force transducer was attached to it (Figure 4.48).[81]

4.3.3.3 Instrumented Shoe

The ground force acting on the foot can be measured by attaching instruments to the foot. The shoe designed for such a measurement is called the instrumented shoe. The instrumented shoe is advantageous in foot force measurement because the measurement can be performed during natural locomotion without restricting the subject to a walkway where force plates or other instruments such as the videocamera are installed.

Figure 4.49 shows an example of the instrumented shoe.[82] It has two load cells attached at the toe and heel. Each load cell consists of an end-support spring element on which strain-gauges are mounted and provides outputs corresponding to the anterior-posterior and medial-lateral shears, axial compression, and torque. An instrumented shoe having eight or nine strain-gauge force transducers has been reported by Kljajic and Krajnik.[83] Each transducer consists of a steel beam with a protuberance in the middle which makes contact with the ground. By employing many transducers, it can provide not only the vertical component of the ground force but also the point of application of the vertical force along the shoe.

A simple instrumented shoe was reported by Miyazaki and Iwakura[84] in which two vertical load transducers were attached at the metatarsal part and the heel. Each transducer consists of a stainless-steel beam with strain-gauges. Also studied was a capacitive transducer, as shown in Figure 4.50, in which a foam rubber sheet is sandwiched between two copper sheets so that the capacitance between the copper sheets varies when the rubber sheet is compressed by the applied load.[85] While the sensitivity may change when the load is not uniform but localized, the observed error was limited to ±10%.

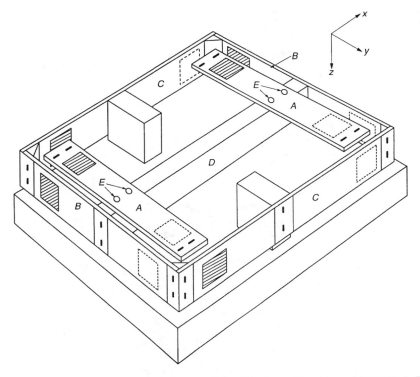

Figure 4.48 A force plate for studying the posture control system in small quadruped animals. A, B, and C are rigidly end- and center-mounted beams, and the centers of beams A are mounted onto the rigid beams D by bolts E. (Modified from Lywood, D. W. et al., *Med. Biol. Eng. Comput.*, 25, 698, 1987.)

4.3.3.4 Foot Force Distribution Measurements

The distribution of the ground force exerted to the sole, which is also called the foot-ground pressure pattern, has been measured for investigating the function of the foot relating to the postural control of the body.[86] In early studies, a printing technique has been used which provides a pressure-dependent footprint. For example, one technique called the kinetograph consists of a corrugated rubber mat and inked fabric so that a difference in pressure is recognized by the different density.[87] Many studies have been carried out using similar printing techniques.[86] Quantitative force distribution can be estimated from the obtained footprint by using the calibrated pressure-sensitive sheet.[88]

A direct visualization technique called the barograph was developed by Elftman.[89] It comprised a black rubber mat having many small protrusions of pyramidal contours and was placed on a glass plate. When viewed from below, the footprint appears as a matrix of black dots. In order to enhance the contrast of the image, a white opaque fluid was introduced between the mat and the glass. After the original barograph, many attempts have been made using different materials and shapes for the deformable mat, as well as the optical observation techniques.[86] Video-to-computer interfacing was also introduced to provide presentations of the processed data.[90] The photo-elastic sheet was employed for optical detection of the force distribution.[91] As shown in Figure 4.51, the applied load is discretized by many hemispherical solids having contacts with a layer consisting of the reflector, photo-elastic sheet, and polarizers, which are placed on a glass plate. When viewed from below, a circular interference pattern appears. The diameter of the circle is a function of the force exerted by the solid to the layer. By processing with a computer, quasi-three-dimensional plots of the ground force distribution could be obtained.[92]

The ground force distribution has also been measured by a matrix force plate consisting of many cells. A force plate has 128 strain-gauge transducers, each 14 mm square and arranged in a 16×8 matrix.[93] The use of a capacitive force transducer consisting of a compressible elastic dielectric material was presented by Nichol.[94] Square-shaped transducers between 3×3 mm and 30×30 mm were used in a matrix arrangement. Matrices of up to 256×128 (32,768) transducers were evaluated.

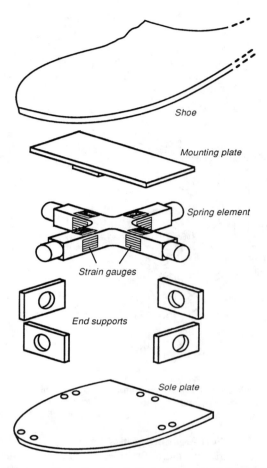

Shoe

Mounting plate

Spring element

Strain gauges

End supports

Sole plate

Figure 4.49   An instrumented shoe. (From Spolek, G. A. and Lippert, F. G., *J. Biomech.,* 9, 779, 1976. With permission.)

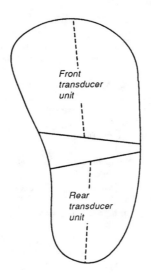

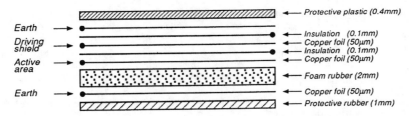

Figure 4.50 A capacitive transducer for a simple instrumented shoe. (From Miyazaki, S. and Iwakura, H., *Med. Biol. Eng. Comput.*, 16, 429, 1978. With permission.)

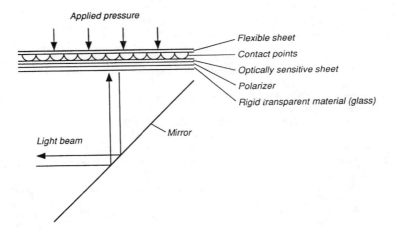

Figure 4.51 A method of direct visualization of the foot-ground pressure pattern. (Modified from Arcan, M. and Brull, M. A., *J. Biomech.*, 9, 453, 1976.)

REFERENCES

1. Åstrand, P. A. and Rodahl, K., *Textbook of Work Physiology*, 3rd ed., McGraw-Hill. New York, 1986.
2. Morris, J. R. W., Accelerometry — a technique for the measurement of human body movements, *J. Biomech.*, 6, 729, 1973.
3. Padgaonkar, A. J., Krieger, K. W., and King. A. L., Measurement of angular acceleration of a rigid body using linear accelerometers, *Trans. ASME J. Appl. Mech.*, 97, 552, 1975.
4. Clemedson, C.-J. and Jönsson, A., Dynamic response of chest wall and lung injuries in rabbits exposed to air shock waves of short duration, *Acta Physiol. Scand.*, 62 (Suppl. 233), 3,1964.
5. Clemedson, C.-J. and Jönsson A., A mechanoelectric transducer for recording transient motion in biological experiments, *J. Appl. Physiol.*, 24, 430, 1968.
6. Contini, R. and Drills, R., Kinematic and kinetic techniques in biomechanics, in *Advances in Bioengineering and Instrumentation*, Alt, F., Ed., Plenum Press, New York, 1966, 3.
7. Ketterkamp, D. B., Johnson, R. J., Smidt, G. L., Chao, E. Y. S., and Walker, M., An electrogoniometric study of knee motion in normal gait, *J. Bone Joint Surg.*, 52-A, 775, 1970.
8. Johnson, R. C. and Smidt, G. L. Measurement of hip-joint motion during walking, *J. Bone Joint Surg.*, 51-A, 1083, 1969.
9. Peat, M., Grahame R. E., Fulford, R., and Quanbury, A. O., An electrogoniometer for the measurement of single plane movements, *J. Biomech.*, 9, 423, 1976.
10. Zangemeister, W. H., Jones, A., and Stark, L., Dynamics of head movement trajectories: main sequence relationship, *Exp. Neurol.*, 71, 76, 1981.
11. McIntosh, M., Duggan, D. E., Watt, D. D., and Goodson, L. H., A photopotentiometric transducer for detecting *in vitro* muscle contractions, *J. Appl. Physiol.*, 20, 349, 1965.
12. Goldstein, S. R., Friauf, W. S., and Wells, J. B., A novel instrument for dynamic and static measurement of large length changes in muscle, *J. Appl. Physiol.*, 36, 128,1974.
13. Wolfendale, P. C. F., Capacitive displacement transducers with high accuracy and resolution, *J. Phys. E*, 1, 817,1968.
14. Lentini, E. A. and Guyton, W., Electronic micrometer, *J. Appl. Physiol.*, 18, 636, 1963.
15. Nalecz, M., Hoffman, M., Maniewski, R., Nowicka, J. Rydlewska-Sadowska, W., and Tarlowski, A., Hall effect transducer for apexcardiography and sphygmography, *J. Biomed. Eng.*, 4, 313, 1982.
16. Nalecz, M., Measurements of mechanical displacements of the order of 10 angstroms using the Hall generator, *Bull. Polish Acad. Sci.*, 17, 1, 1969.
17. Woltjen, J. A., Timm, G. W., Waltz, F. M., and Bradley, W. E., Bladder motility detection using the Hall effect, *IEEE Trans. Biomed. Eng.*, BME-20, 295, 1973.
18. Bando, E., Fukushima, S., Kawabata, H., and Kohno, S., Continuous observation of mandibular positions by telemetry, *J. Preosthet. Dent.*, 28, 485, 1972.
19. McCall, Jr., W. D. and Rohan, E. J., A linear position transducer using a magnet and Hall effect devices, *IEEE Trans. Instrum. Meas.*, IM-26, 133, 1977.
20. Peterson, Jr., C. V. and Otis, A. B., A Hall effect transducer for measuring length changes in mammalian diaphragm, *J. Appl. Physiol. Respirat. Environ. Exercise Physiol.*, 55, 635, 1983.
21. Sweetman, B. J., Jayasinge, W. J., Moore, C. S., and Anderson, J. D. A., Monitoring work factors relating to back pain, *Postgrad. Med. J.*, 52 (Suppl. 7), 151, 1976.
22. Sydenham, P. H., Microdisplacement transducers, *J. Physics, E*, 5, 721, 1972.
23. Feldstein, C. F., Meerbsum, S., Lewis, G., and Culler, V., Transducers for myocardial research, *Med. Instrum.*, 14, 277,1980.
24. Jones, R. D., Adams, R. B., and Luria, M. H., A miniature segment-length strain gauge arch for the assessment of myocardial function, *Med. Instrum.*, 11, 244, 1977.
25. Kolin, A. and Culp, G. W., An intra-arterial induction gauge, *IEEE Trans. Biomed. Eng.*, BME-18, 110, 1971.
26. van der Schee, E. J., de Bakker, J. V., Zwamborn, A. W., Transducer for *in vivo* measurement of the inner diameter of arteries in laboratory animals, *Med. Biol. Eng. Comput.*, 19, 218,1981.
27. Pieper, H. P. and Paul, L. T., Catheter-chip gauge for measuring blood flow velocity and vessel diameter in dogs, *J. Appl. Physiol.*, 24, 259, 1968.
28. Horwitz, L. D., Bishop, V. S., Stone, H. L., and Stegall, H. F., Continuous measurement of internal left ventricular diameter, *J . Appl. Physiol.*, 24, 738, 1968.
29. Stegall, H. F., Kardon, M. B., Stone, H. L., and Bishop, V. S., A portable, simple sonomicrometer, *J. Appl. Physiol.*, 23, 289, 1967.
30. Goodman, C. A. and Castellana, F. S., A digital sonomicrometer for two-point length and velocity measurements, *Am. J. Physiol.*, 243, H634, 1982.
31. Lee, R. D. and Sandler, H., Miniature implantable sonomicrometer system, *J. Appl. Physiol.*, 28,110, 1970.
32. Hunt, J. W., Arditi, M., and Foster, F. S. Ultrasound transducers for pulse-echo medical imaging, *IEEE Trans. Biomed. Eng.*, BME-30, 453, 1983.
33. Roelandt, J., van Dorp, W. G., Bom, N., Laird, J. D., and Hugenholtz, P.G., Resolution problems in echocardiology: a source of interpretation errors, *Am. J. Cardiol.*, 37, 256, 1976.

34. Hokanson, D. E., Mozersky, D. J., Sumner, D. S., and Strandness, Jr., D. E., A phase-locked echo tracking system for recording arterial diameter changes *in vivo*, *J. Appl. Physiol.*, 32, 728, 1972.
35. Tayler, K. D., Mottier, F. M., Simmons, D. W., Cohen, W., Pavlak, Jr., R., Cornell, D. P., and Hankins, G. B., An automated motion measurement system for clinical gait analysis, *J. Biomech.*, 15, 505, 1982.
36. Woltring, H. J., New possibilities for human motion studies by real-time light spot position measurement, *Biotelemetry*, 1, 132, 1974.
37. Macellari, V., CoSTEL, a computer peripheral remote sensing device for 3-dimensional monitoring of human motion, *Med. Biol. Eng. Comput.*, 21, 311, 1983.
38. Morasso, P. and Tagliasco V., Analysis of human movements: spatial localisation with multiple perspective views, *Med. Biol. Eng. Comput.*, 21, 74, 1983.
39. Reed, D. J. and Reynolds, P. J., A joint angle detector, *J. Appl. Physiol.*, 27, 745, 1969.
40. Mead, J., Peterson, N., Grimby, G., and Mead, J., Pulmonary ventilation measured from body surface movements, *Science*, 156, 1383, 1967.
41. Rolfe, P., A magnetometer respiration monitor for use with premature babies, *Biomed. Eng.*, 6, 402, 1971.
42. Ashutosh, K., Gilbert, R., Auchincloss, J. H., Erlebacher, J., and Peppi, D., Impedance pneumograph and magnetometer methods for monitoring tidal volume, *J. Appl. Physiol.*, 37, 964, 1974.
43. Stagg, D., Goldman, M., and Davis, J. N., Computer-aided measurement of breath volume and time components using magnetometers, *J. Appl. Physiol. Respirat. Environ. Exercise Physiol.*, 44, 623, 1978.
44. Saunders, N. A., Kreitzer, S. M., and Ingram, Jr., R. H., Rib cage deformation during static inspiratory efforts, *J. Appl. Physiol. Resp. Environ. Exer. Physiol.*, 46, 1071, 1979.
45. Robertson, Jr., C. H., Bradley, M. E., and Homer, L. D., Comparison of two-and four-magnetometer methods of measuring ventilation, *J. Appl. Physiol. Respirat. Environ. Exercise Physiol.*, 49, 355, 1980.
46. Raab, F. H., Blood, E. B., Steiner, T. O., and Jones, H. R., Magnetic position and orientation tracking system, *IEEE Trans. Aerospace Electron. Syst.*, AES15, 709, 1979.
47. Logan, S. E. and Groszevski, P., Dynamic wrist motion analysis using six degree of freedom sensors, *Biomed. Sci. Instrum.*, 25, 213, 1989.
48. Russell, P., Weld, A., Pearcy, M. J., Hogg, R., and Unsworth, A., Variation in lumbar spine mobility measured over a 24-hour period, *Br. J. Rheumatol.*, 31, 329, 1992.
49. Pearcy, M. J., Twisting mobility of the human back in flexed postures, *Spine*, 18, 114, 1993.
50. Boyce, E. S. Dawes, G. S., Gough, J. D. and Poore, E. R., Doppler ultrasound method for detecting human fetal breathing *in utero*, *Br. Med. J.*, 2, 17, 1976.
51. Wells, P. N. T., A range-gated ultrasonic Doppler system, *Med. Biol. Eng.*, 7, 641, 1969.
52. Doebelin, E. O., *Measurement System Application and Design*, 4th ed., McGraw-Hill, New York, 1990, chap. 4.
53. Krippke, D. F., Mullaney, D. J., Messin, S., and Wyborney, V. G., Wrist actigraphic measures of sleep and rhythms, *Electroencephalogr. Clinical Neurophysiol.*, 44, 674, 1978.
54. Webster, J. B., Messin, S., Mullaney D. J., and Kripke, D. F., Transducer design and placement for activity recording, *Med. Biol. Eng. Comput.*, 20, 741, 1982.
55. Rudolf, F., Jornod, A., Bergqvist, J., and Leuthold, H., Precision accelerometer with μg resolution, *Sensors Actuators*, A21-23, 297, 1990.
56. Suzuki, S., Tuchitani, S., Sato, K., Veno, S., Yokota, Y., Sato, M., and Esashi, M., Semiconductor capacitance-type accelerometer with PWM electrostatic servo technique, *Sensors Actuators*, A21-23, 316, 1990.
57. Rosa, G. N., Some design considerations for liquid rotor angular accelerometers, Instrument notes, 26, Stratham Instrumentation, Inc., Oxnard, CA, 1954.
58. McLaughlin, R. J., Systematic design of cantilever beams for muscle research, *J. Appl. Physiol. Respirat. Environ. Exercise Physiol.*, 42, 786, 1977.
59. Chapman, R. A., High sensitivity isometric force transducers made with piezo-electric or piezo-resistive strain gauges, *J. Physiol. (Lond.)*, 210, 4P-6P, 1970.
60. Hamrell, B. B., Panaanan, R., Trono, J., and Alpert, N. R., A stable, sensitive, low-compliance capacitance force transducer, *J. Appl. Physiol.*, 38, 190, 1975.
61. Hellam, D. C. and Podolsky, R. J., Force measurements in skinned muscle fibers, *J. Physiol. (Lond.)*, 200, 807, 1969.
62. Canaday, P. G. and Fay, F. S., An ultrasensitive isometric force transducer for single smooth muscle cell mechanics, *J. Appl. Physiol.*, 40, 243, 1976.
63. Minns, H. G. and Franz, G. N., A low-drift transducer for small forces, *J. Appl. Physiol.*, 33, 529, 1972.
64. Sutfin, D. C. and Lefer, A. M., A modified strain gauge arch for measurement of heart contractile force, *Med. Electron. Biol. Eng.*, 1, 371, 1963.
65. Jacoby, H. I., Bass, P., and Benett, D. R., *In vivo* extraluminal contractile force transducer for gastrointestinal muscle, *J. Appl. Physiol.*, 18, 658, 1963.
66. Bass, P. and Wiley, J. N., Contractile force transducer for recording muscle activity in unanesthetized animals, *J. Appl. Physiol.*, 32, 567, 1972.
67. Lambert, A., Eloy, R., and Grenier, J. F., Transducer for recording electrical and mechanical chronic intestinal activity, *J. Appl. Physiol.*, 41, 942, 1976.
68. Lanyon, L. E. and Smith, R. N., Bone strain in the tibia during normal quadrupedal locomotion, *Acta Orthop. Scandinav.*, 41, 238, 1970.

69. Lanyon, L. E., Strain in sheep lumbar vertebrae recorded during life, *Acta Orthop. Scandinav.*, 42, 102, 1971.

70. Lanyon, L. E., *In vivo* bone strain recorded from thoracic vertebrae of sheep, *J. Biomechan.*, 5, 277, 1972.

71. Cochran, G. V. B., Implantation of strain gauges on bone *in vivo*, *J. Biomech.*, 5, 119, 1972.

72. Cochran, G. V. B., A method for direct recording of electromechanical data from skeletal bone in living animals, *J. Biomech.*, 7, 563, 1974.

73. Baggott, D. G. and Lanyon, L. E., An independent "post-mortem" calibration of electrical resistance strain gauges bonded to bone surfaces "*in vivo*", *J. Biomech.*, 10, 615, 1977.

74. Carter, D. R., Smith, D. J., Spengler, D. M., Daly, C. H., and Frankel, V. H., Measurement and analysis of *in vivo* bone strains on the canine radius and ulna, *J. Biomech.*, 13, 27, 1980.

75. Carter, D. R., Vasu, R., Spengler, D. M., and Dueland, R. T., Stress fields in the unplated and plated canine femur calculated from *in vivo* strain measurements, *J. Biomech.*, 14, 63, 1981.

76. Cunningham, D. M. and Brown, G. W., Two devices for measuring the forces acting on the human body during walking, *Proc. Soc. Exper. Stress Anal.*, 9, 75,1951.

77. Cavagna, G. A., Force platforms as ergometers, *J. Appl. Physiol.* 39, 174, 1975.

78. Murray, M. P., Seireg, A., and Scholz, R. C., Center of gravity, center of pressure, and supportive forces during human activities, *J. Appl. Physiol.*, 23, 831, 1967.

79. Ishida, A. and Imai, S., Responses of the posture-control system to pseudorandom acceleration disturbances, *Med. Biol. Eng. Comput.*, 18, 433, 1980.

80. van Eyken, A., Perlin, S., Lywood, D. W., and Macpherson, J. M., Robotic force platform for the study of posture and stance in the quadruped, *Med. Biol. Eng. Comput.*, 25, 693, 1987.

81. Lywood, D. W., Adams, D. J., van Eyken, A., and Mac Pherson, J. M., Small, triaxial force plate, *Med. Biol. Eng. Comput.*, 25, 698, 1987.

82. Spolek, G. A. and Lippert, F. G., An instrumented shoe — a portable force measuring device, *J. Biomech.*, 9, 779, 1976.

83. Kljajic, M. and Krajnik, J., The use of ground reaction measuring shoe in gait evaulation, *Clin. Phys. Physiol. Meas.*, 8, 133, 1987.

84. Miyazaki, S. and Iwakura, H., Foot-force measuring device for clinical assessment of pathological gait, *Med. Biol. Eng. Comput.*, 16, 429, 1978.

85. Miyazaki, S. and Ishida, A., Capacitive transducer for continuous measurement of vertical foot force, *Med. Biol. Eng. Comput.*, 22, 309, 1984.

86. Lord, M., Foot pressure measurement: a review of methology, *J. Biomed. Eng.*, 3, 91, 1981.

87. Morton, D. J., Foot biomechanics: functional disorders and deformities, in *Medical Physics*, Glasser, O., Ed., Year Book Publishers, Chicago, 1961, 457.

88. Aritomi, H., Morita, M., and Yonemoto, K., A simple method of measuring the footsole pressure of normal subjects using prescale pressure-detecting sheets, *J. Biomech.*, 16, 157,1983.

89. Elftman, H. O., A cinematic study of the distribution of pressure in the human foot, *Anat. Record.*, 59, 481,1934.

90. Betts, R. P., Franks, C. I., Duckworth, T., and Burke, J., Static and dynamic foot-pressure measurements in clinical orthopaedics, *Med. Biol. Eng. Comput.*, 18, 674, 1980.

91. Arcan, M. and Brull, M. A., A fundamental characteristic of the human body and foot, the foot-ground pressure pattern, *J. Biomech.*, 9, 453, 1976.

92. Cavanagh, P. R. and Ae, M., A technique for the display of pressure distribution beneath the foot, *J. Biomech.*, 13, 69, 1980.

93. Dhanendran, M., Hutton, W. C., and Parker, Y., The distribution of force under the human foot — an on-line measuring system, *Meas. Contr.*, 11, 261, 1978.

94. Nichol, K., A new capacitive transducer system for measuring force distribution statically and dynamically, *Proc. Transducer Tempcon.*, 81, 1, 1981.

Temperature, Heat Flow, and Evaporation Measurement

5.1 OBJECT QUANTITIES

5.1.1 UNITS OF THERMAL QUANTITIES

In the SI system, thermodynamic temperature or absolute temperature is expressed in Kelvins (K), and temperature is expressed in degrees Celsius (°C). The Celsius temperature is defined as

$$0°C = 273.15 \text{ K} \tag{5.1}$$

and the temperature interval or difference expressed in Kelvins and degrees Celsius is identical; 0°C is 0.01 K below the triple point of water. The temperature coefficient is expressed in reciprocal Kelvin (K^{-1}).

The unit of heat is joule (J) in the SI system. While calorie has been widely used in medical fields, it should be converted to joule in the SI system. The conversion factor depends on the definition. 15°C calorie (cal_{15}) is defined as the heat required to warm 1 g of water from 14.5°C to 15.5°C at a pressure of 101,325 kPa; that is,

$$1 \text{ cal}_{15} = 4.1855 \text{ J} \tag{5.2}$$

and an IT calorie (International Table calorie, cal_{IT}) is defined as

$$1 \text{ cal}_{IT} = 4.1868 \text{ J}. \tag{5.3}$$

Units of other quantities relating to temperature, heat, and heat flow are given as follows:

Heat flow rate	W (watt); 1 W = 1 J/s
Density of heat flow rate	W/m^2 (watt per square meter)
Thermal conductivity	W/(m·K) (watt per meter Kelvin)
Heat capacity	J/K (joule per Kelvin)

The content of water vapor in a gas can be expressed in various ways, such as absolute humidity, water vapor pressure, relative humidity, or dew point. Absolute humidity is defined as the water vapor concentration usually expressed by grams per cubic meter. Water vapor pressure is expressed in Pascals (Pa) in the SI system. Relative humidity, RH, is defined as

$$RH = \frac{P_{H_2O}}{P_{H_2O(Sat)}} \times 100 \ (\%) \tag{5.4}$$

221

where p_{H_2O} is water vapor pressure, and $p_{H_2O(Sat)}$ is saturated water vapor pressure at the gas temperature. Dew point is the temperature at which relative humidity becomes 100% when the gas is cooled. When the gas is cooled below the dew point, water condensation may occur. The evaporation rate of water vapor from the body is usually expressed in grams per square meter and hour or in milligrams per square centimeter per hour.

5.1.2 REQUIREMENTS FOR MEASUREMENT RANGES

Temperature is measured at many different sites of the body for clinical diagnosis and patient monitoring. In humans and other in homeothermic animals, the temperature of the central part of the body is stabilized by a physiological thermoregulatory function. The deep tissue temperature at the central part of the body is called core temperature or deep body temperature. The term "body temperature" is often used to indicate core temperature, even though the temperature of the body is not uniform but can vary from site to site.

The core temperature always remains in a range from 35 to 40°C. Most physiological and pathological temperature variations occur in this range, from the lowest temperature in early morning or in cold weather to the highest one during febrile disease or hard exercise.[1] In the case of therapeutic or accidental hypothermia or hyperthermia, a wider measurement range is required.

A temperature resolution of 0.1°C is generally required in core temperature measurement, and that of 0.05°C is sometimes required such as in basal temperature measurement. An absolute accuracy of 0.1°C is acceptable for most purposes.[2]

Skin temperature is also measured in physiological studies, clinical diagnoses, and patient monitoring. To estimate heat exchange between the body and its environment, the mean skin temperature is considered. Mean skin temperature, $\overline{T}_s$, is defined as the sum of the products of the area of each regional surface element, A_j, and its mean temperature, $\overline{T}_{sj}$, divided by the total area of the body surface, A_b:[3]

$$\overline{T}_s = \sum_j \left(A_j \overline{T}_{sj}\right)/A_b. \tag{5.5}$$

Practically, the simplified weighted sum of 3 to 15 points of skin temperatures is accepted as an approximation of mean skin temperature, as in

$$\overline{T}_s = \sum_j W_j \overline{T}_{sj}. \tag{5.6}$$

While many different weighting systems have been proposed in which different measurement sites and corresponding weights, W_j, are assigned, a comparative study showed that most of them agreed within 1°C over 85% of measurements in a wide range of controlled environments.[4]

In clinical patient care, skin temperature has been measured for monitoring peripheral circulation. The temperature of the big toe has commonly been measured for this purpose.[5-8]

The temperature of the skin can vary at least between ambient temperature and core temperature. The temperature of sweating skin sometimes drops below ambient temperature and, at the lowest, falls to the dew point. Skin temperature may vary much more widely when the skin is cooled or warmed externally. A thermometer for skin temperature measurement should, therefore, have a wider range, e.g., 0 to 50°C, although a higher resolution within a narrower range is required in some cases. The resolution, absolute accuracy, and response time of the thermometer required for measuring skin temperature vary according to purpose, but comparable performance by a thermometer for core temperature measurement may be acceptable in most applications.

Anomalies of skin temperature distribution are often observed with abnormal circulation, vascularization, or heat production in the underlying tissue. Visualization of the skin temperature distribution by thermography is a convenient method for discovering these anomalies. Thermography should cover the whole range of skin temperature with sufficient resolution to detect thermal anomalies of physiological or pathological origin. For example, in breast cancer screening, the conventional criterion of temperature difference for a positive anomaly is about 1°C[9] or 2.5°C.[10]

Local temperature measurements in the tissue are sometimes required. At thermal equilibrium, the temperature of a tissue is determined by local heat production and heat transport into and from the site.

Metabolically active tissues have a higher temperature than other sites and can maintain temperatures higher than the arterial blood temperature. However, when the temperature of the tissue increases to a level higher than the arterial blood temperature, the blood cools the tissue. To increase metabolism, the oxygen supply from arterial blood has to be increased, hence the cooling effect by the arterial blood is enhanced. Since 1 liter of arterial blood contains about 0.2 l of oxygen and metabolic heat production is about 21 J (5 kcal) per 1 liter oxygen, the increment of temperature in metabolically active tissue is not more than 1°C, except when an effective counter-current heat exchange exists between arteries and veins, as is seen in some animals.[11] Actual observations of muscle temperatures during exercise have shown that the maximum temperature rise of the quadriceps muscle during bicycle exercise is 0.95°C above rectal temperature.[12] In malignant tumors, temperatures 1 to 2°C higher than arterial blood temperature have been observed.[13,14]

In hyperthermia cancer therapy, local temperature measurement of tissue is required. To achieve maximum therapeutic effect, the temperature is maintained at about 43°C, which is close to the limit of survival of normal cells, while cancer cells can survive at a slightly lower temperature. An error of temperature control of 0.5°C would therefore have serious consequences. Temperature measurement should be accurate enough to keep the tissue temperature within this critical range. Christensen[15] suggested that requirements of thermometry during hyperthermia therapy should include an accuracy and resolution of 0.1°C and a spacial resolution of 1 cm within a range of 20 to 55°C. During the electromagnetic heating used in hyperthermia therapy, temperature measurement in the presence of a strong electromagnetic field is required.

Heat flow measurement is required to estimate heat dissipation from the body surface to the environment. The heat dissipation from the body surface depends on many factors, such as skin temperature, moisture on the skin surface, environmental temperature and humidity, air velocity near the surface, and condition of the covering of the surface. A model experiment showed that, at an environmental temperature of 22°C, density of heat flow rate from a body surface at 35 to 37.5°C covered by a dry cloth in still air was 3.7 W/m^2, whereas when the same surface was covered by a wet cloth and air blown onto it, the density of heat flow rate increased to 47 W/m^2.[16]

In a cold environment, much higher heat dissipation will occur. Heat dissipation measurement from the body surface requires a range from a few watts per square meter to several hundred watts per square meter.

The range required for humidity measurement is from zero up to 100% in relative humidity, saturated water vapor pressure, or the concentration at the highest temperature suspected during measurement. At 37°C, absolute humidity is 43.83 g/mm^3 when water vapor is saturated, and saturated water vapor pressure is 6279 Pa (47.1 mmHg).

When humidity is expressed by relative humidity, temperature should also be measured with sufficient accuracy. Near the body temperature, saturated water vapor pressure varies largely with temperature so that a temperature change of 1°C may cause changes of more than 5% in relative humidity. The rate of evaporation from the skin ranges between a few $g \cdot m^{-2} \cdot h^{-1}$ to several tens of $g \cdot m^{-2} \cdot h^{-1}$ for normal skin, while it increases to 100 $g \cdot m^{-2} \cdot h^{-1}$ or more in extreme pathological situations such as severe burns.

5.2 TEMPERATURE TRANSDUCERS

Many different kinds of temperature sensors are available, which are used by themselves or installed in surface probes, catheters, or needles making contact with or introduced to the object site of the body. To choose one suitable for medical thermometry, it may be worthwhile to compare temperature sensors based on various principles. There are always several alternatives in the choice of sensing device.

5.2.1 THERMISTORS

A thermistor is a semiconductor resistive temperature sensor made by sintered oxides of metals such as manganese, cobalt, nickel, iron, or copper. The resistance of a thermistor has a negative temperature coefficient, typically about –0.04/K. Compared with a platinum wire resister, which has a temperature coefficient of about 0.0039/K, the sensitivity of a thermistor is about 10 times that of the platinum wire temperature probe, hence a thermistor is suitable for use in physiological temperature measurement where relatively higher resolution is required in a narrow temperature range.

The resistivity of a thermistor material, ρ, at an absolute temperature, T, is generally expressed as

$$\rho \propto \exp\left(E_g / 2kT\right) \tag{5.7}$$

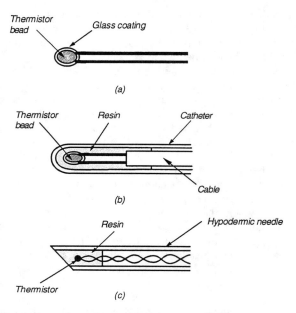

Figure 5.1 Examples of thermistor probes: **(a)** glass-coated thermistor bead with bare wires, **(b)** catheter-type probe, and **(c)** needle-type probe.

where E_g is the band gap energy of the semiconductor, and k is the Boltzmann constant. Thus, if the resistance of a thermistor is R_0 at a temperature T_0, then the resistance at temperature T is expressed as

$$R(T) = R_0 \exp\left(\left(\frac{1}{T} - \frac{1}{T_0}\right)B\right) \qquad (5.8)$$

where $B = E_g/2k$ is a constant that depends on the thermistor material and has a dimension of temperature. B always remains in a range from 1500 to 6000 K.

The temperature coefficient, α, of a thermistor is derived from the above expression as

$$\alpha = \frac{1}{R}\frac{dR}{dT} = \frac{d}{dT}\left(\frac{B}{T}\right) = -\frac{B}{T^2}. \qquad (5.9)$$

This means that the temperature coefficient is negative and is temperature dependent. If $B = 4000$ K, α is about –0.416 at 37°C.

Commercial thermistors for general use have resistance ranging from 6 to 60 kΩ at 0°C, or from 15 to 150 Ω at 37°C. Thermistors having much higher resistances are sometimes required for use in instruments to be operated at lower power. For this purpose, thermistors having a resistance of about 1 MΩ at room temperature are available.

Most commercial thermistors are at least stable enough for clinical use. It has been shown that drift rates of different types of commercial thermistors were around 0.1 mK per 100 days and were unaffected by thermal cycling, while mechanical shock and strain appear to be the cause of high drift rates in some thermistors.[17]

Various types of thermistor probes for general use or for medical use are commercially available. Figure 5.1 shows examples of thermistor probes. Figure 5.1(a) is a glass-coated thermistor bead with bare lead wires. Beads of small diameters down to about 0.3 mm are readily available. Figure 5.1(b) is a catheter-type probe in which the thermistor is connected to a flexible insulated cable; the connected part is also insulated and completely waterproof. The needle-type probe shown in Figure 5.1(c) is also available.

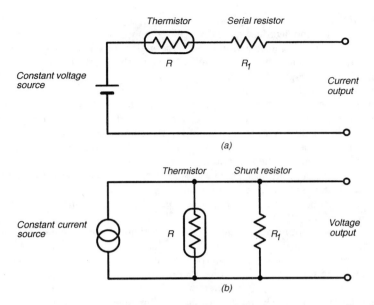

Figure 5.2 Linearization circuits for thermistors: **(a)** using a serial resistor and a constant voltage source, and **(b)** using a parallel resistor and a constant current source.

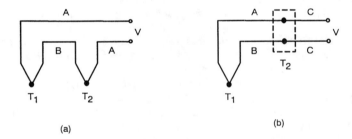

Figure 5.3 Thermocouple circuit with **(a)** two dissimilar metals, A and B, and **(b)** with an additional third metal, C.

The response time of a thermistor probe depends on its shape, size, and covering material, as well as on the surrounding medium. Fine catheter- or needle-type probes have response times of about 0.1 s or less in water, while response times increase to about 3 s or more in air.

While the characteristic of the thermistor is nonlinear, as expressed in Equation (5.8), many techniques have been proposed in order to obtain linear output to temperature. In a narrow temperature range, linearization can be achieved by adding only one resistor, as shown in Figure 5.2. When a constant voltage source is used, a resistor is connected in series, as in (a); when a constant current source is used, a shunt resistor is connected, as in (b). To minimize measurement error in the required temperature range, resistance, R_1, to be connected to the thermistor as circuit (a) or (b) is determined as

$$R_1 = R(B - 2T)/(B + 2T) \tag{5.10}$$

where T is the midpoint of the required temperature range, and R is the resistance of the thermistor at T.[18,19] For example, when $B = 3000$ K and a measurement range from 290 to 310 K is required, the departure from linearity is estimated as 0.03 K. If an error of 0.1 K is permissible, a measurement range from 285 to 315 K can be covered.[18]

5.2.2 THERMOCOUPLES

The thermocouple is a thermoelectric sensor. A circuit composed of two dissimilar metals, A and B, as shown in Figure 5.3(a) provides an electromotive force which depends on the temperature difference between two junctions. This phenomenon is known as the Zeebeck effect.

In the circuit shown in Figure 5.3(a), when the temperature of the reference junction, T_2, is kept constant, electromotive force varies only with the temperature of the measurement junction, T_1. In the circuit shown in Figure 5.3(b), a third metal, C, is connected to both metals A and B. As long as the two new junctions are at the same temperature, this circuit provides the same electromotive force as that of the circuit shown in Figure 5.3(a), regardless of the material of the third metal.

Even though the temperature of the reference junction is kept constant, the electromotive force is nonlinear with the temperature of the measurement junction. However, departure from linearity is small within the temperature range used in medical thermometry and the error is always compensated for in commercial thermometers. Sensitivities of typical thermocouples in a temperature range of 20 to 40°C are about 41 µV/K for copper/constantan, about 40 µV/K for chromel/alumel, and about 6.1 µV/K for platinum/platinum rhodium (10%).

To achieve accurate measurement by a thermocouple, the temperature of the reference junction should be stable enough. The triple point of water can be an accurate reference temperature and is 0.01 ± 0.0005°C. A conventional ice bath containing pure water with ice provides 0°C with an accuracy of about 0.05°C. Automatic ice baths are available that use the thermoelectric cooling element so that an external supply of ice is unnecessary. These systems use expansion of freezing water in a sealed bellows so as to sense the amount of the ice and control the cooling element.

When a high absolute accuracy is not required, the constant temperature bath can be eliminated by employing a method of compensating for the reference junction temperature. A convenient method is to use the circuit shown in Figure 5.3(b) so that the temperature of the input terminals to which the thermocouple is connected is used as the reference. The temperature of the measurement junction is measured as the reference junction temperature plus the temperature difference between junctions estimated by the electromotive force. While the sensitivity of a thermocouple is not constant for different reference junction temperatures, the error is small for narrow temperature ranges and can be easily compensated for.

The integrated circuit (IC) module amplifier shown in Figure 5.4 has a compensation circuit for the reference temperature assembled in the same package (Analog Device AD594/595). The nominal absolute accuracy at 25°C is ±1°C, and the effect of package temperature is less than ±0.025°C per 1°C. Thus, a stability of ±0.25°C will be attained by limiting the package temperature in a range of ±10°C.

For measuring local temperature, many different types of thermocouple probes, such as needle, insulated, or catheter, are available. Bare, fine thermocouple wires down to about 10 µm in diameter are also available commercially.

While ready-made junctions are commonly used, a simple method of welding by capacitor discharge is possible. According to a report, constantan and nickel-chromium wires of 30 µm in diameter are connected to a 0.25-µF capacitor charged with 120 V and brought into contact end-to-end, and then the ends are butt welded.[20]

Microthermocouples, as shown in Figure 5.5, with a junction size of about 1 µm can be made using the technique of fabricating glass-coated microelectrodes.[21] To fabricate this, a tip of thin platinum wire 25 µm in diameter is tapered by electropolishing, and a very thin glass coating on the platinum is made, leaving an exposed cone of platinum. Then, a thin film of tellurium is formed by vapor deposition and insulated by a thin film coating of negative photo-resist. Finally, a gold film is formed by vapor deposition that serves to shield the thermocouple from electromagnetic interference. The response time of this thermocouple is less than 50 ms in water.

Figure 5.6 shows an example of a thermocouple tip that was fabricated by thin-film technique.[22] A quartz rod was tapered, and a tip about 10 µm in diameter was formed. Then a nickel layer was vacuum-deposited on one side and a copper layer on the other, with the two metals overlapping only on the tip end. The probe was finally coated with polymer for insulation. When the tip was placed in the vitreous of the eye and irradiated by a laser light, it reached an equilibrium temperature in about 0.2 ms.

5.2.3   WIRE AND THIN FILM THERMORESISTIVE ELEMENTS

Because a thermoresistive element made from a pure metal has advantages — it has a constant temperature coefficient, linear output can be obtained in a wide temperature range, and a probe having a relatively large contact area with small heat capacity can easily be fabricated — it has been used widely

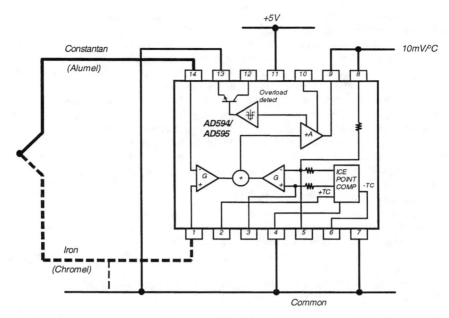

Figure 5.4   An example of an IC-module thermocouple amplifier having a compensation circuit for the reference temperature (Analog Devices AD594/595).

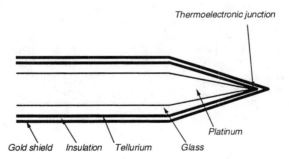

Figure 5.5   Microthermocouple. (Modified from Guilbeau, E. J. and Mayall, B. I., *IEEE Trans. Biomed. Eng.,* BME-28, 301, 1981.)

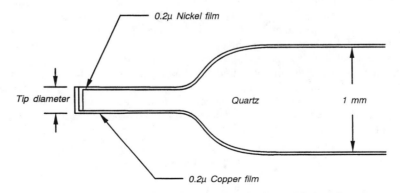

Figure 5.6   An example of a thermocouple tip fabricated by thin film technique. (From Cain, C. P. and Welch, A. J., *IEEE Trans. Biomed. Eng.,* BME-21, 421, 1974. With permission.)

in industrial applications. These advantages are less significant in most biomedical applications, where very local temperature has to be measured in a narrow temperature range; however, it is useful in special situations where extremely fast response is required, such as hot wire or hot film anemometry.

The most common material used as the wire or thin film thermoresistive element is platinum. The platinum thermoresistive element has higher stability and smaller nonlinearity than thermistors, but its temperature coefficient of about 0.0039/K is about one tenth that of the thermistor.

Various types of probes have been made for surface and fluid temperature measurement. Flat grid winding probes are used for measuring the surface temperature of solids, and probes having winding wires encased in protective tubes are used for fluid temperature measurement. The metal wire or film can also be cemented onto the surface for which the temperature is to be measured. However, the resistance change can occur not only by a temperature change but also by a strain, hence it may exhibit spurious output due to a mechanical load or differential thermal expansion.[23]

5.2.4 p-n JUNCTION DIODES AND TRANSISTORS

The voltage across a p-n junction at constant forward-bias current exhibits excellent linear temperature dependency over a wide temperature range; thus, any diode or transistor having a p-n junction can be a temperature sensor.[24,25] It is also advantageous that a p-n junction can be fabricated on a chip with interfacing circuits by integrated circuitry technology, and many convenient IC temperature sensors are commercially available.

The voltage-current characteristics of the forward-bias p-n junction is approximated as

$$I = A\exp\left[\left(qV - E_g\right)/kT\right] \tag{5.11}$$

where I is the forward-bias current, A is a constant depending on the geometry of the junction, q is the electron charge, V is the voltage across the junction, E_g is the bandgap energy, k is Boltzman's constant, and T is the absolute temperature.[26] When the current, I, is held constant, $(qV - E_g)/kT$ is kept constant; thus, the voltage across the junction, V, is a linear function of the absolute temperature, T. The temperature coefficient of V is always negative, and the observed dV/dT in typical small-signal silicon p-n junction diodes ranges from -1.3 to -2.4 mV/K at a current level of about 100 μA.[27] The base-emitter p-n junction in the transistor in which the base is connected to the collector is also used because of the fact that the nonlinearity in temperature dependency is less than that of most diodes.[28]

If the p-n junction in a diode or transistor is driven by different forward current levels, I_1 and I_2, and voltages V_1 and V_2 are developed at these current levels, then, from Equation (5.11)

$$V_1 - V_2 = (kT/q)\ln(I_1/I_2). \tag{5.12}$$

Thus, the difference in voltages corresponding to different current levels maintained at a constant ratio is proportional to the absolute temperature, without any offset. By this principle, a thermometer providing output proportional to the absolute temperature can be realized either by applying a square-wave current to a p-n junction,[29] or by using two matched devices operating at different current levels.[30]

An integrated circuit device that includes a pair of matched transistors and output operational amplifiers has been supplied by National Semiconductor Corp. (LX5600/5700).[31] It provides linear output proportional to the absolute temperature with a sensitivity of 10 mV/K.

The conventional two-terminal current-output device is also based on a similar technique.[32] Figure 5.7 shows an idealized scheme of the device. If transistors Q_1 and Q_2 are assumed to be equal and have a large gain, their collector currents are equal and constrain the collector current of Q_3 and Q_4. Q_3 has r base-emitter junctions, and each one is identical to that of Q_4. Then, from Equation (5.12), the voltage across R is obtained as

$$R \cdot I = (kT/q)\ln r \tag{5.13}$$

thus, the total current $2I$ is proportional to the absolute temperature. Figure 5.8 shows an equivalent circuit of a practical monolithic IC device (Analog Device, Inc., AD590).[32] This device provides output current of 1 μA times absolute temperature for the supply voltage ranging from +4 ~ +30 V. A voltage

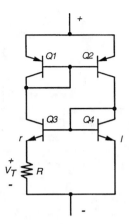

Figure 5.7 Idealized scheme of a two-terminal current-output device. (From Timko, M. P., *IEEE J. Solid State Circ.*, SC-11, 784, 1976. With permission.)

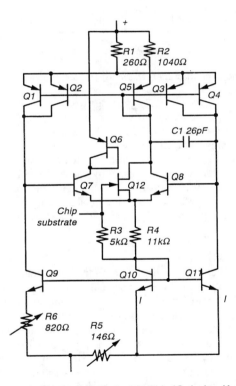

Figure 5.8 An equivalent circuit of a practical monolithic IC device (Analog Devices AD590).

output proportional to the absolute temperature can be obtained by connecting a resistor in series, and by trimming the resistor the error in temperature reading can be adjusted to zero at any desired temperature. The maximum error depends on the temperature span. In the highest-grade device at room temperature, the maximum error is less than 0.1, 0.2, and 0.3 K for temperature spans of 10, 25, and 50 K, respectively.[33]

Devices that produce Celsius, Fahrenheit, or any arbitrary scale output can be realized using a similar technique. For example, a single-chip IC device was fabricated which provides current output with a temperature coefficient of 1 μA/°C and zero output at 0°C.[34] A voltage output device was also fabricated which has a sensitivity of 100 mA/°C and provides zero output at 0°C.[35] A further improved device has higher stability and lower power consumption.[34]

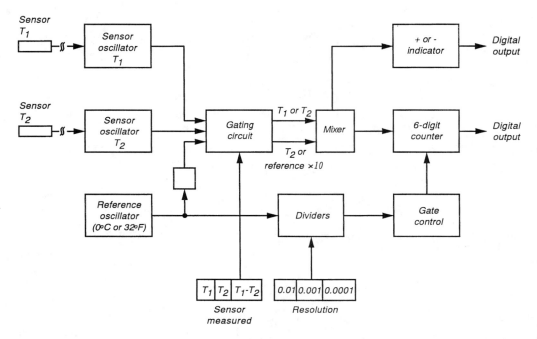

Figure 5.9 Block diagram of a crystal thermometer (Hewlett-Packard HP 2804A).

An IC device (Analog Devices, Inc., AD537) provides frequency output proportional to temperature. In the device, temperature input is converted to voltage and then converted to frequency by a voltage-to-frequency converter fabricated on the same IC chip. By connecting a 1000-pF capacitor externally, a square-wave output with a sensitivity of 10 kHz/K is obtained.[33]

The IC technology provides a method of fabricating diode temperature sensor arrays for temperature distribution measurements.[36,37] Diodes were fabricated by conventional IC technology, and interconnecting wires were also fabricated of evaporated gold. Then a polyimide layer was formed on the wafer, and by etching the back side of the wafer separate silicon islands interconnected by a flexible polyimide layer were formed without bonding. Arrays 3.5, 10, and 20 cm long and 1 mm wide with 20 diodes have been fabricated.[37]

5.2.5 CRYSTAL RESONATORS

The resonant frequency of a quartz resonator has a temperature coefficient, and it can be used as a temperature sensor. The temperature coefficient of typical crystal temperature sensor ranges between 10 and 100 ppm/K. The crystal temperature sensor has an almost uniform temperature coefficient in a wide temperature range, and hysteresis is negligible in medical thermometry.

Figure 5.9 shows the block diagram of a commercial crystal thermometer (Hewlett-Packard, HP 2804A). It has two crystals operated at about 28 MHz and provides the output of each temperature, or the difference between them, with a resolution of 10^{-4} K. Although temperature resolution is limited by stability of the resonator, a short-term stability of 5×10^{-6} K was attained by this system.[38]

A crystal resonator can be excited by ultrasound, and its resonant frequency can also be determined by using an ultrasonic coupling. When an ultrasound near the resonant frequency is applied, the resonator crystal is excited. After the applied wave is terminated, the excitation decays, emitting an ultrasonic wave of the resonant frequency into the surrounding medium. The induced ultrasound can be detected at a distant site, and thus this technique provides remote sensing of temperature in a medium in which the sound can propagate.

A crystal resonator for this purpose is supplied commercially (Tokyo Communication Equipment Co., Ltd.; Kanagawa, Japan). The crystal having a resonant frequency at about 40 kHz is encapsulated in a metal rod, 2 mm in diameter and 7 mm in length. The Q value of the resonator is about 20,000, and a temperature resolution of 0.01 K and absolute accuracy of 0.1 K are attained. In a preliminary experiment, intragastric temperature could be monitored across the abdominal wall using this crystal resonator, a temperature sensor, and the thermometer system with an ultrasonic probe.[39]

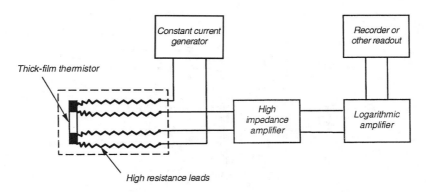

Figure 5.10 Temperature measurement system using high-resistance leads. (Modified from Bowman, *IEEE Trans. Microwave Theor. Tech.,* MTT-24, 43, 1976.)

5.2.6 TEMPERATURE SENSORS FOR USE IN STRONG ELECTROMAGNETIC FIELDS

Tissue temperature measurement in a strong electromagnetic field is sometimes required, especially during hyperthermia cancer therapy using electromagnetic heating. Conventional thermometer probes are erroneous due to the interaction of the field with the metals in the sensor element and its connecting lead. These interactions also distort the field structure and may produce intense heating. To avoid these undesirable interactions, many kinds of thermometers have been studied.

One solution is the use of high-resistance leads. Bowman[40] used a small high-resistance thermistor and plastic high-resistance lead. Employing a four-lead system as shown in Figure 5.10, in which separate lead pairs were used for current supply and to provide a readout of the voltage drop, the thermistor resistance could be measured accurately despite the large and unstable lead resistance. A thin film line fabricated by vacuum deposition of nichrome onto a Mylar substrate has also been employed as a high-resistance lead.[41]

A small thermocouple with very fine wires can also be used. When the diameter of the wire is about 50 μm or less, the interaction between the thermocouple wire and a microwave field is practically negligible.[42] A nonmetallic thermocouple consisting of a carbon-bearing fluorcarbon strand with a carbon-bearing silicone sheath has also been tried.[43] The resistance of this probe, 20 to 40 cm long, was about 25 kΩ, and its output was about 20 mV/K.

Fiberoptic coupling temperature sensors have been developed to solve the problem of electromagnetic interference. Several techniques have been studied for sensing temperature and converting it into optical signals. Among these are liquid crystals,[44,45] phosphors and photoluminescence,[46-49] birefringent crystal with polarizers,[50] semiconductor band-edge absorption shift,[51] and liquid meniscus.[52]

Figure 5.11 shows a configuration of the liquid crystal fiberoptic probe.[45] The liquid crystal housed in a sealed enclosure at the distal end of a fiberoptic catheter is illuminated by a light transmitted through optical fibers, and the reflected light is led to the photodetector through other optical fibers. With a gradual decay in the reflectance due to temperature cycling, a temperature resolution of 0.1°C was attained.

Figure 5.12 shows the configuration of a fluoroptic probe.[47] A pair of rare earth phosphors (two components among La_2O_2S, Gd_2O_2S, and Y_2O_2S) encapsulated at the tip of a fiberoptic catheter are excited by ultraviolet radiation through the fiber, and resulting visible fluorescence is returned by the same fiber. The temperature is determined by the ratio of fluorescence intensities of the two phosphors. By adequate selection of phosphors, absolute accuracy of 0.1°C can be attained over a temperature range required for hyperthermia cancer therapy.

Temperature dependency on the fluorescent decay time is used in a fiberoptic temperature sensor. A magnesium fluorogermanate phosphor excited in an ultraviolet or blue-violet region provides fluorescence in the deep red, and its decay time varies from about 5.3 ms at −200°C to less than a millisecond at 450°C. Near the body temperature, the sensitivity is about .01 ms for a temperature change of 1.2°C. A single-sensor probe of 0.5 mm and a four-sensor probe with a 0.9-mm outer diameter were fabricated.[48] A commercial model of this type is available (Model 3000, Luxtron Corporation; Mountain View, CA).

A fiberoptic temperature probe using the shift of the optical absorption edge of a gallium arsenide semiconductor in the near-infrared is available.[49] Figure 5.13(a) and (b) show configurations of the probes with single sensors and with a four-sensor array. The band edge is shifted toward longer wavelengths

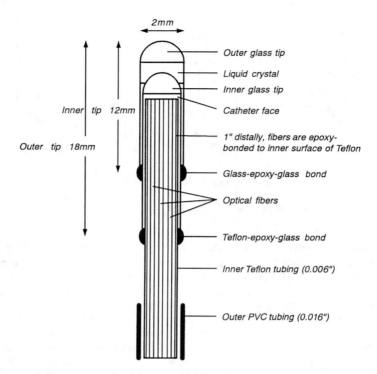

Figure 5.11 A liquid-crystal fiberoptic temperature probe. (From Livingston, G. K., *Radiat. Environ. Biophys.*, 17, 233, 1980. With permission.)

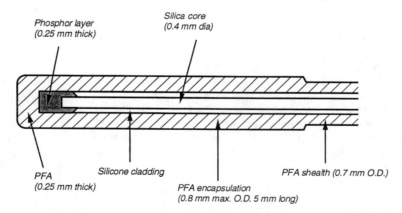

Figure 5.12 A fluoroptic temperature probe. (Modified from Wickersheim, K. A. and Alves, R. V., *Prog. Clin. Biol. Res.*, 107, 547, 1982.)

as temperature increases, as shown in Figure 5.14. When the appropriate light emitting diode (LED) having a center wavelength at the band edge as the light source is used, the intensity of the light reflected by a reflector prism made by gallium arsenide varies with temperature.

Gallium arsenide also exhibits efficient photoluminescence phenomena, and the photoluminescence spectra has a temperature dependency as shown in Figure 5.15.[53] A fiberoptic medical thermometer using this principle has four sensing points, 10 mm apart from each other, in a 0.7-mm diameter probe (FTP-5, ASEA Fiber Optic Sensors; Stockhom, Sweden). A vapor pressure thermometer has also been tested in which about 5 ml of trichlorofluoromethane is confined at the tip of a probe, and the pressure is transmitted by a plastic tube filled with ethandiol.[54]

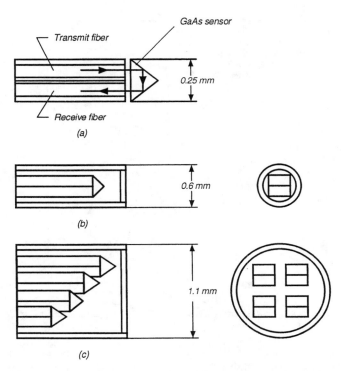

Figure 5.13 Fiberoptic temperature probes using the absorption band edge shift in a GaAs semiconductor. (From Vaguine, V. A. et al., *IEEE Trans. Biomed. Eng.*, BME-31, 168, 1984. With permission.)

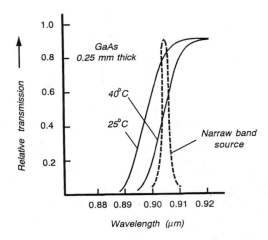

Figure 5.14 Absorption band edge shift in GaAs and the emission spectrum of appropriate light emitting diode to detect the change in the absorption spectrum. (From Vaguine, V. A. et al., *IEEE Trans. Biomed. Eng.*, BME-31, 168, 1984. With permission.)

5.3 NONCONTACT TEMPERATURE MEASUREMENT TECHNIQUES

Noncontact temperature measurement can be realized using radiation heat transfer. Infrared radiation thermometers and thermographies have been used for skin temperature measurement. Microwave radiometers can detect thermal radiation in microwave regions, and they can be used for deep tissue temperature measurements. Imaging of temperature distribution in deep tissue has been attempted even though it is extremely difficult.

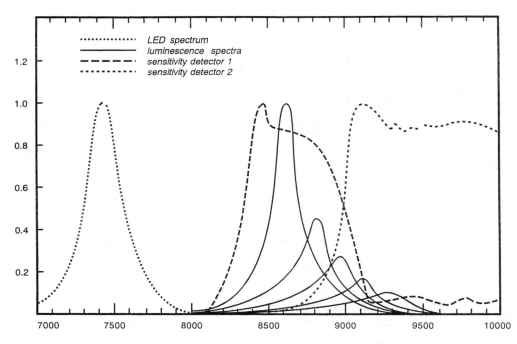

Figure 5.15 Photoluminescence spectra of GaAs crystal at different temperatures, emission spectrum of the light source, and spectral responses of detectors in a photoluminescence-type fiberoptic temperature probe. (From Ovrén, C. et al., *Int. Conf. on Optical Techniques in Process Control, Hague,* Paper B2, 67, 1983. With permission.)

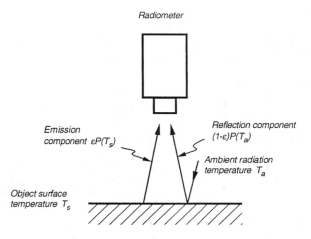

Figure 5.16 Radiation power entering into a radiometer directed to an object surface. It consists of two components: the emission from the object and the reflection of the ambient radiation.

5.3.1 INFRARED MEASUREMENTS
5.3.1.1 Infrared Radiation Thermometers

The radiation thermometer is essentially an instrument that measures thermal radiation power emitted from the object surface. Near the human body temperature, the peak of the thermal radiation is in the far-infrared region; thus, infrared radiometers are used in medical thermometry.

When a radiometer is directed at an object surface as shown in Figure 5.16, both the emission from the object and the ambient radiation reflected at the object surface will enter into the radiometer. Thus, the total power, W, which enters the thermometer can be expressed as the sum of the emission and reflection components as

$$W = \varepsilon P(T_s) + (1 - \varepsilon)P(T_a) \tag{5.14}$$

where ε is the emissivity of the object surface, and T_s and T_a are surface and ambient radiation temperatures. $P(T)$ is Planck's radiation formula at absolute temperature, T, i.e.,

$$P(T) = \int \frac{C_1 \lambda^{-5}}{\exp(C_2/\lambda T) - 1} d\lambda \tag{5.15}$$

where λ is wavelength, and C_1 and C_2 are universal constants given as

$$C_1 = 3.74 \times 10^{-16}\,\mathrm{Wm}^2$$

$$C_2 = 1.44 \times 10^{-2}\,\mathrm{mK}.$$

When the radiometer has a sensitivity in a range from λ_1 to λ_2, the integral in Planck's formula should be taken in that wavelength range.

From Equation (5.14), the surface temperature, T_s, can be obtained when the emissivity, ε, and the ambient radiation temperature, T_a, are known.

While the emissivity of the skin had been considered as almost unity, which means that the skin is essentially a black body, it was shown that the skin emissivity is around 0.97 in the wavelength range of 8 to 14 μm.[55] This implies that about 3% of the reflection may occur at the skin surface; thus, the ambient radiation temperature can affect a temperature measurement by a radiation thermometer.

Temperature readings may vary when the optical axis is not perpendicular to the skin, but it has been shown that the apparent temperature readings do not fall for angles of inclination of less than 70°. A significant fall only occurs above 80° when the emissivity is 0.98, $\lambda = 5$ μm, surface temperature is 30°C, and background ambient radiation temperature is 20°C.[56,57]

When a temperature gradient across the skin exists, the temperature measured by a radiation thermometer will differ from that measured at the surface by means of a contact thermometer probe, because radiation from the body emanates not only from the external surface but also partly from the underlying layer.[58,59] The amount of radiation from the underlying layer may depend on the transmittance of the epidermis and dermis. Anderson and Parrish[60] estimated the approximate depth of penetration at which radiation is attenuated to 1/e of the incident energy. The depth of penetration for $\lambda = 1.2$ μm was estimated to be about 2.2 mm. In longer wavelength ranges, the transmittance is somewhat lower than that for 1.2 μm, but a systematic study of skin penetration depth is still lacking.

Watmough and Oliver[59] estimated the effective temperature difference in an extreme situation in which blood at 310 K was perfused beneath the epidermis at 305 K. They showed that a temperature reading about 2.5°C higher than the surface temperature would be expected using a radiation thermometer with a sensitivity peak at 5 μm. If a much smaller, but still relatively large, temperature gradient of 4°C/cm is assumed and the penetration depth is assumed to be 1 mm, the difference between the radiation thermometer reading and the true surface temperature will be about 0.4°C. This analysis suggests that accurate skin surface temperature measurement using a radiation thermometer requires a correction for skin emissivity and penetration depth; however, in a practical situation, such accurate measurement is rarely required. If an error of about 0.5°C is acceptable, a radiation thermometer can be used for skin temperature measurement without such corrections.

5.3.1.2 Infrared Thermography

The term "thermography" generally implies techniques of thermal imaging of the object surface. Practical techniques of thermography are infrared thermography based on radiation measurement and contact thermography using thermochromic liquid crystals. Infrared thermography can provide thermal images

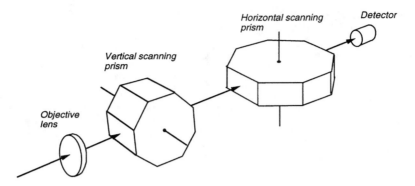

Figure 5.17 The scanning system of a thermography in which two rotating polygonal prisms are used.

in electric signals, which are convenient for image processing and data storage, and it can also realize noncontact, quick, and accurate measurement.

To obtain a thermal image of an object, surface temperature measurements at many points on the object should be performed quickly. This requirement can be achieved by infrared radiometers with either mechanical scanning by means of moving mirrors or prisms, or with electronic scanning in an image tube or a solid-state image device. In conventional medical thermography systems, the mechanical scanning system has been widely used.

Medical infrared thermography systems consist of vertically and horizontally moving mirrors or prisms and an infrared detector, typically HgCdTe at liquid nitrogen temperature. Figure 5.17 shows an example of the scanning system employed in the AGEMA Thermovision System Model 870 (Stockholm, Sweden). Using rotating polygonal prisms, it realizes very high scanning speeds up to 25 frames per second with 280 lines for vertical scanning ranges. A temperature resolution of 0.05 to 0.1°C is attained in most commercial thermography systems.

The thermal image is displayed on a monitor screen as either discrete temperature levels represented by colors or a continuous gray tone. The color image allows quantitative assessment throughout the entire temperature range, while the gray tone image provides a clear picture which is much more accurate for identification of vascular changes.[61]

Conventional thermography systems have an internal temperature reference, so that radiation from the object can be compared with that from the reference body at every line scan. To calibrate a thermography system, a set of discrete temperature references consisting of black bodies at different temperatures[62] or a continuous temperature calibrator consisting of a metal rod with a uniform axial temperature gradient can be used.[63]

Infrared television, making use of a pyroelectric vidicon as the camera tube, has been used. It is less expensive than the mechanical scanning thermography systems and can be operated at room temperature. Spatial resolution of the system is comparable or higher than that of the mechanical scanning system at higher image contrasts, while resolution is lower when the contrast is less.[64] The frame rate of about 25 frames per second is comparable with that of high-speed mechanical scanning. However, temperature resolution is about 1.0°C, which is far below that obtained by ordinary mechanical scanning thermography systems, and its medical application is limited.

Electronic scanning solid-state devices in the infrared region have been developed extensively. To read out signals from an infrared detector array, scanning circuits such as the CCD (charge coupled device) can be used, as shown in Figure 5.18(a). Hybrid devices can be fabricated in which the infrared detector array of InSb or CdHgTe is electrically connected to a CCD by wire bonding or multiple flexible contacts.[65]

Optical signals from a detector strip may be read by a simplified method, which consists of a strip of n-type CdHgTe. As shown in Figure 5.18(b), a constant bias current is applied through the two ohmic end contacts, and the signal is read out from the third ohmic contact.[66] In the CdHgTe strip, carriers are generated, and drift along the strip to the negative end contact due to the electric field is developed in the strip. If the lifetime of the carrier is longer than the traveling time along the strip, carriers originated by illumination reach the readout contact, and thus the optical image along the strip can be scanned. In an example of the actual device, eight CdHgTe filaments, each of which is 700 μm in length and 62.5 μm wide, having a 12.5-μm gap between filaments, are fabricated on a sapphire substrate and operated at

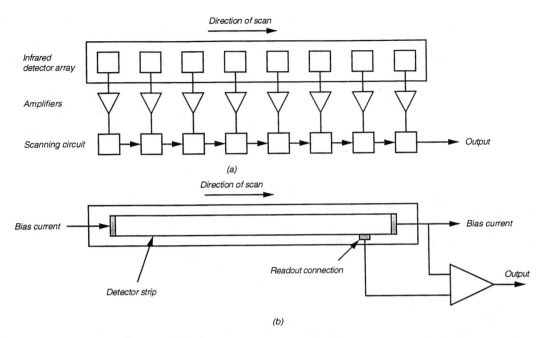

Figure 5.18 Two serial scanning systems for use in thermal imaging: **(a)** detector array with scanning circuit, and **(b)** a method using a detector strip.

liquid nitrogen temperature. This device has been supplied commercially as the SPRITE detector.[67] While the device in itself is not sufficient for obtaining a complete thermal image, it has been used in medical thermography systems combined with a mechanical scanning method.

Monolithically integrated image devices for thermal imaging have also been studied, and image devices operating at a wavelength band around 10 μm were realized utilizing Schottky-barrier infrared detectors, or heterojunction internal photoemission infrared detectors. While the quantum efficiencies of these detectors are lower than that of CdHgTe, their advantage is that they are fabricated on silicon substrates by standard integrated circuit processing techniques. As an example, a monolithic 128 × 128-element IrSi Schottky-barrier detector array was reported.[68] It had a cutoff wavelength of about 9.4 μm, and a temperature resolution of 0.3 K was attained when it was operated at 50 K. A 400 × 400-element heterojunction internal photoemission detector array consisting of Ge_xSi_{1-x} alloy was also reported.[69] Its cutoff wavelength was about 9.3 μm, and a temperature resolution of 0.2 K was attained at an operating temperature of 53 K.

5.3.2　MICROWAVE RADIOMETERS AND IMAGING SYSTEMS

The microwave radiometer detects thermal radiation in the microwave range. At the normal temperature of the human body, much less energy is emitted thermally in the microwave region (many orders of magnitude) than in the infrared region; nevertheless, the thermal emission in the microwave region is detectable and can be a promising method of subsurface temperature measurement. Microwaves of 3 GHz penetrate about 1 cm into tissue, while far-infrared radiation penetrates only 0.1 mm or less.

Microwave radiometers, originally developed for radioastronomy, are sensitive enough for medical thermometry. As an example, a radiometer of 1 to 6 GHz with a temperature-stabilized tunnel diode input stage has a temperature sensitivity of about 0.1°C.[70]

When using the radiometer to measure local subsurface tissue temperature, a contact or remote-sensing aerial is used. The contact type aerial shown in Figure 5.19(a) is used at relatively long wavelengths, typically between 5 and 30 cm. The remote-sensing type of aerial has a large reflector which focuses the radiation from a local site in the body into the horn at the front of the receiver, as shown in Figure 5.19(b). Because the size of the reflector must be greater than several wavelengths, the remote-sensing system is practical only at shorter wavelengths, from about 0.4 to 3 cm.[71]

Using a remote-sensing system and by scanning the reflector over the object, thermal imaging can be achieved. The spatial resolution is limited by the operating wavelength, and higher resolution can be

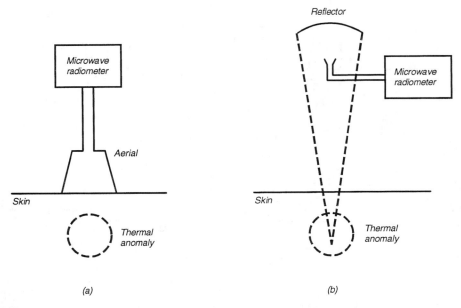

Figure 5.19   Subsurface temperature measurement systems by microwave radiometers: **(a)** contact type, and **(b)** remote-sensing type.

obtained in shorter wavelengths. At the wavelength of 9 mm, spatial resolutions of about 9 mm were achieved, while at the longer wavelength of 3 cm, spatial resolutions were about 2.6 cm.[71]

The multifrequency microwave radiometer provides information about the temperature-vs.-depth profile. A study using a three-band radiometer system operating at 1.5 GHz, 2.5 GHz, or 3.5 GHz showed that the temperature profile can be obtained over a depth of up to 5 cm with an error of about ±0.5°C in meat phantoms and animals.[72]

5.3.3   NONINVASIVE THERMAL IMAGING IN DEEP TISSUE

Even though noninvasive thermal imaging in deep tissue is extremely difficult, many attempts have been made using different techniques such as microwave radiometer, ultrasound, X-ray computer tomography (CT), and magnetic resonance imaging (MRI). Active microwave imaging has been attempted in which local temperatures in the body are estimated from a reconstructed image of the temperature-dependent dielectric constant of the tissue. An experiment using a phantom filled with water showed that a temperature resolution of 1°C with a spatial resolution of 6 mm at 3 GHz is attainable.[73] It was also shown that this technique can be applied to inhomogeneous bodies.[74]

Ultrasonic computed tomography has also been attempted for thermal imaging in the body, in which local temperatures are estimated from temperature dependencies of sound velocity and the attenuation constant.[75] However, absolute temperature measurement is very difficult, and even the measurement of relative change in a regional temperature is not an established technique.

Clinical X-ray CT can detect small changes in X-ray absorption due to thermal expansion of water. It has been shown that a relative precision of 0.25°C can be attained.[76]

Thermal imaging by MRI is based on the temperature dependency of the magnetization relaxation parameters. The longitudinal relaxation time, T_1, is commonly used. In the blood, it has a temperature coefficient of about 1.4%/K. In a study with a phantom, taking blood samples at different temperatures, it has been shown that a temperature resolution of 2°C with a spatial resolution of 6 mm can be obtained from a 5-min scan.[77]

5.4   CLINICAL THERMOMETERS

Since the mercury-in-glass thermometer was introduced in medicine in the early years of the 20th century, the measurement of body temperature has become easy and accurate.[78] The mercury-in-glass clinical

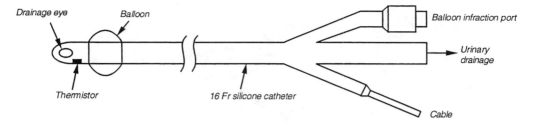

Figure 5.20   The thermistor-tipped Foley's bladder catheter.

thermometer is appreciated in clinical thermometry for its remarkable reliability, as well as its convenience of handling and low cost. Even today, no other thermometer can fully replace it.

There are, however, many situations in which the mercury-in-glass clinical thermometer is unacceptable because of its slow response and large heat capacity and size. It is inconvenient for continuous monitoring of body temperature and data processing by computer. Mercury contamination is also a serious problem in hospitals.[79] Many kinds of thermometers for medical use, therefore, have been developed and used successfully.[80]

5.4.1   INDWELLING THERMOMETER PROBES

The mouth is the most convenient site for routine measurement of core temperature. For continuous monitoring, however, placement of the probe in the mouth for a long period of time is uncomfortable, and oral temperature becomes unstable when the mouth is opened. Body core temperature is, therefore, usually monitored at deeper sites of the body such as the rectum, esophagus, and bladder using indwelling thermometer probes.

5.4.1.1   Rectal Temperature Measurement

The rectal probe has been commonly used for patient monitoring. The probe is simply a flexible catheter having a thermistor at the tip. To avoid the effect of ambient temperature change, the tip of the probe is placed about 8 to 15 cm from the anal sphincter. Rectal temperature has been considered as a reliable index of core temperature, if body temperature is stable.

Rectal temperature is always 0.2 to 0.3°C higher than that obtained in any other part of the body.[81-83] The possibility of the effect of bacterial heat production on rectal temperature has been pointed out;[84] however, doubt is cast on this effect by an experimental observation in which temperature change at bowel sterilization by orally administered antibiotics was insignificant.[85,86]

When body temperature varies, changes in rectal temperature are delayed, compared with other temperatures in the central part of the body. During hypothermia, for example, rectal temperature is higher than esophageal temperature during cooling, and the difference is inverted during rewarming.[87] Due to such large thermal inertia, rectal measurement is not reliable for patient monitoring in anesthesia. Rectal measurement is not recommended in infants and small children, because of the possibility of rectal perforation.[88]

5.4.1.2   Esophageal Temperature Measurement

Esophageal temperature is measured by inserting flexible probes through the mouth or nose, primarily for body temperature monitoring during anesthesia. In the upper part of the esophagus, significant influence of tracheal air temperature has been observed.[89] It is therefore recommended that measurement of esophageal temperature should be taken 24 to 28 cm below the corniculate cartilages or at heart level. At this level, esophageal temperature is intermediate between oral and rectal temperature and rapidly follows internal temperature changes.[87] Esophageal measurement can be tolerated postoperatively if the probe is introduced through the nose as a gastric tube.[88]

5.4.1.3   Bladder Temperature Measurement

Bladder temperature is monitored by a thermistor-tipped Foley's bladder catheter, i.e., a 16-Fr, triple-lumen all silicone catheter with an attached thermistor. Figure 5.20 shows the schematic diagram of an example. It was shown that bladder temperatures are highly correlated with rectal, esophageal, and pulmonary arterial temperatures and closely follow arterial blood temperature even during rapid cooling

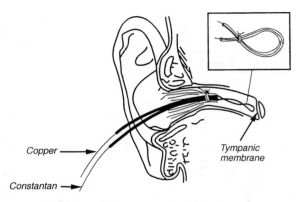

Copper →

Constantan →

Tympanic membrane

Figure 5.21 Tympanic temperature measurement using a brush-type thermocouple probe. (Modified from Benzinger, T. H. and Taylor, G. W., in *Temperature, its Measurement and Control in Science and Industry*, Vol. 3, Herzfeld, C. M., Ed., Reinhold, New York, 1963, 111.)

and rewarming using extracorporeal circulation.[90] Bladder temperature monitoring is recommended particularly for patients in whom Foley's catheterization is indicated.

5.4.2 TYMPANIC THERMOMETERS

Tympanic temperature has been considered as a reliable index of core temperature, and it has been used in physiological studies of thermoregulation. It is also used in patient monitoring during anesthesia and in intensive care units. Noncontact tympanic thermometers have the advantage that body temperature measurement can be performed in a few seconds.

5.4.2.1 Contact Probes

Small thermocouple or thermistor probes have been used for tympanic temperature measurement. Benzinger and Taylor[91] described tympanic probes in which 36-gauge copper-constantan wires were soldered side by side at the tip and drawn into fine polyethylene tubing. A brush probe, as shown in Figure 5.21, was recommended, with the free ends of the bristles facing the interior of the meatus. The authors stated that there was practically no discomfort in the wearing, and minor alterations in hearing were advantageous for ascertaining correct positioning.

The comparative evaluation of temperatures recorded using tympanic and two nasopharyngeal probes at a region that receives its blood supply from an artery supplying the brain showed that differences between these temperatures were 0.05°C or less in environmental temperatures of 24 and 45°C.[91] Tympanic temperature was also compared with esophageal temperature in clinical situations, and it was recognized that both were closely parallel.[92-94] Although the safety and minimum embarrassment offered by this probe have been stressed by many authors, there have been some case reports of tympanic membrane perforation complicating tympanic thermometry using contact probes.[95,96]

5.4.2.2 Noncontact Tympanic Thermometers

The noncontact tympanic thermometer is essentially an infrared radiation thermometer having a probe that can be directed to the tympanic membrane through the external auditory canal. Noncontact measurement of tympanic temperature was first attempted by using a small thermistor with active heating at the proximal end of the probe.[97] Since then, instruments employing fast response infrared detectors have been developed, and different kinds of infrared tympanic thermometers are now supplied commercially.

A commercial tympanic thermometer (First-Temp®, Intelligent Medical Systems; Carlsbad, CA) employs a thermopile detector with a light pipe installed at the tip of the probe. The tip is inserted into the external auditory canal, as shown in Figure 5.22.[98] When the probe is correctly applied to the ear, tympanic temperature can be measured within 2 s. In this model, the probe should be recalibrated after each measurement by placing it on the instrument body on which a reference radiation source is installed. Infrared tympanic thermometers that can be used without recalibration are becoming common.

A clinical observation showed that tympanic temperatures measured by an infrared tympanic thermometer (First-Temp®) were close to pulmonary arterial temperatures over a temperature range of 34.0

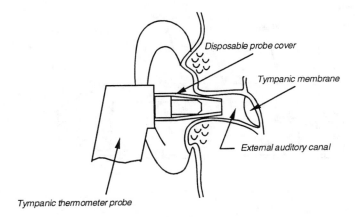

Figure 5.22 Noncontact tympanic temperature measurement with a probe inserted into the external auditory canal. (Modified from O'Hara, G. J. and Phillips, D. B., *U.S. Patent No. 4, 602, 642, 1986.*)

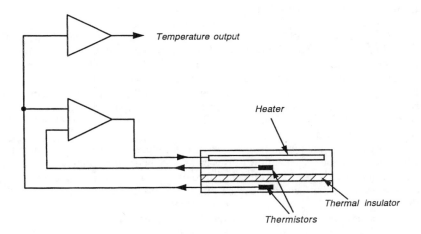

Figure 5.23 The probe and the servo-controlled circuit of the zero-heat-flow thermometer.

to 39.5°C with a correlation coefficient of 0.98, even including the situation of rapid increase in temperature after open heart surgery.[99]

Thermoscan, Inc.® (San Diego, CA) introduced two types of tympanic thermometers, PRO-1 and HM-1, for professional and home use, respectively.[100] Fraden[101] has reviewed the history and the principle of this sensor type. These thermometers differ from the First-Temp® by using a different detection technique. PRO-1 and HM-1 utilize a pyrosensor as the detector element. The infrared heat flow from the tympanic membrane is detected, the measurement time is specified to be less than 1 s, and the minimum time between readings is 8 s. The accuracy meets the ASTM Standard E 1112-86.

The Thermoscan® instant thermometers (and other types) were evaluated clinically by Romano et al.[102] (pediatric intensive care), Weiss et al.[103] (laboratory testing and gynecology patients), and by Wolf and Baker[104] (obstetrics and gynecology). The accuracy and repeatability were found to be in accordance with the specifications given by the manufacturer. One major advantage of the infrared tympanic thermometer is the capability of rapid assessment of fairly accurate body temperatures; therefore, it can save nursing time in busy clinics and on hospital floors.

5.4.3 ZERO-HEAT-FLOW THERMOMETER

A zero-heat-flow thermometer is a device that measures deep tissue temperature from the skin surface. Although skin temperature more or less reflects deep temperature, a large temperature gradient exists across the skin exposed to the outer air. The temperature gradient can be reduced by insulating the skin

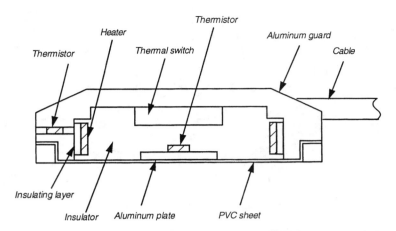

Figure 5.24   Cross-section of the standard probe of a zero-heat-flow thermometer.

surface; however, available insulating materials are insufficient for realizing a condition where the temperature difference between deep tissue and the skin surface can be negligibly small. To ensure complete thermal insulation, a servo-controlled heating system is employed.[105,106]

The probe and the servo-control circuit of the zero-heat-flow thermometer are shown in Figure 5.23. The probe has two thermistors, separated by an insulating layer and a heater. The temperature difference across the insulator is measured and the heating power is controlled in such a way that no temperature gradient exists across the insulating layer. Consequently, no heat flows across this layer so that the layer is equivalent to an ideal insulator. When the probe is applied to the skin surface, it prevents heat loss from the surface and the skin temperature rises to the level of the deep tissue temperature. A commercial unit has been produced (Deep Body Thermometer, Ltd.; Cambridge, U.K.). The probe of the thermometer is a flat, square pad 6×6 cm in size.

In the actual probe, having a configuration as shown in Figure 5.23, some small heat flow remains due to the radial heat flow, and this heat flow causes errors. To reduce the radial heat flow, a metal guard covering the probe and making contact with the surface of the skin at the circumference of the probe is effective, and it was shown that accuracy can be improved by this modification.[107,108] A thermometer employing this type of probe is also produced commercially (Terumo Co.; Tokyo Japan). Disc-shaped probes of different sizes, from 1.5 to 8 cm in diameter, are supplied. Figure 5.24 shows a cross-section of the probe.

To monitor body temperature with a zero-heat-flow thermometer, the probe is attached to the forehead, upper sternum, or abdomen. The initial response time when the probe is applied to the exposed skin is 15 to 20 min. After thermal equilibrium is reached, the probe temperature follows physiological changes in core temperature. It was shown that the temperature measured by a zero-heat-flow thermometer probe placed at the upper sternum did not lag behind auditory canal and intestinal temperatures, even at the onset of fever induced by an intravenous injection of endogenous pyrogen.[106]

Comparative studies of zero-heat-flow probe temperatures with other conventional core temperatures, such as oral, tympanic, nasal, rectal, and pulmonary arterial, show that body temperature monitoring by the zero-heat-flow thermometer is acceptable in most clinical situations.[39,109-111] A study under controlled conditions in a climatic chamber has also shown that zero-heat-flow probe temperatures at the forehead are within 0.1°C of sublingual temperature measured by small temperature sensors with fine stainless-steel leads when the ambient temperature is 20 or 25°C.[112]

The major advantage of the zero-heat-flow thermometer is its convenience in applying probes to patients. This thermometer is particularly suited for long-term body temperature monitoring.

5.4.4   TELEMETERING CAPSULES

Telemetering technique is effective when a sensor can be placed near the object, while direct connection by a cable is unfavorable. For body core temperature measurement in the digestive tract, a temperature sensor and radio transmitter are encapsulated in a swallowable radio telemetering capsule. A passive telemetering technique without using a battery-operated transmitter is also possible and has the advantage

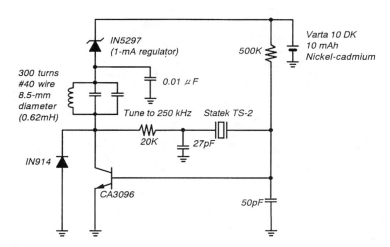

Figure 5.25 The circuit diagram of a radio pill, in which temperature is sensed by a quartz tuning fork. (Modified from Cutchis, P. N. et al., *Johns Hopkins APL Tech. Dig.*, 9, 16, 1988.)

that the operating period is not restricted by battery life. Ultrasonic excitation and detection of a crystal resonator is an example of this type.

5.4.4.1 Radio Pill

The radio pill is a swallowable radio-telemetering capsule which can be used for measurements in the digestive tract.[113,114] The temperature-sensitive radio pill has been used for body core temperature monitoring. One example of a radio pill is 2.2 cm long and 0.9 cm in diameter, is operated at about 350 kHz with a temperature coefficient of about 10 kHz/K, and has a response time of about 4 min for a 97% response in unstirred water.[114] A passive telemetry system in which the transmitter circuit is powered by an externally applied radio frequency excitation is also possible.[115]

A radio pill in which a crystal resonator is employed as the temperature sensor has been developed by Cutchis et al.[116] The crystal oscillator circuit is shown in Figure 5.25. Temperature is sensed by a quartz tuning fork having a sensitivity of about 9 Hz/K operated at 262 kHz. The circuit is powered by a 1.2-V rechargeable battery, and the current drain is about 100 μA. The battery can be charged by electromagnetic induction. The power consumption can be reduced by command mode operation, by which the oscillator is activated for a 10-s readout period when a command signal is applied. A disposable type of this pill is produced commercially (CorTemp®, Human Technologies, Inc.; St Petersburg, FL). It has a silver-oxide battery and operates for 200 hours.

The obvious advantage of the radio pill in temperature measurement is the lack of awareness of it after swallowing. While it is difficult to identify the location of the pill in the intestine, departure from the stomach can be established by observing the effect of a small drink of water. However, to reuse the pill, it must be recovered from the feces, and the procedure may be unpleasant. Disposable pills are preferable, especially in human use.

5.4.4.2 Ultrasonic-Coupling Crystal Resonator

The quartz crystal resonator can be excited by applying an ultrasonic wave near the resonance frequency. Ultrasonic detection of the resonance signal is also possible, as long as the resonator is placed in a medium through which sound propagates.

In one study, a crystal resonator consisted of a quartz tuning fork encapsulated in small metal cylindrical capsule, 2 mm in diameter and 7 mm in length, as shown in Figure 5.26.[117] Its resonant frequency was about 40 kHz with a temperature coefficient of about 3.2 Hz/K. To measure temperature, the resonator was excited for about 0.4 s by applying an ultrasonic wave near the resonant frequency. The damped oscillations in the resonant frequency were measured, and the absolute accuracy was about 0.1°C with a temperature resolution of about 0.01°C. It was shown that when the capsule was swallowed, temperature measurements could be performed from the abdominal skin surface.

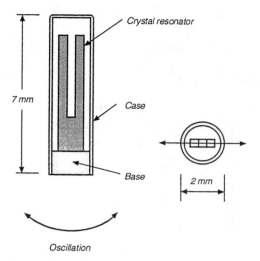

Figure 5.26 An implantable ultrasonic-coupling crystal resonator for local temperature measurement in the body.

5.5 HEAT FLOW MEASUREMENTS

The heat flow can be measured directly by a heat flow transducer, in which the heat flow is converted into an electromotive force. Heat flow measurement is sometimes required when the rate of heat dissipation from the body surface has to be measured. When heat flow at a specific part of the body is to be determined, a small heat flow transducer is attached to that part so that the heat flows through the transducer. The total heat dissipation from the body is measured directly by a calorimeter. While there are many different types of calorimeters, heat flow transducers are employed in the gradient-layer calorimeters so as to measure heat flow across its wall.

5.5.1 HEAT FLOW TRANSDUCERS

A conventional heat flow transducer is a thin, flat plate sensor. If the material of the sensor plate has thermal conductivity k and its thickness is d, then heat flow, Q, which is the amount of heat flow per unit area, is proportional to the temperature difference, ΔT, across the plate so that

$$Q = k\Delta T/d. \tag{5.16}$$

Thus, heat flow Q can be determined by measuring the temperature difference between both sides of the plate.

When a heat flow transducer is attached to the object surface, natural heat flow distribution may be disturbed to some extent. To reduce this effect, the heat flow transducer should be adequately thin, and the material the plate consists of should have high thermal conductivity. As a consequence, very small temperature differences should be measured.

In practical heat flow transducers, the temperature difference is usually measured by thermocouples. Figure 5.27(a) shows an example of a heat flow transducer in which a metal plate is sandwiched by another metal. While the configuration is simple, the sensitivity is low. If a constantan plate of 5-mm thickness is coated by copper on both sides, its sensitivity is only about 0.083 $\mu V/(W/m^2)$. The sensitivity can be increased to some extent by using combinations of metals which provide higher thermal electromotive forces. As an example, a 1.5-mm plate of tellurium-silver alloy thickly coated on its two sides with copper provides about 6.9 $\mu V/(W/m^2)$.[118]

Higher sensitivity can be achieved by using a thermopile in which many thermocouples are connected in series. Figure 5.27(b) shows an example of such a type. The thermocouple wire is wound spirally on a strip having adequate thermal resistance, so that every half-turn of each spiral consists of constantan wire and the other half-turn consists of copper wire. This configuration is known as the Schmidt belt, named for the inventor.[119]

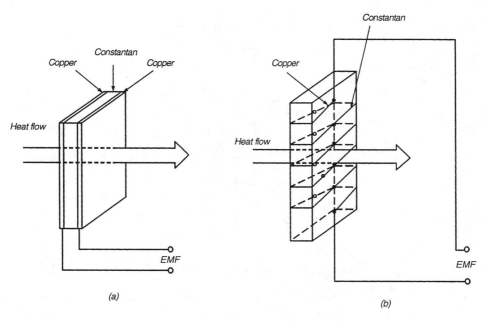

Figure 5.27 A heat flow transducer consists of sandwiched metal plates **(a)** and thermocouples connected in series **(b)**.

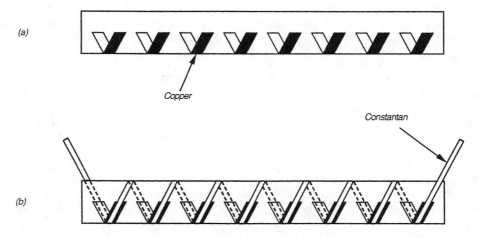

Figure 5.28 A method of fabricating the Schmidt-belt heat flow transducer. (Modified from Cairnie, A. B. and Puller, J. D., *J. Sci. Instrum.*, 36, 249, 1959.)

While heat flow transducers of different shapes and sensitivity are commercially available, small sensitive transducers are required for biomedical applications. For example, a transducer 28 × 28 × 1.6 mm in size having a sensitivity of about 50 μV/(W/m²) is available (H 12, Themonetics Co.; San Diego, CA).

The fabrication procedure of the Schmidt belt heat flow transducer can be simplified by employing a configuration shown in Figure 5.28.[120] A copper ribbon is cut and applied to a paper-board strip as shown in (a). Constantan ribbon is then wound over the copper as shown in (b), and the junctions are soldered. In this configuration, copper and constantan ribbons are connected in parallel at every half-turn; thus, loop current may be generated if a temperature gradient along this half-turn is developed. However, this current does not cause significant reduction of thermal electromotive force, because the resistivity of constantan is about 30 times greater than that of copper.

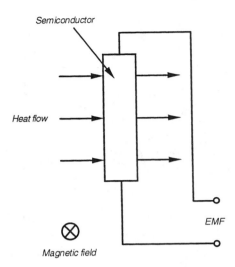

Figure 5.29 The principle of the Nernst effect.

A simplified fabrication technique for a heat flow transducer using nylon net was also reported.[121] The transducer was made by sewing several windings of constantan wire (average number, 36) into a nylon net, and thermoelectric junctions were produced by partly "coppering" the constantan wire in a galvanic bath. Then the net was filled with rubber to keep a fixed distance between the cold and warm sides. With this procedure, the transducer was made electrically insulated, flexible, and robust.

In order to realize very thin heat flow transducers, a technique using the Nernst (transverse thermo-magnetic) effect has been introduced.[122] The principle of the Nernst effect is shown in Figure 5.29. When heat flows through a semiconductor plate to which a transverse magnetic field is applied, electromotive force is developed perpendicular to the direction of heat flow and the magnetic field. As the semiconductor material, InSb-NiSb eutectic and Gd_3As_2-NiAs eutectic were employed. The advantage of applying the Nernst effect is that the output electromotive force is proportional not to the temperature difference but rather to the temperature gradient. For a given heat flow, the temperature gradient is independent of the thickness, so the thermal resistance can be reduced without reducing sensitivity. However, the use of an external magnetic field is a serious disadvantage to most applications and is only used in thermal radiation detectors.[123]

The sensitivity of heat flow transducers cannot be determined accurately by their geometry and material, so calibration is required. In commercial transducers, the sensitivity value of each transducer is attached to it.

Calibration of heat flow transducers is always performed by a standard heat-dissipating surface across which uniform heat flow is maintained. While the application of a heat flow transducer to the standard surface will disturb heat dissipation from the surface, due to the thermal resistance of the transducer, the same situation will occur in actual use. Thus, it is practical to note that the transducer output corresponds to the undisturbed heat flow on the standard surface, as long as the situation of calibration procedure is similar to actual measurement situations.

A simple calibration method for body heat loss measurement in a water immersion experiment was described by Gin et al.[124] The heat flow transducer to be calibrated was applied to a plastic cylinder of uniform thickness in the same way as it was normally applied to the skin. The inside of the cylinder was filled with water, and its temperature was kept at room temperature by a heater. The cylinder was then immersed in stirred water kept at 0°C by crushed ice. The heat flow was calculated as the electric power consumed in the heater divided by the surface area of the cylinder.

5.5.2 HEAT FLOW MEASUREMENT AT THE BODY SURFACE

Heat flow transducers are used for local heat flow measurement on the skin in human and animal studies, e.g., to estimate heat loss from the body or the efficiency of thermal insulation of clothes or animal fur. When a heat flow transducer is attached to a body surface, it should have good thermal contact with the skin. Usually a transducer is attached to the entire inside surface by doublesided adhesive tape. If it is attached only to the edge, then thermal contact at the center can be facilitated by applying a thin layer of thermal-conducting grease.[124]

Total heat loss from the body can be estimated approximately by placing heat flow transducers in different regions and computing the sum of surface area times heat flow at each region. While the accuracy will increase by increasing the number of measuring sites, 5 to 15 regions normally are selected.[125-128]

When heat flow from the skin to the ambient air is measured by a heat flow transducer, the condition of heat exchange at the outer surface of the transducer should be similar to that of the actual skin. In animal studies, the heat flow transducer is sometimes covered with the particular animal's skin or an artificial pelt.[129]

The thermal resistance of the transducer and that between the transducer and the skin may affect the natural heat transfer between the skin and the surrounding medium. If the thermal resistance, R, including transducer and contact resistance, is known by separate calibration, the heat flow corrected for the insulating effect, H_{corr}, can be estimated from the temperature difference between the uncovered skin adjacent to the transducer and an ambient reference, ΔT_1, and that between under the surface of the transducer and the ambient reference, ΔT_2, as

$$H_{corr} = \frac{H_{meas} \cdot \Delta T_1}{\Delta T_2 - R \cdot H_{meas}} \tag{5.17}$$

where H_{meas} is the measured heat flow. According to a study in which this correction is applied to measurements in subjects immersed in water, H_{corr}/H_{meas} ranged between 1.10 and 1.15, and was 1.12 on average when the heat flow transducer Microfoil P/N 20460 (RdF Corp.) was used.[125]

The insulating effect due to the thermal resistance of the heat flow transducer can be compensated for by placing a servo-controlled thermoelectric heat pump on the outer surface of the transducer, as shown in Figure 5.30. By this method, the temperature at the inner surface of the transducer can be equated to the temperature at the uncovered skin adjacent to the transducer. In a steady state, heat flow toward the skin surface where the transducer is placed will be the same as that for uncovered skin, as long as the temperatures at these two sites are the same. This situation will be maintained even though evaporative heat loss exists, because heat transfer in the skin is determined only by the temperature gradient developed in it, which is determined by the temperature difference between the skin surface and the deep tissue. This technique was realized using a heat flow transducer with a 28 × 28-mm thermistor at the center of the inside surface (HA 12-18-5-P, Thermonetics Co.; San Diego, CA) and a thermoelectric heat pump of the same size (CP1.4-71-10, Melcor; Trenton, NJ) with cooling fins at the outer surface.[130]

5.5.3 DIRECT CALORIMETRY

Direct calorimetry means estimation of metabolic rate by measuring heat production in the animal or human subject. Metabolic rate can also be estimated either by the amount of food intake and excretion or by oxygen uptake and carbon dioxide output, both of which are known as indirect calorimetry. (Techniques of oxygen uptake measurement are described in Section 7.3.4.2).

The instrument for calorimetry is called a calorimeter. The essential part of the calorimeter for direct calorimetry consists of a chamber in which the subject is placed. Heat production in the subject is measured as heat flow from the chamber to the outer space.

There are different types of calorimeters as shown in Figure 5.31.[131] The type shown in Figure 5.31(a) is called a gradient-layer calorimeter. The heat produced inside the chamber passes through the layer toward the outer surface. If the layer has uniform thermal conductivity, the average temperature gradient developed in the layer is proportional to the heat flow passing through the layer.

Figure 5.31(b) shows the heat-sink calorimeter, in which heat produced in the chamber is removed by coolant liquid. The rate of heat removal is estimated from the flow rate and temperature rise of the coolant.

Figure 5.31(c) shows the convection calorimeter. Heat flow through the wall is prevented by insulation, and heat production inside the chamber is estimated from flow rate and rise in temperature of the ventilating air.

Figure 5.31(d) shows a differential calorimeter. This type is used for measurements in small animals. Two identical chambers are placed in an air stream. One chamber contains the object of measurement, and the other an electric heater. Heat production is estimated as the heating power when it is adjusted to produce identical temperature increases.

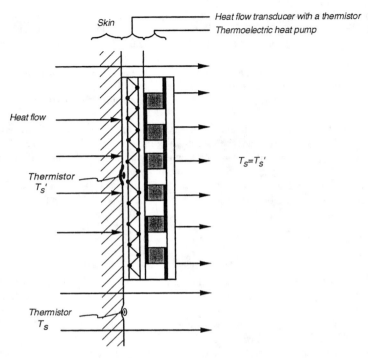

Figure 5.30 A method to compensate for the effect of thermal resistance of the heat flow transducer when it is attached to the body surface. The skin temperature at the inner surface of the transducer is equated to that of the uncovered skin by using a servo-controlled heat pump.

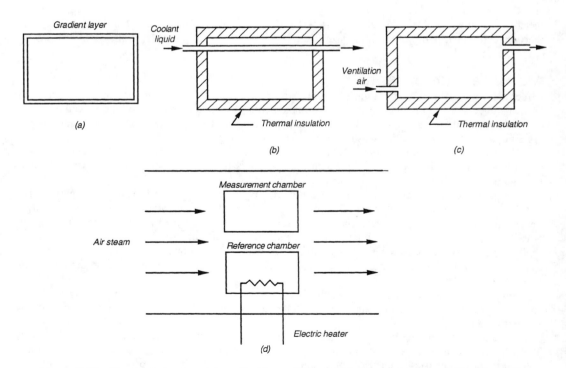

Figure 5.31 Direct calorimeters: **(a)** gradient layer, **(b)** heat sink, **(c)** convective, and **(d)** differential.

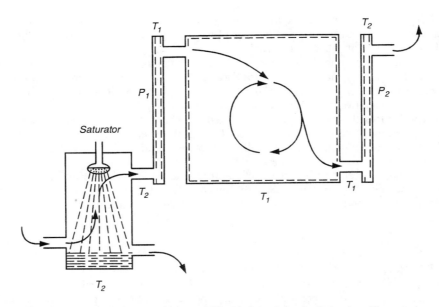

Figure 5.32 A gradient-layer calorimeter with two additional plate-shaped calorimeters. The entering and leaving air is saturated with water vapor at temperature T_1. (Modified from Benzinger, T. H. and Taylor, G. W., in *Temperature, Its Measurement and Control in Science and Industry,* Vol. 3, Hardy, J. D., Ed., Reinhold, New York, 1963, 87.)

Calorimeters for humans and animals require ventilation, but heat transfer can occur through ventilatory air. In a steady state, the ventilatory heat transfer component can be estimated by measuring the temperature and water vapor pressure of air entering and leaving and the air flow rate.

In order to measure ventilatory heat exchange, Benzinger and Kitzinger[132] employed a gradient layer calorimeter chamber with two additional plate-shaped calorimeter chambers, as shown in Figure 5.32. The ventilation air is first saturated with water vapor at temperature T_2 and enters a plate-shaped calorimeter, P_1, maintained at temperature T_1 ($>T_2$). The air entering the calorimeter chamber is warmed to temperature T_1 with dewpoint temperature T_2. The air then passes another plate-shaped calorimeter, P_2, at temperature T_2. The air coming from the calorimeter chamber is cooled, and additional water vapor is condensed out. Finally, the air leaving, P_2, reaches the same condition as the air entering, P_1, which is at temperature T_2 and saturated with water vapor. The total heat produced in the chamber is then measured as the total heat flow through the three gradient calorimeters. If heat flow transducer elements consisting of gradient layers have uniform sensitivity, the total heat flow can be obtained by simply connecting them in series.

The Hannah Research Institute (Ayr, U.K.) large animal calorimeter is an example of the gradient-layer calorimeter.[131] It has a chamber measuring $2.31 \times 2.6 \times 1.52$ m, internally. The gradient layer consists of resin-bonded, paper-board strips 38 mm wide and 0.8 mm thick, which are wound with constantan tape 1 cm wide by 0.8 mm thick. Every half-turn of each constantan spiral is short-circuited by copper tape so that it consists of one thermocouple for every 58 cm^2, making 4500 thermocouples in all. The outer surface of each wall is cooled by circulating water at a constant temperature. The chamber is ventilated at a rate of 2500 l/min. The chamber can operate at any temperature range from 10 to 40°C, with dewpoint temperature air entering the chamber from 8 to 30°C. The accuracy of total heat loss is approximately ±1% or ±1 W, whichever is the greater. The time constant of the gradient layer is 2.04 s, but the presence of objects of high heat capacity inside the chamber can increase this.

5.6 EVAPORATION MEASUREMENT

5.6.1 INTRODUCTION

The water content of the human body is approximately 60% total body weight. Humans and other organisms gain water through the intake of fluids and food and also by oxidation of carbohydrates in the food. Water is lost mainly through urine, feces, respiration, and the so-called insensible water loss via the skin. Total water intake can be as high as 3 l per day in healthy adult subjects, a figure that can be strongly affected by a number of diseases.

For optimal fluid therapy it is necessary to have a thorough understanding of the nature of all the fluid loss mechanisms in the human body. Urine losses are fairly straightforward to measure, whereas respiratory losses and water losses via the skin require more elaborate devices. In this chapter the sensors and methods for evaporative water loss measurements are reviewed.

The total water loss through the skin is mediated via two different physical processes: diffusion through the epidermis[133] and low-level sweat secretion via the sweat glands.[134,135] A common term for these two processes is insensible water loss. If the ambient temperature is raised or if the subject engages in physical work, the evaporative water loss will increase. Most of this increase, which is called sensible perspiration, is due to increased sweat gland activity. The heat loss associated with the evaporative water loss usually constitutes approximately 25% of the total heat loss of the human body.[136] Thus, evaporative heat loss is the most important regulatory mechanism involved in the thermal regulatory system of the human body.

5.6.2   HUMIDITY TRANSDUCERS
5.6.2.1   Electrolytic Water Vapor Analyzer
The water content of a gas can be measured by an electrolytic water vapor analyzer.[137-141] The measurement cell in electrolytic water vapor analyzers is usually a cylindrical tube through which a continuous carrier gas flow is conducted. This gas is usually nitrogen. Inside the tube, two helically wound platinum wires are arranged and supplied from a direct current source. A thin layer of phosphor pentoxide (a very strong dessicant) is deposited between the coils. When the carrier gas flows through the tube, the humidity is absorbed by phosphor pentoxide. A small electrical current passes between the platinum wires, the magnitude of which is related to the water content absorbed in the phosphor pentoxide and thereby also to the water partial pressure of the gas under examination. This measurement principle is restricted to the analysis of gases with low water content.

5.6.2.2   Thermal Conductivity Cell
The heat conduction properties of a gas are related to its composition. This property of a gas can be used for analysis of its composition. An electrically heated wire or a thermistor serves the dual function of heat source and temperature sensor (Figure 5.33). For each gas mixture, the heat source comes into thermal equilibrium with the surrounding gas so that the temperature of the sensor is related to the composition of the gas. The method is limited to the mixture of only two gases in order to get a unique relation between the gas composition and the temperature of the sensing device. To be able to measure the loss of water from the skin, a dry carrier gas such as nitrogen can be used. The sensitivity of a thermal conductivity cell is influenced by the environmental temperature, so the sensor should be kept insulated and temperature controlled.

The gas flow conditions are also very critical for the stability of the measurements. If adequately controlled, this principle can form the basis for a very sensitive and stable humidity recording instrument.

5.6.2.3   Infrared Water Vapor Analyzer
The infrared absorption analyzer is based on the infrared absorption properties of a gas at the 1.37-μm band (water vapor absorption band). The measurement cell is usually divided into two tubes, one containing the gas under study and the other a reference sample. When the analyzer is calibrated, the same dry gas is flushed through both tubes and the infrared sensors record the same intensities. When a humid gas passes through the measurement tube and the reference tube is filled with a dry gas, the detectors will record an imbalance signal which is proportional to the vapor concentration of the gas under study. Sensitivities down to 0.01% of full-scale deflections have been reported. The long-term stability can vary considerably among individual instruments.

5.6.2.4   Dewpoint Hygrometer
The temperature at which the water vapor condenses at a surface can be used for gas water content measurements (dewpoint hygrometry; see Figure 5.34). The light from a light source is reflected in a

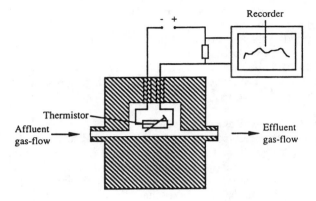

Figure 5.33 The thermal conductivity cell. (From Nilsson, G. E. and Öberg, P. Å., in *Noninvasive Physiological Measurements,* Rolf, P., Ed., Academic Press, London, 1979. With permission.)

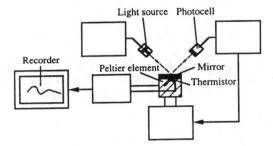

Figure 5.34 Dewpoint hygrometry. (From Nilsson, G. E. and Öberg, P. Å., in *Noninvasive Physiological Measurements,* Rolf, P., Ed., Academic Press, London, 1979. With permission.)

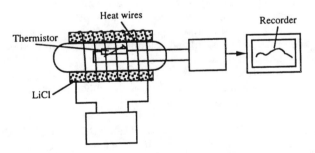

Figure 5.35 The electrohygrometer. (From Nilsson, G. E. and Öberg, P. Å., in *Noninvasive Physiological Measurements,* Rolf, P., Ed., Academic Press, London, 1979. With permission.)

mirror and detected by a photocell. A Peltier element below the mirror controls its temperature. When the temperature of the mirror is lowered, the humidity condenses at some point. The reflectivity drops at this point. The temperature of the mirror is determined by a thermistor, thermo-element, or a platinum wire. The accuracy of an instrument can be ±1% in the temperature interval –60 to +70°C. This instrument is rapid and has good measurement accuracy.

5.6.2.5 Electrohygrometer

Electrical hygrometers are usually based on the variations in resistance of a hygroscopic material when moisture is absorbed in it. Figure 5.35 shows the principle of the method. The sensor consists of a lithium chloride film that can be electrically heated via two wires. When the relative humidity exceeds 11%, vapor is absorbed by the lithium chloride film, causing an abrupt decrease in its resistance which in turn increases the current and thus the temperature of the film. The temperature increase means that part of the moisture absorbed in the film has evaporated. At the crystallization point, the temperature of the lithium salt equilibrium is established. This temperature, which is related to the water content of the ambient air, is measured with a thermistor. The measurement range is usually 12 to 100% relative humidity with an accuracy of about 3%.

5.6.2.6   Capacitive Humidity Sensor

Relative humidity can also be measured with a capacitive transducer. Vaisala Oy (Helsinki, Finland) is the manufacturer of a capacitive element, 4×6 mm, originally developed for meteorological purposes.

This sensor uses an organic polymer dielectric, sensitive only to changes in relative humidity.[142,143] Mechanical solidity was achieved in this sensor by coating a thin polymer film onto a flat glass substrate. An almost linear relation between the sensor capacitance and the relative humidity is obtained in the entire humidity range. A hysteresis effect exists, though, especially after large humidity deflections, but this effect is limited to about 1% of full-scale deflection. The fast response time and the low temperature coefficients of the sensor element make it well suited for rapid measurements in a wide range of temperatures.

5.6.2.7   Impedance Humidity Sensor

A group of organic polymers having constituent ionic monomers, such as sodium styrenesulphonate, exhibit ionic conductivity and are called polymer electrolytes. Their ionic conductivity increases with an increase in water absorption due to increases in ionic mobility and/or charge carrier concentrations. The most well-known sensor of this type is the "Hument" with an accuracy of about 1%. Several copolymers have been investigated with this application in mind. Yamazoe and Shimizu[144] have reviewed the field.

5.6.2.8   Thermoelectric Psychrometer

Ingelstedt[20] presented a thermoelectric micro-psychrometer with the aim of measuring rapid changes in temperature and relative humidity of respiratory air. The thermo-elements were made of constantan and nickel-chromium wires, 0.03 mm in diameter. The wires were joined by an elegant micro-buttwelding technique. The dry and wet thermo-elements were positioned in a cannula of 1.7-mm inner diameter and 2.0-mm outer diameter. The cannula could be introduced into the laryngeal cavity by means of a device attached to the neck by adhesive tape. Recording took place during 5 to 8 respiratory cycles, after which period the cannula was withdrawn and reconditioned for the next recording. The thermo-electric part of the system is straightforward and could be carefully calibrated before use. Recording of the moisture content of the respiratory air was made after a control experiment in a model system. The theory of the dry and wet thermo-elements has been presented in detail.[20] Ingelstedt's method has advantages in studies of airway moisture and can, in a slightly updated form, be very valuable in similar studies.

5.6.3   MEASUREMENT OF EVAPORATIVE WATER LOSS FROM SKIN AND MUCOSA

Several principles exist when it comes to the practical arrangement of humidity sensors in measurement setups. In this section, the most common principles are reviewed.

5.6.3.1   Unventilated Chamber

One of the first measurement principles for evaporative water loss from the skin was the unventilated chamber method.[145-147] A volume of a hygroscopic salt is placed within a closed chamber with its opening facing the skin surface. The hygroscopic salt (8 to 10 g calcium chloride) is desiccated in a linen bag placed in an oven at a temperature over 200°C. The bag containing the salt is then quickly weighed on a sensitive balance and suspended on hooks in a glass tube with the open end facing the skin. To prevent leakage of water vapor, the base of the tube is attached to the skin by double adhesive tape. All the water that evaporates from the circumscribed area of the skin is absorbed by the salt. After 15 to 30 min, the measurement procedure is terminated. The bag containing the salt is again weighed, and the increase in mass, which is equal to the loss of water from the skin surface being investigated, is calculated. The amount of water vapor evaporated from the skin is generally expressed in grams per square meter per hour or in milligrams per square centimeter per hour:

$$WL = \frac{W_{after} - W_{before}}{A_{uv} \cdot \Delta t} \tag{5.18}$$

where WL is the water loss by evaporation (g m^{-2}h^{-1}); W_{after} and W_{before} are the weight of the hygroscopic salt after and before the period of measurement, respectively, in grams; A_{uv} is the area of measurement in the unventilated chamber (m^2); Δt is the period of measurement (hr).

The unventilated chamber method is the least-used method because of some serious disadvantages associated with it. First, the humidity level in the chamber rises during the measurement procedure. It is well known that diffusion of water molecules through the skin is a function of the transepidermal gradient of humidity. When this gradient is disturbed, the transport process is affected. Using the capsule unventilated seriously disturbs the conditions for measurements.

In addition, a sensitive scale is required to obtain good resolution. For a resolution of 10% at a measurement on normal skin, an accuracy on the scale of 0.1 mg is required.[148] The transport of the hygroscopic material between the scale and the chamber can introduce serious errors. The unventilated chamber method cannot be used to study transient conditions.

5.6.3.2 Ventilated Chamber

In the method based on a ventilated chamber, the evaporative water loss from the skin is estimated by passing a continuous flow of gas with a known water content (can be dry) and a known velocity through the measurement chamber. Generally, the humidity and velocity are measured both in the affluent and effluent gas flow. The amount of water lost by the skin surface investigated per unit time and area is calculated according to the equation

$$WL = \frac{F \cdot M \left(p_{H_2O(eff)} - p_{H_2O(aff)} \right)}{R \cdot T \cdot A_v} \tag{5.19}$$

where WL is the water loss per unit of time and area (g m^{-2}h^{-1}); F is the flow of carrier gas (m^{3}h^{-1}); M is the molecular weight of water; R is the gas constant (8.314 J mol^{-1} K^{-1}); T is the temperature (K); and A_v is the area of the measurement in the chamber (m^2). $p_{H_2O(eff)}$ and $p_{H_2O(aff)}$ are the vapor pressures of the effluent and affluent gas, respectively (Pa). Both dry and prehumified gases are used. A critical parameter in the system is the gas flow velocity. If it is too low, condensation in the tubing may occur when the evaporation rate is high. On the other hand, if the gas flow velocity is too high, convective currents may affect the microclimate surrounding the skin, thereby altering the evaporation rate. The water content of the gas is usually measured by an electrolytic water vapor analyzer or by a conductivity cell, as described earlier.

Smallwood and Thomas[148] described a ventilated chamber method aimed especially at measurements on paraplegic patients. The instrument has a traditional design with a measuring head placed on the skin and a mixing chamber in which the measurements take place. The humidity sensor measures surface resistivity in a styrene copolymer layer. The resistance of this layer is inversely proportional to relative humidity. The sensor is linearized by a mathematical formula given by the manufacturer. In their paper, the authors provide an excellent discussion of the measurement errors of the ventilated chamber.

A combination of the unventilated and ventilated chamber is sometimes used for animal studies. Hatting and Luck[149] utilized an unventilated chamber in which the air was circulated by means of a fan. Wang et al.[150] have presented a simple hygrometer for estimation of water loss from rodent skin. Brackenbury et al.[151] addressed the problem of recording respiratory losses of water from exercising birds. They have used an open gas flow system with a mixing chamber in which the sensor (an electrolytic resistor) is positioned.

5.6.3.3 Gradient Estimation Method

This method, when applied to evaporation measurement, was presented for the first time by Nilsson and Öberg.[152] The method has been described in detail by Nilsson[153] and Nilsson and Öberg.[154]

The human body is under undisturbed conditions surrounded by a water vapor boundary layer, about 10 mm in height. If the vapor pressure and temperature distribution of the boundary layer are known, the evaporative water loss and convective heat loss from the skin surface can be calculated. In the case

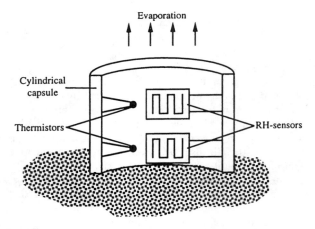

Figure 5.36 The gradient-layer type of Evaporimeter probe. (From Nilsson, G. E. and Öberg, P. Å., in *Noninvasive Physiological Measurements*, Rolf, P., Ed., Academic Press, London, 1979. With permission.)

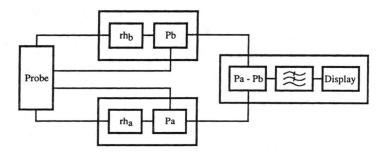

Figure 5.37 Block diagram of the instrument for calculation of the vapor pressure gradient. (From Nilsson, G. E. and Öberg, P. Å., in *Noninvasive Physiological Measurements*, Rolf, P., Ed., Academic Press, London, 1979. With permission.)

of evaporative water loss, the relation between the evaporated amount of water per unit time and area and the vapor pressure gradient is given by the transport equation,[155,156]

$$\frac{1}{A} \cdot \frac{dm}{dt} = -D \cdot \frac{dp}{dx} \tag{5.20}$$

where $(1/A)(dm/dt)$ is the amount of water evaporated per unit time and area (g m^{-2} h^{-1}), D is a constant, and dp/dx is the vapor pressure gradient. This gradient can be calculated if the difference between the vapor pressures at two points is known within the boundary layer on a line perpendicular to the evaporating surface. The way the sensors were arranged is shown in Figure 5.36.

Two thermistors and two relative humidity sensors (Vaisala Oy; Helsinki, Finland) were placed at two levels above the skin surface. The vapor pressure at each point of measurement was calculated as the product of the relative humidity and the saturated vapor pressure. The vapor pressure gradient is calculated according to the block diagram of the instrument, as shown in Figure 5.37.

The accuracy of the instrument was tested by correlating the reduction in weight of a freely evaporating salt solution to the recorded evaporation rate from the same salt solution. Different evaporation rates were obtained by the use of different salt solutions or by heating the salt solution to preselected temperatures. The correlation coefficient between the weight loss and the recorded evaporation rate was calculated to be 0.997. A commercial version of this instrument (Evaporimeter EP1) is manufactured and sold by Servomed AB (Kinna, Sweden).

The Evaporimeter has become the standard method for transepidermal water loss measurements in newborn infants and in dermatology, especially occupational dermatology. The Uppsala Neonatology group headed by Prof. G. Sedin has used the instrument extensively for water loss studies in newborn babies.[156,157] The European Society of Contact Dermatitis has performed a number of studies in order to further standardize the use of the Evaporimeter in clinical measurements.[157-161]

The Evaporimeter has also found many applications within the field of plastic surgery and burns, for which it was originally developed. The study of water permeability of wound coverings is an important problem in burns and plastic surgery. Lamke et al.[162] used an early version of the instrument to study coverings. Erasmus and Jonkman[163] have repeated this study for some modern materials.

At high evaporation rates (over approx. 8 g m^{-2} h^{-1}) the Evaporimeter seems to underestimate the transepidermal water loss. Wheldon and Monteith[164] have analyzed the Evaporimeter theoretically and experimentally. These authors found that the relation between the diffusion resistance of the skin and the diffusion resistance of the probe is important to the accuracy of the method. As long as the probe resistance is small, in comparison with the sum of skin and boundary layer resistance, the Evaporimeter seems to estimate the rate of transepidermal water loss accurately.

Scott et al.[165] compared the Evaporimeter with the ventilated chamber method in measurements at normal and disturbed human skin function. They could verify the results of Wheldon and Monteith.[163] At high transepidermal water loss, occurring when the barrier function of the skin is lost, the gradient method underestimates the water loss over evaporation rates of 7.5 g m^{-2} h^{-1}. At levels higher than this rate, these authors prefer the ventilated chamber method.

REFERENCES

1. DuBois, E. F., The many different temperatures of the human body and its parts, *West. J. Surg. Obstet. Gynecol.*, 59, 476, 1951.
2. Dinovo, J. A., Testing the clinical thermometer, *J. Clin. Eng.*, 7, 119, 1982.
3. Bligh, J. and Johnson, K. G., Glossary of terms for thermal physiology, *J. Appl. Physiol.*, 35, 941, 1973.
4. Mitchell, D. and Wyndham, C. H., Comparison of weighting formulae for calculating mean skin temperature, *J. Appl. Physiol.*, 26, 616, 1969.
5. Ibsen, B., Treatment of shock with vasodilators measuring skin temperature on the big toe: ten years' experience in 150 cases, *Dis. Chest*, 52, 425, 1967.
6. Joly, H. R. and Weil, M. H., Temperature of the great toe as an indication of the severity of shock, *Circulation*, 39, 131, 1969.
7. Henning, R. J., Wiener, F., Valdes, S., and Weil, M. H., Measurement of toe temperature for assessing the severity of acute circulatory failure, *Surg. Gynecol. Obstet.*, 149, 1, 1979.
8. Koholousky, A. M., Sufian, S., Pavlides, C., and Matsumoto, T., Central peripheral temperature gradient: its value and limitations in the management of critically ill surgical patients, *Am. J. Surg.*, 140, 609, 1980.
9. Lloyd Williams, K., Lloyd Williams, F. J., and Handley, R. S., Infra-red thermometry in the diagnosis of breast disease, *Lancet*, 2, 1378, 1961.
10. Amarlic, R., Giraud, D., Altschuler, C., and Spitalier, J. M., Value and interest of dynamic telethermography in detection of breast cancer, *Acta Thermogr.*, 1, 89, 1976.
11. Scholander, P. F. and Schevill, W. E., Counter-current heat exchange in the fins of whales, *J. Appl. Physiol.*, 8, 279, 1955.
12. Saltin, B., Gagge, A. P., and Stolwijk, J. A. J., Muscle temperature during submaximal exercise in man, *J. Appl. Physiol.*, 25, 679, 1968.
13. Lawson, R. N. and Chughtai, M. S., Breast cancer and body temperature, *Can. Med. Assoc. J.*, 88, 68, 1963.
14. Gautherie, M., Thermopathology of breast cancer: measurement and analysis of *in vivo* temperature and blood flow, *Ann N. Y. Acad. Sci.*, 335, 383, 1980.
15. Christensen, D. A., Thermometry and thermography, in *Hyperthermia in Cancer Therapy*, Storm, F. K., G K Hall, Boston, 1983, 223.
16. Kleiber, M., *The Fire of Life, an Introduction to Animal Energetics*, rev. ed., Robert E. Krieger, New York, 1975.
17. Edwards, T. J., Observations on the stability of thermistors, *Rev. Sci. Instrum.*, 54, 613, 1983.
18. Beakley, W. R., The design of thermometers with linear calibration, *J. Sci. Instrum.*, 28, 176, 1951.
19. Hoge, H. J., Comparison of circuits for linearizing the temperature indications of thermistors, *Rev. Sci. Instrum.*, 50, 316, 1979.
20. Ingelstedt, S., Studies on the conditioning of air in the respiratory tract, *Acta Oto-Laryngeol. Suppl.*, 131, 1, 1956.
21. Guilbeau, E. J. and Mayall, B. I., Microthermocouple for soft tissue temperature determination, *IEEE Trans. Biomed. Eng.*, BME-28, 301, 1981.
22. Cain, C. P. and Welch, A. J., Thin-film temperature sensors for biological measurements, *IEEE Trans. Biomed. Eng.*, BME-21, 421, 1974.
23. Kaufmann, A. B., Bonded-wire temperature sensors, *Instrum. Contr. Syst.*, May, 103, 1963.
24. MacNamara, A. G., Semiconductor diodes and transistors as electrical thermometers, *Rev. Sci. Instrum.*, 33, 330, 1962.

25. Cohen, B. G., Snow, W. B., and Tretola, A. R., GaAs p-n junction diodes for wide range thermometry, *Rev. Sci. Instrum.*, 34, 1091, 1963.
26. Sah, C-T., Noyce, R. N., and Shockley, W., Carrier generation and recombination in p-n junctions and p-n junction characteristics, *Proc. IRE*, 45, 1228, 1957.
27. Sclar, N. and Pollock, D. B., On diode thermometers, *Solid State Electron.*, 15, 473, 1972.
28. Davis, C. E. and Coates, P. B., Linearization of silicon junction characteristics for temperature measurement, *J. Phys. E*, 10, 613, 1977.
29. Vester, T. C., p-n junction as an ultralinear calculable thermometer, *Electron. Lett.*, 4, 175, 1968.
30. Ruhle, R. A., Solid-state temperature sensor outperforms previous transducers, *Electronics*, 48(6), 127, 1975.
31. Riezenman, M. J., Integrated temperature transducers, *Electronics*, 47(23), 130, 1974.
32. Timko, M. P., A two-terminal IC temperature sensor, *IEEE J. Solid State Circ.*, SC-11, 784, 1976.
33. Sheingold, D. H., A Guide to analog signal conditioning, in *Transducer Interfacing Handbook*, Norwood, M. A., Ed., Analog Devices, Norwood, MA, 1980, 153.
34. Meijer, G., C., M., An IC temperature transducer, *IEEE J. Solid State Circ.*, SC-15, 370, 1980.
35. DeHaan, G. and Meijer, G. C. M., An accurate small-range IC temperature transducer, *IEEE J. Solid State Circ.*, SC-15, 1089, 1980.
36. Barth, P. W. and Angel, J. B., Thin linear thermometer arrays for use in localized cancer hyperthermia, *IEEE Trans. Electron. Dev.*, ED-29, 144, 1982.
37. Barth, P. W., Bernard, S. L., and Angell, J. B., Flexible circuit and sensor arrays fabricated by monolithic silicon technology, *IEEE Trans. Electron. Dev.*, ED-32, 1202, 1985.
38. Hammond, D. L. and Benjaminson, A., The crystal resonator — a digital transducer, *IEEE Spectrum*, 6, 53, 1969.
39. Tsuji, T., Ohshima, T., Hashimoto, D., and Togawa, T., Human intragastric temperature telemetry with quartz crystal resonator using ultrasonic detection, *Precision Machinery*, 4, 83, 1992.
40. Bowman, R. R., A probe for measuring temperature in radio-frequency-heated material, *IEEE Trans. Microwave Theor. Tech.*, MTT-24, 43, 1976.
41. Larsen, L. E., Moore, R. A., Jacobi, J. H., Halgas, F. A., and Brown, P. V., A microwave compatible MIC temperature electrode for use in biological dielectrics, *IEEE Trans. Microwave Theor. Tech.*, MTT-27, 673, 1979.
42. Szwarnowski, S., Sheppard, R. J., Grant, E. H. and Bleehen, N. M., A thermocouple measuring temperature in biological material heated by microwave at 2.45 GHz., *Br. J. Radiol.*, 53, 711, 1980.
43. Olsen, R. G. and Molina, E. A., The nonmetallic thermocouple: a differential-temperature probe for use in microwave field, *Radiol. Sci.*, 14, 81, 1979.
44. Johnson, C. C. and Rozzell, T. C., Temperature probe for M/W fields, *Microwave J.*, 18, 55, 1975.
45. Livingston, G. K., Thermometry and dosimetry of heat with specific reference to the liquid-crystal optical fiber temperature probe, *Radiat. Environ. Biophys.*, 17, 233, 1980.
46. Samulski, T. and Shrivastava, P. N., Photoluminescent thermometer probes: temperature measurements in microwave fields, *Science*, 208, 193, 1980.
47. Wickersheim, K. A. and Alves, R., V., Fluoroptic thermometer: a new RF-immune technology, in *Biomed Thermology*, Alan R. Liss, New York, 1982, 547.
48. Wickersheim, K. A., A new fiberoptic thermometry system for use in medical hyperthermia, *SPIE Opt. Fibers Med. II*, 713, 150, 1986.
49. Vaguine, V. A., Christensen, D. A., Lindley, J. H. and Walston, T. E., Multiple sensor optical thermometry system for application in clinical hyperthermia, *IEEE Trans. Biomed. Eng.*, BME-31, 168, 1984.
50. Ceta, T. C. and Connor, W. G., Thermometry considerations in localized hyperthermia, *Med. Phys.*, 5, 79, 1978.
51. Christensen, D. A., A new non-perturbing probe using semiconductor band edge shift, *J. Bioeng.*, 1, 541, 1977.
52. Deficis, A. and Priou, A., Non-perturbing microprobes for measurement in electromagnetic field, *Microwave J.*, 20, 55, 1977.
53. Ovrén, C., Adolfsson, M., and Hök, B., Fiberoptic systems for temperature and vibration measurements in industrial applications, *Int. Conf. on Optical Techniques in Process Control, Hague*, Paper B2, 67, 1983.
54. Szwarnowski, S., A thermometer for measuring temperatures in the presence of electromagnetic fields, *Clin. Phys. Physiol. Meas.*, 4, 79, 1983.
55. Togawa, T., Non-contact skin emissivity: measurement from reflectance using step change in ambient radiation temperature, *Clin. Phys. Physiol. Meas.*, 10, 39, 1989.
56. Watmough, D. J., Fowler, P. W., and Oliver, R., The thermal scanning of a curved isothermal surface: implication for clinical thermography, *Phys. Med. Biol.*, 15, 1, 1970.
57. Martin, C. J. and Watmough, D. J., Thermal scanning of curved surfaces, *Acta Thermogr.*, 2, 18, 1977.
58. Watmough, D. J. and Oliver, R., Emissivity of human skin in the waveband between 2μ and 6μ, *Nature*, 219, 622, 1968.
59. Watmough, D. J. and Oliver, R., The emission of infrared radiation from human skin — implications for clinical thermography, *Br. J. Radiol.*, 42, 411, 1969.
60. Anderson, R. R. and Parrish, J. A., The optics of human skin, *J. Invest. Dermatol.*, 77, 13, 1981.
61. Jones, C. H., An evaluation of colour thermography with respect to the diagnosis of breast disease, *Acta Thermogr.*, 4, 59, 1979.
62. Barnes, R., B., Diagnostic thermography, *Appl. Opt.*, 7, 1673, 1968.

63. Hsieh, C. K. and Su, K. C., Design, construction, and analysis of a continuous-temperature infrared calibrator for temperature measurement using an infrared scanner, *Rev. Sci. Instrum.*, 50, 888, 1979.
64. Newman, P., Davison, M., and James, W. B., A prototype, inexpensive thermograph for routine hospital use, *Acta Thermogr.*, 4, 132, 1979.
65. Rode, J. P., Hybrid HgCdTe arrays, *Proc. SPIE*, 443, 120, 1983.
66. Elliott, C. T., New detector for thermal imaging system, *Electron. Lett.*, 17(8), 312, 1981.
67. Blackbarn, A., Blackman, M. V., Charlton, D. E., Dunn, W. A. E., Jenner, M. D., Oliver, K. J. and Wotherspoon, J. T. M., The practical realization and performance of SPRITE detectors, *Infrared Phys*, 22, 57, 1982.
68. Tsauer, B.-Y., McNutt, M. J., Bredthauer, R. A., and Mattson, R. E., 128 × 128-elements IrSi Schottky-barrier focal plane arrays for long-wavelength infrared imaging, *IEEE Electron. Dev. Lett.*, 10, 361, 1989.
69. Tsauer, B.-Y., Chen, C. K., and Marino, S. A., Long-wavelength Ge_xSi_{1-x}/Si heterojunction infrared detectors and 400 × 400-element imager arrays, *IEEE Electron. Dev. Lett.*, 12, 293, 1991.
70. Myers, P. C., Barret, A. H., and Sadowski, N. L., Microwave thermography of normal and cancerous breast tissue, *Ann. N.Y. Acad. Sci.*, 335, 443, 1980.
71. Edrich, J., Jobe, W. E., Cacak, P. K., Hendee, W. R., Smyth, C. J., Gautherie, M., Gros, C., Zimmer, R., Robert, J., Thouvenot, P., Escanye, J. M., and Itty, C., Imaging at centimeter and millimeter wavelength, *Ann. N.Y. Acad. Sci.*, 335, 456, 1980.
72. Hamamura, Y., Mizushina, S., and Sugiura, T., Noninvasive measurement of temperature-versus-depth profile in biological systems using a multiple-frequency-band microwave radiometer systems, *Automedica*, 8, 213, 1987.
73. Bolomey, J. C., Jofre, L., and Peronnet, G., On the possible use of microwave-active imaging for remote thermal sensing, *IEEE Trans. Microwave Theor. Tech.*, MTT-31, 777, 1983.
74. Pichot, C., Jofre, L., Peronnet, G., and Bolomey, J-C., Active microwave imaging of inhomogeneous bodies, *IEEE Trans. Antennas Propagat.*, AP-33, 416, 1985.
75. Greenleaf, J. F. and Bahn, R. C., Clinical imaging with transmissive ultrasonic computorized tomography, *IEEE Trans. Biomed. Eng.*, BME-28, 177, 1981.
76. Fallone, B. G., Moran, P. R., and Podgorsak, E. B., Noninvasive thermometry with a clinical X-ray CT scanner, *Med. Phys.*, 9, 715, 1982.
77. Parker, D. L., Smith, V., Sheldon, P., Crooks, L. E., and Fussell, L., Temperature distribution measurement in two-dimensional NMR imaging, *Med. Phys.*, 10, 321, 1983.
78. Ebstein, E., Die Entwicklung der klinishen thermometrie, *Erg. Inn. Med. Kind.*, 33, 407, 1928.
79. Notani-Sharma, P., Chiva, R. K., and Katchen, M., Little known mercury hazards, *Hospitals*, 54, 76, 1980.
80. Togawa, T., Body temperature measurement, *Clin. Phys. Physiol. Meas.*, 6, 83, 1985.
81. Eichina, L. W., Berger, A. R., Rader, B., and Becker, W. H., Comparision of intracardiac and intravascular temperatures with rectal temperatures in man, *J. Clin. Invest.*, 30, 353, 1951.
82. Cranston, W. I., Gerbrandy, J., and Snell, E. S., Oral, rectal and esophageal temperatures and some factors affecting them in man, *J. Physiol.*, 126, 347, 1954.
83. Isley, A. H., Rutten, A. J., and Runciman, W. B., An evaluation of body temperature measurement, *Anaesth. Intensive Care*, 11, 31, 1983.
84. Pickering, G., Regulation of body temperature in health and disease, *Lancet*, 1:1-9, 59–64, 1958.
85. Rubin, A., Horvath, S. M., and Mellette, H. C., Effect of fecal bacterial activity on rectal temperature of man, *Proc. Soc. Exp. Biol. Med.*, 76, 410, 1951.
86. Durotoye, A. O. and Grayson, J., Heat production in the gastro-intestinal tract of the dog, *J. Physiol.*, 214, 417, 1971.
87. Cooper, K. E. and Kenyon, J. R., A comparison of temperature measured in the rectum, oesophagus, and on the surface of the aorta during hypothermia in man, *Br. J. Surg.*, 44, 616, 1957.
88. Vale, R. J., Monitoring of temperature during anesthesia, *Int. Anesthesiol. Clin.*, 19, 61, 1981.
89. Whitby, J. D. and Dunkin, L. J., Temperature differences in the oesophagus, *Br. J. Anaesth.*, 40, 991, 1968.
90. Lilly, J. K., Boland, J. P. and Zekan, S., Urinary bladder temperature monitoring: a new index of body core temperature, *Crit. Care Med.*, 8, 742, 1980.
91. Benzinger, T. H. and Taylor, G. W., Cranial measurements of internal temperature in man, in *Temperature, Its Measurement and Control in Science and Industry*, Vol. 3, Hardy, J. D., Ed., Reinhold, New York, 1963, 111.
92. Benzinger, M., Tympanic thermometry in surgery and anesthesia, *J. Am. Med. Soc.*, 209, 1207, 1969.
93. Dickey, W. T., Ahlgren, E. W., and Lawes, A. J., Body temperature monitoring via the tympanic membrane, *Surgery*, 67, 981, 1970.
94. Webb, G. E., Comparison of esophageal and tympanic temperature monitoring during cardiopulmonary bypass, *Anesth. Analog*, 52, 729, 1973.
95. Wallance, C. T., Marks, W. E., Adkins, W. Y., and Mahaffey, J. E., Perforation of the tympanic membrane, a complication of tympanic thermometry during anesthesia, *Anesthesiology*, 41, 290, 1974.
96. Tabor, M. W., Blaho, D. M., and Schriver, W. R., Tympanic membrane perforation: complication of tympanic thermometry during general anesthesia, *Oral Surg.*, 51, 581, 1981.
97. Moor, J. W. and Newbower, R. S., Noncontact tympanic thermometer, *Med. Biol. Eng. Comput.*, 16, 580, 1978.
98. O'Hara, G. J. and Phillips, D. B., Method and apparatus for measuring internal body temperature utilizing infrared emission, *U.S. Patent No. 4*, 602, 642, 1986.

99. Shinozaki, T., Deane, R., and Perkins, F. M., Infrared tympanic thermometer: evaluation of a new clinical thermometer, *Crit. Care Med.*, 16, 148, 1988.

100. Thermoscan® Operator's Manual PRO-1 Instant Thermometer, Model IR-1 and IR-1A Instant Thermometer, Thermoscan®, San Diego, CA, 1991

101. Fraden, J., The development of Thermoscan® instant thermometer, *Clin. Pediatr.*, 30, 4(Suppl.), 11, 1991.

102. Romano, M. J., Fortenberry, J. D., Autrey, E., Harris, S., Heyroth, T., Parmeter, P., and Stein, F., Infrared tympanic thermometry in the pediatric intensive care, *Crit. Care Med.*, 21, 1181, 1993.

103. Weiss, M. E., Pue, A., and Smith III, J., Laboratory and hospital testing of new infrared tympanic thermometers, *J. Clin. Eng.*, 16, 137, 1991

104. Wolf, G. C. and Baker, C. A., Tympanic thermometry for recording basal and body temperature, *Fertil. Steril.*, 60, 922, 1993

105. Fox, R. H. and Solman, A. J., A new technique for monitoring the deep body temperature in man from the intact skin surface, *J. Physiol.*, 212 8P, 1971.

106. Fox, R. H., Solman, A. J., Isaacs, R., Fry, A. J., and MacDonald, I. C., A new technique for monitoring deep body temperature from the skin surface, *Clin. Sci.*, 44, 81, 1973.

107. Kobayashi, T., Nemoto, T., Kamiya, A., and Togawa, T., Improvement of deep body thermometer for man, *Ann. Biomed. Eng.*, 3, 181, 1975.

108. Togawa, T., Nemoto, T., Yamazaki, T., and Kobayashi, T., A modified internal temperature measurement device, *Med. Biol. Eng.*, 14, 361, 1976.

109. Togawa, T., Noninvasive deep body temperature measurement, in *Noninvasive Physiological Measurements*, Vol.1, Rolfe, P., Ed., Academic Press, London, 1979, 261.

110. Lees, D. E., Kim, Y. D., and MacNamara, T. E., Noninvasive determination of core temperature during anesthesia, *South. Med. J.*, 73, 1322, 1980.

111. Muravchick, S., Deep body thermometry during general anesthesia, *Anesthesiology*, 58, 271, 1983.

112. Nemoto, T. and Togawa, T., Improved probe for a deep body thermometer, *Med. Biol. Eng. Comput.*, 26, 456, 1988.

113. Mackay, S. and Jacobson, J., Endoradiosonde, *Nature*, 179, 1239, 1957.

114. Rowlands, E. N. and Wolff, H. S., The radio pill, telemetering from the digestive tract, *Br. Commun. Electron.*, 7, 598, 1960.

115. Nagumo, J., Uchiyama, A., Kimoto, S., Watanuki, T., Hori, M., Suma, K., Ouchi, A., Kumano, M., and Watanabe, H., Echo capsule for medical use (a batteryless endoradiosonde), *IRE Trans. Biomed. Electron.*, BME-9, 195, 1962.

116. Cutchis, P. N., Hogrefe, A. F., and Lesho, J. C. The ingestible thermal monitoring system, *Johns Hopkins APL Tech. Dig.*, 9, 16, 1988.

117. Tsuji, T., Oshima, T., Hashimoto, D., and Togawa, T., Thermometry of human organ temperature with quartz crystal resonator using ultrasonic detection, in *Biotelemetry XI*, Uchiyama, A. and Amlander, C. J., Eds., Waseda University Press, Tokyo, 1992, 378.

118. Hatfield, H. S. and Wilkins, F. J., A new heat-flow meter, *J. Sci. Instrum.*, 27, 1, 1950.

119. Schmidt, E., U.S. Patent No. 1,528,383 (1923).

120. Cairnie, A. B. and Pullar, J. D., Temperature controller based on measurement of rate-of-change of temperature, *J. Sci. Instrum.*, 36, 249, 1959.

121. Danielsson, U., Convective heat transfer measured directly with a heat flux sensor, *J. Appl. Physiol.*, 68, 1275, 1990.

122. Goldsmid, H. J., Knittel, T., Savvides, N., and Uher, C., Measurement of heat-flow by means of Nernst effect, *J. Phys. E Sci. Instrum.*, 5, 313, 1972.

123. Goldsmid, H. J. and Sydney, K. R., A thermal radiation detector employing the Nernst effect in Cd_3As_3-NiAs, *J. Phys. D, Appl. Phys.*, 4, 869, 1971.

124. Gin, A. R., Hayward, M. G., and Keatinge, W. R., Method of measuring heat loss in man, *J. Appl. Physiol. Resp. Environ. Exer. Physiol.*, 49, 533, 1980.

125. Strong, L. H., Gee, G. K., and Goldman, R. F., Metabolic and vasomotor insulative responses occurring on immersion in cold water, *J. Appl. Physiol.*, 58, 964, 1985.

126. Toner, M. M., Sawka, M. N., Holden, W. L., and Pandolf, K. B., Comparison of thermal response between rest and leg exercise in water, *J. Appl. Physiol.*, 59, 248, 1985.

127. Sagawa, S., Shiraki, K., and Konda, N., Cutaneous vascular responses to heat simulated at a high altitude of 5,600 m, *J. Appl. Physiol.*, 60, 1150, 1986.

128. Ferretti, G., Veicsteinas, A., and Rennie, D. W., Regional heat flows of resting and exercising men immersed in cool water, *J. Appl. Physiol.*, 64, 1239, 1988.

129. McGinnis, S. M. and Ingram, D. L., Use of heat-flow meters to estimate rate of heat loss from animal, *J. Appl. Physiol.*, 37, 443, 1974.

130. Tamura, T., Nemoto, T., and Togawa, T., Heat flow meter for measuring regional heat flux in man, *Rep. Inst. Med. Dent. Eng.*, 15, 103, 1981.

131. McLean, J. A. and Tobin, G., *Animal and Human Calorimetry*, Cambridge University Press, London, 1987.

132. Benzinger, T. H. and Kitzinger, C., Gradient layer calorimetry and human calorimetry, in *Temperature, Its Measurement and Control in Science and Industry*, Vol. 3, Hardy, J. D., Ed., Reinhold, New York, 1963, 87.

133. Moyer, C. A. and Butcher, H. R., *Burns, Chock and Plasma Volume Regulation*, C.V. Mosby, St. Louis, MO, 1967.

134. Rothman, S., *Physiology and Biochemistry of the Skin*, University Press, Chicago, 1954, 233.

135. Bettley, F. R. and Grice K. A., A method for measuring the transepidermal water loss and a means of inactivating sweat glands, *Br. J. Dermatol.*, 77, 627d, 1965.

136. Hardy, J. D. and Du Bois, E. F., Basal metabolism, radiation, convection and vaporization at temperatures of 22 to 35°C, *J. Nutr.*, 15, 477, 1937.

137. Gasselt, H. R. M. and Vierhout R. R., Registration of insensible perspiration of small quantities of sweat, *Dermatologica*, 127, 255, 1963.

138. Spruit, D. and Malten, K., E., The regeneration rate of water vapour loss of heavily damaged skin, *Dermatologica*, 132, 115, 1966.

139. Thiele, F. A. J. and Senden, K. G., Relation between skin temperature and the insensible perspiration of the human skin, *J. Invest. Dermatol.*, 47, 307, 1966.

140. Grice, K., Satter, H., Sharratt, M., and Baker, H., Skin temperature and transepidermal loss, *J. Invest. Dermatol.*, 57, 108, 1971.

141. Rajka, G. and Thune, P., The relationship between the course of psoriasis and the transepidermal water loss, photoelectric plethysmography and reflex photometry, *Br. J. Dermatol.*, 94, 253, 1976.

142. Misevich, K. W., Capacitive humidity transducers, *IEEE Trans.*, IECI-16, 6, 1969.

143. Suntula, T. and Antson, J., A thin film humidity sensor, in *Vaisala News*, Vol. 59, Lindqvist, A., Ed., Vaisala Oy, Helsinki, 1973, 12.

144. Yamazoe, N. and Shimizu, Y., Humidity sensors: principles and applications, *Sensors Actuators*, 10, 379, 1986.

145. Felsher, Z. and Rothman, S., The insensible perspiration of the skin in hyperkeratotic conditions, *J. Invest. Dermatol.*, 6, 271, 1945.

146. Monash, S. and Blank, H., Location and reformation of the epithelial barrier to water vapour, *Arch. Dermatol.*, 78, 710, 1958.

147. Shahidullah, M., Raffe, E. J., Rimmer, A. R., and Frain-Bell, W., Transepidermal water loss in patients with dermatitis, *Br. J. Dermatol.*, 81, 722, 1969.

148. Smallwood, R. H. and Thomas, S. E., An inexpensive portable monitor for measuring evaporative water loss, *Clin. Phys. Physiol. Meas.*, 6(2), 147, 1985.

149. Hatting, J. and Luck, C. P., A sensitive, direct method for the measurement of water loss from body surface, *A. Afr. J. Med. Sci.*, 38, 31, 1973.

150. Wang, P. Y., Evans, D. W., Samji, N., and Lewellyn-Thomas, E., A simple hygrometer for the measurement of evaporative water loss from rodent skin, *J. Surg. Res.*, 28(2), 182, 1980.

151. Brackenbury, J. H., Avery, P., and Gleason, M., Measurement of water loss in exercising animals using a humidity tester, *Med. Biol. Eng. Comput.*, 20, 433, 1982.

152. Nilsson, G., E. and Öberg, P. Å., A new method for measurement of transepidermal water loss, *3rd Nordic Meeting Med. Biol. Eng*, Tampere, 1975, 53.

153. Nilsson, G. E., Measurement of water exchange through the skin, *Med. Biol. Eng. Comput.*, 15, 209, 1977.

154. Nilsson, G. E. and Öberg, P. Å., Measurement of evaporative water loss: methods and clinical applications, in *Noninvasive Physiological Measurements*, Vol. 1, Rolfe, P., Ed., Academic Press, London, 1979, 279.

155. Chapman, S. and Cowling, T. G., *The Mathematical Theory of Non-uniform Gases*, Cambridge University Press, London, 1953, 244.

156. Eckert, E. R. G. and Drake, R. M., *Heat and Mass Transfer*, 1959, 449.

157. Hammarlund, K., Nilsson, G. E., Öberg, P. A. and Sedin, G., Transepidermal water loss in newborn babies, *Acta. Ped. Scand.*, 66, 553, 1977.

158. Rutter, N. and Hull, D., Water loss from the skin of term and preterm babies, *Arch. Dis. Childhood*, 54, 858, 1979.

159. Pinnagoda, J., Tupker, R. A., Coenraads, P. J., and Nater, J. P., Comparability and reproducibility of the result of water loss measurements: a study of 4 evaporimeters, *Contact Dermatitis*, 20, 241, 1989.

160. Pinnagoda, J., Tupker, R. A. Smit, J. A., Coenraads, P. J., and Nater, J. P., The intra- and interindividual variability and reliability of transepidermal water loss measurements, *Contact Dermatitis*, 21, 255, 1989.

161. Pinnagoda, J., Tupker, R. A., Agner, T., and Serup, J., Guidelines for transepidermal water loss (TEWL) measurements, *Contact Dermatitis*, 22, 164, 1990.

162. Lamke, L-O., Nilsson, G. E., and Reitner, L., The evaporative water loss from burns and the water-vapour permeability of grafts and artificial membranes used in the treatment of burns, *Burns*, 3, 159, 1977.

163. Erasmus, M. E. and Jonkman, M. F., Water vapour permeance: a meaningful measure for water vapour permeability of wound coverings, *Burns*, 15(6), 371, 1989.

164. Wheldon, A. E. and Monteith, J. L., Performance of a skin evaporimeter, *Med. Biol. Eng. Comput.*, 18, 201, 1980.

165. Scott, R. C., Oliver, G. J. A., Dugard, P. H. and Singh, H. J., A comparison of techniques for the measurements of transepidermal water loss, *Arch. Dermatol. Res.*, 274, 57, 1982.

Chapter 6

Bioelectric and Biomagnetic Measurement

6.1 OBJECTS OF MEASUREMENTS

6.1.1 UNITS OF ELECTROMAGNETIC MEASUREMENTS

6.1.1.1 Electrical Units

Signals and bias potentials from electrodes are expressed in volts or fractions of volts (millivolts, microvolts, or nanovolts). In some applications, the power of a physiological signal is used and then expressed in watts. Spectral representations of signals are usually expressed in

$$\frac{(\text{volts})^2}{\text{hertz}} \tag{6.1}$$

or

$$\frac{\text{volts}}{\sqrt{\text{hertz}}} \tag{6.2}$$

where 1 hertz (Hz) is the unit of frequency in the SI-system.

Electrode properties are closely related to signals and noise. A common method of expressing the properties of signals and noise is to regard the power

$$P = \frac{\overline{v^2}}{R} \tag{6.3}$$

where $\sqrt{\overline{v^2}}$ is the root-mean-square (RMS) noise voltage and R is the circuit resistance. Alternatively, it is sometimes more convenient to consider the spectral distribution of the signal or noise present and to evaluate its spectral density, $S(f)$, in a given bandwidth, df. The average power of signals or noise can be evaluated for a load resistance of 1 ohm by the use of integration.

Electrode currents are measured in amperes or a fraction thereof. Resistance to the flow of electrical charges is usually expressed as a complex unit impedance

$$Z = R + jX \tag{6.4}$$

where R is the real part, called the resistance, and is measured in ohms. X is the reactance and is also expressed in ohms. Reactances of physiological electrodes can be capacitive or inductive. Capacitance is measured in farad (F) and inductance is measured in henry (H):

$$\frac{X}{R} = \tan \Theta .$$ (6.5)

The quotient between X and R can be used to express the phase angle between the current through an electrical component and the voltage drop across the same component. Θ is expressed in degrees or radians. Usually electrodes have resistive-capacitive properties and can be modeled by an RC-circuit. On rare occasions, the electrodes have inductive properties, and the reactance is then an inductive one.

6.1.1.2 Magnetic Units

Magnetic flux density or magnetic induction (B) has the dimension:

$$\left[Q^{-1}MT^{-1} \right] .$$

The unit is derived from the equation

$$F = I \times B$$ (6.6)

where F is the force acting on a unit length of a linear conductor carrying the current I in a field B. The unit for B is tesla (T), while I is expressed in amperes and F in newtons per meter.

$$1\,T = 1\,VS/m^2 = 10^4 \text{ gauss.}$$

The magnetic flux has the dimension:

$$\left[Q^{-1}ML^2T^{-1} \right] .$$

Magnetic flux is expressed in webers (Wb) as the surface integral of B and is expressed as:

$$1\,Wb = 1\,T/m^2 .$$

The magnetic field has the dimension:

$$\left[Q\,L^{-1}\,T^{-1} \right] .$$

The magnetic field strength, H, is a vector quantity, the rotation of which is equal to the current density including the displacement current and is expressed as current per unit length, usually in amperes/meter. The magnetic field strength and the magnetic flux density are related as $B = \mu H$, where μ is permeability expressed in henry per meter.

6.1.2 REQUIREMENTS FOR MEASUREMENT RANGES

6.1.2.1 Bioelectric Events

Most electrical events in the human body have amplitudes or levels well below 1 V. The resting potential of a cell may be as high as 0.05 to 0.1 V, whereas voltages recorded from the skull and related to activity of the central nervous system may be as low as a few microvolts. Table 6.1 summarizes the magnitude and frequency domain required for the recording of some common physiological signals.

Table 6.1 Electrophysiological Variables and their Magnitude and Frequency Ranges

Electrophysiological Parameter	Signal Range (mV)	Signal Frequency Range (Hz)
Electrocardiography (ECG)	0.5–4	0.01–250
Electroencephalography (EEG)	0.001–0.1	d.c.-150
Electrocorticography and brain depth	0.1–5.0	d.c.-150
Electrogastrography	—	d.c.-1
	0.1–5	d.c.-1
Electromyography (EMG)	—	d.c.-10,000
Eye potentials		
EOG	0.005–0.2	d.c.-50
ERG	0.0–0.6	d.c.-50
Nerve potentials	0.01–3	d.c.-10,000

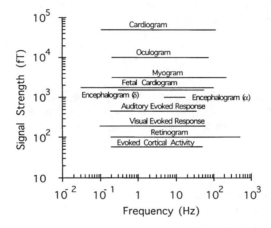

Figure 6.1 Typical signal strengths and frequency ranges for various biomagnetic signals (1 fT = 10^{-15} T).

6.1.2.2 Biomagnetic Events

The magnetic signals from biological organs are extremely weak. At maximum they reach the magnitude of 10^{-10} T, which are extremely low signal levels. As a comparison, the static magnetic field of the Earth is on the order of 10^{-5} to 10^{-6} T. The low signal levels put special demands on the recording techniques used in biomagnetic measurements. As can be seen from Figure 6.1,[1] most biomagnetic signals are limited to the low frequency range.

6.2 ELECTRODE THEORY

6.2.1 THE ELECTRODE-ELECTROLYTE INTERFACE

Bioelectrodes are devices that transform biochemical and physiological phenomena into electrical currents or generate such phenomena from electrical currents. Such devices make use of the fact that electrolytes in biological solutions and body tissues contain charged particles, ions. The electrode's function is to transfer charge between ionic solutions and metallic conductors. By this definition, we can include not only stimulating systems but also those systems that are used for the sensing of electrophysiological information, such as electrocardiograms. In the latter case, the signals to be recorded are electrical events resulting from biochemical activity of nerve and muscle tissues. The presence of ions in a solution indicates that (1) the ions are the carriers of electrical charge and (2) they are atoms or molecules with a finite mass. When we pass currents through either solutions or tissues, we are using ions to conduct the current. We must distinguish between majority carriers and minority carriers in the solution. In the human body, a majority carrier is an ion such as sodium or chloride. Intracellular potassium also contributes to conduct current in higher frequency ranges. Such ions are quite mobile and are present in large enough concentrations so that when current flows they are in the majority and carry most of the transferred charge. Minority carriers are ions in a very low concentration. Minority ions do not normally carry much of the total transfer charge.

Ions interface between metals and electrolytes, between one kind of electrolyte and another, across membranes, etc. They generate potentials independently. They describe to us how batteries act and how electrodes polarize. The amount of the ion in a unit volume of the solution is measured as a concentration. If the concentration is low on one side of a boundary and high on the other side, then there is a force acting on that ion influencing it to move and to equalize the difference in concentrations. The ion must respond to the concentration gradient. Quite independent of this mechanism, the ion is a charged carrier. All other molecules in a solution are not charged. For instance, sugar is readily soluble but does not ionize. An uncharged molecule moves only in response to concentration gradients. Conversely, an electron that is a charged particle with almost zero mass moves only in response to electrical fields. The electrical potentials developed by ions in solution result from their activity following exposure to these two forces at once, so the ion concentration is a function of both the charge and the concentration gradient.

The charge is further described by an equation that says its potential is equal to a Nernst potential, (see Section 7.2.1.1) which is proportional to the gas constant, R, and the absolute temperature, T:

$$E_m = \frac{RT}{nF} \ln(M^+) + E_0. \tag{6.7}$$

The electrode potential is inversely proportional to the number of charges per ion (n, valence) and the Faraday constant, F; M^+ is the activity of the ion, and E^0 is a constant. If we introduce constants for a monovalent ion in the formula, we obtain

$$E_m = 0.0258 \ln(M^+) + E_0. \tag{6.8}$$

Most of physiological electrochemistry is based on the assumption that one ion does not interfere with another ion that will have to compete with it. This means that the solution is dilute and there is enough free space between ions that they will act freely. The activity of such an ion will be linearly related to its concentration, and activity and concentration will be equal. Most solutions in the human body are dilute. For example, in normal saline we have 0.9% by weight of sodium chloride in solution. If we go to very high sodium chloride concentrations, we could not consider their chemical activity proportional to their concentration. The equilibrium of ion movement would then be affected.

6.2.2 LIQUID JUNCTION POTENTIALS

The mobility of individual ions is very different, depending primarily on their size. When ions exist in a water solution, they very often hydrate, or accumulate water molecules. This accumulation of water molecules means that the geometric size of the molecule changes dramatically, which affects the mobility of the ion in solution. A good example is sodium and chloride ions in a water solution. A sodium ion and a chloride ion exist, and the same forces act on both ions. However, the unhydrated chlorine ion moves much more freely than the unhydrated corresponding ion. The difference in mobility of these ions has created the same situation as in a selective membrane. Charge differences will develop, and concentration gradients will oppose the charge gradients. An equilibrium is reached, and we usually have a liquid junction potential. This phenomenon will occur any time two solutions have different concentrations and ions of different mobilities. A liquid junction potential, E, can be of the order of millivolts, maybe up to 50 mV, depending on the concentrations. At a junction of two solutions of ionic concentrations, C_1 and C_2, the liquid junction potential, E, is given as

$$E = \frac{u^+ - v^-}{u^+ + v^-} \frac{RT}{F} \ln \frac{C_1}{C_2} \tag{6.9}$$

where u^+ and v^- are the mobilities of the cations and the anions, respectively. R is the gas constant, T is the absolute temperature, F is the Faraday constant, and n is the number of charges carried by each ion.

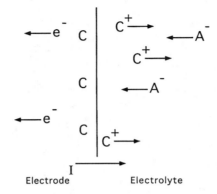

Figure 6.2 Electrode-electrolyte interface with a current passing through the electrode into the electrolyte. The electrode is a metal with metallic atoms C^+, and the electrolyte is a solution containing cations of the electrode metal C^+ and anions A^-.

6.2.3 DOUBLE LAYER

If a metal is introduced in an electrolyte solution, double layers are formed close to the electrode surface. The electrode-electrolyte interface is illustrated in Figure 6.2.[2] A net current that crosses the interface passing from the electrode to the electrolyte consists of

1. Electrons moving in a direction opposite to that of the current of the electrode
2. Cations (denoted by C^+) moving in the same direction as the current
3. Anions (denoted A^-) moving in the direction opposite that of the current in the electrolyte

For a charge to cross the interface, since there are no free electrons in the electrolyte and no free cations or anions in the electrodes, something must occur at the interface that transfers the charge between these types of carriers. What actually occurs are chemical reactions at the interface, which can be represented in general by the following equations:

$$C \rightleftarrows C^{n+} + ne^-$$ (6.10)

$$A^{m-} \rightleftarrows A + me^-$$ (6.11)

where n is the valence of C, and m is the valence of A.

In these equations we are assuming that the electrode is made of the same materials as the cation and that this material in the electrode at the interface can become oxidized to form a cation and a free electron. The cation is discharged into the electrolyte, while the electron remains as a charge carrier in the electrode. The reaction involving the anions is given in Equation (6.11). In this case, an anion coming to the electrode-electrolyte interface can be oxidized to a neutral atom, giving off one or more free electrons to the electrode. It is important to know that these two reactions are reversible and that a reduction reaction going from right to left in the equations can occur, as well. When no current is crossing the electrode-electrolyte interface, these free actions often still occur, but the rate of oxidation reactions equals the rate of the reduction reaction so that the net transfer charge across the interface is zero. When the current flows from the electrode to electrolyte as indicated in Figure 6.2, the oxidation reactions dominate; when the current is flowing in the opposite direction, the reduction reactions dominate.

To further understand the characteristics of the electrode-electrolyte interface we must consider what happens when we place a piece of metal in a solution containing ions of that particular metal. When the metal comes into contact with the solution, the reaction presented in Equation (6.10) begins immediately. The reaction initially goes either to the left or to the right, depending on the concentration of cations in the solution, and at some point the equilibrium conditions are established for that particular reaction. The local concentration of cations in the solution at the interface changes and affects the anion concentration at this point, as well. The net result is that neutrality or charge is not maintained in this region of the electrode. So, the electrode surrounding the metal is at a different electrical potential from the rest of the solution. A potential difference known as a half-cell potential is determined by the metal

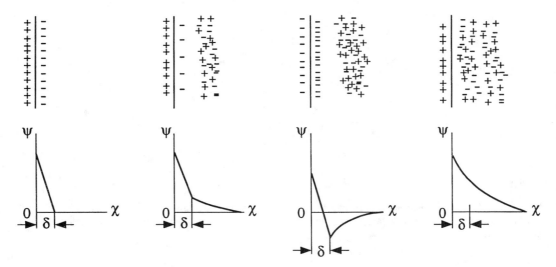

Figure 6.3 Charge and potential distributions at an electrode-electrolyte interface, based upon (from left to right) Helmholtz,[3] Gouy,[5] Stern,[6] and pure Gouy.

involved, the concentration of its ions in the solution, and the temperature, as well as other second-order factors. Knowledge of the half cell potential is important for understanding the behavior of biopotential electrodes.

Several forms of charge distribution are possible. The simplest was conceived by Helmholtz,[3] who described the electrical double layer, later called the Helmholtz layer. He proposed the arrangement of Figure 6.3(a),[3,4] which represents a layer of ions tightly bound to the surface of the electrode and an adjacent layer of oppositely charged ions in the solution. With such a simple arrangement, the potential distribution close to the electrode would be as shown. Because of the thermal motion of ions, it was believed that the simple Helmholtz electrical double layer was not adequate to describe the environment of an electrode. Gouy [5] suggested another charge distribution in which the fixed Helmholtz layer of negative charge was not enough to balance the positive charge of the electrode. To satisfy this requirement, he proposed the existence of a diffused charge distribution adjacent to the Helmholtz layer. Gouy's arrangement of a fixed and diffused layer and the accompanying potential distribution appear in Figure 6.3(b).

Stern[6] believed that a fixed layer could contain more negative charges than are required to balance a positive charge on the electrode. This situation, in combination with a diffuse Gouy layer, is presented in Figure 6.3(c), along with the expected potential distribution. Any combination of a fixed and diffused layer is called a Stern layer. It is also possible for the charge distribution to consist entirely of the diffused Gouy layer and to exhibit the potential distributions shown in Figure 6.3(d). As stated previously, the particular charge distribution that exists depends on the species of metal used for the electrode and the type of electrolyte. It is the ionic distribution that gives an electrode its properties.

Parsons[7] described electrodes in terms of their reaction at the double layer. Electrodes in which no net transfer of charge occurs across the metal electrolyte interface were designated by him as perfectly polarized. Those in which unhindered exchange of charge is possible are called perfectly nonpolarizable. Real electrodes have properties that fall between these idealized limits. It should be apparent by this definition that a truly polarized electrode has all the characteristics of a capacitor.

MacInnis[8] has stated that the term "electrode polarization" is used in two ways: (1) as described previously and (2) referring to the conditions when an electrode-electrolyte potential is altered by the passage of a current. Although the various theories can result in different structures for the ion distribution close to the electrode, they all support the fact that there exists a kind of separation of charges at the metal electrolyte interface that results in an electrical double layer. All theories also result in the suggestion that one type of charge dominates on the surface of the metal and the opposite charge is distributed immediately adjacent to the electrode in the electrolyte.

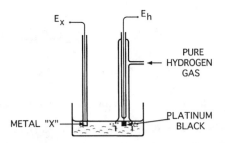

Figure 6.4   The standard hydrogen electrode (SHE).

6.2.4   ELECTRODE POTENTIAL

As a result of the charge distribution close to the electrode surface, the electrode itself acquires a potential. It is not possible to measure the potential of a single electrode with respect to a solution. Therefore, electrode potentials are always measured in relation to a standard electrode, which is simple to reproduce in the laboratory. Such an electrode is the standard hydrogen electrode (SHE). Although the choice of a particular reference electrode fixes the electrode potential, the reference electrode does not affect the difference in potential measured between two electrodes in an electrolyte.

The standard hydrogen electrode (Figure 6.4)[4] consists of a platinum black electrode in contact with a solution containing hydrogen ions of unit activity and dissolved molecular hydrogen. The activity of the latter is specified by requiring it to be in equilibrium with hydrogen at 1 atm in the gas phase. The platinum black has a large capacity for absorbing hydrogen and probably acts as the catalyst to convert the hydrogen to its ionic form. The hydrogen gas is bubbled over the electrode, and the cell is operated at 1 atm. The potential of such an electrode is defined as zero volts at all temperatures, contrary to the physical reality, but can be circumvented by identifying the temperature at which the measurement of the electrode is made.

6.3 SURFACE POTENTIAL ELECTRODES

6.3.1   ELECTROCARDIOGRAM ELECTRODES

Analyses of the electrocardiogram (ECG) are common diagnostic procedures in modern healthcare. Several million diagnostic ECGs are recorded every day. Also, monitoring of the ECG and heart rate in intensive care is providing additional information. The electrode on the skin surface has many important properties for making these measurements reliable and undisturbed.

Electrocardiogram electrodes exist in a variety of models and sizes for various purposes, from disposable simple electrodes for temporary purposes to very advanced high quality designs for quantitative and highly sophisticated measurements. Most electrodes used today are disposable, with an adhesive tape surrounding the electrode's metal surface. Recessed electrodes permit the use of a paste between the skin and the metal, while other electrodes place the metal in direct contact with the skin. Pre-gelled electrodes are delivered with a conducting foam or paste as a medium between the metal and the skin. For a survey of existing design principles, the reader is referred to Geddes[4] and to Webster.[9]

Reusable electrodes are made from silver, zinc alloys, copper, and nickel. Usually they can be utilized for several years and are mainly used for ECG analysis requiring up to 10 electrodes or more on the skin surface. Bulb and plate electrodes of various sizes and shapes are the most common; however, disposable electrodes are used increasingly instead of bulb-plate electrodes.

Artifacts from ECG electrodes are mainly of the power-line frequency or baseline instability type. The half-cell potential and the electrolyte skin impedance are important factors when trying to reduce the stability and noise sensitivity.[10,11] Typical examples of baseline wander and power-line interference are seen in Figures 6.5(a) and (b).[9]

These disturbances can be considerably reduced by the use of properly chosen electrode paste which reduces electrode impedance and the intra-electrode offset potential. The latter reduction is also important from the standpoint of ECG amplifier saturation. Today, pre-gelled silver-silverchloride disposable electrodes are used almost exclusively for monitoring and diagnostic ECG recordings. The surface of the electrode is partially covered with Ag-AgCl. The disposable electrodes have a number of advantages

Figure 6.5 (a) Baseline wander and (b) power-line interference.

compared to the plates: less risk of infection, smaller size, better long-term adhesive properties, and better electrical properties. The chloridization of Ag-AgCl electrodes is an important procedure in regard to electrode stability. Geddes[4] suggests several good methods for this process. Chloridization and pre-gelling of electrodes improve the electrical properties substantially.

6.3.1.1 The Silver-Silverchloride Electrode
At the surface of an Ag-AgCl electrode, a reversible electrochemical reaction takes place. If the electrode is operated as a half-cell, chloride is deposited on the electrode when it is the anode, and AgCl on the electrode surface is reduced to Ag. The Cl^- ions leave the electrode and enter the electrolyte solution when it is operating as a cathode.

The Ag-AgCl electrode is generally described as[12]

$$Ag^+|AgCl|Cl^-$$

and consists of a metal silver substrate coated with AgCl. When such an electrode comes into direct contact with an electrolyte containing chloride ions, an exchange takes place between the electrode and the solution. The electrode can pass a current of limited strength without changes in the chemical composition of the solute close to the electrode. As indicated earlier, the potential of a silver-silverchloride electrode can be described by two equilibrium reactions:

$$Ag\,(solid) \rightleftarrows Ag^+(solution) + e^-(metal\ phase) \tag{6.12}$$

$$Ag^+\,(solution) + Cl^-(solution) \rightleftarrows AgCl\,(solid). \tag{6.13}$$

The reactions above are valid under the assumption that the product of the activity coefficients for Ag^+ and Cl^- are constant (see Section 7.1.1).

$$\alpha_{Ag^+} \times \alpha_{Cl^-} = k^s. \tag{6.14}$$

The ionic product equals the solubility product. When an electrolyte solution containing Cl^- at physiological concentration is used, α_{Cl^-} is larger than α_{Ag^+} by several order of magnitude, thus the change in α_{Cl^-} due to the electrode reaction is a negligible fraction. On the other hand, α_{Ag^+} is correlated with α_{Cl^-} by Equation (6.14), so α_{Ag^+} cannot also be changed. Consequently, α_{Ag^+} is practically a constant under these circumstances.

The Nernst equation gives

$$E = E_{0\,Ag,Ag} + \frac{RT}{ZF}\ln \alpha_{Ag^+}. \tag{6.15}$$

As long as α_{Ag^+} is constant, E does not change even when an electrode reaction occurs.

6.3.1.2 Stainless-Steel Electrodes
Stainless-steel electrodes are commonly used for recording an ECG. Stainless steel of the standard types 304 or 316 have been used extensively. The gel used in this connection creates a low half-cell potential

Table 6.2 Offset and Polarization Voltages and Impedances of Stainless-Steel and Ag-AgCl Electrodes

Electrode Type	Offset (mV)	Impedance (10 Hz, ohms)	Polarization (mV)	
			5 sec	35 sec
Stainless steel	1–50	800–2200	600–1200	400–900
Ag-AgCl	0.1–50	70–300	1–30	1–10

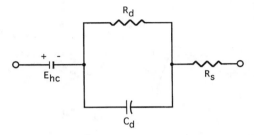

Figure 6.6 Equivalent circuit for a biopotential electrode in contact with an electrolyte. E_{hc} denotes the half-cell potential. R_d and C_d are associated with the electrode-electrolyte interface and polarization effects. R_s is the series resistance in the electrolyte and the lead wire.

if potassium citrate EDTA (ethylene diaminetetra acetate) or potassium sulfate is used. The stainless-steel electrode will polarize far more than the silver-silverchloride electrode. Table 6.2 shows polarization data for both types of electrodes after discharges of 2 millicoulomb (mC) (200 V). Polarization mechanisms are O_2 reduction and H_2 development at the cathode. At the anode, O_2 evolution, gel impurity, and formation of metal oxides dominate the creation of an almost permanent polarization.

6.3.1.3 Electrode Impedance

Electrocardiogram electrodes have been characterized in several ways. Electrical impedance and polarization properties can be described in terms of diagrams and models based on the RC-characteristics over a particular frequency range. A number of authors have reviewed the impedance properties of bioelectrodes and have described extensively the electrode properties.[4,13,14] Generally, the electrodes can be described as models similar to the one given in Figure 6.6.[2] In this circuit, R_d and C_d represent the resistive and reactive components. C_d represents the capacitance across the double layer of the electrode-electrolyte interface. The parallel resistance, R_d, represents the leakage resistance across this double layer. The component values of this equivalent circuit are strongly influenced by the electrode material and, to a lesser extent, by the properties of the electrolyte. It can be seen from the equivalent circuit that the magnitude of the impedance of an electrode is frequency dependent. At high frequency, the reactance of the capacitor, C_d, is usually much lower than the resistance of R_d, and the total impedance will be determined by R_s. At low frequencies, $1/\omega\, C_d \gg R_d$, which means that the total impedance is approaching $R_s + R_d$. In between these frequencies, the magnitude of the impedance will vary with the frequency. The battery, E_{hc}, symbolizes the half-cell potential of the electrode. This component is usually omitted for nonpolarizable electrodes, such as the silver-silverchloride electrode.

The component values for the model are strongly related to the way the electrode has been fabricated. Figure 6.7[15] shows an example taken from platinum electrodes. In this example, the electrode consists of pure platinum. After the initial measurements, the electrodes are platinized, i.e., covered with a layer of platinum black which changes the electrode properties and thus also dramatically changes the parameter values of the model.

The simple electrode model presented in Figure 6.7 is not an exact model of the electrode properties. For instance, both the resistance and capacitance vary with the current density of the electrode surface. Figure 6.8[4] shows an example of stainless-steel electrodes. The increase in current density increases the capacitance of the electrode model, and a decrease in the resistance follows an increased current density. This phenomenon illustrates the complexity of electrode-electrolyte interfaces.

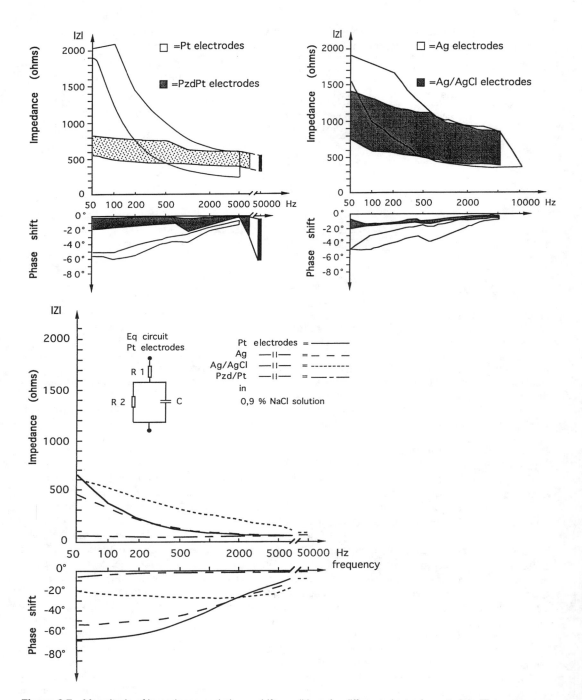

Figure 6.7 Magnitude of impedance and phase-shift conditions for different electrode materials. The upper part of the figure shows results from an animal model study and the lower part from a model/equivalent circuit. For Pt electrodes, $R_1 = 50\ \Omega$, $R_2 = 600\ \Omega$, and $C = 3.3\ \mu F$. For Pzd/Pt electrodes, $R_1 = 49\ \Omega$, $R_2 = \infty\Omega$, and $C = 1000\ \mu F$.

The skin has electrical properties that can be modeled in terms of R and C components. It is also important to consider this impedance because it comprises a significant part of the total impedance between the generator (the heart) and the amplifier input terminals. The skin impedance properties add to those of the electrode, and the skin equivalent circuits may affect the signal as much as the electrode properties.

Figure 6.9[16] shows an attempt to study the electrical properties of the upper skin layers. The stratum corneum and the other upper skin layers have been removed by stripping the skin with adhesive tape.

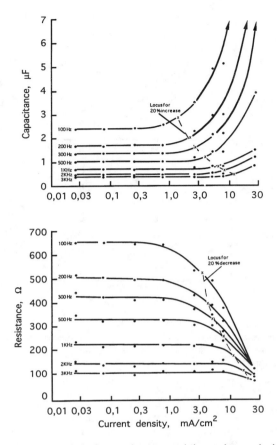

Figure 6.8 Dependence of the series-equivalent resistance and the series-equivalent capacitance of a stainless-steel/0.9% saline interface on current density and frequency.

Every time the tape is removed, parts of the skin are attached to the tape, and in this way the skin changes its electrical properties. As can be seen from Figure 6.9, the impedance of the skin can be closely modeled by RC parallel circuits connected in series. Stripping of the outer cell layers corresponds to the removal of a few of the RC-circuits.

Skin properties can be changed dramatically from an electrical viewpoint by either stripping or rubbing the skin surface with sandpaper. This technique is used for lowering the skin impedance and thereby improving the quality of ECG recordings. A similar technique, the skin drilling technique, was suggested by Schackel.[17] Here the skin is abraded with a dental burr. Schackel[17] obtained a reduction in resistance 1/5 to 1/10 that of the undrilled sites and ECG recordings of very high quality with this technique.[17]

ECG paste applied on the electrode can have the same effect as skin abrasion or skin drilling (Geddes[4] has covered this topic extensively). A variety of gel or paste compositions has been manufactured to improve the conductive properties of the skin. When using an electrode preparation with surface electrodes it is important to remember that the electrode impedance is initially high but decreases when the applied paste diffuses into the skin. Roman[18] and Almasi and Schmitt[19] have studied the 10-Hz impedance. The latter authors reported that the decrease in impedance with time was exponential in nature with a time constant of 6.9 min. The final impedance values were reached about 30 min after application of the paste. They also noted major differences in the skin impedances at various sites on the human skin.

6.3.1.4 Motion Artifacts

As described earlier in this chapter, polarizable electrodes in direct contact develop a double layer of charges adjacent to the electrode surface. If the electrode is moving with respect to the electrolyte, the double layer is disturbed and the result is a change in the half-cell potential. When the motion stops,

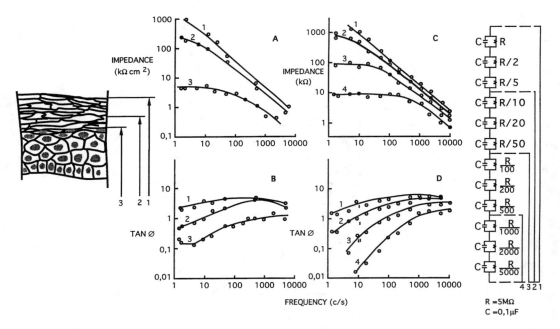

Figure 6.9 The impedance and phase of human skin under a dry silver electrode **(A)** and **(B)** after various degrees of stripping (see left-hand diagram), and of an electrical circuit **(C)** and **(D)** with successive elements removed (see right-hand diagram).

the double layer is re-established and the electrode develops the initial half-cell potential. When recordings are made between two electrodes and one of the electrodes is moving, a potential difference is generated between the electrodes during the movement. This type of potential difference is called a motion artifact, the existence of which is a serious problem in most types of electrode measurements. It is reasonable to assume that nonpolarizing electrodes are better in this respect than polarizable ones. Neuman[2] has shown this in a simple experiment. By moving a metallic silver electrode in a saline solution before and after chloridization, he was able to demonstrate the difference in motion sensitivity between two electrode preparations.

The double layer is not the only source of motion artifacts in ECG recordings. Tam and Webster[10] have shown that variations in the electrolyte paste-skin potential can also cause motion artifacts if this potential varies with the motion of the electrode. They indicated that this effect can be significantly reduced by removing stratum corneum with abrasive paper. Ödman[20] found that dielectric/piezoelectric properties of the stratum corneum can cause motion artifacts. If the stratum corneum is stretched mechanically, a redistribution of charges occurs. Such a redistribution will result in a motion artifact at the electrode. This mechanism is closely related to the dielectric/piezoelectric properties of the stratum corneum. Removal of this part of the skin eliminates this motion artifact generator. However, removal of the body's outer protective barrier opens a way for infectious diseases to enter the body. Cautious use of this method is recommended.

6.3.1.5 Dry Contactless or Capacitive Electrodes

The invention of the field effect transistor made it possible to design amplifiers with a very high input impedance. Using such an amplifier as the input stage, it is possible to apply electrodes directly on the skin without paste or low-ohmic paths between the electrode and the tissue. Such electrodes are called dry or capacitive electrodes.

A common type of capacitive electrode functions in the following way: A metal plate is placed against the stratum corneum and a metal electrolyte interface is created. The impedance is mainly resistive because the stratum corneum works as an isolating medium. The deeper parts of the skin have higher conductivity, so we have a capacitor with the metal plate and the dermis serving as conducting plates and the stratum corneum serving as the dielectric.

The distance between the plates is much larger than the corresponding distance in an electrode double layer, and the capacitor formed has a much lower capacitance. The resistance is much higher, however, in comparison with the wet electrodes.

By using an impedance-converting amplifier in direct contact with the electrode, we can detect the biopotential even if the source impedance is very high. Usually the input impedance of the amplifier exceeds 1 GΩ. Examples of such an electrode have been described by Ko and Hynech.[21]

Richardson et al.[22,23] suggested a type of insulated electrode in which an oxide layer (Al_2O_3, $SiO_{2)}$ constitutes the isolation layer between the metal and the tissue. A field effect source follower was mounted close to the electrode and used as an impedance transforming device. Variations of the same theme have been presented by David and Portnoy,[24] Matsuo et al.,[25] and Yon et al.[26] As mentioned earlier, the dry electrodes have the advantage of not requiring the use of electrode paste. There are also some disadvantages associated with this electrode type. The very high impedance amplifier is very sensitive to electric fields around the electrode. In addition, capacitive changes in the stratum corneum may affect the low frequency response of the electrode.[27]

Bergey et al.[28] suggest the use of a pasteless electrode during long-term applications such as, for instance, space missions and intensive care. High-impedance amplifiers have to be used in connection with these types of electrodes because of the high skin-to-electrode impedance.

6.3.2 ELECTROMYOGRAM ELECTRODES

Two electrode types are used to record the electromyographic (EMG) signal. Surface electrodes are used on the surface of a muscle or the skin above the muscle under study. Needle or wire electrodes are inserted into the muscle for signal extraction. Electrodes for EMG can be used singularly (monopolar) or in pairs (bipolar).

6.3.2.1 Surface Electrodes

The most common surface electrode used for EMG analysis is a variant of the ordinary disc Ag-AgCl electrode. Coupling between the muscle and the electrode is improved by using gel between the electrode surface and the tissue. The EMG Ag-AgCl electrode has the advantages and drawbacks described earlier in this chapter. Special types of electrodes are manufactured and sold; in comparison with the ECG electrodes, they are usually small and lightweight. In addition, dry or isolated electrodes are used in electromyography. As described earlier, this electrode type has the advantage of not requiring paste or other conducting media between the electrode surface and the tissue. However, the dry electrode for EMG has a higher inherent noise level than the disc types. The dry electrodes do not have the long-term stability required because of changes in the dielectric properties caused by perspiration. As a result, they have been little used in electromyography.[29]

Godin et al.[30] has modeled the noise characteristics of stainless-steel surface electrodes. The effect of electrode surface area on electrode impedance and noise was studied using circular stainless-steel electrodes of varying diameters. The results were formulated in terms of mathematical models. The primary noise components were thermal and amplifier impedances generated. Electrode impedance was found to be a power function of both electrode diameter and frequency.

The surface electrodes can only be used for EMG recordings from superficial muscle groups. Selective recordings from small muscles are also difficult to perform. However, the noninvasiveness of the surface electrode is advantageous for child measurements, controlling external prostheses, and psychophysiological measurements where relaxation is important.

Recently, several new electrode design ideas have been described for a variety of applications. Cram and Engstrom[31] and Cram and Steger[32] introduced a muscle scanning technique that can be used for EMG assessment of multiple muscle sites. The technique involves the use of "post-style" silver-silver chloride electrodes positioned 2 cm apart in a hand-held adapter. The electrode setup is positioned at each muscle site for 2 s to get an average of integrated EMG activity. Usually 6 to 10 s are allowed for the EMG signal to stabilize before the 2-s sample is taken. Using this technique, the authors could scan many bilateral muscle sites in 10 to 15 min to get what is called a "snapshot view" of the pattern of muscle activity. Thompson et al.[33,34] have evaluated this technique and found that 2-s intervals were sufficient to ensure the same recording quality with fixed skin surface electrodes; however, 10-s recording intervals further improved the quality of the recording.

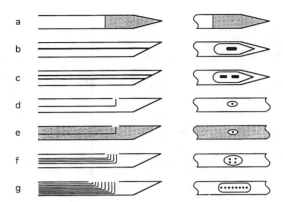

Figure 6.10 Different types of needle electrodes for EMG recordings.

6.3.2.2 Needle and Wire Electrodes

These types are very common in EMG recordings and exist in many variants. de Luca[29] has described the most common types (see Figure 6.10).[9] The concentric needle electrode is the most common one with a monopolar single electrode and a concentric ring needle case. The relatively small pickup area allows recording of individual motor unit action potentials. The electrode can easily be repositioned within the muscle.

A number of specially designed types of electrodes have been used by authors with special interests. Wire electrodes have the advantage of being flexible and are therefore well suited for kinesiology studies. Small diameter wires made from alloys of platinum, silver, nickel, or chromium are preferred. Insulations such as nylon, polyurethane, and Teflon® are used.

Two electrodes are inserted in a cannula, which is introduced into the muscle. The cannula is then withdrawn, leaving the wire in position in the muscle; 1 to 2 mm of the distal tip are deinsulated and bent in the form of two hooks. The wires can easily be withdrawn if required.

Giroux and Lamontagne[35] compared surface electrodes with the performance of intramuscular electrodes under isometric and dynamic conditions. They found that surface electrodes were more reliable than the intramuscular wire electrodes in day-to-day investigations.

Nandedkar and Sanders[36] have studied the recording characteristics of monopolar needle electrodes. Disposable monopolar electrodes were found to be somewhat more selective than were selective electrodes. The geometrical properties — in terms of recording sensitivity as a function of distance — differ between the monopolar and the concentric electrode. Nandedkar et al.[37] have also described the recording properties of disposable concentric needle electrodes which were compared with reusable ones. The two electrode types were found to be very similar in model and *in vivo* studies, although minor differences in the electrical properties were found.

6.3.3 ELECTROENCEPHALOGRAM ELECTRODES

Small (7 to 10 mm) Ag-AgCl, silver, or gold-plated electrodes are used for measurements of the electroencephalographic (EEG) signal. Because of the low signal levels obtained in these measurements, the application of the electrodes on the scalp requires care. The electrodes are "glued" on the scalp by using collodion. Ordinary ECG paste is used for improvement of the conductive properties close to the electrode. It should be noted that this way of applying electrodes is a time-consuming process. Some reports on special applications of EEG electrode techniques have been published.

Wyllie et al.[38] evaluated subdural electrode systems used for evaluation before epilepsy surgery in both adults and children. The electrode systems had between 21 and 100 electrodes of 3-mm stainless steel mounted on a Silastic® sheet. The electrodes were placed in various locations to cover as completely as possible the expected area of epileptogenicity. The main contribution of the subdural electrode system was a more precise definition of the epileptogenic area to be resected. Kramer[39] studied the influence of noise on bioelectric signal recording. He found that electrode noise very often has a low frequency character which covered the same frequency range as the signal of interest, which makes signal interpretation difficult. Silver-silverchloride electrodes were found to be superior to other types. However, poorly prepared electrodes created an enhanced noise level which often coincided with the delta activity of the EEG.

Maxwell et al.[40] have evaluated an implantable EEG electrode during long-term implantation (2600 hours) in seven patients. The electrodes are constructed of polyurethane tubing with stainless-steel contacts. The electrode positioning was easy to carry out with precision using stereotaxic techniques. No electrode migration was observed. The implantable stereo depth electrode had an advantage over surface electrode techniques in selected cases of epilepsy.

6.4 MICRO- AND SUCTION ELECTRODES

Microelectrodes have been developed to study the electrical properties and activity of membranes of living cells. Two types of electrodes are used:

1. A fluid-filled glass pipette with a tip size diameter of 0.1 to 1 μm for membrane and intracellular studies
2. A metal-wire electrode with an uninsulated area of 1 to 1000 μm² which is used to record potentials in the extracellular space

The impedances of both types of electrodes are very high and on the order of 1 to 1000 MΩ. Electrodes of these types require amplifiers with a very high input impedance and input capacitance neutralization to ensure undisturbed recording of electrophysiological events at the cellular level. The electrode should not disturb the normal function of the cell. The electrode must be able to penetrate the cell membrane without leakage of cytoplasm, and the electrode content should not poison the cell.

Glass microelectrodes consist of fine glass capillaries heated and drawn to very small diameters. They are filled with a highly conductive solution (commonly 3 M KCl). An equilibrium is established between the electrode solution and the intracellular content which makes it possible to record the intracellular potential. Metal electrodes are used primarily for the recording of dynamic events such as the action potential. Suction electrodes have been used for a very long time to record electrophysiological events. Suction (or negative pressure) as a way of attaching the electrodes to the tissue under study has recently gained popularity in new ECG machines for rapidly obtaining precordial recordings.

6.4.1 GLASS MICROELECTRODES

Figure 6.11[13] illustrates the penetration of a cell membrane by a glass microelectrode. The fine electrode tip penetrates the membrane without disturbing its function or creating leakage to the outside. An Ag-AgCl electrode is positioned in the filling solution to make contact with the interior of the electrode. A circuit model providing the major electrical properties of the electrode is seen in Figure 6.11(b). This circuit can be simplified to an RC-filter constituting a low-pass filter. The distributed resistance, R, appears in the electrolyte of the tip. This resistance is very high (on the order of 10 to 100 MΩ), and other resistances associated with the Ag-AgCl electrode can be neglected in comparison with the tip resistance. The tip is also associated with two potentials: tip and liquid junction. The tip potential decreases with increasing tip diameter and also increases with increased resistivity of the solution surrounding the tip. The potential arises from the hydrated glass layer of the tip which behaves like a cation exchange membrane.

Agin and Holtzman[41] and Agin[42] have studied the tip potential as a function of the molarity of the test solution in the electrode. They found a linear relationship between the electrode-electrolyte concentration and the tip potential over a wide range of variations.

The resistance calculation in the tip depends on the electrolyte resistivity and the top angle of the truncated cone formed by the electrodes. Cobbold[13] gives the expression for the tip resistance as

$$R = \rho \frac{\cot(\phi/2)}{\pi \cdot r_t} \tag{6.16}$$

where ρ is the resistivity of the electrolyte, ϕ the total angle of the tip, and r_t the radius of the tip.

At small angles

$$R \approx \frac{2\rho}{\pi r_t \phi}. \tag{6.17}$$

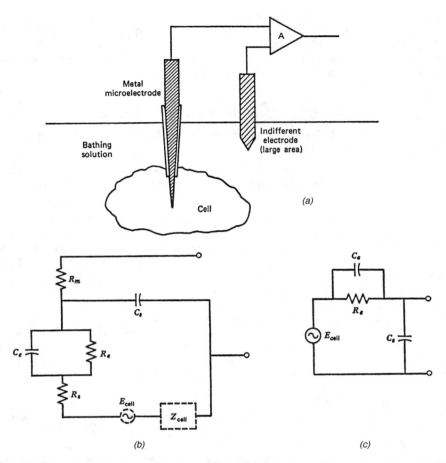

Figure 6.11 Circuit models of a glass microelectrode: **(a)** physical picture, **(b)** circuit model, and **(c)** simplified model.

The capacitance of the tip is generated by the glass wall separating the inner and outer solution. If the wall thickness is constant throughout the tip length, the capacitance can be expressed as

$$C = \frac{0.55 \cdot \varepsilon_r}{\ln(R/r)} \tag{6.18}$$

where ε_r is the relative permitivity of the glass, and R and r are the outer and inner radii of the tip, respectively. C is measured in pF/cm.

In addition, distributed, wiring, and input capacitances together increase the total capacitive load. The combination of a high tip resistance and a distributed capacitance forms a rather high RC time constant that affects the signal under study; however, by means of a positive feedback circuit this capacitance can be neutralized.

6.4.2 METAL MICROELECTRODES

Geddes,[4] in his excellent monograph, has reviewed manufacturing techniques and electrical properties of metal microelectrodes. Ferris[14] covers the same topic extensively. Microelectrodes made from metal are simply metal wires or needles that are isolated from their surroundings except for a small tip that is in contact with the cell or tissue under study. The coating is usually a dielectric resin or an insulating sheet. The materials used are stainless steel, tungsten, or platinum. Electropointing is used for shaping the metal tip. Detailed advice on parameters can be found in Geddes.[4]

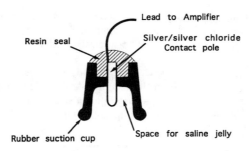

Figure 6.12 Shackel's rubber suction cup electrode.

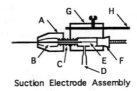

Suction Electrode Assembly

Figure 6.13 The components of the suction electrode assembly: **(A)** tapped hypodermic needle chuck, **(B)** silicone rubber seal, **(C)** stainless-steel or aluminium threaded mounting, **(D)** suction port, **(E)** 30-gauge Ag-AgCl wire, **(F)** rubber stopper, **(G)** plastic fitting, and **(H)** holder mounted to a manipulator.

Various materials and techniques exist for applying the insulating sheath to metal electrodes. Vinyl lacquer and enamels can be used for this purpose. Ferris[14] recommends several brands. Dielectric coating is accomplished by dipping the needle into the coating material and slowly removing it. However, each electrode material requires a special technique that must be developed experimentally.

6.4.3 SUCTION ELECTRODES

Suction electrodes were described long ago by Roth[43] and Ungerleider.[44] Hoffman et al.[45] compared microelectrode techniques with suction electrodes and concluded that "if the suction electrode is properly used the monophasic action potentials recorded with it may be taken as a reliable index of the shape of the action potential during the entire phase of repolarization."

Reports on a number of different types of suction electrodes have been published over the years. Andrews[46] has described a small suction electrode with a diameter of 1 cm. This electrode was originally intended for ECG and EEG recordings. The electrode cup was filled with electrode jelly, which partly served as a means for tightening it against the skin and partly as a contact bridge between the skin and the recessed electrode. In this way, movement artifact can be considerably reduced. The electrode construction gave a rather high impedance. Shackel[17] presented a suction electrode (Figure 6.12) based on a silver-silverchloride electrode which is recessed and has characteristics similar to the electrode described by Andrews.[46]

Microelectrodes have been used with suction applied to the interior electrolyte. Action potentials have been recorded from axons on which the electrodes are fixed via the pressure difference (Figure 6.13). Abraham[47] has presented an ECG suction electrode for precordial ECG mapping. The electrode consists of a 6 × 8 matrix of electrodes to which suction is supplied from a pump.

Suction electrodes have been used extensively in experimental cardiology but also to some extent in clinical studies of the heart. This type of electrode allows studies of the time relations of intraventricular action potentials from the inside of the myocardium. Korsgren et al.[48] described a new suction electrode for the recording of monophasic action potentials from dogs' hearts. It is assumed that this type of electrode used on the myocardium can form a reliable index of both the "time of arrival" of the excitation as well as the shape of the transmembrane potential.[45]

Sjöstrand[49] and Shabetai et al.[50] reported on the development of a suction electrode. Samuelsson and Sjöstrand[51] optimized the electrode design by evaluating the impedance properties for different types of electrode materials. Platinum black (platinized platinum) had the lowest impedance but showed mechanical instability, so Ag-AgCl electrodes were used in the final design. In his thesis work, Olsson[52] developed the suction technique for intracardiac measurements further and made the technique clinically usable.

Downey et al.[53] have designed a suction cup electrode for EMG measurements using nickel-copper as the metal base for the electrode. They compared the suction electrodes with recordings from monopolar needle electrodes and found very good agreement between the intramuscular electrodes and those of the surface (suction).

Pope et al.[54] suggested the use of a suction electrode to study esophageal and gastrointestinal muscular function by the use of a suction electrode in the esophagus. The electrode configuration was varied. Agar bridge electrodes, spike electrodes, and Ag-AgCl electrodes were compared. The use of suction to keep the electrode in position was found to be important, especially when using the spike electrode.

6.5 BIOMAGNETISM

6.5.1 BIOMAGNETIC FIELDS

Biomagnetism is the general term for the investigation of biological organisms by measurement of the very weak magnetic fields they generate. So far the major clinical interest has been the study of the electrical activity of the brain and the heart. Biomagnetism can be defined as the study of the magnetic phenomena associated with the electrical ones (mostly electrocardiograph and electroencephalograph). There are several reasons why these types of measurements are of great diagnostic interest. The magnetic measurements are noninvasive and therefore clinically attractive. The magnetic measurements often give information complementary to their electrical counterparts. The magnetic measurements are also of value, as they increase our understanding of the corresponding electrical phenomena. The frequency content of the magnetic signals is usually in the low frequency range. Figure 6.1[1] shows the frequency interval and typical magnitude of some magnetic fields produced by the human body.

Studying the magnetic activity of the brain (magnetoencephalography, MEG) and the magnetic field from the heart (magnetocardiography, MCG) are the main applications in the medical field today. From the distribution of the magnetic field around the brain it has been possible to determine accurately the source of epileptic seizures (the inverse problem) which had been difficult from the corresponding electrical field, i.e., electroencephalography. The magnetic fields are not attenuated or distorted by the tissues of the head. Cohen[55] was first to record the spontaneous rhythms of the brain in the frequency range 8 to 12 Hz. Several review articles cover the progress of biomagnetism of the brain.[56,57]

The electrical and the magnetic activity of the heart have similar curve forms. In the magnetic recording, the P, QRS, and T waves can be easily identified. Cohen and Chandler[58] were the first to map the magnetic field of the heart using a induction coil magnetometer. In the myocardium, ectopic focii causing cardiac arrythmia can better be localized by analyzing the magnetic activity of the heart. Nakaya and Mori[59] have recently reviewed the field of magnetocardiography. Malmivuo and Plonsey[60] have recently presented a textbook which partly covers the theoretical background of this field.

6.5.2 MAGNETOPNEUMOGRAPHY

The lung function can also be analyzed by using the magnetic field that comes from inhaled particles. When industrial workers and miners inhale large quantities of dust, a proportion of the inhaled dust is retained in the lungs and can be magnetized with external fields. Cohen[61] used the magnetic properties of the inhaled dust to assess the dust content of the lungs. A magnetic scanning technique known as magnetopneumography has been developed for this specific problem. Kalliomäki et al.[62] and Stroink[63] have written review articles about this particular subfield of biomagnetics. Many industrial workers inhale dust, some of which may have negative long-term effects. The contaminating particles that are inhaled and retained in the lungs of industrial workers contain magnetic particles that can set up a small magnetic field that can be measured noninvasively with the sensitive equipment available today. The information from these types of measurements can be used to assess the total dust load of the lungs. A common component in respirable dust is magnetite Fe_3O_4.

Figure 6.14[63] shows the magnetization procedure and the detection of magnetic particles in the lungs. In the presence of a magnetic field with a magnitude of 30 to 100 mT, the dust particles become magnetized and are aligned with the direction of the external magnetic field. The rate at which the dust particles align with the field is dependent on the magnitude. Usually a few (up to 20) seconds are enough to align the particles. The mapping of the magnetic field is carried out by using a SQUID magnetometer (see Section 6.5.4).

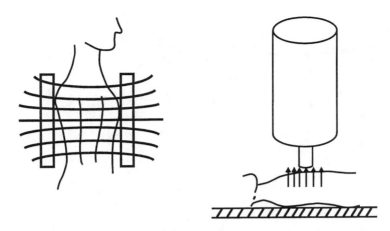

Figure 6.14 The magnetization and detection of magnetic dust in the lungs.

6.5.3 INDUCTION COIL MEASUREMENTS

If an induction coil used at room temperature is exposed to a time-varying magnetic field, the induced voltage is proportional to the time derivative of the flux and the number of turns:

$$e_{ind} = -N\frac{d\theta}{dt} \qquad (6.19)$$

where e_{ind} is the induced voltage, N is the number of turns, and $\frac{d\theta}{dt}$ is the time derivative of magnetic flux. Induction coil-based measurements have had only limited use in biomagnetic measurements. The sensitivity is limited but can be somewhat improved if a high permeability core is used. Baule and McFee[64] used a magnetic coil arrangement in their first studies of the magnetic field of the heart. They used two coils, each containing two million turns wound on a core of magnetic material. The output voltage was on the order of 30 μV and was amplified after 60-Hz noise suppression. The curve form these authors obtained was similar to that of the ECG and it was named the MCG (magnetocardiogram). Wikswo et al.[65] and Wikswo[66] present recordings from a single giant axon of the crayfish. This group used both closed, one-piece induction coils, or clip-on toriods originally designed for oscilloscope measurements. The sensitivity with this room-temperature equipment is a current noise at the input of 0.15 nA/Hz$^{1/2}$. This neuromagnetic technique allows multiple measurements along the nerve.

6.5.4 SQUID SYSTEMS

Cohen et al.[67] introduced the superconducting quantum interference magnetometer (SQUID) and also the shielded room for reproducible real-time measurement of the MCG. SQUID utilizes the Josephson effect to measure very small variations in the magnetic flux.[68-70] In 1962, Josephson[71] suggested the possibility of electron tunneling between two superconducting regions separated by a resistive barrier (the "weak link"). A current magnitude, characteristic for the weak link and smaller than a typical critical value, can penetrate the resistive barrier without causing a voltage drop.

The sensing element of a SQUID device is a superconducting ring interrupted by Josephson junctions. Because of its superconducting nature, a SQUID must be operated at a lower temperature than the transition temperature, T_c. For SQUIDs constructed from niobium ($T_c = 9.3$ K), this means an operating temperature of 0.5 T_c or 4.7 K.

SQUIDs are usually operated in an RF or d.c. mode. The difference between the two modes has to do with the way the Josephson junction is biased. RF refers to a system in which the flux changes are detected by a resonant tank circuit which is inductively coupled to the SQUID loop. In the d.c. variant, the SQUID loop contains two Josephson junctions, and the loop is fed by a d.c. current. The flux changes can be detected as voltage drops across the junctions.

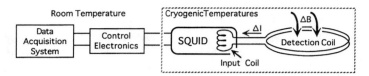

Figure 6.15 Block diagram of a SQUID magnetometer.

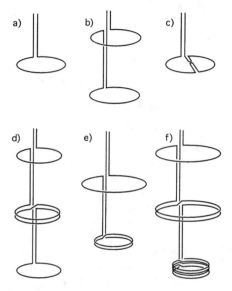

Figure 6.16 Different types of gradiometer coils for SQUID systems: **(a)** Magnetometer, **(b)** first derivative gradiometers, **(c)** planar gradiometer, **(d)** second-derivative gradiometer, **(e)** first-derivative asymmetric gradiometer, and **(f)** second-derivative asymmetric gradiometer.

Figure 6.15[1] shows how a SQUID system is built. The detection coil senses the changes in external magnetic fields and transforms them into an electric current. The input coil transforms the currents into a magnetic flux. Control electronics and data acquisition systems are commonly part of the setup. The SQUID amplifier and the detection coil are superconducting, and liquid helium is used to maintain the system in a superconducting state.

The detection coils of a SQUID can be arranged in a number of ways; Figure 6.16[1] gives a few examples. The choice of configuration depends on a number of factors, such as desired sensitivity, the type of magnetic field source under study, and the matching between the detection coil and the SQUID coil. Spatial resolution and sensivity are often important parameters that are not independent from each other. Improved sensitivity through coil diameter increases results in decreased spatial resolution.

Gradiometer coils can be used in SQUIDs to improve the signal-to-noise levels. Since the field strength of a magnetic dipole is inversely proportional to the cube of the distance between the coil and the field source, the field becomes uniform in direction and magnitude at a distant point. If two coils in the field with opposite winding directions are separated by a distance, then we can get a system that rejects uniform fields.

The first SQUID systems for biomagnetic applications were single-channel systems. With such a system, measurements at multiple sites were time consuming and tedious. Multiple channel systems were introduced to facilitate multiple site measurements and to improve reliability of data. For instance, in heart measurements it is advantageous if the whole organ is mapped with several channels to allow real-time measurements.

SQUIDs act like current-to-voltage converters with extremely high gain. The noise properties are excellent. Very high linearity is a typical property of a SQUID system. The input sensitivity can be on the order of 10^{-12} A/Hz. SQUID theory and various systems for biomagnetic measurements are described in detail in the literature.[67,69]

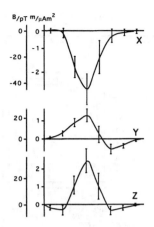

Figure 6.17 *x, y, z* components of the mean normal MCG during the QRS-complex. The amplitude of the equivalent dipole, m/μAm² and the corresponding magnetic induction, B/pT, 120 mm from the center of the heart are given in the figure.

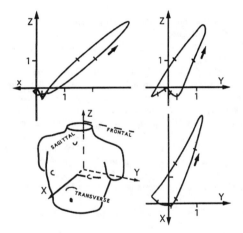

Figure 6.18 Vector loop projections of the mean normal MCG in the left sagittal (*x,z*), frontal (*y,z*), and transverse (*y,z*) planes.

A normal magnetocardiogram with its *x*, *y*, and *z* components from 60 normal subjects is seen in Figure 6.17.[72] The total component of the field can also be expressed in vector form, the vectorcardiogram (VMCG, Figure 6.18).[73]

6.5.5 MAGNETIC NOISE AND SHIELDING

SQUID magnetometers are very sensitive devices and have enough sensitivity for most biomagnetic measurements; however, a number of external magnetic fields may disturb a biomagnetic measurement. The d.c. magnetic field of the Earth is close to 50 μT, which is about 1 million times greater than the magnetic field generated by the heart. The brain signals are about two orders of magnitude lower than those of the heart. The noise in laboratory buildings is of the magnitude 1 to 10 nT/Hz$^{1/2}$ at 1 Hz.

Part of the noise can be eliminated by using gradiometer coils in the SQUID equipment. Usually 2- or 3-order gradiometers are used without a loss in sensitivity. Magnetically shielded rooms can also reduce the background noise levels. The walls are made of Mumetall® and aluminum in several layers. At low frequencies the shielding factor can be as high as 10^4, but this figure rapidly decreases at higher frequencies. Figure 6.19[9] shows the design of a typical shielded room.

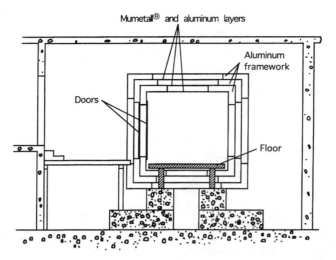

Figure 6.19 The construction of a magnetically shielded room.

REFERENCES

1. Fagaly, R. L., Neuromagnetic instrumentation, in *Advances in Neurology,* Vol. 54, *Magnetoencephalography,* Sato, S., Ed., Raven Press, New York, 1990, chap. 2.
2. Neuman, M. R., Biopotential electrodes, in *Medical Instrumentation, Application and Design,* Webster, J. G., Ed., Houghton Mifflin, Boston, 1978, chap. 5.
3. Helmholtz, H., Studien über Elektrischen Grenzschichten, *Ann. Phys. Chem.,* 7, 337, 1879.
4. Geddes, L. A., *Electrodes and the Measurements of Bioelectric Events,* Wiley-Interscience, London, 1972.
5. Gouy, M., Sur la constitution de la charge électrique à la surface d'un électrolyte, *J. Phys. (Paris),* 9, 457, 1910.
6. Stern, O., Zur theorie der elektrolytischen Doppelschicht, *Z. Electrochem.,* 30, 508, 1924.
7. Parsons, R., Electrode double layer, in *The Encyclopedia of Electrochemistry,* Hampel, C. A., Ed., Reinhold Publishing, New York, 1964.
8. MacInnes, D. A., *The Principles of Electrochemistry,* Reinhold, Dover, New York, 1961.
9. Webster, J. G., Ed., *Encyclopedia of Medical Devices and Instrumentation,* Vol. I, John Wiley & Sons, New York, 1988.
10. Tam, H. W. and Webster, J. G., Minimizing electrode motion artifact by skin abrasion, *IEEE Trans. Biomed. Eng.,* BME-24, 134, 1977.
11. Huhta, J. C. and Webster, J. G., 60-Hz interference in electrocardiography, *IEEE Trans. Biomed. Eng.,* BME-20, 91, 1973.
12. Ives, D. J. G. and Janz, G. J., Eds., *Reference Electrodes. Theory and Practice,* Academic Press, London, 1961.
13. Cobbold, R. S. C., *Transducers for Biomedical Measurements: Principles and Applications,* John Wiley & Sons, New York, 1974.
14. Ferris, C. D., *Introduction to Bioelectrodes,* Plenum Press, London, 1974.
15. Öberg, P. Å. and Sjöstrand, U., Studies of blood-pressure regulation. III. Dynamics of arterial blood pressure on carotid-sinus nerve stimulation, *Acta Physiol. Scand.,* 81, 96, 1971.
16. Tregear, R. T., *Physical Functions of Skin,* Academic Press, London, 1966.
17. Shackel, B., A rubber suction cup surface electrode with high electrical stability, *J. Appl. Physiol.,* 13, 153, 1958.
18. Roman, J., Flight research program. III. High impedance electrode techniques, *Aerospace Med.,* 37, 790, 1966.
19. Almasi, J. J. and Schmitt, O. H., Systematic and random variations of ECG electrode system impedance, *Ann. N.Y. Acad. Sci.,* 170, 509, 1970.
20. Ödman, S., Potential and impedance variations following skin deformation, *Med. Biol. Eng. Comput.,* 19, 271, 1981.
21. Ko, W. H., and Hynech, J., Dry electrodes and electrode amplifiers, in *Biomedical Electrode Technology,* Miller, H. A. and Harrison, D. C., Eds., Academic Press, New York, 1974, 169.
22. Richardson, P. C., Coombs, F. K., and Adams, R. M., Some new electrode techniques for long-term physiologic monitoring, *Aerospace Med.,* 39, 745, 1968.
23. Richardson, P. C., Progress in long-term physiologic sensor development, in *Proc. Biomed. Eng. Symp.,* San Diego, 39, 1967.
24. David, R. M. and Portnoy, W. M., Insulated electrocardiogram electrodes, *Med. Biol. Eng.,* 10, 742, 1972.
25. Matsuo, T., Iinuma, K., and Esashi, M., A barium-titanate-ceramics capacitive-type EEG electrode, *IEEE Trans. Biomed. Eng.,* BME-20, 299, 1973.

26. Yon, E. T., Neuman, M. R., Wolfson, R. N., and Ko, W. H., Insulated active electrodes, *IEEE Trans. Ind. Electron. Control. Instrum.*, IECI-17, 195, 1970.
27. Webster, J. G., Interference and motion artefact in biopotentials, in *IEEE Region 6 Conference Record*, Institute of Electrical and Electronics Engineers, 1977, 53.
28. Bergey, G. E., Squires, R. D., and Sipple, W. C., Electrocardiogram recording with pasteless electrodes, *IEEE Trans. Biomed. Eng.*, BME-18, 206, 1971.
29. de Luca, C. J., Electromyography, in *Encyclopedia of Medical Devices and Instrumentation*, Vol. 2, Webster, J. G., Ed., John Wiley & Sons, 1988, 1111.
30. Godin, D. T., Parker, P. A., and Scott, R. N., Noise characteristics of stainless steel surface electrodes, *Med. Biol. Eng. Comput.*, 29, 585, 1991.
31. Cram, J. R. and Engstrom, D., Patterns of neuromuscular activity in pain and non-pain patients, *Clin. Biofeedback Health*, 9, 106, 1986.
32. Cram, J. R. and Steger, J. C., EMG scanning in the diagnosis of chronic pain, *Biofeedback Self-Regulation*, 8, 229, 1983.
33. Thompson, J. M., Erickson, R. P., and Offord, K. P., EMG muscle scanning: stability of hand-held surface electrodes, *Biofeedback Self-Regulation*, 14, 55, 1989.
34. Thompson, J. M., Madsen, T. J., and Erickson, R. P., EMG muscle scanning: comparison to attached surface electrodes, *Biofeedback Self-Regulation*, 16, 167, 1991.
35. Giroux, B., and Lamontagne, M., Comparisons between surface electrodes and intramuscular wire electrodes in isometric and dynamic conditions, *Electromyogr. Clin. Neurophysiol.*, 30, 397, 1990.
36. Nandedkar, S. D. and Sanders, D. B., Recording characteristics of monopolar EMG electrodes, *Muscle Nerve*, 14, 108, 1991.
37. Nandedkar, S. D., Tedman, B., and Sanders, D. B., Recording and physical characteristics of disposable concentric needle EMG electrodes, *Muscle Nerve*, 13, 909, 1990.
38. Wyllie, E., Lüders, H., Morris, III, H. H., Lesser, R. P., Dinner, D. S., Rothner, A. D., Erenberg, G., Cruse, R., and Friedman, D., Subdural electrodes in the evaluation for epilepsy surgery in children and adults, *Neuropediatrics*, 19, 80, 1988.
39. Kramer, G. S., Influence of electrode noise on bioelectric signal recording, *J. Clin. Eng.*, 8, 243, 1983.
40. Maxwell, R. E., Gates, J. R., Fiol, M. E., Johnsson, M. J., Yap, J.C., Leppik, I. E., and Gumnit, R. J., Clinical evaluation of a depth electroencephalography electrode, *Neurosurgery*, 12, 561, 1983.
41. Agin, D. and Holtzman, D., Glass microelectrodes: origin and elimination of tip potentials, *Nature*, 211, 1194, 1966.
42. Agin, D., Electrochemical properties of glass microelectrodes, in *Glass Microelectrodes*, Lavallée, M., Schanne, O. F., and Hebert, N. C., Eds., John Wiley & Sons, New York, 1969, 62.
43. Roth, I., A self-retaining skin contact electrode for chest leads in electrocardiography, *Am. Heart J.*, 9, 526, 1933–1934.
44. Ungerleider, H. E., A new precordial electrode, *Am. Heart J.*, 18, 94, 1939.
45. Hoffman, B. L., Cranefield, P. F., Lepeschkin, E., Surawicz, B., and Herrlich, H. C., Comparison of cardiac monophasic action potentials recorded by intracellular and suction electrodes, *Am. J. Physiol.*, 196, 1297, 1959.
46. Andrews, H. L., A new electrode for recording bioelectric potentials, *Am. Heart J.*, 17, 599, 1939.
47. Abraham, N. G., A suction electrode net for precordial ECG mapping, *J. Med. Eng. Technol.*, 7, 285, 1983.
48. Korsgren, M., Leskinen, E., Sjöstrand, U., and Varnauskas, E., Intracardiac recording of monophasic action potentials in the human heart, *Scand. J. Clin. Lab. Invest.*, 18, 561, 1966.
49. Sjöstrand, U., Intrakardiellt registrerade monofasiska aktionspotentialer (paper given at Fysiologföreningen, Stockholm, Nov. 22, 1966).
50. Shabetai, R., Surawicz, B., and Hammill, W., Monophasic action potentials in man, *Circ. Res.*, 38, 341, 1968.
51. Samuelsson, R. and Sjöstrand, U., Endocardial recording of monophasic action potentials in the intact dog, *Acta Soc. Med. Upsal.*, 76, 191, 1971.
52. Olsson, S. B., *Monophasic Action Potentials of Right Heart*, Elanders Boktryckeri AB, Gothenburg, 1971.
53. Downey, J. M., Belandres, P. V., and Di Benedetto, M., Suction cup ground and reference electrodes in elelctrodiagnosis, *Arch. Phys. Med. Rehab.*, 70, 64, 1989.
54. Pope, II, C. E., Ask, P., and Tibbling, L., Evaluation of intraluminal EMG electrodes for the oesophagus and gastrointestinal tract, *Med. Biol. Eng. Comput.*, 22, 461, 1984.
55. Cohen, D., Magnetoencephalography: evidence of magnetic fields produced by alpha-rhythm currents, *Science*, 161, 784, 1968.
56. Weinberg, H., Brickett, P., Coolsma, F., and Baff, M., Magnetic localisation of intracranial dipoles: simulation with a physical model, *Electroencephalogr. Clin. Neurophysiol.*, 64, 159, 1986.
57. Weinberg, H., Stroink, G., and Katila, T., Biomagnetism, in *Encyclopedia of Medical Devices and Instrumentation*, Vol. I, Webster, J. G., Ed., John Wiley & Sons, New York, 1988, 303.
58. Cohen, D. and Chandler, L., Measurements and simplified interpretation of magnetocardiograms from humans, *Circulation*, 39, 395, 1969.
59. Nakaya, Y. and Mori, H., Magnetocardiography, *Clin. Phys. Physiol. Meas.*, 13, 191, 1992.
60. Malmivuo, J. and Plonsey, R., *Bioelectromagnetics*, Oxford University Press, New York, 1995.
61. Cohen, D., Ferromagnetic contamination in the lungs and other organs of the human body, *Science*, 180, 745, 1973.

62. Kalliomäki, P.-L., Aittoniemi, K., and Kalliomäki, K., Occupational health applications and magnetopneugraphy, in *Biomagnetism: An Interdisciplinary Approach*, Williamson, S. J., Romani, G.-L., Kaufman, L., and Modena, I., Eds., Plenum Press, New York, 1983, 561.

63. Stroink, G., Magnetic measurements to determine dust loads and clearance rates in industrial workers and miners, *Med. Biol. Eng. Comput.*, 23(Suppl. 1), 45, 1985.

64. Baule, G. M. and McFee, R., Detection of the magnetic field of the heart, *Am. Heart J.*, 66, 95, 1963.

65. Wikswo, Jr., J. P., Barach, J. P., and Freeman, J. A., Magnetic field of a nerve impulse: first measurement, *Science*, 208, 53, 1980.

66. Wikswo, J. P., Magnetic measurements on single nerve axons and nerve bundles, *Med. Biol. Eng. Comput.*, 23(Suppl. Part I), 3–6, 1985.

67. Cohen, D., Edelsack, E. A., and Zimmerman, J. E., Magnetocardiograms taken inside a shielded room with a superconducting point-contact magnetometer, *Appl. Phys. Lett.*, 16, 278, 1970.

68. Giffard, R. P., Webb, R. A., and Wheatley, J. C., Principles and methods of low-frequency electric and magnetic measurements using an RF-biased point-contact superconducting device, *J. Low Temp. Phys.*, 6, 533, 1972.

69. van Duzer, T. and Turner, C. W., *Principles of Superconducting Devices and Circuits*, Elsevier, New York, 1981.

70. Clarke, J., Ultrasensitive measuring devices, *Physica*, 126 B, 441, 1984.

71. Josephson, B. D., Possible new effect in superconductive tunneling, *Phys. Lett.*, 1, 251, 1962.

72. Nousiainen, J., Oja, O. S., Tuominen, T., and Malmivuo, J., Normal magnetocardiogram, *Med. Biol. Eng. Comput.*, 23(Suppl. Part 1), 69, 1985.

Chapter 7

Chemical Measurement

7.1 OBJECTS OF MEASUREMENTS

7.1.1 UNITS OF CHEMICAL QUANTITIES

Usually, the amount of substance is expressed by mass. The unit of mass is kilograms (kg) in the SI unit system. The amount of substance can also be expressed by mole (mol). The number of elementary entities in 1 mol is the Avogadro constant, which is usually denoted by N_A and is 6.0221367×10^{23}. While the term "mole" comes from "molecule", this unit can also be applied to atoms, ions, or any elementary entities of substance, if adequately defined. In the SI unit system, a mole is defined as the number of atoms in 0.012 kg of ^{12}C.

The amount of a specific substance in a solution can be expressed by its mass, volume, or mole. The concentration is the mass of the substance in unit volume of the solution, and the unit of concentration in the SI system is kg/m^3. In physiological texts, related units such as g/l, mg/dl, or mg/l have also been used.

The molar concentration is the amount of substance expressed in moles in a unit volume of the solution, and the unit in the SI system is mol/m^3. In physiological texts, mol/l, mmol/l, μmol/l, and nmol/l have also been used. The concentration and molar concentration are usually denoted by c, but molar concentration of a specific substance is often denoted by a chemical formula with brackets. For example, $[H^+]$ and $[HCO_3^-]$ are the molar concentrations of H^+ and HCO_3^-, respectively.

The ratio of quantities of substances in a mixture is usually expressed either by a mass fraction or molar fraction. Both quantities are nondimensional values. Molar concentration, mass fraction, and molar fraction are also called concentrations. Fractions can be expressed in percent. To express very small fractions, ppm (parts per million) and ppb (parts per billion) are also used.

In ordinary chemical reactions, the reaction rate depends on the concentrations of the substances involved in the reaction. However, the tendency to facilite a reaction is not strictly proportional to the concentration of the substance, but rather to a quantity called activity. The concept of activity is especially important in estimating the condition of equilibrium. The activity is a dimensionless quantity and is unity for a pure substance. When the concentration is low enough, the activity is equivalent to a molar fraction. Activity is denoted by a, or a with a subscript which denotes the substance.

In an electrolyte solution, the dissociation and association occurs as

$$AB \rightleftarrows A^+ + B^- \tag{7.1}$$

and the rates of both reactions are balanced in equilibrium. The degree of dissociation can be described as

$$\frac{\left[A^+\right]\left[B^-\right]}{[AB]} = k \tag{7.2}$$

or

$$\frac{a_{A^+} \cdot a_{B^-}}{a_{AB}} = K \tag{7.3}$$

where a_{A^+}, a_{B^-}, and a_{AB} are activites of A^+, B^-, and AB, respectively. Both k and K are called equilibrium constants. When a molar concentration is used, as in Equation (7.2), the equilibrium constant k depends on the concentration of the solution; however, when activities are used as in Equation (7.3), the equilibrium constant, K, is independent of the concentration. K is called the thermodynamic dissociation constant. The unit of k is molar concentration, but K is dimensionless. In weak electrolytes, the dependency of k on the concentration is negligible.

To describe the concentration of a hydrogen ion, pH has been used and is defined as

$$pH = -\log_{10}\left[H^+\right] \tag{7.4}$$

where $[H^+]$ is expressed by mol/l. In a neutral solution, pH = 7.0.

Partial pressure of dissolved gas is denoted by "p" followed by the chemical formula of the gas. For example, partial pressures of oxygen and carbon dioxide are denoted as pO_2 and pCO_2, respectively. The unit of partial pressure is pascal (Pa) in the SI system, but mmHg is still commonly used.

The total amount of gas dissolved in a solution is expressed by the ratio of the equivalent volume of the gas at standard condition (0°C, 1 atm) to the volume of the solution. It is a dimensionless quantity and commonly expressed by volume percent (vol%). The amount of dissolved gas is also expressed by a molar concentration such as mmol/l.

7.1.2 OBJECTS OF CHEMICAL MEASUREMENT

Chemical quantities in the body are estimated by analyzing a sampled material such as the blood or urine. However, the concentrations of some chemical quantities may vary rapidly so that continuous monitoring *in vivo* is required.

Blood is the most common object for chemical measurement, because many important chemical substances are transported by means of blood circulation. Actually, concentrations of many substances in the blood reflect those in the whole body.

Conventionally, blood samples are taken from a vein and analyzed by chemical analyzers. In order to monitor chemical quantities in the blood continuously, either continuous drainage of the blood or placing a chemical transducer in a blood vessel is required. However, if the object substance is highly permeable to the capillary wall, the concentration in the interstitial fluid is considered to be equated to that in the blood. Thus, the concentration of the substance in the blood can be monitored by a transducer placed in the interstitial fluid space. Urine is also a common object in clinical laboratory analysis. Although the available information from the urine is smaller compared to that from the blood, urine analysis is advantageous because of the ease of obtaining samples. During continuous drainage of urine, continuous monitoring of chemical quantities is conceivable. However, the ranges of concentrations of substances in the urine can vary widely when the urine output varies. Actually, urine pH normally ranges from 4.6 to 8.0, and urine osmotic pressure, which corresponds to the molar concentration of the solute, ranges from 200 to 850 mmol/l. To monitor the rate of excretion of a substance into urine, both the amount of urine and the concentration of the substance should be determined.

On the other hand, glucose, protein, or hemoglobin in the urine is not detected in normal urine, thus the appearance of such a substance in the urine can be a sensitive sign of abnormality. Components of expiratory gas are also important clinically and are sometimes monitored during anesthesia and in intensive care. Oxygen and carbon dioxide are also measured to determine the metabolic rate.

7.1.3 REQUIREMENTS AND LIMITATIONS IN CHEMICAL MEASUREMENT

Chemical transducers are used in various clinical situations such as during anesthesia, in an intensive care unit, in an emergency situation, and in ambulatory monitoring of postoperative or chronic patients. It is also necessary to use a closed-loop artificial organ, in which chemical quantities in the body are servo-controlled to a desired level.

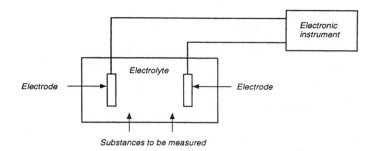

Figure 7. 1 Diagram of an electrochemical measurement system.

To monitor chemical quantities in subjects, there are different approaches to applying a transducer to the body, such as attaching a transducer to the body surface, inserting it into the tissue or blood vessel, implanting it in the body, attaching it to the extracorporeal circulation circuit, or installing it in the artificial organ. When totally implanted devices are desired for chronic use, long-term stability must be a requirement. The use of a catheter through which a transducer can be inserted and removed for recalibration or replaced by a new one can be a possible solution when ultimate stability cannot be attained. However, the risk of infection will still remain if there is a gap between the outer surface of the catheter and the skin tissue. Although several attempts with percutaneous connectors which form a tight contact between the artificial material and the skin have been made, most artificial materials were unsatisfactory for this purpose.[1] Only some materials, such as hydroxyapatite, could be used to form a fairly good contact with the skin tissue. It is expected that by using such a material, reliable transcutaneous sockets or connectors for long-term use can be effected.[2,3]

When a chemical transducer is used *in vivo*, some general requirements summarized below need to be considered:

- Initial calibration and recalibration, if neccessary, should be performed in sterile conditions.
- A chemical reaction occurring at the transducer should not cause toxicity to the tissue.
- The temperature coefficient of the transducer should be compensated for so that it can be used at an *in situ* temperature.
- When the transducer is placed in the blood vessel, its surface should be anticoagulant, and it should be small enough not to obstruct blood flow.
- The transducer should be electrically insulated from the body fluid so that leakage current, which may cause an electrical hazard, never occurs even if a power-line voltage is accidentally applied between the transducer cable and the body.
- The size and shape of the transducer should be designed so that it does not cause any mechanical hazard to the surrounding tissue.

7.2 CHEMICAL TRANSDUCERS

7.2.1 ELECTROCHEMICAL TRANSDUCERS

The electrochemical transducer is a device that converts a chemical quantity into electric potential or current, making use of the electrochemical principle. Figure 7.1 shows the fundamental electrochemical measurement system. It consists of two electrodes, the electrolyte solution, and the electronic instrumentation for potential or current measurement. When the electrolyte is dissolved in water, it dissociates into ions. The ion moves when an electric field is developed in the solution and causes ionic current. However, the ionic current cannot be measured directly by the ordinary electronic instrument in which the current is sustained by the movement of free electrons in the electric conductor. Therefore, in order to measure ionic current with an electronic instrument, a pair of interfacing parts are required to convert an ionic current into an electronic one. In electrochemistry, the electrode is defined as the interface between the electrolyte and the electric conductor. The system comprised of electrodes and electrolytes can be an electrochemical transducer where the concentration of the substance to be measured is reflected as the potential between the electrodes or current through the electrodes. The electrochemical transducer, in which the object quantity is reflected to the potential difference, is called the potentiometric transducer. That in which the object quantity is reflected to the current drain is called the amperometric transducer.

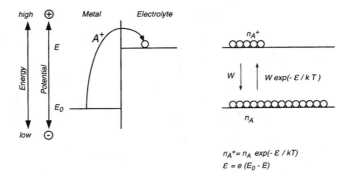

$$n_{A^+} = n_A \ exp(-\varepsilon/kT)$$
$$\varepsilon = e \ (E_0 - E)$$

Figure 7. 2 Representation of the concept of energy levels involved in an electrode reaction (left), and transition probabilities between these two states (right). At thermal equilibrium, populations of the entity at two energy levels obey Boltzman's law. e = an electron charge.

7.2.1.1 Electrode Potential and Reference Electrodes

At the surface of a metal placed in an electrolyte solution, a potential difference is developed between the electrode material and the electrolyte solution, and that potential difference is called the electrode potential. However, the electrode potential of each electrode cannot be measured individually, because a potential measurement requires at least two electrodes. Only the sum of both electrode potentials can be measured. If the electrode potential of one electrode is stable enough, the change in electrode potential of another electrode can be measured as the change in potential difference between two electrodes. As described later in detail, the electrode potential is usually defined as the relative potential difference from a hydrogen electrode whose electrode potential is assumed to be zero.

When no current flows through an electrode, the electrode potential will reach an equilibrium. The electrode potential in equilibrium depends on the concentration (activity in rigorous treatment) of the substance involved in the electrode reaction and the temperature. Generally, the electrode reaction that contributes to the electrode potential involves charge transfer across the boundary between the electrode material and the solution. Oxidation or reduction of a metal is a typical electrode reaction. Consider a reaction

$$A \rightleftarrows A^+ + e^- \tag{7.5}$$

where e^- is an electron. In this reaction, a substance, A, is oxidized to A^+ when the reaction occurs from left to right, and A^+ is reduced to A when the reaction occurs in the opposite direction. In a metal electrode, A corresponds to the neutral metal atom bounded on the solid electrode surface, and A^+ corresponds to its ion suspended in the solution.

If a potential difference between the metal and the electrolyte exists, as shown in Figure 7.2, the above reaction corresponds to the movement of a charge between two states at different potential levels, and thus it accompanies a gain or loss of electrical potential energy. In an oxidation process, A splits into A^+ and e^-, and then A^+ is dissolved in the solution, whereas e^- remains in the metal electrode, resulting in movement of the positive charge from the metal to the solution. If the potential of the solution is higher than that of the metal, as shown in Figure 7.2, A^+ gains energy because, in this process, the positive charge is moved from a lower potential level to a higher one. If the potential levels of the metal and the solution are E_0 and E, the energy difference, ε, between the two states is given by

$$\varepsilon = e\left(E_0 - E\right) \tag{7.6}$$

where e is the electron charge.

In thermal equilibrium, populations of atoms in both states, expressed as n_A and n_{A^+}, should obey Boltzman's distribution so that

$$n_{A^+} = n_A \cdot \exp\left(-\varepsilon/kT\right) \tag{7.7}$$

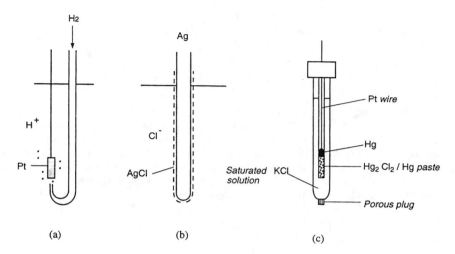

Figure 7.3 Reference electrodes: **(a)** hydrogen electrode, **(b)** Ag-AgCl electrode, and **(c)** calomel electrode.

where T is absolute temperature, and k is Boltzman's constant (1.38×10^{-23} J/K). On the right side of Figure 7.2, transitions between two energy levels are shown. At thermal equilibrium, the transition rates of both directions are equal so that Equation (7.7) is valid.

Substituting Equation (7.6) with Equation (7.7) and replacing populations n_A and n_{A^+} by activities of both states expressed as a_A and a_{A^+} one obtains

$$a_{A^+} = a_A \cdot \exp\left(-e(E - E_0)/kT\right). \tag{7.8}$$

Thus, E is expressed as

$$E = E_0 + (kT/e)\ln\left(a_{A^+}/a_A\right). \tag{7.9}$$

If e and k are multiplied by N_A, k/e becomes R/F, where R is the gas constant (8.31 J/mol/K) and F is Faraday's constant (9.65×10^4 C/mol). Then, Equation (7.9) can be written as

$$E = E_0 + (RT/F)\ln\left(a_{A^+}/a_A\right). \tag{7.10}$$

If a_A is unity, and a_{A^+} is ten times a_A at 300 K, the potential difference, $E - E_0$, is estimated to be about 0.059 V = 59 mV.

If z electrons are involved in the electrode reaction, instead of one electron being involved in the reaction in Equation (7.4), F in Equation (7.9) should be replaced by zF so that

$$E = E_0 + (RT/zF)\ln\left(a_{A^+}/a_A\right). \tag{7.11}$$

This expression is called the Nernst equation, which is the fundamental equation in electrochemistry. More detailed and rigorous derivation of the Nernst equation can be found in textbooks on electrochemistry or physical chemistry (for example, Bockris and Reddy,[4] Bard and Faulkner,[5] and Brett and Brett[6]).

There is a kind of electrode in which the electrode potential is sufficiently stable and is not affected by the change in the concentration of substance which is involved in the electrode reaction. Such an electrode is useful as a reference electrode by which potential change is measured. The hydrogen electrode, the silver-silverchloride electrode, and the calomel electrode are typical electrodes of this kind. Figure 7.3 shows the configurations of these electrodes.

The hydrogen electrode consists of a platinized platinum electrode immersed in an acid solution. Hydrogen gas bubbles continuously onto the electrode surface as shown in Figure 7.3(a). The electrode reaction can be expressed as

$$H_2 \rightleftarrows 2H^+ + 2e^-. \tag{7.12}$$

Due to a continuous supply of hydrogen gas, dissolved hydrogen gas is saturated and its activity is maintained at unity. If the concentration of the hydrogen ion H^+ is unity, its activity also becomes unity, thus the electrode potential becomes E_0 according to Equation (7.11). Such a condition is realizable in a strong acid in which pH = 0. Because the second term of Equation (7.11) vanishes in this condition, the temperature dependency of the electrode potential due to this term can be eliminated. The electrode potential of the standard hydrogen electrode is assumed to be zero at all temperatures, and the hydrogen electrode is used as a standard in determining the electrode potentials of other electrodes.

The silver/silverchloride electrode consists of pure silver with a porous layer of silver chloride on its surface, as shown in Figure 7.3(b). The electrode reaction is represented by

$$Ag \rightleftarrows Ag^+ + e^-. \tag{7.13}$$

The metallic silver is a pure substance so that the activity is unity. Thus, the electrode potential is determined only by the concentration of Ag^+ in the solution. However, if solid AgCl and a high concentration of Cl^- exist near the electrode surface, the concentration of Ag^+ is maintained at an almost constant level because the solid AgCl partly dissociates into ions as

$$AgCl \rightleftarrows Ag^+ + Cl^-. \tag{7.14}$$

In thermal equilibrium, the product of concentration of both ions is maintained at a constant level, so that

$$\left[Ag^+\right]\left[Cl^-\right] = 1.7 \times 10^{-10} \quad (mol/l)^2 \tag{7.15}$$

at 25°C. When an electrode reaction by Equation (7.13) occurs, the change in concentration of Ag^+ is compensated for by the reaction of Equation (7.14), so that the dissociation equilibrium expressed as Equation (7.15) is maintained. As long as $[Cl^-]$ is large enough, $[Ag^+]$ is maintained as constant because the fraction of change in $[Cl^-]$ due to the reaction of Equation (7.13) is negligibly small if $[Cl^-] \gg [Ag^+]$; consequently, the electrode potential is maintained at a constant level. However, in the silver-silverchloride electrode, the second term in Equation (7.13) is not zero, thus the electrode potential depends on the temperature.

Silver-silverchloride electrodes can be easily prepared by electrolysis. An example of a preparation procedure is as follows: Clean a silver plate or wire by immersing in 3 mol/l HNO_3, wash in water, and then place in 0.1 mol/l of HCl solution. Apply d.c. current with a density of about 0.4 mA/cm^2 for about 30 min toward the electrode from a silver metal anode. A silver chloride layer is then formed on the electrode surface.

The electrode potential of a silver-silverchloride electrode depends on the concentration of the chloride ion. The electrode potentials in saturated, 3.5 mol/l, or 1.0 mol/l KCl are 0.199, 0.205, and 0.235 V, respectively, at 25°C. The temperature coefficient of electrode potentials is about –0.14 mV/°C in saturated KCl, and +0.250 mV/°C in 1.0 mol/l KCl.

The calomel electrode consists of mercurous chloride (Hg_2Cl_2) and pure mercury. Figure 7.3(c) shows an example of its configuration. Saturated, 3.5, 1.0, or 0.1 mol/l KCl is commonly used for the inner solution of the electrode. The electrode reaction is expressed as

$$2Hg \rightleftarrows Hg_2^{2+} + 2e^-. \tag{7.16}$$

In equilibrium, the concentration of Hg_2^{2+} is kept constant by the same mechanism as that in the silver-silverchloride electrode, hence the electrode potential is maintained at a constant level. The electrode

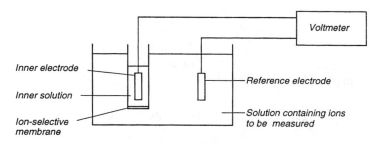

Figure 7.4   Diagram of an ion-selective electrode and the measurement system.

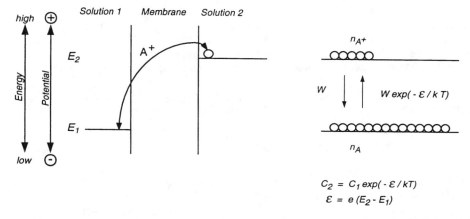

Figure 7.5   Representation of the energy levels at an ion-selective membrane (left) and transition probabilities between these two states (right). A^+ = a cation.

potential of a calomel electrode is about +0.241 V at 25°C when a saturated KCl is used as its inner solution, and the temperature coefficient is about +0.22 mV/°C.

7.2.1.2   Potentiometric Sensors

The ion-selective electrode is a typical potentiometric transducer. It consists of two electrodes and an ion-selective membrane placed in between them, as shown in Figure 7.4. The change in ionic concentration is reflected upon the potential difference across the membrane. The change in potential difference is detected by two electrodes placed in the solutions at both sides of the membrane.

An ideal ion-selective electrode consists of a membrane through which only a specific ion is permeable. While actual membranes are not ideal, a membrane that is highly permeable to a specific ion and less permeable to other ions can be used as the ion-selective membrane when concentrations of other ions are relatively low.

The principle of the ion-selective electrode can be explained as similar to the origin of the electrode potential described previously. Suppose a membrane is permeable to ion A^+ but is impermeable to other ions and is placed between solutions 1 and 2, as shown on the left side of Figure 7.5. If the potentials of solution 1 and 2 are E_1 and E_2, respectively, then ion A^+ gains energy $\varepsilon = e(E_2 - E_1)$ when it is moved from solution 1 to solution 2. In thermal equilibrium, concentrations of A^+ in solutions 1 and 2, denoted by c_1 and c_2, respectively, are related to the potential across the membrane as

$$c_2 = c_1 \exp\left(-e\left(E_2 - E_1\right)/kT\right). \tag{7.17}$$

Then the potential across the membrane is solved as

$$E_2 - E_1 = \left(kT/e\right)\ln\left(c_1/c_2\right) \tag{7.18}$$

or

$$E_2 - E_1 = (RT/F) \ln(c_1/c_2) \tag{7.18'}$$

In rigorous treatment, activities, a, should be used instead of concentrations:

$$E_2 - E_1 = (RT/F) \ln(a_1/a_2). \tag{7.19}$$

If the concentration or activity of A^+ in solution 2 is constant, the membrane potential is given as

$$E_2 - E_1 = \text{const.} + (RT/F) \ln a_1 \tag{7.20}$$

where a_1 and a_2 are activities of A^+ in solutions 1 and 2. The change in concentration, or activity, of A^+ in solution 1 can be determined by the change in the potential across the membrane. Consequently, it is measured as the change in potential difference between two electrodes placed in these two solutions. If the activity of A^+ in solution 1 is increased 10 times, membrane potential increases by about 59 mV at 25°C.

Equation (7.20) is valid when the membrane is permeable only to a specific ion A^+. If the membrane is also permeable to other ions such as B^+, C^+,···, the resulting membrane potential is expressed as

$$E_2 - E_1 = \text{const.} + (RT/F) \ln\left(a_{A^+} + K_B a_{B^+} + K_C a_{C^+} + \cdots\right) \tag{7.21}$$

where K_B and K_C are selectivity coefficients of B^+ and C^+, and a_{A^+}, a_{B^+}, a_{C^+} denote activites of A^+, B^+, and C^+, respectively. If the selectivity coefficient of an ion is small enough, the influence of the change in activity of that ion upon the potential across the membrane is small. However, if the activity of that ion is high, the effect on the membrane potential will become significant even if the selectivity coefficient is small. For example, in pH measurement in a range of high pH where [H^+] is fairly small, the membrane potential may be strongly influenced by other ions.

In the above consideration, it is assumed that only cations are permeable through the membrane. Similar results are expected for a membrane in which only anions are permeable, while the sign of the developed potential is reversed. However, in a membrane in which both cations and anions are permeable, potentials generated by cations and anions are cancelled out, and thus such a membrane cannot be used as an ion-selective electrode. Current studies on ion-selective electrodes are to be found in reviews.[7,8]

The structure of typical ion-selective electrodes is shown in Figure 7.6. The potential change developed at the ion-selective membrane is measured as the change in potential difference between the working electrode and the reference one. The characteristics of ion-selective electrodes depend on the electrical property of the ion-selective membrane. The glass electrode has a large internal resistance, which is usually several hundreds or several thousands of MΩ; therefore, an amplifier having high input impedance has to be used. The input impedance of conventional amplifiers for glass electrodes is usually higher than 10^{13} Ω, whereas the resistance of the liquid ion-exchange membrane is below 30 MΩ, and that of a pressed-pellet membrane is below 100 kΩ.

The response time of the ion-selective electrode depends on the structure, size, and geometry of the electrode and the electrical property of the membrane material. A 95% response of typical ion electrodes ranges from several seconds to several minutes.

The pH is commonly measured by the glass electrode. The configuration of a typical pH electrode is shown in Figure 7.6(a). It has a Ag-AgCl electrode and is filled with a buffer solution. The potential developed at the pH-sensitive glass membrane is measured by comparison with a reference electrode.

When the pH has to be measured in a small sample, the capillary-type pH electrode has been employed, e.g., in clinical blood gas analyzers.[9] It consists of a fine capillary made of pH-sensitive glass surrounded by the reference solution, as shown in Figure 7.7. For the measurement, the sample is introduced into the capillary. The potential that builds up across the glass membrane is measured between an electrode in the reference buffer solution outside the capillary and another reference electrode connected to the sample via a salt bridge.

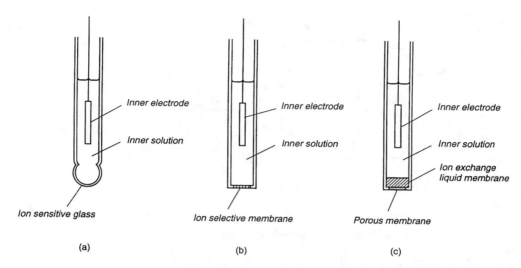

Figure 7.6 Ion-selective electrodes: **(a)** glass electrode, **(b)** pressed-pellet membrane electrode, and **(c)** liquid ion-exchange membrane electrode.

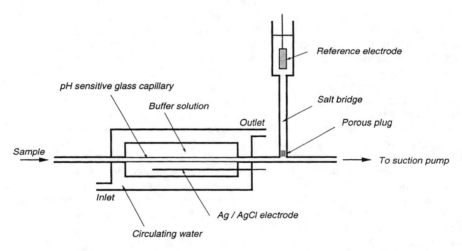

Figure 7.7 Diagram of pH measurement using a capillary-type glass electrode.

Some kinds of glass have ion selectivity for different cations such as sodium and potassium, and ion-selective electrodes can be formed by using materials such as the ion-selective membrane. For example, the sodium glass NSA 11-18 is used in sodium electrodes. While this material is influenced by the hydrogen ion to some extent, the effect is serious only in lower pH ranges where the hydrogen concentration is high.

Many other solid materials have been used as ion-selective membranes, such as solid inorganic materials, single crystals of fluolides of rare-earth elements, polycrystalline silver sulfide, and fused mixtures of silver sulfide and silver chloride or bromide.

A liquid membrane electrode is formed by an absorbent material so that the ion-exchange material dissolved in a lipophilic solvent is absorbed in it. While liquid membranes of this type are not stable enough, their lifetime can be extended by introducing polymer-matrix membranes such as polyvinyl chloride (PVC). A typical ion-selective electrode of this type is the Ca^{2+} electrode.

Other types of liquid membranes are used in potassium and ammonium electrodes. For example, the valinomycin membrane is used in the potassium electrode, and the nonactin, narasin, or monensin membranes are used in the ammonium electrode. Various applications of ion-selective electrodes are well reviewed by Koryta.[10]

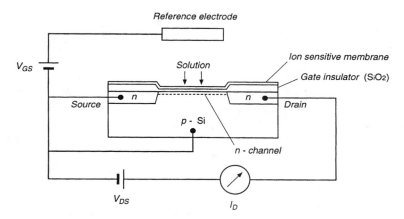

Figure 7.8 Ion-selective, field-effect transistor (ISFET).

Ion-selective electrodes, especially pH electrodes, are used in gas sensors and biosensors in which the specific substance induces a chemical reaction in a layer attached to the electrode surface so that particular substances are detected by the ions produced, (as described in Sections 7.2.1.4 and 7.2.4).

An ion-selective electrode can be formed by coating the metal electrode surface directly with the ion permeable material. Such an electrode has no inner solution and is called the coated-wire electrode (CWE).[11] Due to its simple configuration, the coated-wire electrode is suitable for miniaturization.

The ion-selective field-effect transistor (ISFET) consists of a field-effect transistor (FET) and an ion-selective membrane that covers the silicon dioxide layer on the conductive channel.[12-16] As shown in Figure 7.8, the ISFET is usually fabricated on a p-type silicon substrate similar to the n-channel MOSFET (metal-oxide semiconductor field-effect transistor). However, the insulating layer on the n-channel is covered by the ion-selective membrane and is exposed to the solution, instead of the gate, with an electric lead connection in the ordinary FET. By placing a reference electrode in the solution, a bias voltage, V_{GS}, is applied onto the insulating layer on the conductive channel similar to applying gate bias voltage to an FET. The potential developed at the membrane by the existing ion in the solution is then measured by the change in current through the channel.

ISFET has several advantages when it is compared to ordinary ion-selective electrodes: ISFET provides low-impedance output so that the high-impedance amplifier required in glass electrodes is unnecessary. It can be fabricated by conventional IC technology and is simple to miniaturize. Because the ion-selective membrane can be thinner than that in ordinary ion-selective electrodes, the response can be faster. Besides that, it can be produced by ordinary IC technology which suits mass production; consequently, it is expected to be supplied for single use.

An example of the construction of an ISFET is shown in Figure 7.9.[16] It was fabricated by selective etching of a silicon wafer strip. Surface layers were formed by CVD (chemical vapor deposition). Because the whole surface is covered by the insulating layer, it can be immersed in a solution. In pH ISFETs, SiO_2, Si_3N_4, Al_2O_3, SnO_2, and Ta_2O_5 are used as ion-sensitive membranes. NAS glass in Na+ ISFET, polyvinyl chloride containing valinomycin or crown ether in K+ ISFET, and dodecyl phosphate plus dioctylphenyl phosphates in Ca2+ ISFET are used as ion-selective membranes. If a hydrophobic material such as polystyrene is used as the membrane, instead of an ion-selective one, the device has no ion sensitivity and can be a reference electrode.[17]

Using IC fabrication technology, different ISFETs can be fabricated on the same wafer. For example, a combined pH and Na+ ISFET and combined pH and reference ISFET have been fabricated.[18]

While the sensitivity of ISFET is comparable to that of the ordinary ion-selective electrodes, the response time is shorter. According to a study by Matsuo and Esashi,[18] the sensitivity of the pH ISFET with the Ta_2O_5 membrane was 56 to 57 mV/pH in a range from 1 to 13 pH, which is close to the value of 59 mV/pH predicted by the Nernst equation. The 95% response time was less than 0.1 s with a drift of less than 0.2 mV/h, while the response time of a typical glass pH electrode is more than a few seconds.

7.2.1.3 Amperometric Sensors

Polarography uses a typical amperometric transducer in which the rate of a chemical reaction is detected by the current drained through an electrode. If the chemical reaction of a substrate involves charge

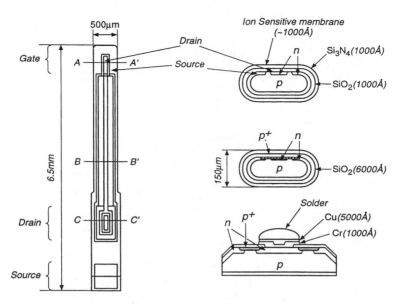

Figure 7.9 An example of the construction of ISFET. (From Matsuo, T. and Esashi, M., *Sensors Actuators*, 1, 77, 1981. With permission.)

transfer between the electrode surface and the substrate, the reaction will cause some electric current which is detectable by an externally connected electronic instrument. The rate of such a reaction is governed by the concentration (rigorous activity) of the substrate and the potential between the electrode and the solution. Therefore, by applying constant voltage at an adequate level, the concentration of the substrate can be measured by the current drain. In biomedical applications, dissolved oxygen and hydrogen peroxide are commonly measured by polarographic electrodes, and these electrodes are also used in various kinds of enzyme and microbial electrodes.

The configuration of the polarographic oxygen electrode is shown in Figure 7.10. It consists of a platinum cathode and a silver anode. When a constant voltage of about −0.6 V is applied to the platinum side, a reaction

$$O_2 + 2H_2O + 4e^- \rightarrow 4OH^- \tag{7.22}$$

occurs as long as oxygen is present at the electrode surface. If the amount of oxygen near the electrode surface is limited, this reaction ceases when all the oxygen is consumed. But if oxygen is supplied continuously to the electrode surface, a continuous current remains. By covering the electrode with an oxygen permeable membrane so that the oxygen tension outside the membrane determines the oxygen supply to the electrode surface, the oxygen tension outside the membrane reflects the current at the equilibrium.

If oxygen is supplied by diffusion through a membrane of thickness d and area A, the current I is approximated by

$$I = 4FAD\alpha\, p/d \tag{7.23}$$

where F is Faraday's constant, D and α are the diffusion coefficient and the solubility of oxygen in the membrane, respectively, and p is the partial pressure of the oxygen.[19] In this expression, the factor 4 appears due to the fact that four electrons are involved in a reaction of one oxygen molecule. As long as this relation is valid, the current is directly proportional to the partial pressure of oxygen, p, but it also depends on the geometry of the electrode and the solubility and diffusion constant of the oxygen in the membrane. To maintain a constant sensitivity, these parameters should be kept at constant levels.

When the solution surrounding the electrode does not move, the concentration gradient of oxygen is built up in the medium outside the membrane due to the oxygen flux onto the electrode surface;

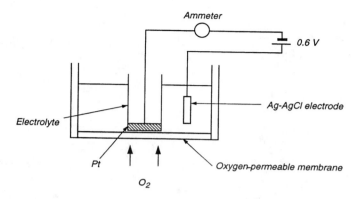

Figure 7.10   Diagram of the polarographic oxygen electrode.

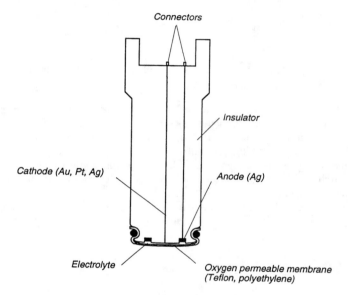

Figure 7.11   Clark oxygen electrode.

consequently, the sensitivity is reduced. But if the medium is stirred, the concentration gradient disappears and the sensitivity apparently increases. This effect causes unwanted flow dependency. The use of a membrane having a lower diffusion constant than that of the medium reduces flow dependency. Also, the use of an electrode size smaller than the thickness of the membrane is effective in reducing flow dependency, because the diffusion field is spread widely at the membrane surface.[20] When the outside of the membrane is exposed to a gas instead of a liquid, the concentration gradient is not a problem.

A conventional polarographic oxygen electrode is shown in Figure 7.11. Because this type of oxygen electrode was first introduced by Clark,[21] it is called the Clark oxygen electrode. It consists of a fine platinum wire, and the anode is silver or silver-silverchloride. The membrane is generally made of polypropylene or polyethylene and is usually about 20 μm thick. In order to keep an electrolytic connection between electrodes, a thin layer of electrolyte between the membrane and the electrode surface, typically about 5 to 10 μm, is maintained.

The polarographic principle has also been used for detecting hydrogen peroxide. The electrode reaction involved at the working electrode is expressed as

$$H_2O_2 \rightarrow 2H^+ + O_2 + 2e^-. \tag{7.24}$$

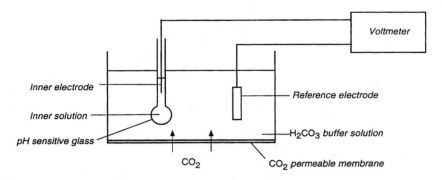

Figure 7.12 Diagram of the carbon dioxide electrode.

This reaction is anodic, and the working electrode is usually maintained at about +0.6V. The construction of the electrode is the same as that of the oxygen electrode except for the polarity of the applied potential. Because hydrogen peroxide is produced by some enzymatic reactions, such as the reaction of glucose oxidase, this principle is used in enzyme electrodes, as described later.

7.2.1.4 Electrochemical Gas Sensors

Several kinds of gas sensors based on electrochemical principles have been developed and are used in many different fields, such as biotechnology, agriculture, environmental monitoring, or the food and drug industries. In biomedical applications, the carbon dioxide electrode based on the potentiometric principle and the oxygen electrode based on the amperometric principle are commonly used in blood gas measurement. The solid electrolyte oxygen sensor is used in respiratory gas measurement. Electrochemical gas sensors are also used in enzyme and microbial electrodes to detect the product of enzymatic reactions and microbial metabolism.

The carbon dioxide electrode consists of a glass pH electrode covered with a gas permeable membrane such as Teflon and a sodium bicarbonate solution filler between the membrane and the glass membrane.[22] Figure 7.12 shows the construction of a carbon dioxide electrode. This type of electrode is called the Severinghaus electrode. The dissolved carbon dioxide penetrates into the inner solution through the membrane. A part of the dissolved carbon dioxide combines with water and forms carbonic acid (H_2CO_3), and a part of the carbonic acid dissociates into H^+ and HCO_3^-. Then the increase of H^+ can be detected as a decrease in pH by using a glass pH electrode or some kind of pH electrode such as the pH ISFET. Details of the chemical reaction and the characteristics of the Severinghaus electrodes are found in References 23 and 24. Carbon dioxide electrodes for *in vivo* and transcutaneous measurements are described in Sections 7.3.1 and 7.3.3.

Many kinds of gas sensors can be formed by covering pH electrodes with appropriate gas permeable membranes. Such gas-sensing probes as CO_2, NO_2, H_2S, SO_2, HF, HCN, and NH_3 are available from commercial sources. Figure 7.13 shows a typical configuration of such an electrode. Solid-state devices can also be used as the electrode. For example, a hydrogen sensor can be made of a metal oxide semiconductor (MOS) covered by palladium, and an ammonia gas sensor can be covered by iridium.[25] Each sensor has a sensitivity of about 1 ppm in air.

To measure oxygen tension, the Clark oxygen sensor (described in the preceding section) is commonly used, both for dissolved oxygen in a solution and oxygen in gas phase. In blood gas analyzers, oxygen tension in sampled blood is measured by introducing the sample into a cuvette where a Clark oxygen sensor is installed, but the standard gases are also introduced into the cuvette when the sensor is calibrated. For oxygen measurement in gas phase, the galvanic-cell oxygen sensor and the solid electrolyte sensors are also used.

The galvanic-cell oxygen sensor is a device based on the same electrochemical reaction at the cathode as that in the Clark oxygen sensor, but it consists of an anode having a higher ionization tendency in order to form a battery and does not need the external voltage supply required for the cathode reaction. This type of oxygen sensor has been used widely in simple direct-reading oxygen meters for monitoring oxygen content in the air. The configuration of a typical galvanic-cell oxygen sensor is shown in Figure 7.14.[26] It consists of a cathode made of silver with a surface area of about 30 cm^2 and an anode

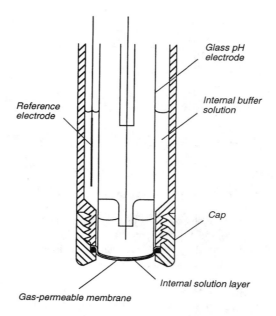

Figure 7.13 Typical configuration of the gas-sensing probe consisting of a glass pH electrode covered by a gas-permeable membrane.

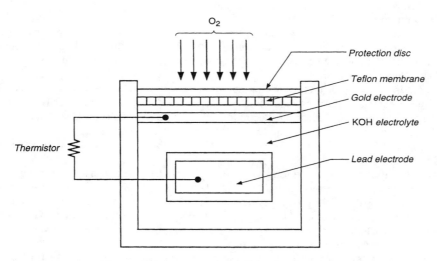

Figure 7.14 A galvanic-cell type oxygen sensor.

made of lead, which has a polyethylene membrane with a thickness of 80 μm. It produces an output current of 200 μA in air at 15°C.[27] Potassium hydride is commonly used as the electrolyte. In the galvanic cell, the anode material is consumed by oxidation, thus its lifetime is limited by the amount of anode material. A typical lifetime for a galvanic-cell, direct-reading oxygen meter is about 9000 h in atmospheric air (Rexnord Electronic Products Div., BioMarine Industries Inc.; Malvern, PA).

Nitrous oxide (N_2O) can be measured by a Clark-type electrode. In an aqueous electrolyte, N_2O is reduced to N_2 at a silver anode according to the reaction

$$N_2O + H_2O + 2e^- \rightarrow N_2 + 2OH^-. \tag{7.25}$$

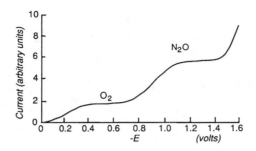

Figure 7.15 Voltammogram of a Clark electrode in the presence of O_2 and N_2O. (From Harn, C. E. W., *J. Phys. E. Sci. Instrum.*, 13, 470, 1980. With permission.)

Reduction of N_2O occurs at a larger negative potential level, as shown in Figure 7.15, and O_2 and N_2O can be clearly separated.[23]

A technique was proposed in which O_2 and CO_2 can be measured simultaneously.[24] The configuration of the electrode is the same as the conventional Clark electrode, but a nonaqueous solvent such as dimethyl sulphoxide is used as the inner electrolyte. When a negative pulse is applied to the working electrode, O_2 is reduced as

$$O_2 + e^- \rightarrow O_2^- \tag{7.26}$$

and a positive pulse is applied. If CO_2 is not present, an oxidizing reaction occurs. However, if CO_2 is present, O_2^- is consumed by the following reaction

$$\begin{aligned} O_2^- + CO_2 &\rightarrow CO_4^- \\ CO_4^- + CO_2 &\rightarrow C_2O_6^- \\ C_2O_6^- + O_2^- &\rightarrow C_2O_6^{2-} + O_2. \end{aligned} \tag{7.27}$$

Therefore, partial pressures of O_2 and CO_2 can be determined by negative and positive currents.

Some solids exhibit ionic conduction at high temperatures, and such a material is called a solid electrolyte. Solid electrochemical sensors can be fabricated by using solid electrolytes. A typical solid electrolyte is a ceramic of zirconium oxide (ZrO_2) called zirconia. The oxygen ion (O^{2-}) can diffuse in it efficiently at a high temperature, typically above 200°C, thus this material can be used as an ion-selective membrane in the ion-selective electrode. Some materials, such as lantanum trifluoride (LaF_3), can be solid electrolytes even at room temperatures.[28,29]

If a solid electrolyte forms a wall which separates two gases having different oxygen contents, oxygen transport occurs through the electrolyte. At the side where oxygen partial pressure is higher, a reaction

$$O_2 + 4e^- = 2O^{2-} \tag{7.28}$$

occurs, and the resulting oxygen ion diffuses to the opposite side. A reaction

$$2O^{2-} - 4e^- = O_2 \tag{7.29}$$

then occurs at another side where the oxygen partial pressure is lower. This process will reach an equilibrium when an electric field is built up so that the diffusion of oxygen ion is obstructed. As a result, a potential difference appears across the wall in a manner similar to the ion-selective electrode, and the potential difference is given as

$$E = \frac{RT}{4F} \ln \frac{\rho_1}{\rho_2}. \tag{7.30}$$

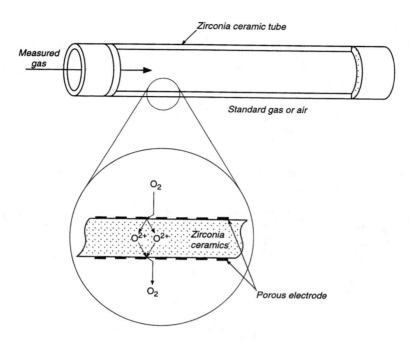

Figure 7.16   A flow-through, zirconia oxygen sensor.

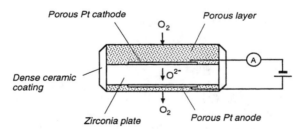

Figure 7.17   Limiting-current type of solid-electrolyte oxygen sensor. (Modified from Saji, K. et al., *Proc. 4th Sensor Symp.,* Institute of Electrical Engineers of Japan, Tokyo, 1984, 147.)

Because four electrons are involved in the transport of one oxygen molecule, the factor of 4 appears in the denominator, and the ratio of the activities in Equation (7.19) is replaced by the ratio of oxygen partial pressures at both sides. Near the atmospheric oxygen partial pressure, a 1% difference in oxygen concentration generates a potential difference of approximately 1 mV. As seen in this equation, the generated potential is proportional to the absolute temperature, and thus the sensor temperature should be kept constant. In actual instrumentations, the sensor temperature is stabilized within ±1°C.

Figure 7.16 shows an example of the configuration of the zirconia oxygen sensor. Microporous platinum electrodes are attached on the inside and outside wall of a zirconia cylinder, which operates at a temperature above 700°C. The sample gas flows through the cylinder, and the outside of the cylinder is exposed to the atmospheric air. The baseline drift is always small so that recalibration is practically unnecessary. The response is faster than that of conventional electrochemical sensors used in a solution.

Typical characteristics of commercially available oxygen sensors of this type are an operating temperature of 700°C, a measurement range of oxygen concentration from 0 to 100%, an error ±2% of full scale, and a response time of less than 3 s at a sample flow rate of 50 ml/min (LC-700 Toray; Tokyo Japan). Inflammable gases such as SO_2, HCl, and CO_2 affect the output, but halothane does not.

There is a different type of solid electrolyte oxygen sensor, called the limiting current type, which provides current output.[30] Figure 7.17 shows its configuration. It consists of a zirconia plate with electrodes at both sides and a porous material layer attached on one side. When an adequate potential is applied between two electrodes, so that the side having the porous layer is negative to the other side,

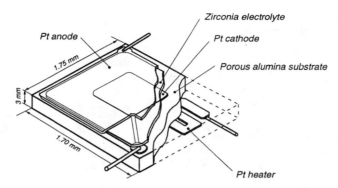

Figure 7.18 Configuration of a practical oxygen sensor of limiting-current type. (From Kondo, H. et al., *Proc. 6th Sensor Symp.*, Institute of Electrical Engineers of Japan, Tokyo, 1986, 251. With permission.)

the oxygen ion which has a negative charge is transported from the negative to the positive side. Finally, the oxygen partial pressure under the porous layer becomes practically zero, and the current drain ceases as long as no oxygen is supplied to the electrode under the layer. However, if oxygen diffuses through the porous layer, current flows in proportion to the diffused oxygen flux. Consequently, this sensor provides output current proportional to the oxygen partial pressure in the surrounding space. Figure 7.18 shows the configuration of a practical oxygen sensor of this type.[31] The sensing element, $1.7 \times 1.75 \times 0.3$ mm, consists of zirconia with electrodes formed by deposited thin platinum film, an alumina porous substrate, and a platinum heater. The sensor operates at 690°C, and the power consumption for maintaining the operating temperature is about 0.8 W.

7.2.2 OPTICALLY BASED CHEMICAL TRANSDUCERS
The simple observation that different substances have different colors indicates the possibility of applying optical techniques to identify chemical species. Actually, spectrophotometric techniques have been used widely in chemical analysis. By introducing fiberoptic techniques, optical measurement can be performed even *in vivo*. Many other techniques are also introduced to form optically based chemical sensors for biomedical use.

7.2.2.1 Spectrophotometric Chemical Analysis
Spectrophotometers, as well as optically based chemical sensors, utilize the interaction between electromagnetic radiation and molecules, ions, or atoms. According to quantum theory, every molecule, ion, or atom has a unique set of discrete energy levels. When a transition from lower to upper energy levels occurs, an energy equal to the difference between two energy levels has to be supplied. At a transition from upper to lower level, the same energy is released. The energy of light is also quantized, and the quantized energy, which is the energy of a photon, is given as $h\nu$, where h is Planck's constant (6.63×10^{-34} Js) and ν is frequency. Different optical phenomena such as absorption, emission, fluorescence, or the Raman effect are used in spectrophotometric analysis. Figure 7.19 shows transition schemes of such phenomena.

In absorption measurement, a light beam of selected wavelength passes a layer of solution, and the attenuation of the beam is measured. The power of a light beam is attenuated exponentially in a uniform media according to the relation known as Beer-Lambert's law, which is expressed as

$$A = \log \frac{I_o}{I} = \varepsilon l c \tag{7.31}$$

where A is the absorbance, ε is the molar absorptivity, l is the length of the optical path, c is the concentration, and I_o and I are the incident and transmitted light intensities, respectively. Beer-Lambert's law is valid only for media that are nonscattering. Deviation from Beer-Lambert's law is observed in a solution of higher concentration of the absorbing substance.

For *in vivo* measurements, absorption measurement usually cannot be performed by detecting transmitted light through the object. In such situations, measurements have to be performed by detecting

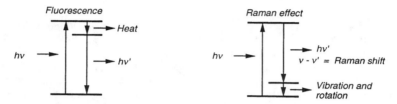

Figure 7.19 Transition schemes in optical phenomena employed in spectrophotometric chemical measurements.

reflected or scattered light to estimate the absorption of the specified materials. Such a situation is common in fiberoptic sensors, as described later.

There are two kinds of reflection: specular and diffuse. Specular reflection takes place in the interface between two optical media. This type of reflection obeys Fresnel's theory. Diffuse reflection will take place when light penetrates into the medium and is absorbed and scattered and partly returns to the surface.

The diffuse reflectance is dependent on the optical properties of the materials used. The diffuse reflectance properties of a material can be described by the Kubelka-Munk theory, which is based on the material properties, K, the absorption coefficient, and the scattering coefficient, S. The reflectance, R, is a function of the concentration of the absorbing medium, c, which, according to Kubelka-Munk, can be expressed as

$$F(R) = \frac{\varepsilon c}{S} = Kc \tag{7.32}$$

where ε is the molar absorptivity and $F(R)$ is a measure of the reflectance known as the Kubelka-Munk function. Reflectance properties can be standardized by using highly reflecting materials such as $BaSO_4$.

Scattering is a phenomenon in which the direction of light is altered by interacting with objects such as small particles. Rayleigh and Mie scattering are examples of elastic scattering for which the wavelength of the radiation is not changed by the scattering phenomenon. If the scattering particle is small in comparison with the incident wavelength, Rayleigh scattering, which is generated at atomic and molecular levels, dominates. Mie scattering occurs when the scatterer size is large compared to the wavelength used. The scattered light component is strongly forward directed and related to the concentration of the scattering particles.

The Raman effect is a phenomenon in which components having wavelengths slightly different from that of the incident light are observed when a light beam is scattered by matter. The shift in wavelength or frequency in the scattered light occurs due to the transition to the vibrationally or rotationally excited energy level, as shown in Figure 7.19, and is called the Raman shift. It contains a set of different wavelength or frequency components corresponding to many different energy levels of molecular vibration and rotation, thus the spectrum of the Raman shift provides information about the molecular structure. A laser source is commonly used in Raman spectroscopy, because it is monochromatic which means it is convenient for detecting small shifts in wavelength in the scattered light. The technique of Raman spectrometry, using the Fourier transform spectrometer, is becoming a powerful tool in analytical chemistry and has been used to analyze many biochemical species.[32]

In emission measurement, molecules, ions, or atoms are excited to higher energy levels by appropriate processes such as heating in a flame or an arc, bombardment of electrons or ions, or exposure to electromagnetic radiation. Some substances such as luminol can be excited chemically; the light emission caused by such chemical excitation is called chemiluminescence.[33]

Fluorescence is a kind of emission in which molecules, ions, or atoms are excited optically. In fluorescence measurements, the medium under study is excited at a short wavelength and the fluorescence occurs at a longer wavelength. While species that exhibit strong fluorescence are limited, fluorescence is detectable even it is weak by the use of a sensitive detector and a filter that rejects the excitation light beam. Actually, many biochemical species can be analyzed by fluorescence spectroscopy.[34] When the lifetime of the fluorescence is longer, the phenomenon is called phosphorescence.

At low concentrations of the analyte under study, a linear response can be obtained according to the equation

$$I = a \cdot I_o \cdot \varepsilon \cdot b \cdot l \cdot c \tag{7.33}$$

where I is the intensity of the measured fluorescence, I_o is the incident light intensity, a is the instrument constant, ε is the molar absorbtivity, l is the optical path length in the sample, and b is the quantum yield of fluorescence

Absorption, reflection, scattering, and fluorescence measurements require a radiation source. A continuous spectrum of visible and near-infrared region is obtained by a tungsten filament lamp or halogen lamp, and that of the ultraviolet region is obtained by a hydrogen or deuterium lamp. In the far-infrared region, a heated inert solid can be used as a source. To obtain a monochromatic beam, a prism or a reflection grating is commonly used. By modulating the source and applying the Fourier transform to the detected signal, an absorption spectrum can be obtained without using a monochromator. This technique, useful in the infrared region, is known as the Fourier transform spectrometer.[35]

Spectrophotometers also include photodetectors. In ultraviolet, visible, and near infrared up to about 1 μm, the photomultiplier tube and photodiode are commonly used. In a photomultiplier tube, an incident photon produces a photoelectron, and the emitted electron is amplified by producing many secondary electrons at electrodes arranged in a cascade. Finally, 10^6 to 10^7 electrons are collected at the anode. Noise can be reduced by cooling the tube, and an ultimate sensitivity can be attained so that every single photon is countable. Such a device is called a photon counter.

The photodiode is a device in which current through the reverse biased p-n junction is proportional to the incident light intensity. Higher sensitivity is attained by employing a process called avalanche multiplication. When a p-n junction is reverse biased at near breakdown conditions, an electron generated at the junction is accelerated and creates electron-hole pairs, each electron and hole produces an additional electron-hole pair, and after continuing this process many times the number of electrons or holes is multiplied up to several hundred times. The photodiode in which avalanche multiplication occurs is called the avalanche photodiode. The dark current can be reduced by cooling. By fabricating either linear or two-dimensional configurations, photodiodes can be used as an image sensor. Further details on spectrophotometric chemical analyzers can be found in texts of instrumental analysis and analytical chemistry (for example, Skoog and Leary[36] and Skoog et al.[37]).

7.2.2.2 Fiberoptic Chemical Transducers

The monitoring of chemical variables by using optical fibers is mainly carried out by studying changes in suitable selected optical parameters of a medium with well-known properties. There are a number of optical properties that can be measured: absorbance, reflectance, scattering, fluorescence, and phosphorescence, as described previously.

A wide range of chemical parameters can be measured with fiberoptic probes; Harmer and Scheggi[38] have presented an overview (Figure 7.20). The instrumentation associated with each of the chemical parameters varies substantially and is mostly designed to match the specific characteristics or requirements determined by each problem area. However, a few groups of approaches can be identified.

The extrinsic sensor is taken to mean a probe where the optical fibers lead light to and from a volume in which the measurand is located. The fibers are passive light conductors. They are not active in the transduction process, only in transporting photons to and from the medium under study. In the extrinsic class of sensors we can find many chemical sensors. Medical probes for oxygen saturation measurements are typical examples of extrinsic sensors.

	Absorption transmission	Emission fluorescence	Fluorescence quenching
pH	O	O	
pO_2		O	O
pCO_2	O	O	
Glucose		O	
UO_2^{++} (uranyl)		O	
Plutonium(III)			O
Beryllium(II)		O	
Sodium		O	
Metal cations		O	
Halide ion		O	O
Iodine			O
H_2O_2		O	
Ammonia	O		
HCN (vapor)	O		
Halothane			O
Immunoassay	O	O	
Phosphate esters	O		
Moisture	O		

Figure 7.20 Overview of fiberoptic chemical sensors.

In the intrinsic sensor, on the other hand, the fiber itself is the active part of the transduction process. In chemical sensing, the evanescent mode sensors are typical examples. Typically, a cladding layer of part of the optical fiber is removed, leaving the core exposed to the measurand. The light guided by the probe has an evanescent wave component extending into the medium surrounding the optical fiber. Chemical reactions take place in the medium and in a surface layer, such as an immense assay. Evanescent-wave nepholometry is a typical example of this kind of sensor. Lieberman[39] gives an extensive overview of this type of sensor.

The instrumentation part of the fiberoptic chemical sensor systems can vary substantially, depending on the particular parameter that is measured. A number of light sources are used. Gas lasers, a tungsten lamp combined with filters, light-emitting diodes, and semiconductor laser diodes are all examples of well-known light sources.

The detector part of an analysis system can also be designed in various ways. Photodiodes, photomultipliers, photoconductive cells, and thermal sensitivity cells have different sensitivity characteristics in different parts of the spectrum under study.

A pair of optical fibers or fiber bundles can be used for transmitting excitation and detection. In some measurements, both the exciting light and the signal can be transmitted along the same optical fiber. Fluorescence measurement is suited for this type because the signal has a wavelength different from that of the exciting light so that the signal can easily be separated from the exciting light by using a filter.

The variations in optical fiber attenuation must sometimes be taken into account. If the fiber lengths are short, this is generally not a problem. Silica fibers have good transmission from the ultraviolet region to the near infrared (1.7 to 1.8 μm).

In comparison with other sensor principles, the fiberoptic sensor for chemical sensing, or optrode, has a number of advantages:

1. The optical fibers are thin and flexible. They permit the design of small, lightweight sensors by which invasive measurements can be performed at various sites in the human body.
2. Optical fibers do not corrode and can be implanted in the body for long periods without changing their optical properties. This property is of importance in experimental medicine.
3. Electrical and magnetic noise can easily be eliminated by using optical fibers as transmission lines for sensor information. The materials of an optical fiber are isolators and thus do not allow the induction of currents from external fields. This property is especially important in sensor use in connection with magnetic resonance imaging (MRI).
4. Fiberoptic sensors are available for response to chemical analytes or parameters for which electrodes are not available.
5. The isolating nature of the fiber material means that the patient is galvanically insulated from the monitoring devices. This contributes to the electrical safety of the patient.
6. Optical fiber probes can be introduced into the tissue of deeply located organs with very little trauma. A number of interesting parameters can be studied with this type of probe.
7. The low cost of optical fibers will make it likely that future fiberoptic sensors for medical monitoring applications can be disposable and of a low price.

Optical fiber sensors can also have certain disadvantages:

1. They can be sensitive to ambient light, for instance, surgical lamps. This problem can be solved by coding the optical information. Another way is to cover the fiber with a light yet tight layer.
2. The optical fibers have impurities that can give additional absorption, fluorescence, and Raman scattering distortion of the signal spectrum.
3. Immobilized indicators often suffer a reduction in sensitivity after immobilization. Sterilization agents can interfere with the properties of the indicators.

Extensive reviews of fiberoptic probes for chemical analytes can be found in Culshaw and Dakin[40] and Wolfbeis.[41]

7.2.3 CHEMICAL TRANSDUCERS OF ACOUSTIC AND THERMAL PRINCIPLES

The amount of a substance can be measured by its mass. To weigh a small amount of substance, a microbalance is commonly used, and measurement by weighing a mass is called gravimetry. A mass can also be measured by acoustic techniques. If a substance is deposited on a mechanical resonator, the mass will be estimated from the change in its resonance frequency. If a substance is uniformly deposited on the surface of a mechanical element through which sound propagates, the propagation pattern will be changed, and thus the mass is estimated from a change in the propagation parameter, such as the sound velocity. A mechanical device in which the amount of a substance is measured by its mass is called the mass, or gravimetric, sensor. There are two types of mass sensors: the bulk acoustic wave (BAW) sensor and the surface acoustic wave (SAW) sensor. Although both types of mass sensor have been developed mostly for use in gas phase, there are also many studies of use in liquid phase.

The BAW sensor consists of a piezoelectric crystal resonator in which bulk acoustic waves propagate. The resonance frequency varies according to the mass loaded onto the resonator surface. While any piezoelectric material can be used as a BAW sensor, the single-crystal quartz has been widely used because of its long-term stability and high mechanical Q value. Such a sensor is called the quartz crystal microbalance (QCM or QMB). The QCM consists of a quartz crystal resonator coated by a compound that selectively adsorbs some specific substance. QCM was first developed as a detector for gas chromatography, and has since been applied to chemical sensors.[42,43]

In QCM, a quartz crystal resonator oscillating in shear mode or thickness shear mode is used. When a substance is adsorbed on it, the mass is loaded to the resonator so that the resonance frequency decreases. The shift in resonant frequency is described by the Sauerbrey equation:[44]

$$\Delta f = -2f_0^2\left(\rho_q v_q\right)^{-1}\Delta m \tag{7.34}$$

where f_0 is resonance frequency, Δm is mass load per unit area, and ρ_q and v_q are density and shear wave velocity, respectively. Substituting the material parameters of a quartz crystal, the frequency shift is given as

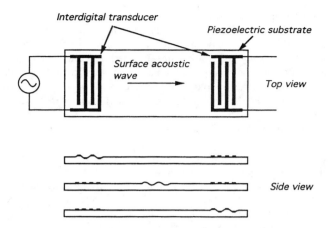

Figure 7.21   Interdigital transducer, the surface acoustic wave.

$$\Delta f = -2.26 \times 10^{-6} f_0^2 \Delta m. \tag{7.35}$$

QCM has extremely high sensitivity (typically 500 to 2500 Hz/µg can be attained[45]) and has been widely used for detecting gases such as sulfur dioxide, carbon monoxide, hydrogen chloride, and aromatic and aliphatic hydrocarbons. However, most QCM gas sensors have to be operated in low and constant relative humidity, thus their biomedical applications are limited.

QCM can also be used in liquid phase or in contact with a solution at one side of the resonator. QCM can be a biosensor if a selectivity is realized using substances of biological origin such as enzymes, antibodies, or genes, as described in the section on biosensors.

QCM can also be used combined with an electrochemical method. If a thin electrode forms on a quartz crystal and is exposed to an electrolyte solution, the mass change at the electrode surface caused by electrosorption can be detected. This type of QCM is called the electrochemical QCM, or EQCM. A resonator 320 µm thick oscillates around 5 MHz, and a mass change of 18 ng/cm^2 causes a frequency change of 1 Hz.[46] For example, using a gold-plated quartz crystal, adsorption of bromide and iodide can be detected.[47]

The SAW sensor also consists of a piezoelectric material, but unlike the BAW sensor, an elastic wave propagating along the surface of a solid is generated by the electrodes deposited on the same side of the crystal in the form of an interdigital transducer, as shown in Figure 7.21. The different types of elastic waves shown in Figure 7.22[48] can be excited selectively using various crystalline materials, interdigital electrode separation, and the operating frequency. The Rayleigh wave has a surface-normal component which is strongly affected by the viscosity of the adjacent medium and is not suited for use in liquid phase. The horizontally polarized shear wave has no surface-normal component and propagates efficiently at a solid-liquid interface. The Lamb wave propagates in a thin membrane and is potentially more sensitive to mass loading than a Rayleigh wave at the same frequency. The acoustic plate mode (APM) wave propagates by reflecting both surfaces, as shown in Figure 7.23. It is used in gas sensors, viscosity sensors, and biosensors.

In SAW sensors, a dual device configuration as shown in Figure 7.24 is commonly used. It consists of two SAW delay lines and amplifiers so as to oscillate at frequencies determined by delay times. One delay line has a surface coated by a sorptive film, while the other remains uncoated. Using the dual delay line configuration, the effect of temperature fluctuation can be compensated for and the mass loading can easily be detected by the difference in frequency of both delay lines, usually on the order of kHz.

The oscillating frequency of a feedback loop consisting of a delay line and an amplifier is affected by the mass loading and the change in the elastic properties of the delay line. However, the latter effect can be ignored when the coating material is soft. The frequency shift in a Rayleigh wave SAW device made of an ST-cut quartz substrate can be described approximately as

$$\Delta f = -1.26 \times 10^{-6} f_0^2 h \rho \tag{7.36}$$

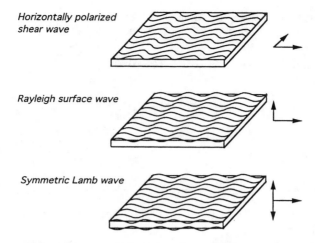

Figure 7.22 Different types of elastic waves. (Modified from Ballantine, D. S. and Wohljen, H., *Anal. Chem.*, 61, 704A, 1989.)

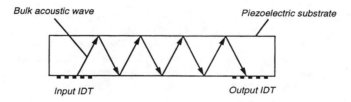

Figure 7.23 Propagation of the acoustic plate mode (APM).

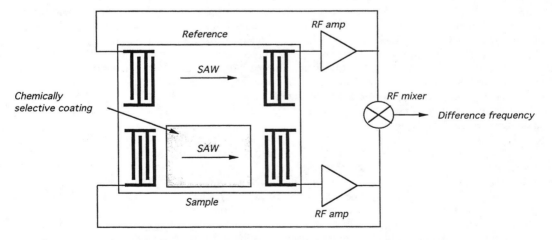

Figure 7.24 Dual device configuration in the surface acoustic wave measurement.

where h and ρ are the thickness and the density, respectively, of the coating film. The product of the thickness and the density corresponds to the mass loading per unit area in Equation (7.36). The sensitivity of a SAW device is less than that of a QCM at the same frequency, but due to the smaller numerical factor appearing in Equation (7.36) compared to Equation (7.34), the SAW device can be used at a higher frequency so that it is potentially more advantageous than the QCM device.

The thermal principle has been applied to chemical sensors. If some heat is produced by a chemical reaction, the occurrence of the reaction can be detected by an elevation in temperature. Enzymatic

Table 7.1 Biological Elements and
Transducing Principles that May Be
Used To Construct a Biosensor

Biological elements	Transducing principles
Organic molecules	Electrochemical
Enzymes	Optical
Antibodies	Mechanical
Nucleic acids	Thermal
Receptors	
Cells	
Tissues	
Organs	
Microorganisms	

reactions always produce heat in the range of 20 to 100 kJ/mol, and many substances can be detected by thermometric techniques.[49,50]

The flow-through system is commonly employed to measure heat production associated with a chemical reaction continuously. The temperature difference between the inlet and outlet of a reactor column is measured by small temperature sensors such as thermistors. A smaller sized apparatus was also made with a silicon chip.[51]

7.2.4 BIOSENSORS

The term "biosensor" commonly means a device incorporating a biological sensing element either intimately connected to or integrated within a transducer.[52] The purpose of incorporating a biological element is to realize higher selectivity to the substance to be measured in a mixture containing many other substances. By using such a highly selective element, the separation and purification processes which are always required in ordinary chemical analysis become unneccessary so that *in situ* or *in vivo* measurement can be carried out.

Many different biological elements and transducing principles, as listed in Table 7.1, have been employed in biosensors. The selectivity to a substance to be measured is generally realized by biological elements such as enzymes or antibodies. Some biological elements are sufficiently stable so that they can be extracted and incorporated in sensing devices. However, if the biological element having the desired function is not stable enough when it is extracted from the cell or tissue, there are different approaches in which cells, tissues, or even living organisms can be incorporated in a device.

The possible applications of biosensors are not limited to medical use alone but are also spreading into other fields of industry such as fermentation control and environmental applications for monitoring pollution. In this section, only biosensors for medical use are described briefly. More details on biosensors for clinical use, as well as those for use in other fields, can be found in many textbooks.[52-55] A summary of current information is available in reviews entitled "Chemical Sensors" which appear bi-annually in the journal *Analytical Chemistry*.[56-59]

7.2.4.1 Enzyme-Based Biosensors

The enzyme-based biosensor is a chemical transducer in which the catalytic property of an enzyme is utilized. Because enzyme reactions are highly specific, superior to any synthetic catalysts, it is expected that enzyme-based biosensors will be used even though many other substances exist which have to be removed by tedious pretreatment in ordinary analytical methods.

In biosensors, immobilized enzymes are commonly used. Although enzyme activity is reduced to some extent by immobilization, immobilized enzymes have advantages in that they can be used in continuous operation without supplying enzymes. Many different techniques regarding the immobilization of enzymes have been proposed, as illustrated in Figure 7.25.[60] For one particular enzyme, it is important to find an adequate method of immobilization so as to retain higher enzyme activity and better stability.

Enzyme-based biosensors can be formed by a number of transducing principles: electrochemical, optical, thermal, or gravimetric. The enzyme-based biosensor based on the electrochemical principle is called the enzyme electrode. The enzyme electrode measures either the formation of a product or the consumption of a substrate, if the product or substrate is electroactive.

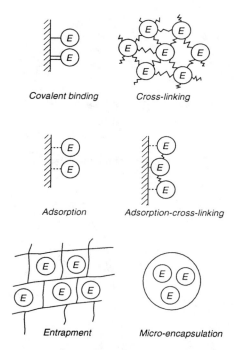

Covalent binding *Cross-linking*

Adsorption *Adsorption-cross-linking*

Entrapment *Micro-encapsulation*

Figure 7.25 Different techniques of immobilization of enzymes. (From Barker, S. A., in *Biosensors, Fundamentals and Applications,* Turner, A. P. F., Karube, I., and Wilson, G. W., Eds., Oxford Science, London, 1987, 85. With permission.)

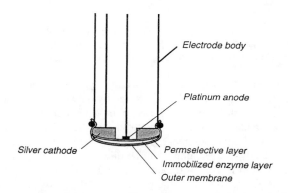

Electrode body

Platinum anode

Silver cathode

Permselective layer
Immobilized enzyme layer
Outer membrane

Figure 7.26 An example of polarographic glucose electrode.

The amperometric principle has been used widely in enzyme electrodes in which either the consumption of oxygen or the formation of hydrogen peroxide is measured by a polarographic electrode. For example, glucose is oxidized by the enzymatic reaction of glucose oxidase (GOD) as

$$\text{Glucose} + O_2 \xrightarrow{\text{GOD}} \text{Gluconolactone} + H_2O_2. \tag{7.37}$$

Oxygen as the substrate and hydrogen peroxide as the product both can be detected by the polarographic principle. The glucose electrode consists of a polarographic electrode with an immobilized glucose oxidase layer covered by a glucose-permeable membrane, as shown in Figure 7.26. If the rate of the above reaction is limited by the glucose supply through the membrane by diffusion, then the glucose concentration is measured by the change in current of either the oxygen electrode or the hydrogen peroxide electrode. When oxygen measurement is employed, differential operation of two oxygen

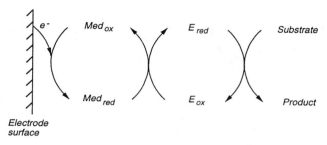

Figure 7.27 Reaction scheme of the mediated enzyme electrode.

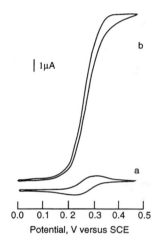

Figure 7.28 Voltammogram of a mediated electrode in the presence of glucose: **(a)** and **(b)** are without and with glucose oxidase, respectively. (From Hill, H. A. O. and Sanghera, G. A., in *Biosensors, A Practical Approach,* Cass, A. E. G., Ed., Oxford University Press, London, 1990, 19. With permission.)

electrodes, with and without GOD, is needed to measure oxygen consumption by the above reaction. Hydrogen peroxide production can, however, be measured by only one electrode, which is advantageous because of the simplicity. This process has been used more often than the oxygen-measurement type.

When enzymes catalyze oxidation of the substrate so that the reaction requires oxygen, fluctuation in the oxygen content affects the measurement, especially when the concentration of the substrate is high. In ordinary chemical analysis, the sample can be diluted by an oxygenated buffer; however, such a procedure is unacceptable for *in situ* measurement. This problem can be solved by the use of a mediator, which mediates charge transfer between enzyme molecules and the working electrode, and an electrochemical technique called cyclic voltammetry. The enzyme electrode using a mediator is called the mediated enzyme electrode, whereas the one not using a mediator is called the unmediated or direct-electron-transfer enzyme electrode.

The reaction scheme depicting the roll of a mediator is shown in Figure 7.27.[61] Ferrocene (bis(η^5-cyclopentadienyl) iron, $FeCp_2$) is an excellent mediator for oxidase enzymes. It alternates between oxidized and reduced forms by accepting and releasing an electron when negative and positive potentials are applied to the electrode. In cyclic voltammetry, a triangular potential is applied so as to increase and decrease potential at a constant rate between two limits. A closed-loop voltage-current diagram, called a voltammogram, as shown in Figure 7.28(a), is obtained. However, when the catalytic reaction occurs, current due to oxidation of the mediator is increased, as shown in Figure 7.28(b). The increment of current is proportional to the concentration of the substrate, and thus the concentration can be determined if it is calibrated adequately.

Many kinds of amperometric enzyme electrodes have been developed which are specific to different substances such as glucose, sucrose, lactate, uric acid, acetylcholine, creatinine, and cholesterol, as well as alcohols, amines, or amino acids. While most of them can be used only under laboratory conditions, glucose electrodes for the analysis of whole blood were developed and one of them has been launched commercially as a disposable test strip (MediSense, Inc.; Cambridge, MA).

Enzyme electrodes are also formed by either the ion-selective electrode (ISE) or ion-selective field-effect transistor (ISFET), both of which are based on the potentiometric principle and called potentiometric enzyme electrodes. They consist of an ISE or ISFET covered by an immobilized enzyme layer and a protective membrane. In a potentiometric enzyme electrode, the product of the enzymatic reaction is measured by the potentiometric method. For example, the hydrolyzation of urea is catalyzed by urease as

$$\text{Urea} + 2H_2O + H^+ \xrightarrow{\text{Urease}} 2NH_4^+ + HCO_3^- \tag{7.38}$$

and then NH_3 is formed as

$$NH_4^+ + OH^- \rightarrow NH_3 + H_2O. \tag{7.39}$$

Because of these reactions, urea can be measured by the NH_3 electrode which is based on the potentiometric principle. Other possible potentiometric enzyme electrodes such as amino acids, penicillin, and amygdalin are found in the literature.[62]

The thermometric principle is also employed in some enzyme-based biosensors. Enzymatic reactions always produce heat in the range of 20 to 100 kJ mol[-1]. The amount of the substrate can be estimated from the produced heat. By injecting the temperature-stabilized sample into a small column filled with an immobilized enzyme with a constant flow rate, the produced heat can be measured by the temperature difference between the inlet and outlet of the column. Typical substances which are analyzed by the enzyme-based thermometric method are ethanol, glucose, lactate, oxalic acid, penicillin, sucrose, and urea.[49] A thermistor coated with an immobilized enzyme can be a simple biosensor and is called the enzyme-bound thermistor.[63]

7.2.4.2 Immunosensors

The immunosensor is a type of chemical transducer in which the antigen-antibody reaction is utilized so as to realize highly specific and sensitive measurement, similar to the analytical technique called immunoassay. The antigen (A_g) and the antibody (A_b) form an antigen-antibody complex (A_gA_b) as

$$A_b + A_g \leftrightarrow A_bA_g \tag{7.40}$$

where the reaction is reversible so that the ratio, K, is defined as

$$K = \left[A_bA_g\right]/\left[A_b\right]\left[A_g\right] \tag{7.41}$$

and is constant at equilibrium; this constant is called the affinity constant. When A_g is introduced and the amount of A_b remains unchanged, the amount of introduced A_g can be determined by the increment of A_gA_b. If the antibody, A_b, is immobilized and fixed on the sensor surface, as shown in Figure 7.29(a), formation of A_gA_b will cause some change in electrode potential or mass, which can be measured by electrochemical or gravimetric techniques.

The potentiometric detection of the antigen-antibody reaction was demonstrated by a study in which human chorionic gonadotropin (hCG), or an antibody against hCG, was immobilized onto a titanium wire, and potential changes were observed when either the respective antibody or antigen was added.[64] To detect the change due to the antigen-antibody reaction, an ion-selective, field-effect transistor can be used; such a device is called the immunochemically sensitive FET or IMFET.[65] Gravimetric detection was also demonstrated by a study in which an antibody, goat antihuman IgG (immunoglobulin G), was immobilized on a surface-acoustic-wave device (SAW). The antigen, human IgG, could be successfully detected as the oscillation frequency shift due to the mass load of adsorbed antigen.[66]

Highly sensitive immunosensors can be formed by utilizing an enzymatic reaction, in which a fixed amount of enzyme-labeled antigen is introduced. As shown in Figure 7.29(b), labeled or unlabeled antigen can be bound onto an immobilized antibody. After removing free antigen, enzyme activity is

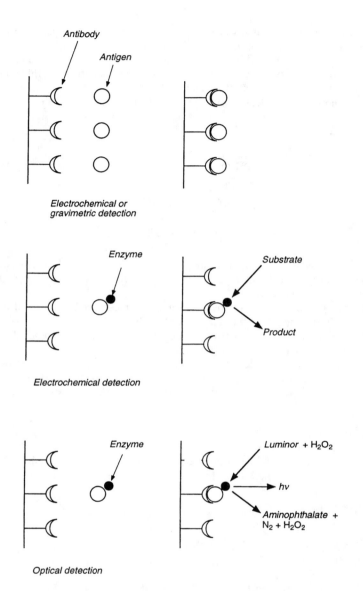

Figure 7.29 Immunosensors of gravimetric, electrochemical, and optical detection.

measured by introducing a substrate and detecting the change of the product by an appropriate technique, such as the electrochemical or optical method. The binding reactions of labeled and unlabeled antigen are competitive: When the concentration of unlabeled antigen is increased, more antibodies are occupied by unlabeled antigen so that the amount of antibodies bound by labeled antigen decreases; consequently, after separating free antigen, the resulting enzyme activity will decrease. Due to the catalytic function of the enzyme, one molecule of enzyme can produce a great number of product molecules, and thus extremely high sensitivity can be attained.

For the labeling enzyme, catalase and glucose oxidase have been commonly used. Because catalase facilitates the reaction

$$2H_2O_2 \rightarrow 2H_2O + O_2 \qquad (7.42)$$

the activity of catalase can be measured by the production of oxygen when hydrogen peroxide is introduced. Aizawa et al.[67] constructed an immunoelectrode to monitor human chorionic gonadtropin (hCG) using catalase and an oxygen electrode. The antibody for the antigen hCG was immobilized onto a cellulose membrane and was placed on a Clark-type oxygen electrode. The membrane was exposed to labeled and nonlabeled hCG so as to compete for the antibody, washed to remove free hCG, and exposed to the hydrogen peroxide solution. The amount of nonlabeled hCG was determined by the rate of increase in oxygen tension.

Catalase is advantageous because of its high turnover number, which is the number of catalytic reactions occurring per unit of time; however, the use of hydrogen peroxide as the substrate is inconvenient because it is not sufficiently stable. Glucose oxidase is also used in enzyme-linked immunoelectrodes. It catalyzes the reaction

$$\text{Glucose} + O_2 \rightarrow \text{Gluconolactone} + H_2O_2. \tag{7.43}$$

While the turnover number of GOD is less than that of catalase, it is advantageous for the reason that the substrate glucose is stable enough.

The enzyme-labeled antigen (or antibody) can also be detected optically if the enzyme catalyzes a chemoluminescent reaction as shown in Figure 7.29(c). For example, peroxidase catalyzes the reaction

$$\text{Luminol} + H_2O_2 \rightarrow \text{Aminophthalate} + N_2 + H_2O + h\nu \tag{7.44}$$

so that the enzyme activity can be measured by measuring the intensity of chemoluminescence. The emitted light can be measured by a photomultiplier through an optical fiber. By this technique, human serum albumin, β_2 microglobulin, and insulin could be measured.[68-71]

7.2.4.3 Microbial Sensors

The microbial sensor is a type of chemical transducer in which immobilized microorganisms are incorporated so that microbial-catalyzed reactions can occur. While the selectivity of a microbial sensor for substrates is realized by the enzymes in incorporated microorganisms, there are particular advantages in using microorganisms vs. using isolated enzymes: (1) microbial sensors are less sensitive to inhibition by solutes and are more tolerant of suboptimal pH, (2) they have a longer lifetime than enzyme electrodes, and (3) they are less expensive because an active enzyme need not be isolated. On the other hand, there are disadvantages: (1) some microbial sensors have a longer response time than enzyme electrodes, (2) they need more time to return to the baseline level after use, and (3) microorganisms contain many enzymes and care must be taken to ensure selectivity.[72]

Most microbial sensors are classified as either a respiration-measuring or metabolite-measuring type. As shown in Figure 7.30(a), the respiration-measuring microbial sensor consists of immobilized aerobic microorganisms and an oxygen electrode. When a substrate, which can be metabolized by the microorganism, is contained in a solution saturated by oxygen, a metabolic reaction occurs by the consumption of dissolved oxygen, thus the substrate can be measured from the decrease in oxygen tension.

By means of respiration-measuring microbial sensors, many substances such as glucose, assimilable sugars, acetic acid, ammonia, and alcohols are measured selectively by using different microorganisms.[73,74] The biochemical oxygen demand (BOD) can also be measured by this type of microbial sensor, and it has been used in environmental control.[75,76]

Figure 7.30(b) shows the metabolite-measurement microbial sensor. It consists of immobilized microorganisms and a sensor that detects the metabolite produced by the reaction catalyzed by that microorganism. Using different types of gas and ion sensors, many kinds of substances can be measured by detecting different metabolites. For example, the fuel cell electrode is used to detect H_2 for measuring formic acid, the CO_2 electrode for glutamic acid or lysine, and the pH electrode for cephalosporin and nicotinic acid.[75]

Some bacteria exhibit luminescence. If these luminous bacteria are immobilized and combined with a photodetector, substances that affect the bioluminescence can be measured by detecting the change in luminescence with a sensor called the photomicrobial sensor. While some substances such as glucose may increase luminescence, many toxic substances such as benzalkonium chloride, sodium dodecyl sulfate, chromium, and mercury decrease luminescence; thus, the photomicrobial sensors are applicable to environmental monitoring.[77,78]

The combination of a microbial sensor and an immobilized enzyme membrane (a hybrid biosensor) is effective in some cases. For example, a urea sensor can be formed by combining an immobilized

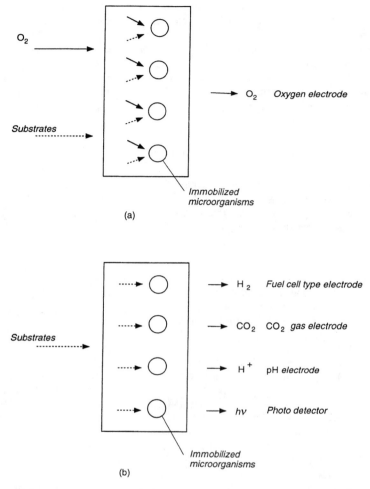

Figure 7.30 · Microbial sensors: **(a)** respiration-measuring and **(b)** metabolite-measuring. (Modified from Karube, I. and Nakanishi, K., *IEEE Eng. Med. Biol.,* 13, 364, 1994.)

urease membrane and a microbial NH_3 sensor using nitrifying bacteria.[79] In this sensor, higher selectivity is realized by using an immobilized enzyme. The microbial NH_3 sensor is superior to the potentiometric ammonium electrode in the sense that interference from ions or volatile compounds such as amines is less common.

A microbial sensor has been proposed for detecting mutagenic chemicals. Because mutagenicity correlates to carcinogenicity, this type of sensor is expected to be used for the screening of carcinogens.[80,81] The sensor consists of aerobic recombination-deficient bacteria and the oxygen electrode. The recombination-deficient strain, *Bacillus subtilis* M45 Rec⁻, and the wild strain, *B. subtilis* H17 Rec⁺, were employed for experiments. Both bacteria were immobilized onto membranes and fixed to the oxygen electrodes. If mutagen was applied to these Rec⁻ and Rec⁺ electrodes at steady state, the current of the Rec⁻ electrode began to increase because Rec⁻ was killed and oxygen consumption was reduced, while the current of the Rec⁺ electrode remained unchanged because Rec⁺ is resistant to mutagenic agents due to the existing DNA recombination enzyme system.

7.2.5 INSTRUMENTAL ANALYSIS

In analytical chemistry various kinds of instruments, such as spectrometers and chromatographs, are used to determine the amount of chemical species contained in a sample. Such instrumentations are called chemical analyzers. Although most chemical analyzers are sophisticated systems which may not be regarded as chemical transducers, the aim of each analyzer is the same as that of a chemical transducer

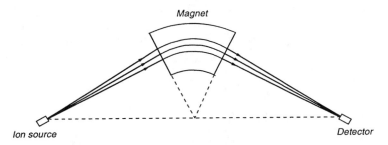

Figure 7.31   Mass spectrometer with a magnetic sector.

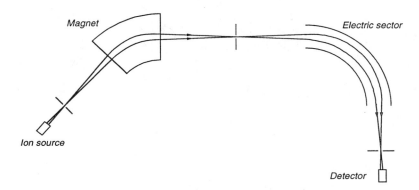

Figure 7.32   Mass spectrometer with a double-focusing magnetic sector.

that converts a chemical quantity into an electrical signal. The principles of various chemical analyzers, not mentioned in the previous sections, are briefly introduced in the following sections.

7.2.5.1   Mass Spectrometry

Mass spectrometry is an analytical technique in which atoms or molecules are separated according to their mass. The sample is always vaporized, ionized in a vacuum, and then separated into different mass species by using a mass analyzer.

Ionization of a sample is usually achieved by electron bombardment. Gas molecules entering the ionizing region interact with an electron beam and lose an electron to become positively ionized. Ionization is also achieved by ion-molecule interaction in which a sample molecule is positively ionized by its reacting with positively charged ions. Such a technique is called chemical ionization. There are several forms of mass analyzers such as the magnetic sector analyzer, the quadrupole analyzer, the ion trap, the time-of-flight analyzer, and ion cyclotron resonance instruments.

The principle of the magnetic sector analyzer is shown in Figure 7.31. The ions formed in the source are accelerated, pass through a static magnetic field, and reach the ion detector. The radius, r, of the circular motion of a charged particle of mass, m, and velocity, v, is determined by the force developed by the presence of a magnetic field:

$$mv^2/r = Bzev \tag{7.45}$$

where B is the magnetic field strength, and ze is the charge. The mass, or mass to charge ratio (m/ze), is estimated from the velocity, v, which is determined by the potential through which the ions are accelerated. By combining it with an electrostatic sector, as shown in Figure 7.32, focusing characteristics can be improved. This type of instrument is called the double-focusing magnetic sector, and the use of only one magnetic field is called the single-focusing magnetic sector. More details can be found in textbooks or reviews (for example, Duckworth et al.,[82] Chapman,[83] and Gaskel[84]).

The quadrupole analyzer consists of four parallel rods of hyperbolic or circular cross-section, as shown in Figure 7.33. A voltage having a d.c. component, V_d, and an a.c. component, $V_a \cos \omega t$, is applied

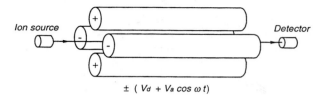

$$\pm \left(V_d + V_a \cos \omega t \right)$$

Figure 7.33 Quadrupole mass spectrometer.

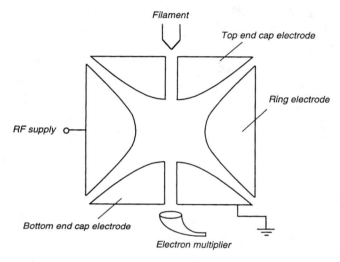

Figure 7.34 Ion-trap mass spectrometer.

between adjacent rods. The ion injected into the center of the rods travels along the axis and oscillates due to the applied a.c. electric field. The oscillation is stable only when a certain condition is fulfilled between V_d, V_a, and the mass, so that the ions reach the ion detector. Mass separation can then be achieved by scanning applied voltages and keeping the ratio V_d/V_a constant. Since the quadrupole instrument does not require magnets, it can be built compactly and inexpensively. More detail can be found in textbooks (for example, Dawson[85]).

The ion trap consists of three cylindrically symmetrical electrodes, as shown in Figure 7.34. An RF voltage is applied to the ring electrode relative to the end-cap electrodes. The ions injected into the cell are trapped in it, being constrained in a stable orbit oscillating both axially and radially. By increasing the amplitude of the RF voltage, each ion gains kinetic energy so that it is ejected from stable orbit and some ions strike the detector. By a gradual increase of the RF field amplitude, instability of ionic motion occurs in order to increase the mass-to-charge ratio, m/z, and then the mass scan is achieved. The presence of a low pressure of helium gas improves sensitivity and mass resolution due to the effect of collisions with helium reduce kinetic energy of the ion and causes migration toward the center of the trap.

The time-of-flight analyzer is an instrument in which the mass of an ion is determined from the time to travel a constant distance when the ion is accelerated by a constant voltage.[86] If the mass is m; the charge, ze; the acceleration voltage, V; and traveling distance, L, then the time of flight, t, is given as

$$t = \left(m/2zeV \right)^{1/2} L. \tag{7.46}$$

To measure the time of flight, ions should be formed in a moment, and that is achieved by a pulsed electron beam in the ionization region. The advantage of this method is its high sensitivity, due to the absence of a beam slit for spatial separation.

The ion cyclotron resonance instrument consists of a magnet, transmitter, and receiver electrode plates as shown in Figure 7.35. An a.c. voltage is applied to the transmitter plates. An ion generated in the

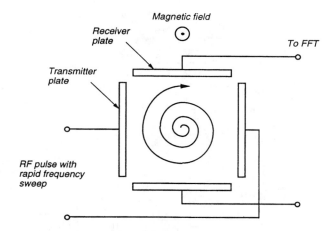

Figure 7.35 Cyclotron resonance mass spectrometer.

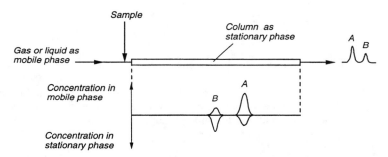

Figure 7.36 The principle of chromatography. Chemical components A and B are separated based on the difference in speeds of moving through a column.

field moves in a circular path perpendicular to the magnetic field direction. If the cyclotron resonance condition

$$\omega = Bze/m \qquad (7.47)$$

is fulfilled, where ω is the angular frequency of applied a.c. potential, B the magnetic field strength, ze the charge, and m the mass, then the ion gains energy from the field so that the radius of circular motion increases and finally reaches the receiver plate. If the applied frequency is swept rapidly, all ions having different masses are excited and cause a signal containing many different frequency components corresponding to ionic species of different masses. The mass spectrum can then be obtained by applying the Fourier transform to the output signal. More details can be found in the reviews.[87]

7.2.5.2 Chromatography
Chromatography is an analytical technique in which the components in a mixture are separated based on differences in their speeds of moving through a column when a gas or liquid is forced to flow through it, as shown in Figure 7.36. The technique of using a gas is called gas chromatography, and that of using a liquid is called liquid chromatography. A liquid chromatograph, in which higher pressure and a column packed with smaller particles are employed to achieve higher resolution as well as to reduce separation time, is called high-performance liquid chromatography (HPLC).

The chromatographic method consists of a stationary phase which is a solid or liquid held in the column, and a mobile phase, which is a gas or a liquid forced into the column. Even if each component in the sample can be dissolved in both phases, the amount dissolved in each phase depends on the chemical species. Consider a small section of the column with the volume of the mobile phase, V_m, and

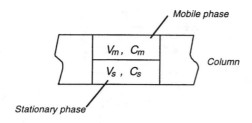

Figure 7.37   A small section of the column.

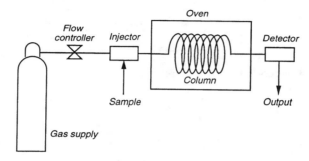

Figure 7.38   Gas chromatography system.

that of the stationary phase, V_s, and assume a component is dissolved in the mobile phase and the stationary phase with concentrations of c_m and c_s, respectively, at equilibrium, as shown in Figure 7.37. Then the fraction, R, of the component existing in the mobile phase is given as

$$R = \frac{c_m V_m}{c_m V_m + c_s V_s} = \frac{1}{1 + c_s V_s / c_m V_m} \qquad (7.48)$$

Because only a component contained in the mobile phase can move, the resulting velocity of the component is proportional to R. While V_s/V_m is constant, R depends on c_s/c_m, which is called the partition ratio or the partition coefficient and is different for different components.

Figure 7.38 shows the basic components of the gas chromatography (GS) system. In gas chromatography, a carrier gas such as helium, nitrogen, or hydrogen is used as the mobile phase. The sample is vaporized and introduced rapidly so as to form a plug of sample vapor. A hollow tube several meters or several tens of meters long and having an inside diameter of a few tenths of a millimeter is commonly used as a column. Such a column is called a capillary column. The stationary phase consists of a material, such as siloxane polymer, and is coated on the inside wall of the capillary tube. Many types of detectors, such as the flame ionization detector, the thermal conductivity detector, the electron-capture detector, and the photo-ionization detector, are used in gas chromatography. Besides these, the mass spectrometer is an excellent detector, so that gas chromatography-mass spectrometry (GC-MS) can be a convenient technique for analyzing complex organic mixtures. The essential limitation of gas chromatography is that it is applicable only for substances that are volatile or can evaporate at oven temperature. Further details of gas chromatography can be found in textbooks (for example, Baugh[88]).

Figure 7.39 shows the basic scheme for high-performance liquid chromatography. The mobile phase is a solvent in which the analyte is soluble. For water-soluble substances, a polar solvent such as water is used; for fat-soluble substances, a nonpolar solvent such as hydrocarbons is used as the mobile phase. The stationary phase consists of fine particles, typically 5 to 10 μm in diameter, packed in a column. Due to the very high flow resistance of the column, the mobile phase is driven by a nonpulsatile pump which generates a pressure of over hundred atmospheres. As a detector, the ultraviolet detector is commonly used, while other detectors such as the refractive index, fluorescence, or conductivity detector are also used. Further details on high-performance liquid chromatography can be found in textbooks (for example, Meyer[89]).

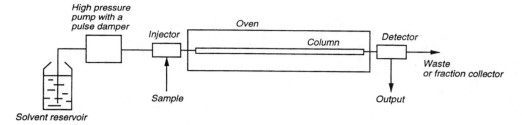

Figure 7.39 High-performance liquid chromatography (HPLC).

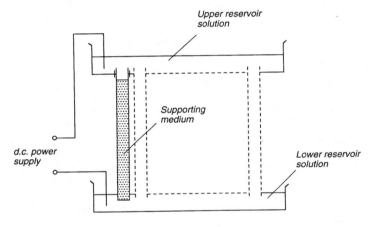

Figure 7.40 Electrophoresis unit.

7.2.5.3 Electrophoresis

The electrophoretic technique is an analytical method in which the migration of ions in an electric field causes the separation of ions exposed to the field. If an ion with a charge, q, is exposed to an electric field, E, a force, qE, arises and drives the ion. In a viscous medium, a frictional force will arise and reaches an equilibrium at which the driving and the frictional forces are in balance so that the ion moves at a constant velocity, v, given as

$$v = qEf \tag{7.49}$$

where f is called the frictional coefficient, which is proportional to the solvent viscosity and the effective radius of the ion. Thus, a smaller molecule moves faster and a larger one moves more slowly if the charges are identical.

The electrophoresis unit consists of a column filled with a supporting medium connected to two reservoirs as shown in Figure 7.40. Agarose or polyacrylamide gels are commonly used as the supporting medium. The column usually consists of two glass plates, separated by a 1- to 2-mm gap filled with gels, forming identical parallel slots for multiple samples. Samples are loaded at the upper part of the slot, and a d.c. potential is applied between the two reservoirs to produce a uniform electric field in the medium. After the molecules in the sample are separated according to their mobilities, the gels are stained by a dye or analyzed by autoradiography if the sample is radiolabeled.

The capillary electrophoresis shown in Figure 7.41 has also been used. It involves narrow-bore tubes, usually 50 μm in diameter and 50 to 100 cm long, filled with a supporting medium. The molecules close to the end of the capillary are detected by an ultraviolet detector similar to that in HPLC. The advantage of using a capillary is the enhancement of heat dissipation due to a large surface-to-volume ratio which permits the use of a high voltage, which decreases the analysis time. By applying 10 to 50 kV, the typical analysis time can be reduced to 10 to 30 minutes, as opposed to the many hours required in the original

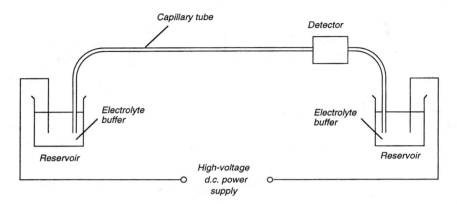

Figure 7.41   Capillary electrophoresis.

apparatus. Further details on electrophoresis can be found in textbooks and reviews (for example, Andrews,[90] Ewing et al.,[91] and Wilson and Walker[92]).

7.2.5.4   Magnetic Resonance

Magnetic resonance is an analytical technique based on the measurement of absorption of electromagnetic radiation by either electrons or nuclei in the presence of an external magnetic field. When the absorption of electrons is measured, the technique is called electron spin resonance (ESR), and when that of nuclei is measured, it is called the nuclear magnetic resonance (NMR). An alternative to the term electron spin resonance — electron paramagnetic resonance (EPR) — is more rigorous because the resonance condition is determined not only by the electron spin but also by the change in electron orbit.

The principle of magnetic resonance can be explained by a simple classical model in which an electron or a nucleus can be regarded as a spinning charged particle. Due to the spinning motion, the particle has an angular momentum. At the same time, it causes a magnetic moment induced by the effective ring current due to the rotating charge. Because angular momentum is quantized on an atomic scale, the emerging magnetic moment is also quantized; consequently, the particle has a magnetic moment at discrete levels only, which is proportional to the quantized angular momentum. The proportionality constant is called the magnetogyric ratio (or gyromagnetic ratio) and is denoted by γ. The magnetic moment vector, μ, is written as

$$\mu = \gamma l \tag{7.50}$$

where l is the angular momentum vector. In a free electron and a nucleus having a spin of 1/2 (such as ^{1}H, ^{13}C, and ^{31}P), only two states of the angular momentum are allowed. In an external magnetic field, $\mathbf{B}$, a particle having a magnetic moment, μ, has a magnetic energy, U, expressed as

$$U = -\mu \cdot \mathbf{B}. \tag{7.51}$$

When the magnetic moment is quantized, only discrete states can be realized. For a particle having a spin of 1/2, there are two states of magnetic moments corresponding to parallel and antiparallel to the external magnetic field. The energy difference between these two states is then given as

$$\Delta U = -\gamma \hbar \mathbf{B} \tag{7.52}$$

where $\hbar$ is one quantum unit of angular momentum and equal to $h/2\pi$, where h is Planck's constant. Transition between these two states can occur when a photon having an energy, $h\nu$, equal to ΔU is absorbed. Thus,

$$h\nu = \gamma \hbar \mathbf{B}. \tag{7.53}$$

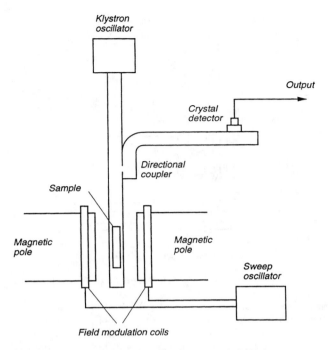

Figure 7.42   Electron spin resonance (ESR) instrument.

The coefficient $\gamma\hbar$ is often expressed as a product of a dimensionless factor, g, called the Zeeman splitting constant, and a factor, β, called the magneton so that

$$h\nu = g\beta\,\mathbf{B}. \tag{7.54}$$

The Zeeman splitting constant of a free electron is

$$g_e = 2.0023193 \tag{7.55}$$

and the factor β for a free electron, called the Bohr magneton, is

$$\beta_e = 9.274153 \times 10^{-24}\,\mathrm{JT^{-1}}. \tag{7.56}$$

The Zeeman splitting constant is different for different nuclei: 5.56948 for ^{1}H, 1.40483 for ^{13}C, and 2.26322 for ^{31}P. The factor, β, for a nucleus is called the nuclear magneton and is

$$\beta_n = 5.0507866 \times 10^{-27}\,\mathrm{JT^{-1}}.$$

The resonant frequency of the ordinary ESR measurement is in the microwave range, while that of the major NMR measurement is in the radio frequency range, due to the difference of β_e and β_n for about three orders of magnitude. The instrumentation of ESR and that of NMR are shown in Figure 7.42 and Figure 7.43. Both instruments have magnets with field modulation coils by which an external magnetic field is applied and is swept in a range in which the resonant spectrum can be observed. In ESR, the resonance is observed in the microwave range. Thus, the sample is placed in a microwave cavity. The microwave is produced by a klystron oscillator and detected by a crystal detector. In NMR, a radio-frequency transmitter and a receiver with two sets of transmitter and receiver coils are used.

In ESR measurement, species having unpaired electrons, such as free radicals and transition metal ions, are analyzed. These species consist of only a small part of the sample. In NMR, the resonance of

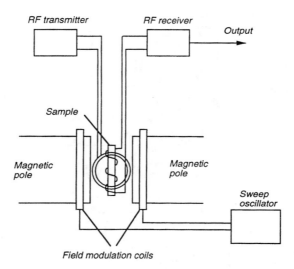

Figure 7.43 Nuclear magnetic resonance (NMR) instrument.

protons (^{1}H) is measured most commonly. Because ^{1}H exists in water and any organic substance, observation of the strength of absorption does not provide much information about the chemical composition of a sample. However, the resonant frequency is affected by the local field in a molecule. Local fields in hydrogen nuclei at various sites in a molecule are different, leading to differences in resonant frequencies called chemical shifts. This principle can be applied to chemical analysis. Other nuclei having a magnetic moment, such as ^{13}C, ^{15}N, and ^{31}P, are also used in biochemical studies. Further details of ESR and NMR for chemical analysis can be found in textbooks (for example, Skoog and Leary,[93] Weil et al.,[94] and Wilson and Walker[92]).

7.2.5.5 Other Analytical Methods Based on Physical Material Properties
Physical properties, such as the mechanical, thermal, electromagnetic, or optical properties, of a matter depend on its chemical composition. Thus, measurement of each physical material property can be potentially applicable to chemical analysis. However, for a mixture of many different chemical species, no single measurement of a physical material property, such as density, thermal conductivity, refraction index, etc., provides sufficient information to identify and determine the constituents. A possible situation in which such a measurement can be useful is one in which only the chemical species of interest dominantly contributes to the material property. For example, the content of plasma protein can be estimated by the refraction index, because the contribution of other constituents is small enough. Another possible application is a combination of a separating device, such as a chromatographic column, and a nonspecific sensing element. For example, a sensing element based on thermal conductivity can be a convenient detector of gas chromatography.

7.2.5.5.1 *Density.*   The density or specific gravity of a solution provides an estimate of the amount of the solutes. The specific gravity of a fluid can be measured by a float gauge based on the principle that buoyancy exerted on a matter in a fluid is simply the weight of the fluid excluded when the matter enters. This principle has been used in the clinical examination of urine using an instrument called the urinometer, which is a float with a scale in specific gravity.

The density of a fluid can also be estimated from sound velocity. For gases and liquids, sound velocity, c, is given by

$$c = \sqrt{K/\rho} \tag{7.57}$$

where K is the bulk modulus of elasticity and ρ is the density.

7.2.5.5.2 *Viscosity.*   The viscosity of a liquid is commonly measured by either a capillary viscometer or a rotational viscometer. The capillary viscometer consists of a narrow capillary tube of known

diameter and length. The sample is forced into it, and viscosity is determined from the pressure drop and flow rate using Poiseuille's law. In the rotational viscometer, the sample is filled in a narrow gap between two surfaces, one rotating and the other stationary, and viscosity is determined from the torque of the rotating element.

7.2.5.5.3 Thermal Conductivity.
The thermal conductivity of a gas, liquid, or solid depends on its composition. Thus, in principle, its measurement can be used in chemical analysis. An example is gas chromatography, as mentioned in Section 7.2.5.2. In this application, the instrument detects any chemical species separated by the column if its thermal conductivity is different from that of the carrier gas. In the detector, differences in thermal conductivity are detected as a change in heat dissipation from the heated filament.

7.2.5.5.4 Refractive Index.
In a solution, the refractive index may vary with the concentration of solutes. If the refractive index varies with only one known solute, its concentration can be estimated by refractometry. For example, the serum protein content has been measured by a simple refractometer, in which the refractive index is measured by measuring the critical angle of the total reflection. Total reflection occurs when a light beam passes across an interface of two media from one having larger refractive indices to another having smaller refractive indices. When the refractive indices of two media are n_1 and n_2 ($n_1 > n_2$), then the critical angle, θ_c, for total reflection is given as

$$\sin \theta_c = n_2 / n_1. \qquad (7.58)$$

7.2.5.5.5 Optical Rotation.
Optical rotation is a phenomenon that occurs when a plane-polarized light beam is transmitted through a certain medium and the plane of polarization rotates. When optical rotation is measured in a sample of constant length, the amount of optical rotation is proportional to the rotatory power and concentration of the substance which possesses the property of optical rotation. Thus, the concentration can be estimated if the rotatory power of the substance is known. This technique has been used for measurement of sucrose concentrations. It is also applicable, in principle, for measurement of the glucose concentration in a body fluid because of the fact that only d-glucose (dextroglucose, dextrose) exists in biological organisms. Actually, glucose measurement in the aqueous humor of the eye was attempted using a contact lens to which a light source, polarizers, and a photodetector were attached (Rabinovitch et al.,[95] March et al.,[96] and King et al.[97]).

7.2.5.5.6 Electric Conductivity and Permittivity.
The conductivity of a solution depends on its ionic concentration and is proportional to that ionic concentration, as long as the concentration is low. While the proportionality coefficients vary for different ions, the conductivity of a solution provides information about the ionic concentration if the relative ionic compositions of the solution are known. For example, conductivity can be a convenient measure of the quality of purified water.

The permittivity (dielectric constant) can be determined by measuring the capacitive component of the impedance. If an a.c. electric field is applied to a solution that contains molecules with permanent dipole moments, the dipoles will reorient themselves in the field direction at every cycle, causing a displacement current. Because permittivity is frequency dependent, and the frequency characteristics depend on the chemical species, permittivity measurement in an adequate frequency range can be used as an analytical technique.[98]

7.2.5.5.7 Magnetic Susceptibility.
Magnetic susceptibility is a quantity that determines magnetization relative to magnetic field strength. When a specific substance in a sample exhibits strong magnetism, but others do not, the concentration of that substance can be analyzed by measuring magnetic susceptibility. The paramagnetic oxygen analyzer is based on this principle. Oxygen gas is paramagnetic so that its magnetic susceptibility is positive and large, while most other gases in medical use (except nitrous oxide, N_2O) are diamagnetic so that their magnetic susceptibility is negative and small.

Different types of paramagnetic oxygen analyzers have been used. Figure 7.44(a) shows the principle of an oxygen analyzer with wedge-shaped magnetic pole pieces and a glass dumbbell containing a diamagnetic gas and suspended by an elastic wire so that the dumbbell stays in equilibrium at the region where the magnetic field is strongest. When the dumbbell is surrounded by a paramagnetic gas, the gas tends to move into the region where the magnetic field is stronger, pushing the dumbbell out from that region. Consequently, the oxygen content of the surrounding gas can be measured as the rotation angle of the dumbbell system that is detected optically.[99]

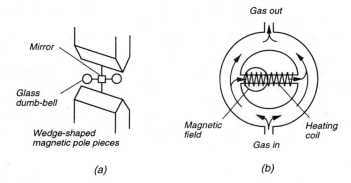

Figure 7.44 Paramagnetic oxygen analyzers: **(a)** dumbbell type and **(b)** gas-flow-detecting type.

Figure 7.44(b) shows the principle of another type of paramagnetic oxygen analyzer. It has a central tube with a heating coil, and a strong magnetic field is situated at one side, as shown in the figure. According to Curie's law, the magnetization of a paramagnetic substance is inversely proportional to the absolute temperature. Thus, a paramagnetic gas flows from the cold side to the hot side. The oxygen content is then measured by the gas flow rate in the tube.[100]

7.3 CONTINUOUS MEASUREMENT OF CHEMICAL QUANTITIES

In routine clinical situations, chemical quantities in body fluids are usually determined by samples taken from the patient. However, when the chemical quantities vary rapidly, continuous measurement using chemical transducers is preferable. In order to monitor chemical quantities of the body, a transducer can be inserted; such a transducer is called the indwelling transducer. On the other hand, if the body fluid can be drained or sucked continuously from the body, measurement can be performed using a transducer outside the body; such a method is called *ex vivo* measurement.

While the above-mentioned techniques require an invasive procedure, noninvasive measurements which can be performed across the skin are desirable. These are possible when the substances to be detected are permeable through the skin. Unfortunately, most clinically important substances are impermeable through the skin. Only gases such as oxygen and carbon dioxide are fairly permeable, and their partial pressures can be measured transcutaneously. When the stratum corneum, the outermost layer of the skin, is removed, interstitial fluid can be withdrawn by applying negative pressure on the skin surface. Some species, such as glucose in the body fluid, can be measured transcutaneously. Optical techniques can be applied if the substance to be analyzed exhibits selective absorption or other specific optical properties in the wavelength range in which skin absorption is small.

Some chemical quantities, such as the carbon dioxide content in mixed venous blood, can be estimated by respiratory gas analysis. Respiratory gas analysis also provides quantitative information about gas exchange and metabolism.

7.3.1 MEASUREMENTS BY INDWELLING TRANSDUCERS

Intravascular transducers have been studied extensively, and some of them have been used clinically. Transducers placed in the tissue are also used when a less intrusive method is preferred or a local change has to be measured; oxygen, carbon dioxide, and pH can be measured by such indwelling transducers. Ions such as sodium, potassium, calcium and catabolites such as glucose and urea have also been measured. Different kinds of transducers (electrochemical, fiberoptic, and mass spectrometric) have been developed and evaluated.

7.3.1.1 Intravascular Measurements

Although many attempts at intravascular measurements of chemical quantities have been carried out, clinically accepted methods are limited. Among them, intravascular blood gas measurements, especially oxygen measurements, have been studied extensively because of the need for the monitoring of neonates in whom blood oxygenation may often fail to achieve an adequate level.[101] Blood gas monitoring is also performed in adults in intensive care.[102]

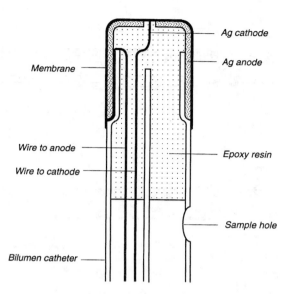

Figure 7.45 A Clark-type oxygen electrode for intravascular measurement. (Modified from Conway, M. et al., *Paediatrics*, 57, 244, 1976.)

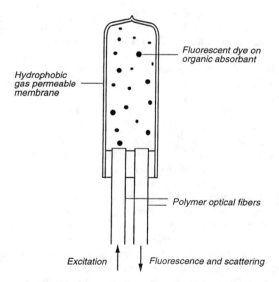

Figure 7.46 A fiberoptic intravascular catheter for pO$_2$ measurement. (Modified from Peterson, J. I. et al., *Anal. Chem.*, 56, 62, 1984.)

In neonates, the catheter is usually inserted via the umbilical artery. By this approach, a catheter about 0.5 to 1.0 mm in diameter, or even slightly larger, can be used despite the body size being extremely small. In adults, catheters are inserted into the radial or brachial artery or into the central vessels via the femoral artery or vein.

Figure 7.45 shows an example of the Clark-type oxygen electrode.[103] It has a silver cathode and anode at the tip. The electrolyte is initially deposited on the tip, which is dip coated by PVC so as to form a membrane of about 25 μm. When it is immersed in blood, water vapor diffuses through the membrane and forms a liquid electrolyte layer in 10 to 45 min. Many attempts using amperometric oxygen electrodes for intravascular monitoring have been reported.[104-108]

Both oxygen partial pressure (pO$_2$) and hemoglobin oxygen saturation (SO$_2$) can be measured by fiberoptic intravascular catheters. To measure pO$_2$, a method based on fluorescence, as shown in

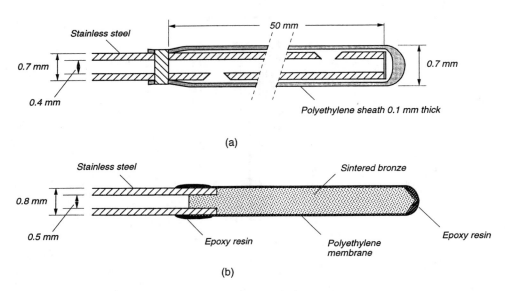

Figure 7.47 Catheters for mass spectrometric blood gas measurement. (Modified from Pinard, E. et al., *Med. Biol. Eng. Comput.*, 16, 59, 1978; Lundsgaard, J. S. et al., *J. Appl. Physiol. Resp. Environ. Exer. Physiol.*, 48, 376, 1980.)

Figure 7.46, was proposed.[109] It consists of a fluorescent dye, such as perylene dibutyrate, adsorbed in organic beads and contained within a hydrophobic membrane. The dye is excited with blue light at 486 nm and emits fluorescence at 514 nm. The fluorescence is quenched by the presence of oxygen. The effect depends on pO_2, and pO_2 can be determined by the intensity ratio between excitation and the fluorescence. SO_2 can be determined by the absorption spectra of oxyhemoglobin and deoxyhemoglobin, and the measurement can be performed intravascularly using fiberoptic catheters.[110] Also used is a combined catheter in which optical fibers for oxygen saturation measurement, a polarographic pO_2 electrode, and a blood sampling lumen are situated in a catheter.[111]

Intravascular measurement of pCO_2 has also been attempted using a potentiometric pH electrode surrounded by a CO_2 permeable membrane. For example, Parker et al.[112] reported a catheter-tip pCO_2 and pO_2 combined electrode. Coon et al.[113] reported an intravascular evaluation of a combined pH and pCO_2 electrode. Shimada et al.[114] described a catheter-tip pCO_2 electrode using a pH ISFET. Despite these efforts, an electrochemical approach to intravascular pCO_2 measurement is more problematic than for pO_2, and it still has not been accepted in clinical blood gas monitoring.

Blood gas partial pressure can also be monitored intravascularly by a mass spectrometer using a catheter which provides gas diffusion through a membrane at the tip. By using a mass spectrometer, various gases such as O_2, CO_2, and N_2 can be measured simultaneously. Figure 7.47(a) shows an example of the configuration of a catheter-type sampler.[115] The catheter consists of a stainless steel tube of 0.4 mm o.d. and 0 2 mm i.d., and the tip is covered by a polyethylene membrane. While the internal diameter is small, the membrane conductance is always sufficiently large so that the membrane is the limiting factor for steady gas flow.[116] Figure 7.47(b) shows another configuration in which a sintered bronze plug is used as a membrane support.[117] As the membrane, polyethylene and silicone rubber are commonly used, because O_2 and CO_2 are selectively permeable but water is practically impermeable. A stainless-steel tube is advantageous to prevent the diffusion of gases into the catheter, but it is too inflexible. Thus, flexible catheters made of a high-density polyethylene tube[118] or a nylon tube coated with polyurethane[119] have also been used.

Oxygen saturation and blood gas partial pressures are also measured with fiberoptic catheters. Such catheters have been used for intravascular monitoring of blood gases in neonates and intensive care patients. Earlier attempts mostly were concerned with oxygen saturation measurement by reflection spectrometry.[120-122] Taylor et al.[123] described an intravascular fiberoptic catheter which measures oxygen saturation and a dye dilution curve using four different wavelengths and having a lumen for pressure measurement. Wilkinson et al.[110] described a disposable fiberoptic catheter for newborn infants. The catheter is made of 4-F (1.3 mm o.d.) polyurethane dual-lumen tubing, where one lumen is for infusion, blood sampling, and pressure measurement. It was reported that it could be used for patient monitoring for up to 180 hours. In reflection measurement in flowing blood, reflection intensity changes with blood flow velocity. A study showed that differences in blood flow velocity of 0 to 30 cm/s resulted in changes of light reflection of about 50%, which corresponds to about a 10% change in oxygen saturation.

Gehrich et al.[124] described a fiberoptic catheter with fluorescence indicators for pO_2, pCO_2, and pH. It has three optical fibers and a thermistor and can be introduced into the artery through a 20G radial artery catheter. A fluorescence indicator is attached at the tip of each fiber. For pO_2 and pH measurement, oxygen-sensitive and pH-sensitive dyes are used, and pCO_2 is measured by the pH change in a buffer solution encapsulated within a silicone membrane as in the carbon dioxide electrode (see Section 7.2.1.4). A catheter for pO_2, pCO_2, and pH measurement with a bent fiber and side window arrangement was also reported by Zimmerman and Dellinger.[125] Wolthuis et al.[126] described a fiberoptic pO_2 catheter using a viologen indicator which becomes strongly absorbent after ultraviolet light stimulation. The rate at which the indicator returns to transparency is proportional to pO_2.

Intravascular measurements of pH; other ions such as K^+, Ca^{++}, and Na^+; or catabolites such as glucose have been available using electrochemical or fiberoptic transducers. However, such ions and small molecules are fairly permeable through the capillary wall, thus their intravascular concentrations may reflect tissue concentrations. In clinical situations, tissue measurements are easier, less invasive, and safer. Thus, at least in a clinical situation, intravascular measurement should be considered only when it is absolutely necessary.

7.3.1.2 Tissue Measurements

When a chemical transducer is positioned in the tissue, it will have direct contact with interstitial fluid. The concentration of chemical species in this fluid can be measured. Because capillaries are exposed to the interstitial fluid, smaller molecules and ions, which can freely diffuse through the capillary wall, will exist in the interstitial fluid at almost the same concentrations as those in the circulating blood. Thus, tissue measurement can be a supplement to direct intravascular measurement. Actually, tissue measurement is emerging as a possible method for glucose monitoring.

Tissue measurement is also used to study very local changes in some chemical species. An example is intramuscular measurements in which pH may change when the muscle works. Nowadays, special attention has been paid to the tissue measurement of nitric oxide (NO), which is an important bioregulatory molecule. Such a local tissue measurement is sometimes performed in an area as small as cellular size using microtransducers.

An example of clinical tissue measurement is the monitoring of the pH of the interstitial tissue fluid in newborn infants and surgical patients. Small-tipped pH glass electrodes have been developed for this purpose. Figure 7.48 shows an example of the tissue pH electrode.[127] It has a pH-sensitive glass tip 1 mm long and 1.3 mm in diameter. For a measurement, it is penetrated into the skin at least 3 mm so that the tip is placed in the subcutaneous tissue. In neonatal monitoring, the probe is applied to the thigh. It is applicable even on a fetus. Using an amnioscope, it is inserted into the vagina and the electrode probe is fixed onto the fetal scalp by means of a double-helical spiral.

The pH of a muscle can also be monitored in physiological studies or in a clinical situation. Khuri et al.[128] studied intramyocardial pH monitoring using a steel-jacketed glass electrode for assessing the adequacy of myocardial preservation. Covington et al.[129] reported on a method of monitoring intramyocardial pH during open heart surgery using a pH ISFET. In this study, the pH ISFET, 1.25 × 2.00 mm, was mounted with epoxy in a 2.0-mm probe with a collar to control the depth of insertion. The pH probe was applied next to a temperature probe in order to adjust the result according to its temperature coefficient. In this type of surgery, temperature drops significantly due to the regular surgical procedure of inducing elective cardiac arrest by perfusing a cold cardioplegic solution. An Ag-AgCl reference electrode was attached to the skin.

Glucose monitoring in the tissue has been studied extensively. Most of these studies use the amperometric enzymatic glucose sensor in which H_2O_2, O_2 or redox reaction of a mediator such as dimethylferrocene is measured (see Section 7.2.4.1). One difficulty in tissue glucose measurement is that the oxygen level in the tissue is low and unstable. Thus, the O_2-measuring type needs differential operation so the enzymatic electrode signal has to be compared with the O_2 level measured by another electrode. The H_2O_2-measuring type is advantageous because of simplified configuration. However, it still consumes oxygen in the electrode reaction, and the oxygen supply becomes insufficient when the glucose concentration is high and the oxygen level is low. This problem has been solved by using a membrane by which the ratio of glucose-to-oxygen flux can be decreased. A study showed that by using a polyurethane

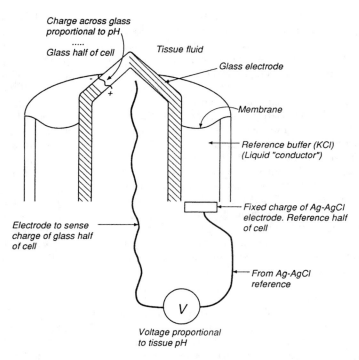

Figure 7.48 A pH glass electrode for tissue pH measurement. (From Hochberg, H. M., *J. Clin. Eng.*, 6, 17, 1981. With permission.)

membrane on a H_2O_2 enzymatic glucose sensor, a linear response could be obtained in a glucose concentration from 0 to 500 mg/dl without being affected by oxygen tensions above 25 mmHg.[130] The effect of oxygen level can also be eliminated by using a mediator (see Section 7.2.4.1). This technique was introduced in glucose sensors for tissue measurements.[131]

For long-term monitoring, needle-type glucose sensors of small diameters, down to 0.5 mm or less, have been developed. For example, a H_2O_2-based glucose sensor having the shape and size equivalent to a 26-gauge needle (0.45 mm o.d.) was developed.[132] This sensor can be implanted into the subcutaneous tissue through a 16-gauge cannula. The cannula is then removed, leaving the sensor in place, and secured with plaster. Animal experiments with this type of sensor showed that the sensitivity and response time 10 days after implantation were almost the same as those 3 days after implantation.[133]

For long-term implantation of a sensor, biocompatibility is absolutely necessary. It is desirable to achieve this without reducing sensitivity and other performances. A study showed that good results were obtained by the encapsulation of perfluorosulfonic acid (Nafion®).[134] It was shown that even after 3 months of implantation, the fibrous connective tissue was relatively thin (<100 µm), and linearity and precision were comparable to conventional glucose electrodes.

Tissue measurement of different chemical species has also been studied extensively. A study showed that a graphite-epoxy capillary electrode coated with a thin film of Nafion® exhibits selectivity to cationic species. The primary neurotransmitters — dopamine, norepinephrine, and 5-hydroxitryptamine (5-HT) — could be detected by the amperometric principle, when the electrode was implanted in the brain tissue of a rat.[135] The Nafion®-coated carbon electrode can be used to detect nitric oxide (NO). In a study, a thermally sharpened carbon fiber with a tip diameter of about 0.5 µm, operated in either the amperometric or voltanmetric mode, was applied to detect NO in a single smooth muscle cell[136] and in the rat brain.[137] A NO-selective electrode was also made with a Pt/Ir alloy coated with a nitrocellulose polymer film, as illustrated in Reference 138.

7.3.2 *EX VIVO* MEASUREMENTS AND MEASUREMENTS BY MICRODIALYSIS

The *ex vivo* measurement is a technique in which a vital object is measured outside the body. A typical *ex vivo* chemical measurement is blood analysis either by continuous or intermittent blood drainage. The *ex vivo* measurement has many advantages: It does not need an indwelling chemical transducer,

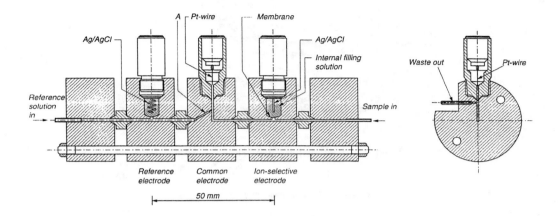

Figure 7.49 A flow-through electrode system for continuous potassium measurement during heart-lung bypass. (From Ossward, H. F., in *Proc. Int. Conf. on Monitoring of Vital Parameters during Extracorporeal Circulation, Nijmegen, 1980,* Karger, Basel, 1981, 149. With permission.)

thus the transducer can easily be calibrated and replaced. If the withdrawn fluid is discarded, contamination of the sampled fluid is not a serious problem, and any chemical processes to the sample can be applicable as in ordinary chemical analysis. However, if the withdrawn fluid is discarded continuously, it causes a loss of the body fluid, and thus the measurement system should be designed so as to minimize the flow rate of continuous drainage or the amount of intermittent sampling. Using advanced silicon fabrication technologies, micro-miniaturization of a chemical transducer system has been achieved, and *ex vivo* measurements in clinical situations are now becoming a reality. Microdialysis is also a kind of *ex vivo* technique, but, instead of withdrawing whole body fluid, it extracts only small molecules by employing the dialysis principle.

7.3.2.1 Measurement by Blood Drainage

By withdrawing blood continuously, chemical measurements can be achieved using a flow-through cell electrode system. For example, Ossward et al.[139] attempted continuous potassium measurement using an ion-selective potassium electrode and applied it to on-line-measurement during open heart surgery with a heart-lung machine. During heart lung bypass, the circulating blood is heparinized, and thus blood clotting is not a serious problem. Blood was withdrawn from the extracorporeal circuit and entered an ion-selective flow-through electrode system (Figure 7.49).[140] The rate of continuous blood drainage was 17.1 ml/h, and the time delay was about 2 min.

Shibbald et al.[141] reported an on-line patient-monitoring system for simultaneous analysis of blood K^+, Ca^{2+}, Na^+, and pH using a flow-through cell with a chemFET (chemically sensitive, field-effect transistor) made on a printed-circuit substrate, as shown in Figure 7.50. The volume of the cell was 30 µl, and an overall dead space was about 460 µl. Blood sampling was achieved by a dual-lumen cannula in which heparin-saline was pumped into the external lumen, and the heparinized blood was withdrawn at a rate of 1.46 ml/min. The sampling period of a measurement was 20 to 25 s, and sampling was performed at a frequency of 12 times per hour for monitoring renal dialysis.

Efforts have been made to reduce sample volume, and it was reduced to 10 µl using a chemFET system as shown in Figure 7.51(a).[142] The flow-through cell was a sandwich arrangement of a chip carrier, a chemFET chip, and a PMMA slide with an engraved flow-through channel. For measurement of an ion, two ISFETs with the same kind of membrane were located in such a way that the meander flow-through channel separated both ISFETs. The cell was connected to the inner lumen of a dual-lumen cannula, and the calibration solution with heparin was supplied through the outer lumen. Interruption of the stream of calibration solution allowed blood to enter the cell. After reaching the first ISFET, as shown in Figure 7.51(b), the differential output gave the measurement signal. When the cycle time was 2 min, blood consumption was 7.2 ml per 24 h. The system was tested in pigs to monitor K^+, Ca^{2+}, and pH.

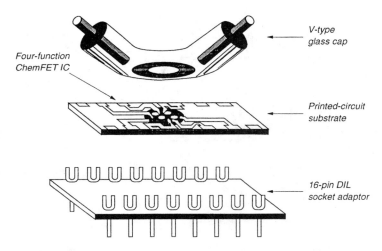

Figure 7.50 A flow-through cell with chemically sensitive field-effect transistor fabricated on a printed-circuit substrate. (From Shibbald, A. et al., *Med. Biol. Eng. Comput.,* 23, 329, 1985. With permission.)

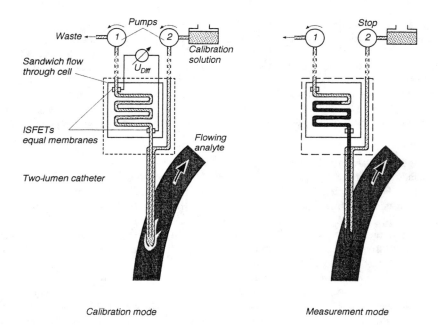

Figure 7.51 A small flow-through cell with two ISFETs. (From Gumbrecht, W. et al., *Sensors Actuators,* B1, 477, 1990. With permission.)

7.3.2.2 Measurement by Microdialysis

Microdyalysis is a technique for extracting small molecules into a perfused solution by interposing a dialysis hollow fiber implanted in the tissue. Using microdialysis, glucose and other small molecules (such as neurotransmitters in the body fluid) can be analyzed continuously in connection with conventional chemical analyzers.

The microdialysis probe consists of a double lumen cannula with a dialysis membrane situated at the tip, as shown in Figure 7.52. The probe tip is placed in the body fluid space, and a buffer solution is perfused via the outer lumen of the cannula. The solutes that are permeable through the dialysis membrane enter into the perfusion fluid at a rate determined by the concentration gradient, the membrane permeability, and the surface area of the membrane. The fluid is then led to the analyzer via the inner

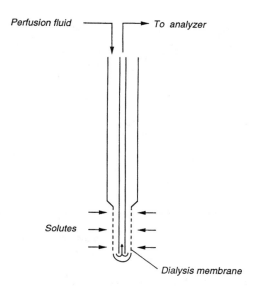

Perfusion fluid ⟶ ⟵ To analyzer

Solutes ⟶

Dialysis membrane

Figure 7.52 Microdialysis probe.

lumen of the cannula. Microdialysis probes having different cut-off molecular weights are available commercially (for example, CMA/Microdyalysis AB; Stockholm, Sweden).

Glucose concentration in the extracellular fluid space can be measured using a tissue microdialysis technique. Pfeiffer et al.[143] used a microdialysis probe with a polycarbonate-polyether co-polymer membrane. The probe has a diameter of 0.6 mm and consists of two concentric steel cannulas. A tubular dialysis membrane 0.5 mm in diameter is glued to two cannulas at the tip. The probe was implanted in the para-umbilical tissue and perfused with phosphate buffer saline at a rate of 4.5 μl/min. Glucose concentration in the dialysate was measured by a H_2O_2-type enzymatic glucose sensor. The probe was calibrated by comparing the glucose level in the dialysate and the actual blood glucose obtained by sampling blood. It was shown that the obtained subcutaneous glucose corresponded well with blood glucose both in healthy subjects and diabetic patients. Laurell[144] reported in a similar study that an enzymatic reactor and a Clark-type oxygen electrode were used to measure glucose concentration in the dialysate.

Microdialysis has been successfully applied in neuroscientific research. Clemens and Phebus[145] reported a measurement of the release of ascorbate for dopaminergic stimulation in conscious rats using microdialysis in the brain. The probe consisted of two parallel stainless-steel cannulae with a U-shaped loop of dialysis tubing 250 μm in diameter and 5 mm long at its tip. The probe was placed surgically in the brain tissue in the anesthetized animal. After recovery, saline was perfused at a rate of 8 μl/min. The dialysate was collected at 15-min intervals, and the ascorbate concentration was assayed by high-performance liquid chromatography.

Flentge et al.[146] reported on-line continuous brain dialysis. In the study, choline and acetylcholine in the extracellular fluid space in the brain were continuously measured using a setup shown in Figure 7.53. The probe consisted of a microdialysis tube of saponified cellulose ester and was placed in the brain tissue of a rat. A buffer solution was perfused at a rate of 1 μl/min. The dialysate was led into an enzyme reactor (preoxidator), and released H_2O_2 was detected by an electrochemical detector using the amperometric principle. Microdialysis has also been applied to analyze amino acids and biogenic amines using liquid chromatography-electrochemistry techniques and to analyze lactate, pyruvate, purines, and some drugs using liquid chromatography-ultraviolet absorbance detection.[147]

7.3.2.3 Measurement by Effluent Fluid Analysis

A group from the National Defence Medical College in Saitama, Japan, under the leadership of Dr. Kikuchi, has developed methods for the determination of glucose in effluent fluid.[148-153] Effluent fluid is defined as the interstitial fluid that can be collected if the outermost layer of the skin is broken and fluid from the extracellular space is exposed. For example, if the stratum corneum is removed by stripping,

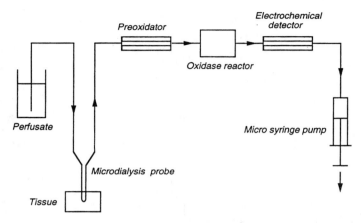

Figure 7.53 An example of continuous monitoring of neurotransmitters in the brain using microdialysis technique. (Modified from Flentge, F. et al., *Anal. Chem.*, 204, 305, 1992.)

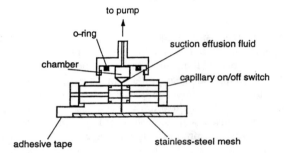

Figure 7.54 The effluent-fluid collecting chamber. (Modified from Kayashima, S. et al., *IEEE Trans. Biomed. Eng.*, 38, 752, 1991.)

fluids in small quantities can be collected. This fluid is, in many respects, a mirror of the corresponding blood concentrations of a variety of substances — for instance, glucose and creatinine, with the exception of protein and lipids.[152] This means that the effluent fluid can be used as a medium for glucose determination in blood. The concentrations in blood and effluent fluids are not time synchronous, though. A delay on the order of 5 to 10 min exists between the two compartments. Effluent fluid can be collected by applying weak suction on the skin (400 mmHg absolute pressure). The use of a collecting probe attached to the skin facilitates the collection (see Figure 7.54). In this procedure, the corneum layer of the epidermis has been removed by repetitive stripping. Analysis of the glucose content either can be done by an ISFET or can be accomplished by using ATR techniques. Ito et al.[153] have demonstrated that it is possible to apply both methods for the glucose determination.

A good correlation exists between what is read in terms of glucose concentration with ISFET and a clinical analyzer. The Japanese group has studied diabetic patients exposed to a 75-g glucose loading. They found that the glucose concentration increased in both the effluent fluid and serum. The effluent fluid glucose concentration changes in proportion to the serum glucose concentration but with a 5- to 10-min delay.

The advantages of this method are that it is noninvasive and is a quasi-continuous method for blood glucose monitoring. One problem is the very small amount of fluid (average 11.9 ± 3.7 ml/h/cm^2) that can be extracted with this method. This technique is an interesting contribution to the problem of continuous analysis of glucose concentration in whole blood.

7.3.3 TRANSCUTANEOUS MEASUREMENTS

Blood gas determination can provide important diagnostic information concerning the efficiency of pulmonary gas exchange and the status of alveolar ventilation, blood gas transport, and tissue oxygenation. Today, relatively simple techniques have been developed that can monitor blood gases from skin

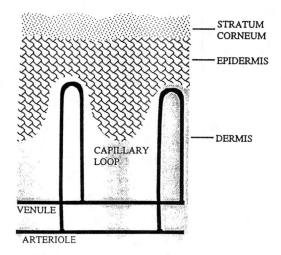

Figure 7.55 A cross-section of the human skin.

measurements. These techniques make it possible to avoid blood sampling, which is painful and associated with risks. Blood sampling techniques are, by definition, discontinuous and do not provide information on blood gas status in periods between the samples.

In neonatology, blood gas monitoring is an important method for monitoring alveolar ventilation, which provides information about early changes in oxygenation. Regarding transcutaneous monitoring of blood gases, several noninvasive techniques have been developed, some of which have found widespread use in the intensive care of critically ill patients. Methods include:

1. Transcutaneous monitoring of pO_2, pCO_2, and pH, using electrode techniques
2. Oxygen saturation measurement by pulse oximetry
3. The use of mass spectrometry in gases that diffuse through the skin

7.3.3.1   The Skin as a Membrane in Transcutaneous Measurements

The skin forms a barrier between the interior of the human body and its surroundings. In transcutaneous measurements, the skin always interacts with the measurement system. For instance, in optical measurements skin absorbs, refracts, and scatters photons. Skin constitution, anatomy, and pigmentation strongly affect the photon transport and thereby the measurement. Anderson and Parrish[154] have reviewed the optical properties of human skin. In electrode measurement, skin thickness and composition affect the oxygen diffusion time.

Oxygen permeability was measured by Tregear[155] and Scheuplein and Blank.[156] In connection with mass spectrometry determination of blood gases,[157] they found a diffusion rate of 25×10^{-6} ml cm^{-2} s^{-1} for heated adult skin, with data being affected by temperature and thickness/structure of the skin.

The examples given above lead to a somewhat closer look at skin anatomy and physiology. The interested reader who wants more extensive treatment of this topic is referred to Tregear[155] and Anderson and Parrish.[154] Several review articles discuss the optics of the skin from a more theoretical, physical perspective.[158-161]

The human skin consists of three separate layers: stratum corneum, epidermis, and dermis. The layers vary in thickness from 0.2 to 2 mm on different parts of the body. Figure 7.55 shows a cross-section of the human skin.

The stratum corneum is a nonliving layer of dehydrated cells which gives the skin many of its barrier properties. For instance, the electrical resistance of this part of the skin is rather high and reduces the risk of hazardous electrical currents entering the body. The epidermal layer is a living layer consisting of proteins, lipids, and melanin-producing cells which give the skin its typical color. The thickness of the epidermis is usually 0.1 to 0.2 mm. The dermis is perfused by capillaries arranged in vertical loops, 0.2 to 0.4 mm in length. The capillary density varies with skin site but is usually in the range of 1/mm^2.

The capillaries receive blood from arterioles, which form a network parallel to the skin surface. The skin blood supply is drained by venules and larger veins. In parallel with the arterial-capillary-venule system, arterio-venous anastomoses exist between the arterial and the venous blood vessel systems.

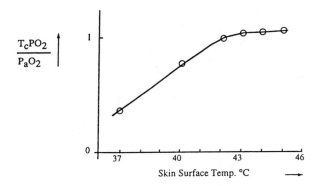

Figure 7.56 The relation between t_cpO_2 and PaO_2 at skin temperatures of 37 to 45°C. (Modified from Rolfe, in *IEE Medical Electronics Monographs,* 18–22, Hill, D. W. and Watson, B. W., Eds., Peter Peregrinus, Steveange, 1976, 126.)

These shunts are frequently found in local skin areas especially in the palms, feet, and face, the acral regions of the body. The shunts are muscular regions innervated by the sympathetic nervous system. These shunting vessels are part of the thermal regulatory system of the body, and the shunts are controlled by information about the body temperature. At higher body temperatures, these channels are wide open.

Normally the skin is not very permeable to gases, but at higher temperatures, the skin's ability to transport gases is improved. Baumberger and Goodfriend[157] positioned a finger in electrolytes at 45°C and found that the oxygen level rose in the liquid to values close to the value present in arterial blood. The heat dilates the blood vessels, blood flow increases, and gas transport becomes facilitated. Figure 7.56 shows the relation between t_cpO_2 and PaO_2 at skin temperatures from 37 to 45°C in a newborn infant.[118] Complete agreement is not reached for temperatures lower than 43°C in infants and for even higher temperatures in adults.

7.3.3.2 Transcutaneous Blood Gas Measurements
7.3.3.2.1 *Transcutaneous Measurement of pO_2.* It is important to estimate tissue oxygenation levels because they reflect the function of several physiological processes forming the links of the oxygen transport chain, e.g., gas exchange in the lung, blood circulation at various levels in the circulatory system, gas exchange at the cellular level, etc. The earlier techniques for blood partial oxygen pressure (pO_2) determination used blood samples from peripheral arteries followed by measurements in laboratory blood gas analyzers.

The idea to measure arterial pO_2/PaO_2 from the skin surface (t_cpO_2) was developed in 1951 by Baumberger and Goodfriend,[157] as described earlier in this chapter. The vasodilatation and the associated increased perfusion are prerequisites for t_cpO_2 measurements; Evans and Naylor[162] have shown that pO_2 on the skin surface is close to zero. Skin heating increases the permeability of oxygen in the skin but also increases the oxygen partial pressure at a given value of the oxygen saturation, SO_2.

Oxygen transducers for transcutaneous electrochemical measurements are based on polarography (see Section 7.2.1.3). The Clark-type oxygen electrode (see Section 7.2.1.4) inspired many ideas for the further development of the original concept into a commercial transcutaneous sensor for t_cpO_2 measurements.[163,164] Figure 7.57 shows the sensor head of a miniaturized polarographic-heated pO_2 electrode. In general, the cathode is made of platinum or gold. The anode is usually a silver ring coated with silver chloride (Ag-AgCl). The electrolyte is usually potassium chloride with a buffering agent added.

The sensor can be calibrated conveniently at two points by using gases with known oxygen concentrations. "Low point" or "zero" calibration is performed by exposing the sensor to 100% nitrogen or a sodium sulfite solution. "High point" calibration is performed with a gas mixture containing 10% oxygen. Manufacturers recommend calibration of the low point on a monthly basis. Calibration in room air is recommended before each measurement. The Clark-type pO_2 electrode gives an accurate estimate of arterial oxygen in infants.[165] In adults, the method usually underestimates PaO_2 by 5 to 15%.[166]

Today, we have a wealth of information concerning the clinical use of the t_cpO_2 electrode. Several monographs have been published;[167-169] in addition, many articles concerning experiences in neonatology have been published. Rithalia[170] has reviewed the most important literature.

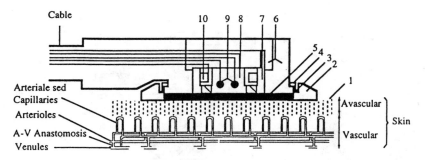

Figure 7.57 Transcutaneous pO_2 electrodes.

The principal application of t_cpO_2 has been in the monitoring of newborn infants. This patient group is suitable because of its thin epidermal layer. In premature infants, and especially in those with respiratory disturbances, there is a need for frequent O_2 determinations. Administration of O_2 to infants requires close control of arterial concentrations, as too high p_aO_2 levels can lead to retinal and pulmonary injuries. Concentrations too low can result in brain damage. In this connection, t_cpO_2 monitoring is an important method in the care of newborn infants with initial respiratory problems.

Transcutaneous assessment of arterial blood gas tensions is, however, influenced by a number of physiological, technical, and methodological parameters that are impossible to predict or control.[171] Transcutaneous t_cpO_2 measurements assess the oxygen tension in the skin under the electrode and not in the systemic circulation. This means that the skin measurement always measures local pressure level, not oxygen pressures at the systemic level. Blood flow changes in the skin are the major source of variations in skin pO_2 measurements.

To "arterialize" capillary blood flow, skin heating to 44 to 45°C is necessary. Long-term applications may cause burns,[172] and frequent changes in the sensor position are necessary to avoid burns. After each repositioning, a period of 15 to 20 min is required before readings become stable and new data can be collected. Transcutaneous partial pressure sensors are very sensitive pieces of equipment and must be handled with care. Equipment should be checked on a regular basis through the calibration procedure given above or through direct comparison with direct blood sampling.

7.3.3.2.2 Transcutaneous pCO_2 Sensor.

Huch et al.[173] were the first group to report successful development of a t_cpCO_2 sensor. Such a sensor has many similarities to a t_cpO_2 probe. The electrode is a pH-sensitive glass electrode with an Ag-AgCl reference electrode positioned in a concentric manner around the pH electrode. The buffer electrolyte is used, and a CO_2-permeable Teflon membrane forms a barrier between the interior of the sensor and its surroundings, similar to the Severinghaus electrode (see Section 7.2.1.4). The response time in t_cpCO_2 measurements is usually longer than for corresponding pCO_2 measurements. In addition, the response time is temperature dependent.[174]

Newer developments combine t_cpO_2 and t_cpCO_2 in the same sensor. Such a sensor is presented in Figure 7.58 (Roche-Kontron) where a Clark polarographic electrode is combined with a Stow-Severinghaus pCO_2 electrode.[175] A gold or platinum cathode is used for pO_2 measurements, and a pH-sensitive glass electrode for the pCO_2 recording.

A heated Ag-AgCl reference anode is common to the two systems, which is also true for the electrolyte and the diffusion membrane which is permeable for both oxygen and carbon dioxide. Several commercial instruments are available on the market: Hellige (Germany), Novametric (U.S.), Radiometer (Denmark), and Roche-Kontron (Switzerland). Transcutaneous pCO_2 monitoring has found major application in neonatology. Arterial pCO_2 varies inversely with alveolar ventilation and gives instant information about the effectiveness of ventilation.[175]

7.3.3.2.3 Mass Spectrometry in Transcutaneous Blood Gas Analysis.

Human skin is permeable to gases to a limited extent. The rate at which oxygen is delivered to adult skin can be calculated as 1.5×10^{-5} ml $\times$ cm^{-2} $\times$ s^{-1} and for neonates' skin as 0.5×10^{-3} ml $\times$ cm^{-2} $\times$ s^{-1}. If one assumes that no consumption of oxygen takes place in the tissue, these figures agree well with the ones that have been determined experimentally.[176] These authors state that if the measurement system is to avoid causing depletion in the surrounding tissue, it must consume less than 10^{-7} ml cm^{-2} s^{-1} of oxygen. Because the mass spectrometer can measure gases at the incoming gas levels of 10^{-7} to 10^{-8} ml s^{-1}, gas sampling can be attained with a sampling area of about 1 cm^{-2}.

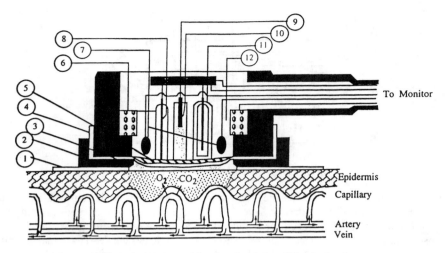

Figure 7.58 A combined t_cpO_2 and t_cpCO_2 electrode.

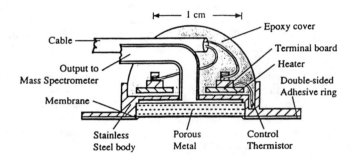

Figure 7.59 A gas sampler for transcutaneous blood gas measurement using the mass spectrometer.

The inlet to the mass spectrometer is an important part of the system because of the very small gas volumes that enter the instruments. The inlet cannula must be made from fine-bore metal tubing. The pumping system must be clean and leak-free. The stability of the pump system is very important. The gas sampler is a heated chamber made from gas-impermeable material in which one wall is exchanged for a gas-permeable membrane. Figure 7.59 illustrates the probe design of Delpy and Parker.[176] A porous metal disc acts as the membrane support medium. The probe is heated and temperature controlled by a thermistor. The diffusion membrane is mounted on the metal support and applied in direct contact with the skin.

The mass spectrometer is a well-known instrument for analysis of several gas components in liquids or gas volumes (see Section 7.2.5.1). This instrument is used routinely in clinical practice to analyze respiratory gases in connection with cardiac-output measurement (Fick's method). In the instrument, gas molecules are separated according to their molecular mass. The instrument is expensive and technically complex, which means that frequent adjustments are necessary.

7.3.3.2.4 *Gas Chromatography in Transcutaneous Blood Gas Analysis.* The gas chromatograph is an analyzer having the sensitivity required for transcutaneous measurement (see Section 7.2.5.2). Several manufacturers have tried to apply gas chromatography to the monitoring of blood gases, and evidence shows that gas chromatographs can provide near continuous arterial blood gas sampling.[177] The gas is collected from the skin surface through a spiral groove in a metal disc in close contact with the skin surface. The collection of gases diffusing through the skin is interrupted by helium gas intervals.

Existing equipment has not reached clinical acceptance because of the complexity, cost, and volume of the measurement setup. Microminiaturization of gas chromatographs by micromachining has been

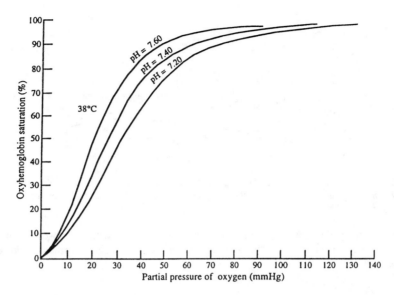

Figure 7.60 Oxygen dissociation curve of hemoglobin. (From Mendelson, Y., in *Encyclopedia of Medical Devices and Instrumentations,* John Wiley & Sons, New York, 1988, 448. With permission.)

suggested, and a combination of the present concept together with the features of new microtechnology may change the usability of this interesting concept dramatically.

Transcutaneous blood gas monitoring can never replace arterial blood sampling as a method for blood gas status monitoring but may contribute to analysis because of its relative simplicity, analysis speed, and the continuous recording that is of little harm to the patient. Newer methods such as pulse oximetry and capnography are slowly displacing the transcutaneous methods. Electrochemical sensors, though, are in a majority as devices for transcutaneous measurement of blood gases. Improvements such as the combined O_2/CO_2 sensor may increase the interest for transcutaneous electrode-based sensors again. Training and education for medical personnel regarding the proper handling of devices is imperative for future acceptance of these types of devices.

7.3.3.3 Transcutaneous Arterial Oxygen Saturation Monitoring

Oxygen is transported from the lungs to individual cells in the tissue volume via two different routes. About 98% is chemically bound to the hemoglobin molecule in the erythrocyte as oxihemoglobin (HbO_2). The additional 2% is physically dissolved in blood plasma. There is a nonlinear relation between the partial pressure of oxygen and the amount of oxygen in HbO_2. The curves are sigmoid-shaped and dependent on both blood temperature and the pH of blood (Figure 7.60).[178] As can be seen from the shape of this curve, pO_2 is a sensitive indicator of the blood oxygen level in the upper, right-hand part of the curve, and SO_2 is a more sensitive indicator in the left, steeper part of the curve below 80% saturation. In principle, SO_2 can be derived from the pO_2 value from the dissociation curve, but errors may occur, especially under abnormal physiological conditions. Under normal conditions, the pO_2 of arterial blood is about 100 mmHg. The corresponding SO_2 is about 98%.

7.3.3.3.1 *The Basis of Oximetry.* Kramer[179] and Matthes[180] have worked with hemoglobin oxygen saturation measurement in tissue. Their work is based on the fact that the absorbance of whole blood varies strongly with the wavelength used and also with the level of oxygenation. Oximetry is based on the differences in the optical transmission spectra between oxygenated and deoxygenated hemoglobin in the visible and near-infrared parts of the spectrum. The differences are illustrated by Figure 7.61.

The two spectra cross at a wavelength of 805 nm, a point called the isobestic point. Here the absorption is independent of the oxygenation level of the blood. The point is often used as a reference. The largest difference between the two curves can be found around 650 to 660 nm.

In this interval, the transmission of oxygenated hemoglobin is much higher than for deoxygenated hemoglobin. Transmission measurements in hemoglobin solutions at two wavelengths make it possible to assess the amount of hemoglobin present and its state of oxygenation.

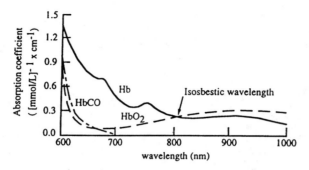

Figure 7.61 Optical transmission spectra of oxygenated and deoxygenated hemoglobin. (From Mendelson, Y., in *Encyclopedia of Medical Devices and Instrumentations,* John Wiley & Sons, New York, 1988, 448. With permission.)

A solution of hemoglobin can be characterized by the degree to which it absorbs optical radiation. Beer-Lambert's law can be written as the relation between the incident and transmitted light intensity levels for a fixed geometry:

$$\frac{I_t}{I_o} = e^{-acd} \tag{7.59}$$

where I_t is the transmitted light intensity, I_o is the incident light intensity, a is the absorption coefficient, c is the concentration of absorbing material, and d is the length of the optical path.

This law is only valid for conditions free from light scattering. This in turn means that only hemolyzed blood can be modeled with this equation. The light scattering occurring in whole blood must be compensated for.

Beer-Lambert's law is often written as an expression for optical density, *OD*:

$$OD = \log_{10}\frac{I_o}{I_t} = a \cdot c \cdot d \tag{7.60}$$

where *OD* is the optical density of the blood. If the optical density is determined at two wavelengths (usually spproximately 660 and 805 or 940 mm), the optical oxygen saturation, SO_2, level can be calculated from the equation

$$SO_2 = A - B \cdot \frac{OD_1}{OD_2} \tag{7.61}$$

where A and B are empirically determined constants and OD_1 and OD_2 are the optical densities at the wavelengths λ_2 λ_1 and λ_2.

7.3.3.3.2 Early Oximeters.

Beer-Lambert's law was first used by Wood and Geraci[181] to determine arterial SO_2 in tissue. As pointed out earlier, this model is not good for scattering media. However, the quotient between density measurement at two wavelengths eliminates many of the errors caused by multiple scattering in tissue and blood.

The first measurements were performed on the ear lobe, which was compressed to 200 mmHg to record transmission in the bloodless tissue of the ear pinna. After this first measurement, the clamp was released and a second recording was made.

The SO_2 can then be calculated as the difference between the two subsequent measurements; however, many problems resulting in inaccuracy were associated with the clamping-declamping procedure.

Hewlett-Packard released the first commercial oximeter in 1970. This instrument was based on a high-intensity tungsten lamp and narrow bandfilter for eight wavelengths between 650 and 1050 nm. The filters are mounted on a rotating wheel in the light beam. The beam is focused onto the end of an

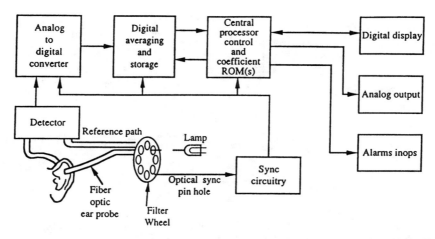

Figure 7.62 Block diagram of an ear oximeter using eight wavelengths. (From Mendelson, Y., in *Encyclopedia of Medical Devices and Instrumentations*, John Wiley & Sons, New York, 1988, 448. With permission.)

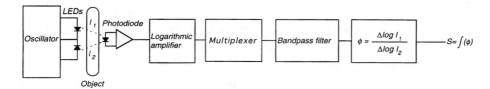

Figure 7.63 Principle of the pulse oximeter.

optical fiber which leads the light to the ear. Photons that have passed the ear are collected and brought to the instrumental part of the system (a block diagram is presented in Figure 7.62). A heater arterializes the blood in the volume under study for about 20 s. Oxygen saturation is determined by comparing the transmissions at the eight wavelengths used. From Beer-Lambert's law, we can determine

$$\%SO_2 = \frac{A_o + \sum_{n=1}^{8} A_n \log T_n}{B_o + \sum_{n=1}^{8} B_n \log T_n} \cdot 100 \tag{7.62}$$

where A_n and B_n are constants that can be determined from experiments in which gases of known concentrations have been inhaled. The Hewlett-Packard oximeter has an accuracy of about 3% and a response time close to 1.5 s. The major disadvantage with this product is the heavy fiberoptic cable. The oximeter is no longer being manufactured.

7.3.3.3.3 Pulse Oximeter.

Aoyagi[182] was the first to suggest a different approach to saturation measurements; he suggested that the optical absorption differences seen at every heartbeat are caused by an influx of arterial blood only. This phenomenon is often illustrated as in Figure 7.63, in which the various attenuations in tissue are illustrated. Usually 660 nm is used as one of the wavelengths, and 940 nm (or 805, the isobestic point) is utilized for the second wavelength.

Pulse oximeters are calibrated empirically by correlating the optical changes in tissue with corresponding arterial SO_2 values. The standard pulse oximetry probe consists of red and infrared light-emitting diodes (LEDs) and a photodetector. Mendelson and Ochs[183] have pointed out the importance of a large detector area in order to improve the signal-to-noise ratio.

Pulse oximetry signals can be recorded on most parts of the human skin. Fingertips, earlobes, and toes are the most frequently used areas for pulse oximetry measurements, but all the areas of the skin

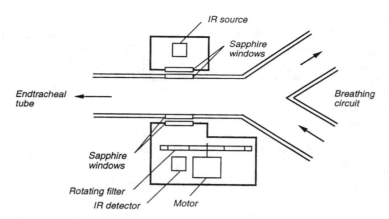

Figure 7.64   Diagram of a main-stream capnometer. (Modified from Olsson, S. G. et al., *Br. J. Anaest.*, 52, 491, 1980.)

give good signals. Tur et al.[184] have mapped the amplitude of the photoplethysmographic signal and its dependence on the anatomical site for the probe. Frequent calibration of pulse oximeters is not required. This fact and the convenience of using it to obtain noninvasive and continuous monitoring of oxygen saturation make it extremely advantageous for monitoring. In some countries, it is recommended that this technique be used for monitoring 100% of surgical interventions, for safety reasons. The instrument market has 30 to 40 competing companies today.

7.3.4   RESPIRATORY GAS ANALYSIS

The respiratory airway is a passage of chemical substances between the body and its environment. Aerobic metabolism is sustained by intaking oxygen and excreting carbon dioxide through the lung. Other substances, such as anesthetic agents and poisonous gases may also pass through the lung. Inspired and expired air is analyzed to obtain clinical and physiological information. Once a gas sample is taken from the airway, gas components can be analyzed by ordinary gas analyzers such as gas chromatographies or mass spectrometers. However, frequent or continuous gas analysis is required in patient monitoring and physiological studies, and specially designed transducers or instrumentations are required for such purposes. Actually, respiratory gas monitors are commonly used during anesthesia and in intensive care units. Continuous recording of oxygen consumption and carbon dioxide production is performed for estimation of the metabolic rate in physiological studies and clinical examinations.

7.3.4.1   Monitoring Ventilation

During general anesthesia or in an intensive care unit, where ventilation is maintained artificially, respiratory gas monitors are widely used. Keeping the inspired air at an adequate O_2 level, which is usually above 25%, is essential to maintain respiration. If gas is continuously sampled from the inspiratory stream in the anesthetic circuit, the O_2 content can be monitored by conventional O_2 sensors such as the polarographic O_2 electrode (see Section 7.2.1.4). When a rapid response is required, the solid electrolyte O_2 sensor (see Section 7.2.1.4) or the paramagnetic O_2 sensor (see Section 7.2.5.5) can be used.

   CO_2 content in expired gas provides information as to whether or not the lung is well ventilated. Inspired gas is usually free of CO_2, thus the CO_2 content in the stream near the mouth varies with the breathing cycle. Thus, detailed analysis of CO_2 content in expired gas requires an instantaneous gas monitor, called a capnometer. In ordinary capnometers, CO_2 is measured by infrared absorption at a wavelength of 4.26 μm, but CO_2 can also be monitored by a mass spectrometer.

   There are two types of capnometers: side-stream and main-stream (or in-line). In the side-stream capnometer, expired gas is continuously sampled from the airway and is led to a CO_2 analyzer via a sampling tube. In the main-stream capnometer, a CO_2 sensor, which is always of the infrared absorption type, is situated between the endotracheal tube and the breathing circuit.[185] Figure 7.64 shows the configuration of the airway adapter of a main-stream capnometer. It consists of an infrared source, a rotating filter, and a detector. The radiation passes through the gas to be measured in an airway adapter with sapphire windows, and the absorption is measured. A sealed cell with a known concentration of CO_2 can be a standard filter so that the CO_2 concentration is measured quantitatively compared to it.

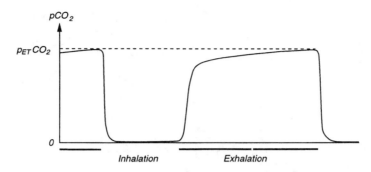

Figure 7.65 An example of a capnogram.

The main-stream capnometer is advantageous in that it does not require gas sampling and thus it has no transit time delay from the sampling site to the analyzer; however, condensation of water vapor on the optical windows can cause false readings. To avoid this, the windows have to be heated above body temperature. Some other factors also affect measurement accuracy, such as concentrations of nitrous oxide and other anesthetic agents, oxygen, and water vapor and atmospheric pressure.

Figure 7.65 shows an example of a capnogram. In the exhalation phase, partial pressure of carbon dioxide (pCO_2) increases rapidly due to the mixing of dead-space gas with alveolar gas and then reaches a plateau. At the end of exhalation, pCO_2 always reaches the maximum, and that value is called the end-tidal CO_2 tension, denoted as $p_{ET}CO_2$. During anesthesia and in intensive care units, monitoring of $p_{ET}CO_2$ is useful because changes in metabolism, pulmonary circulation, and alveolar ventilation reflect on $p_{ET}CO_2$. In commercial capnometers, $p_{ET}CO_2$ is displayed on the monitor screen together with the trend of pCO_2 (for example, Novametrix End Tidal CO_2 Monitor 1260, Novametrix Medical Systems, Inc.; Wallingford, CT). More details about the CO_2 analyzer and the capnometer can be found in reviews.[186,187]

Other than CO_2, optical absorption measurements at different wavelengths can be used for monitoring different gases such as nitrous oxide and halothane. By using two or more wavelengths, simultaneous monitoring can be realized. A study shows that O_2 and CO_2 can be simultaneously measured by absorption measurement, in which CO_2 is measured at 4.3 μm, and O_2 is measured at 147 nm.[188]

For respiratory gas monitoring, mass spectrometers have also been used; the principle and configurations of mass spectrometers are briefly described in Section 7.2.5.1. By using a mass spectrometer, different gas components can be measured simultaneously, according to their molecular weights. Mass spectrometers are fast enough for breath-by-breath on-line monitoring. For example, $p_{ET}CO_2$ and O_2 tension in inhaled gas (p_IO_2) can be monitored.[189] A mass spectrometer can be used for monitoring many patients simultaneously by connecting a multiplexing valve which selects gas samples successively from more than 10 patients.[190,191] One system services 16 patients using long sample lines of up to 30 m (Lifewatch Monitor, Perkin Elmer Co.; Pomona, CA).

7.3.4.2 Estimation of Metabolic Rate

Oxygen consumption and carbon dioxide production can be determined by measuring respiratory gas contents together with respiratory gas flow. Then the metabolic rate can be estimated. For this purpose, the Douglas bag, in which exhaled gas is collected, has been widely used in exercise physiology. However, such a measurement provides only averaged rates of oxygen consumption and carbon dioxide production, while continuous measurements are required both in physiological studies and clinical monitoring.

To measure oxygen consumption continuously, either the closed circuit or the open-circuit system can be used.[192] In the closed-circuit system, oxygen consumption is measured as the rate of decrease in gas volume in a closed circuit having a reservoir filled with pure oxygen and an absorber which absorbs carbon dioxide and water vapor. A spirometer (described in Section 3.4.2) can be used for this purpose; however, measurement by a closed-circuit system can be made only during a short period of time. Thus, open-circuit systems have been widely used to observe the metabolic rate for a long period of time. In the open-circuit system, the subject breathes fresh air from the atmosphere either through a non-return valve attached at a mouthpiece or mask or from an airstream maintained by means of a pump. The latter method is called the flow-through system.

The Oxylog[193] is a portable open-circuit system in which the subject wears a tight-fitting face mask, and inspiratory and expiratory air is separated by two non-return valves. The ventilation flow rate is

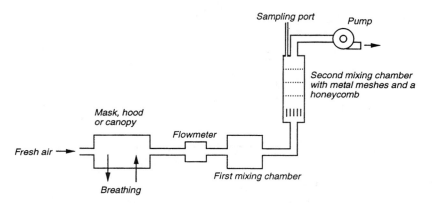

Figure 7.66 Diagram of the flow-through system for continuous oxygen consumption measurement.

measured by a turbine flowmeter incorporated into the inspiratory valve, and the oxygen content in the expiratory air, separated by the expiratory valve, is measured by a polarographic-type oxygen sensor (see Section 7.2.1.4). An evaluation of the Oxylog showed that oxygen consumption values were highly correlated to those obtained by Douglas's bag method with a correlation coefficient of above 0.99, and stable values could be obtained for at least 6 hours.[194]

Figure 7.66 shows the flow-through circuit. The subject breathes at the flow-through circuit using a face mask, hood, or canopy. The airflow rate through the circuit must be higher than the peak inspiratory and expiratory flow rates; otherwise, backflow may occur in the circuit and cause rebreathing. Usually, the airflow rate through the circuit is fixed at about 40 l/min or higher in adult subjects,[195] but at about 1 l/min in babies.[196] The expired gas and room air should be well mixed to eliminate phasic change in gas concentration due to respiration. Usually, two or more mixing chambers are used for this purpose. The use of an electric fan in the first chamber is recommended, and the second chamber should consist of a cylinder containing an aluminum honeycomb and meshed baffle plates.[195,197] The mixed gas is then sampled and gas contents are measured using a paramagnetic analyzer (see Section 7.2.5.5) or solid electrolyte transducer (see Section 7.2.1.4) for oxygen and an infrared analyzer for carbon dioxide.

When a flow-through system is applied during heavy exercise, oxygen consumption and carbon dioxide production increase 10 times or more than at rest, thus the airflow rate in the circuit should be set at a higher level than the maximum flow rate of breathing during exercise. However, if the airflow rate is fixed at such a higher level, the expired gas at rest is mixed with a large amount of fresh air. Thus, the difference in gas content between room air and mixed gas becomes very small and errors may increase. To overcome this difficulty, a servo-controlled system can be applied in which the airflow rate through the circuit is controlled so as to maintain the oxygen content in the mixed gas at a constant level.[198] When the oxygen content of the room air is 20.9%, and if that of the mixed gas is kept at 19.9%, then oxygen consumption is given as

$$\dot{V}_{O_2} = \dot{V}_M(0.209 - 0.199) = 0.01\dot{V}_M \qquad (7.63)$$

where $\dot{V}_M$ is the flow rate of the mixed gas. Oxygen content was measured by a Clark-type polarographic sensor having a time constant of about 3 s, and the time required to reach 63% of total response for a step input was about 4.5 s. A commercial model based on this principle is available (MRM-1 Oxygen Consumption Computer, Waters Instruments, Inc.; Rochester, MN). The subject has to wear a face piece but does not need to wear a mouthpiece or nose clip.

A portable system based on the servo-controlled flow-through method was also reported.[199] The subject wears a hood, as shown in Figure 7.67, and all instruments can be carried in a backpack. In this unit, a limiting-current, solid electrolyte oxygen sensor (see Section 7.2.1.4) is used. It has the advantages of short response time and low operating power; the time to reach 95% of total response for a step input was 4 s, and the power consumption is less than 1 W and can be supplied by a battery in a portable system.

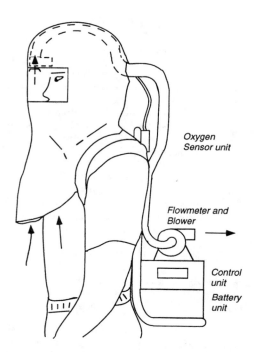

Figure 7.67 A portable instrument of a servo-controlled, flow-through oxygen consumption measurement system.

7.3.4.3 Electronic Noses

Sensor arrays combined with pattern recognition routines have been developed to assess gas mixtures of various kinds. These new sensors are often called electronic noses. A comprehensive review has been published by Gardner and Bartlett.[200] The electronic nose works in a way similar to the human olfactory organ. A sense signal pattern from a sensor array is handled by a computer in which artificial neural networks are used. Olfaction seems to be based on sensors that appear to have a low specificity for each receptor cell, yet the combination of many cells will result in a large information content.

Electronic noses seem to have interesting applications in a variety of fields. It has been possible to classify liquors, tobacco brands, perfumes, and beers.[201-203] Scents from humans have also been investigated by Ödman.[204] His application has been able to find victims in debris following catastrophes.

The sensor mechanisms are of various design principles. Tin oxide sensors have been used by the Warwick group.[205] The active part of the sensor is doped tin oxide, which essentially works as an n-type semiconductor. Odors react with the chemisorbed lattice oxygen. The reaction injects electrons into the conduction band of the semiconductor but also increases the electron mobility through the lowering of the intragranular barriers; Figaro, Inc. (Japan) is manufacturing a commercial version of this sensor type. This type of sensor must be operated at about 3 to 400°C. The response time is usually a few seconds.

Another type of odor sensor is made from electroactive polymers, which change resistance when exposed to various gases. Polymer sensors can operate at room temperatures; and several polymer-odor combinations have been tested.[206] Langmuir-Blodgett films have also shown sensitivity to odors.[207]

Lundström et al.[208] have designed a structure of gas-sensitive field effect transistors with gates of catalytic metals such as Pd, Ir, and P(t). The gas-sensing property is based on the horizontal shift of the I_D-V_g curve that occurs upon exposure to gas mixtures of different kinds. The operational temperature of the field effect devices is usually 50 to 200°C. It is likely that gas sensors in the form of electronic noses can be used in diagnostic analysis of respiratory gases, monitoring of respiratory gases in cases of intoxication, and other healthcare-oriented problem areas in which gas analysis is of importance.

REFERENCES

1. Park, J. B., *Biomaterials — An Introduction*, Plenum Press, New York, 1979.
2. Aoki, H., Akao, M., Shin, Y., Tsuji, T., and Togawa, T., Sintered hydroxiapatite for a percutaneous device and its clinical application, *Med. Progr. Technol.*, 12, 213, 1987.
3. Tsuji, T., Aoki, H., Shin, Y., and Togawa, T., Hydroxiapatite percutaneous devices implanted in forearms of three volunteers for four years, *Artif. Organs*, 14(Suppl. 3), 185, 1990.
4. Bockris, J. O. M. and Reddy, A. K. N., *Modern Electrochemistry*, Plenum Press, New York, 1973.
5. Bard, A. J. and Faulkner, L. R., *Electrochemical Methods — Fundamentals and Applications*, John Wiley & Sons, New York, 1980.
6. Brett, C. M. A. and Brett, A. M. O., *Electrochemistry — Principles, Methods and Applications*, Oxford University Press, London, 1993.
7. Solsky, R. L., Ion-selective electrodes, *Anal. Chem.*, 60, 106R, 1988.
8. Solsky, R. L., Ion-selective electrodes, *Anal. Chem.*, 62, 21R, 1990.
9. Sanz, M. C., Ultramicro-methods and standardization of equipment, *Clin. Chem.*, 3, 406, 1957.
10. Koryta, J., Theory and application of ion-selective electrodes, *Anal. Chem. Acta*, 233, 1, 1990.
11. Martin, C. R. and Freiser, H., Coated-wire ion-selective electrodes and their application to the teaching laboratory, *J. Chem. Educ.*, 57, 512, 1980.
12. Bergveld, P., Development of an ion-selective solid-state device for neurophysiological measurements, *IEEE Trans. Biomed. Eng.*, BME-17, 70, 1970.
13. Bergveld, P., Development, operation, and application of the ion-sensitive field-effect transistor as a tool for electrophysiology, *IEEE Trans. Biomed. Eng.*, BME-19, 342, 1972.
14. Matsuo, T. and Wise, K., An integrated field-effect electrode for biopotential recording, *IEEE Trans. Biomed. Eng.*, BME-21, 485, 1974.
15. Moss, S. D., Janata, J., and Johnson, C. C., Potassium ion-selective field effect transistor, *Anal. Chem.*, 47, 2238, 1975.
16. Esashi, M. and Matsuo, T., Integrated micromulti-ion sensor using field effect of semiconductor, *IEEE Trans. Biomed. Eng.*, BME-25, 184, 1978.
17. Matsuo, T. and Nakajima, H., Characteristics of reference electrodes using a polymer gate, ISFET, *Sensors Actuators* 5, 293, 1984.
18. Matsuo, T. and Esashi, M., Methods of ISFET fabrication, *Sensors Actuators* 1, 77, 1981.
19. Cobbold, R. S. C., *Transducers for Biomedical Measurements: Principles and Applications*, John Wiley & Sons, New York, 1974.
20. Hudson, J. A., Measurement of oxygen tension by the oxygen cathode, *Med. Biol. Eng.*, 5, 207, 1967.
21. Clark, L. C., Monitor and control of blood and tissue oxygen tensions, *Trans. Am. Soc. Artif. Internal Organs*, 2, 41, 1956.
22. Severinghaus, J. W., Blood gas concentrations, *Handbook of Physiology, Respiration*, Vol. II, Fenn, W. O and Rahn, H., Eds., American Physiology Society, Washington, D.C., 1965, 1475.
23. Hahn, C. E. W., Techniques for measuring the partial pressures of gases in the blood. Part 1. *In vitro* measurements, *J. Phys. E. Sci. Instrum.*, 13, 470, 1980.
24. Hahn, C. E. W., Blood gas measurement, *Clin. Phys. Physiol. Meas.*, 8, 3, 1987.
25. Winquist, F. and Danielsson, B., Semiconductor field effect devices, in *Biosensors, A Practical Approach*, Cass, A. E. G., Ed., Oxford University Press, London, 1990, 171.
26. Gramling, J. J., Oxygen analyzers, *Hospital Instrumentation Care and Servicing for Critical Care Unit*, Spooner, R. B., Ed., Instrumentation Society of America, Pittsburgh, PA, 1977, 81.
27. Mackereth, F. J. H., An improved galvanic cell for determination of oxygen concentrations in fluid, *J. Sci. Instrum.*, 41, 38, 1964.
28. Harke, S., Wiemhöfer, H. D., and Göpel, W., Investigations of electrodes for oxygen sensors based on lanthanum trifluoride as solid electrolyte, *Sensors Actuators*, B1, 188, 1990.
29. Lukaszewica, J. P., Miura, N., and Yamazoe, H., A LaF3-based oxygen sensor with perovskite-type oxide electrode operative at room temperature, *Sensors Actuators*, B1, 195, 1990.
30. Saji, K., Takahashi, H., Kondo, H., Takeuchi, T., and Igarashi, I., Limiting current type oxygen sensor, *Proc. 4th Sensor Symp.*, Institute of Electrical Engineers of Japan, Tokyo, 1984, 147.
31. Kondo, H., Takahashi, H., Saji, K., Takeuchi, T., and Igarashi, I., Thin film limiting current type oxygen sensor, *Proc. 6th Sensor Symp.*, Institute of Electrical Engineers of Japan, Tokyo, 1986, 251.
32. Levin, I. W. and Lewis, E. N., Application of Fourier transform Raman spectroscopy to biological assemblies, in *Fourier Transform Raman Spectroscopy*, Chase, D. B. and Rabolt, J. F., Eds., Academic Press, San Diego, CA, 1994, 157.
33. Nakashima, K. and Imai, K., Chemiluminescence, molecular luminescence spectroscopy, in *Chemical Analysis Series*, Vol. 77, Schulman, S. G., Ed., John Wiley & Sons, 1993, 1–23.
34. Meites, L., *Handbook of Analytical Chemistry*, McGraw-Hill, New York, 1963.
35. Griffiths, P. R. and de Haseth, J. A., *Fourier Transform Infrared Spectrometry*, John Wiley & Sons, New York, 1986.
36. Skoog, D. A. and Leary, J. J., *Principle of Instrumental Analysis*, 4th ed., Saunders College Publishing, Philadelphia, PA, 1992.

37. Skoog, D. A., West, D. M., and Holler, F. J., *Analytical Chemistry*, 6th ed., Saunders College Publishing, Philadelphia, PA, 1994.
38. Harmer, A. and Scheggi, A., Chemical, biochemical, and medical sensors, in *Optical Fiber Sensor Systems and Applications*, Culshaw, B. and Dakin, J., Eds., Artech House, Norwood, MA, 1989, 599.
39. Lieberman, R. A., Intrinsic fiberoptic sensors, in *Fiberoptic Chemical Sensors and Biosensors*, Vol. 1, CRC Press, Boca Raton, FL, 1991, chap. 5.
40. Culshaw, B. and Dakin, J., *Optical Fiber Sensor Systems and Applications*, Artech House, Norwood, MA, 1989.
41. Wolfbeis, O. S., in *Fiberoptic Chemical Sensors and Biosensors*, Vol. 1, CRC Press, Boca Raton, FL, 1991.
42. Hlavay, J. and Gilbault, G. G., Applications of the piezoelectric crystal detector in analytical chemistry, *Anal. Chem.*, 49, 1890, 1977.
43. Hlavay, J. and Gilbault, G. G., Detection of ammonia in ambient air with coated piezoelectric crystal detector, *Anal. Chem.*, 50, 1044, 1978.
44. Sauerbrey, G., Verwendung von Schwingquarzen zur Wägung dünner Schichten und zur Mikrowägung, *Zeitschrift für Physik*, 155, 206, 1959.
45. Alder, J. F. and McCallum, J. J., Piezoelectric crystals for mass and chemical measurements, *Analyst*, 108, 1169, 1983.
46. Deakin, M. R. and Byrd, H., Prussian blue coated quartz crystal microbalance as a detector for electroinactive cations in aqueous solution, *Anal. Chem.*, 61, 290, 1989.
47. Deakin, M. R., Li, T. T., and Melroy, O. R., A study of the electrosorption of bromide and iodide ions on gold using the quartz crystal microbalance, *J. Electroanal. Chem.*, 243, 343, 1988.
48. Ballantine, D. S. and Wohljen, H., Surface acoustic wave, *Anal. Chem.*, 61, 704A, 1989.
49. Danielsson, B. and Mosbach, K., Theory and application of calorimetric sensors, in *Biosensors, Fundamentals and Applications*, Turner, A. P. F., Karube, I., and Wilson, G. S., Eds., Oxford Science, London, 1987, 575.
50. Danielsson, B. and Winquist, F., Thermometric sensors, in *Biosensors, A Practical Approach*, Cass, A. E. G., Ed., Oxford University Press, London, 1990, 191.
51. Xie, B., Danielsson, B., Norberg, P., Winquist, F., and Lundström, I., Development of a thermal micro-biosensor fabricated on a silicon tip, *Sensors Actuators*, B6, 127, 1992 .
52. Turner, A. P. F., Karube, I., and Wilson, G. S., in *Biosensors, Fundamentals and Applications*, Turner, A. P. F., Karube, I., and Wilson, G. S., Eds., Oxford Science, London, 1987.
53. Cass, A. E. G., in *Biosensors, A Practical Approach*, Cass, A. E. G., Ed., Oxford University Press, London, 1990.
54. Turner, A. P. F., *Advances in Biosensors*, JAI Press, Hampton Hill, 1993.
55. Alcock, S. J. and Turner, A. P. F., *In Vivo Chemical Sensors: Recent Developments*, Cranfield Press, Bedford, 1993.
56. Janata, J. and Bezegh, A., Chemical sensors, *Anal. Chem.*, 60, 62R, 1988.
57. Janata, J., Chemical sensors, *Anal. Chem.*, 62, 33R, 1990.
58. Janata, J., Chemical sensors, *Anal. Chem.*, 64, 196R, 1992.
59. Janata, J., Josowicz, M., and De Vaney, D. M., Chemical sensors, *Anal. Chem.*, 66, 207R, 1994.
60. Barker, S. A., Immobilization of the biological component of biosensors, in *Biosensors, Fundamentals and Applications*, Turner, A. P. F., Karube, I., and Wilson, G. S., Eds., Oxford University Press, London, 1987, 85.
61. Hill, H. A. O. and Sanghera, G. S., Mediated amperometric enzyme electrodes, in *Biosensors, A Practical Approach*, Cass, A. E. G., Ed., Oxford University Press, London, 1990, 19.
62. Kuan, S. S. and Guilbault, G. G., Ion-selective electrodes and biosensors based on ISEs, in *Biosensors, Fundamentals and Applications*, Turner, A. P. F., Karube, I., and Wilson, G. S., Eds., Oxford University Press, London, 1987, 135.
63. Tran-Minh, C. and Vallin, D., Enzyme-bound thermistor as an enthalpimetric sensor, *Anal. Chem.*, 50, 1874, 1978.
64. Yamamoto, N., Nagasawa, Y., Sawai, M., Sudo, T., and Tsubomura, H., Potentiometric investigations of antigen-antibody and enzyme-enzyme inhibitor reactions using chemically modified metal electrodes, *J. Immunol. Meth.*, 22, 309, 1978.
65. Janata, J. and Blackburn, G. F., Immunochemical potentiometric sensors, *Ann. N. Y. Acad. Sci.*, 428, 286, 1984.
66. Roederer, J. E. and Bastiaans, G. J. Microgravimetric immunoassay with piezoelectric crystals, *Anal. Chem.*, 55, 2333 1983.
67. Aizawa, M., Morioka, A., Suzuki, S., and Nagamura, Y., Enzyme immunosensor. III. Amperometric determination of human chorionic gonadotropin by membrane-bound antibody, *Anal. Biochem.*, 94, 22, 1979.
68. Aizawa, M., Suzuki, S., Kato, T., Fujiwara, T., and Fujita, Y., Solid-phase enzyme immunoassay of IgG and anti-IgG using a transparent and nonporous antibody-bound plate, *J. Appl. Biochem.*, 2, 190, 1980.
69. Aizawa, M., Ikariyama, Y., Matsuzawa, M., and Shinohara, H., Optical biosensors and chemical sensors, *Dig. 4th Int. Conf. Solid-State Sensors and Actuators*, Institute of Electrical Engineers of Japan, Tokyo, 1987, 783.
70. Ikariyama, Y. and Suzuki, S., Luminescence immunoassay of human serum albumin with haemin as labeling catalyst, *Anal. Chem.*, 54, 1126, 1982.
71. Ikariyama, Y., Suzuki, S., and Karube, M., Luminescence catalyst immunoassay of β_2-microglobulin with haemin as a chemically amplifiable label, *Enzyme Microb. Technol.*, 5, 215, 1983.
72. Karube, I. and Nakanishi, K., Microbial biosensors for process and environmental control, *IEEE Eng. Med. Biol.*, 13, 364, 1994.
73. Karube, I., Micro-organism based sensors, in *Biosensors, Fundamentals and Applications*, Turner, A. P. F., Karube, I., and Wilson, G. S., Eds., Oxford Science, London, 1987, 13.

74. Karube, I. and Suzuki, S., Microbial biosensors, in *Biosensors, A Practical Approach,* Cass, A. E. G., Ed., Oxford University Press, London, 1990, 155.
75. Karube, I., Matsunaga, T., and Suzuki, S., A new microbial electrode for BOD estimation, *J. Solid Phase Biochem.,* 2, 97, 1977.
76. Tan, T. C., Li, F., and Neoh, K. G., Measurement of BOD by initial rate of response of a microbial sensor, *Sensors Actuators,* B10, 137, 1993.
77. Lee, S., Sode, K., Nakanishi, K., Marty, J., Tamiya, E., and Karube, I., A novel microbial sensor using luminous bacteria, *Biosensors Bioelectron.,* 7, 273, 1992.
78. Lee, S., Suzuki, S., Tamiya, E., and Karube, I., Sensitive bioluminescent detection of pesticides utilizing a membrane mutant of *Escherichia coli* and recombinant DNA technology, *Anal. Chem. Acta,* 257, 183, 1992.
79. Okada, T., Karube, I., and Suzuki, S., Hybrid urea sensor *Eur. J. Appl. Microbiol. Biotechnol.,* 14, 149, 1982.
80. Karube, I., Matsunaga, T., Nakahara, T., and Suzuki, S., Preliminary screening of mutagens with a microbial sensor *Anal. Chem.,* 53, 1024, 1981.
81. Karube, I., Nakahara, T., Matsunaga, T., and Suzuki, S., Salmonella electrode for screening mutagens, *Anal. Chem.,* 54, 1725, 1982.
82. Duckworth, H. E., Barber, R. C., and Venkatasubramanian, V. S., *Mass Spectroscopy,* 2nd ed., Cambridge University Press, London, 1988.
83. Chapman, J. R., *Practical Organic Mass Spectrometry,* 2nd ed., John Wily & Sons, Chichester, 1993.
84. Gaskel, S. J., Mass spectrometry in medical research, *Clin. Phys. Physiol. Meas.,* 6, 1, 1985.
85. Dawson, P. H., *Quadrapole Mass Spectrometry and Its Applications,* Elsevier Scientific, Amsterdam, 1976.
86. Wiley, W. C. and McLeran, I. H., Time-of-flight mass spectrometer with improved resolution, *Rev. Sci. Instrum.,* 26, 1150, 1955.
87. Wilkins, C. L. and Gross, M. L., Fourier transform mass spectrometry for analysis, *Anal. Chem.,* 53, 1661A, 1981.
88. Baugh, P. J., *Gas Chromatography, A Practical Approach,* Oxford University Press, London, 1993.
89. Mayer, V. R., *Practical High-Performance Liquid Chromatography,* 2nd ed., John Wiley & Sons, Chichester, 1994.
90. Andrews, A. T., *Electrophoresis,* Clarendon Press, Oxford, 1986.
91. Ewing, A. G., Wallingford, R. A., Capillary electrophoresis, *Anal. Chem.,* 61, 292A, 1989.
92. Wilson, K. and Walker, J. M., *Principles and Techniques of Practical Biochemistry,* 4th ed., Cambridge University Press, London, 1994.
93. Skoog, D. A. and Leary, J. J., *Principles of Instrumental Analysis,* 4th ed., Saunders College Publishing, Fort Worth, TX, 1992.
94. Weil, J. A., Bolton, J. R., and Wertz, J. E., *Electron Paramagnetic Resonance,* John Wiley & Sons, New York, 1994.
95. Rabinovitch, B., March, W. F., and Adams, R. L., Noninvasive glucose monitoring of the aqueous humor of the eye. Part I. Measurement of very small optical rotation, *Diabetic Care,* 5, 254, 1982.
96. March, W. F., Rabinovitch, B., and Adams, R. L., Noninvasive glucose monitoring of the aqueous humor of the eye. Part II. Animal studies and the scleral lens, *Diabetic Care,* 5, 259, 1982.
97. King, T. W., Cote, G. L., McNichols, R., and Goetz, M. J., Multispectral polarimetric glucose detection using a single pockels cell, *Opt. Eng.,* 33, 2746, 1994.
98. Kell, D. B. and Davey, C. L., Conductimetric and impedimetric devices, in *Biosensors, A Practical Approach,* Cass, A. E. G., Ed., Oxford University Press, London, 1990, 125.
99. Ellis, F. R. and Nunn, J. F., The measurement of gaseous oxygen tension utilizing paramagnetism: an evaluation of the "Servomex" OA.150 analyzer, *Br. J. Anaesth.,* 40, 569, 1968.
100. Thring, M. W., British instrument exhibition — London 1951, *J. Sci. Instrum.,* 28, 293, 1951.
101. Rolfe, P., Intra-vascular oxygen sensors for neonatal monitoring, *IEEE Eng. Med. Biol. Eng.,* 13, 336, 1994.
102. Rolfe, P., *In vivo* chemical sensors for intensive-care monitoring, *Med. Biol. Eng. Comput.,* 28, B34, 1990.
103. Conway, M., Durbin, G. M., Ingram, D., MacIntosh, N., Parker, D., Reynolds, E. O. R., and Souter, L., Continuous monitoring of arterial oxygen tension using a catheter-tip polarographic electrode in infants, *Paediatrics,* 57, 244, 1976.
104. Kimmich, H. P. and Kreuzer, F., Catheter PO_2 electrode with low flow dependency and fast response, *Prog. Resp. Res.,* 3, 100, 1969.
105. Mindt, W., Sauerstoffsensoren fur *in vivo*-mussungen, Proc. *Jahrestagung D Ges. Biomed. Tech. Erlangen,* 29, 1973.
106. Huxtable, R. F. and Fatt, I., A flexible catheter-type oxygen sensor, *J. Appl. Physiol.,* 37, 435, 1974
107. Goddard, P., Keith, I., Marcovitch, H., Robertson, N. R. C., Rolfe, P., and Scopes, J. W., Use of a continuously recording intravascular oxygen electrode in the newborn, *Arch. Dis. Child.,* 49, 853, 1974.
108. Jansen, T. C., Lafeber, H. N., Visser, H. K. A., Kwant, G., Oeseburg, B., and Zijlstra, W.G., Construction and performance of a new catheter-tip oxygen electrode, *Med. Biol. Eng. Comput.,* 16, 274, 1978.
109. Peterson, J. I., Fitzgerald, R. V., and Buckhold, D. K., Fiberoptic probe for *in vivo* measurement of oxygen partial pressure, *Anal. Chem.,* 56, 62, 1984.
110. Wilkinson, A. R., Phibbs, R. H., and Gregory, G. A., Continuous measurement of oxygen saturation in sick newborn infants, *J. Pediatr.,* 93, 1016, 1978.
111. Parker, D., Continuous measurement of blood oxygen tension and saturation, in *Monitoring Vital Parameters During Extracorporeal Circulation,* Kimmich, H. P., Ed., Karger, Basel, 1981, 23.

112. Parker, D., Delpy, D., and Lewis, M., Catheter-tip electrode for continuous measurement of pO_2 and pCO_2, *Med. Biol. Eng. Comput.,* 16, 599, 1978.

113. Coon, R. L., Lai, N. C. J., and Kampine, J. P., Evaluation of a dual-function pH and pCO_2 *in vivo* sensor, *J. Appl. Physiol.,* 40, 625, 1976.

114. Shimada, K., Yano, M., Shibatani, K., Komoto, Y., Eshashi, M., and Matsuo, T., Application of catheter-tip I.S.F.E.T. for continuous *in vivo* measurement, *Med. Biol. Eng. Comput.,* 18, 741, 1980.

115. Pinard, E., Seylaz, J., and Mamo, H., Quantitative continuous measurement of pO_2 and pCO_2 in artery and vein, *Med. Biol. Eng. Comput.,* 16, 59, 1978.

116. Wald, A., Hass, W. K., Siew, F. P., and Wood, D. H., Continuous measurement of blood gases *in vivo* by mass spectrography, *Med. Biol. Eng.,* 8, 111, 1970.

117. Lundsgaard, J. S., Jensen, B., and Grønlund. J., Fast-responding flow-independent blood gas catheter for oxygen measurement, *J. Appl. Physiol. Resp. Environ. Exer. Physiol.,* 48, 376, 1980.

118. Rolfe, P., Arterial oxygen measurement in the newborn with intra-vascular transducer, in *IEE Medical Electronics Monographs,* Hill, D. W. and Watson, B. W., Eds., Peter Peregrinus, Stevenage, 1976, 18–22, 126.

119. Parker, D., A flexible catheter for continuous measurement of blood-gas tensions *in vivo* by mass spectrometry, in *Digest Int. Conf. Med. Biol. Eng.,* Ottawa, 1976, 60.

120. Polanyi, M. L. and Hehir, R. M., *In vivo* oximeter with fast dynamic response, *Rev. Sci. Instrum.,* 33, 1050, 1962.

121. Enson, Y., Jameson, A. G., and Cournand, A., Intracardiac oximetry in congenital heart disease, *Circulation,* 29, 499, 1964.

122. Cole, J. S., Martin, W. E., Cheung, P. W., and Johnson, C. C., Clinical studies with a solid state fiberoptic oximeter, *Am. J. Cardiol.,* 29, 383, 1972.

123. Taylor, J. B., Lown, B., and Polanyi, M., *In vivo* monitoring with a fiberoptic catheter, *J. Am. Med. Assoc.,* 221, 667, 1972.

124. Gehrich, J. L., Lübers, D. W., Opitz, N., Hansman, D. R., Miller, W. W., Tusa, J. K., and Yafuso, M., Optical fluorescence and its application to an intravascular blood gas monitoring system, *IEEE Trans. Biomed. Eng.,* 33, 117, 1986.

125. Zimmerman, J. L. and Dellinger, R. P., Initial evaluation of a new intra-arterial blood gas system in humans, *Crit. Care Med.,* 21, 495, 1993.

126. Wolthuis, R. A., McCrae, D., Hartl, J. C., Saaski, E., Mitchell, G. L., Garcin, K., and Willard, R., Development of a medical fiberoptic oxygen sensor based on optical absorption change, *IEEE Trans. Biomed. Eng.,* BME-39, 185, 1992.

127. Hochberg, H. M., Current clinical experience with continuous tissue pH monitoring, *J. Clin. Eng.,* 6, 17, 1981.

128. Khuri, S. F., Josa, M., Marston, W., Braunwald, N. S., Smith, B., Tow, D., VanCisin, M., and Barsamian, E. M., First report of intramyocardial pH in man. II. Asssessment of adequacy of myocardial preservation, *J. Thorac Cardiovasc. Surg.,* 86, 667, 1983.

129. Covington, A. K., Vardes-PerezCasga, F., Weeks, P. A., and Brown, A. H., pH ISFETs for intramyocardial pH monitoring in man, *Analusis Mag.,* 21, M43, 1993.

130. Shichiri, M., Kawamori, R., Yamazaki, Y., Hakui, N., and Abe, H., Wearable artificial endocrine pancreas with needle-type glucose sensor, *Lancet,* 2, 1129, 1982.

131. Claremont, D. J., Sambrook, I. E., Penton, C., and Pickup, J. C., Subcutaneous implantation of a ferrocene-mediated glucose sensor in pigs, *Diabetologia,* 29, 817, 1986.

132. Bindra, D. S., Zhang, Y., and Wilson, G. S., Design and *in vitro* studies of a needle-type glucose sensor for subcutaneous monitoring, *Anal. Chem.,* 63, 1692, 1992.

133. Moatti-Sirat, D., Capron, F., Poitout, V., Reach, G., Bindra, D. S., Zhang, Y., Wilson, G. S., and Thévenot, D. R., Toward continuous glucose monitoring: *in vivo* evaluation of a miniaturized glucose sensor implanted for a several days in rat subcutaneous tissue, *Diabetologia,* 35, 224, 1992.

134. Turner, R. F. B., Harrison, D. J., Rajotte, R. V., and Baltes, H. P., A biocompatible enzyme electrode for continuous in vivo glucose monitoring in whole blood, *Sensor Actuator,* B1, 561, 1990.

135. Gerhardt, G. A., Oke, A. F., Nagy, G., Moghaddam, B., and Adams, R. N., Nafion-coated electrodes with high selectivity for CNS electrochemistry, *Brain Res.,* 290, 390, 1984.

136. Malinski, T. and Taha, Z., Nitric oxide release from a single cell measured *in situ* by a porphyrinic-based microsensor, *Nature,* 358, 676, 1992.

137. Malinski, T., Bailey, F., Zhang, Z. G., and Chopp, M., Nitric oxide measured by a porphyrinic microsensor in rat brain after transient middle cerebral artery occlusion, *J. Cereb. Blood Flow Metab.,* 13, 355, 1993.

138. Ichimori, K., Ishida, H., Fukahori, M., Nakazawa, H., and Murakami, E., Practical nitric oxide measurement employing a nitric oxide-selective electrode, *Rev. Sci. Instrum.,* 65, 2714, 1994.

139. Ossward, H. F., Asper, R., Dimai, W., and Simon, W., On-line continuous potentiometric measurement of potassium concentration in whole blood during open-heart surgery, *Clin. Chem.,* 25, 39, 1979.

140. Ossward, H. F., Continuous monitoring of electrolyte concentrations by direct potentiometry during extracorporeal circulation, in *Proc. Int. Conf. on Monitoring of Vital Parameters during Extracoporeal Circulation,* Nijmegen, 1980, Karger, Basel, 1981, 149.

141. Shibbald, A., Covington, A. K., and Carter, R. F., Online patient monitoring system for the simultaneous analysis of K^+, Ca^{2+}, Na^+ and pH using a quadruple-function chemFET integrated-circuit sensor, *Med. Biol. Eng. Comput.,* 23, 329, 1985.

142. Gumbrecht, W., Schelter, W., Montag, B., Rasinski, M., and Pfeiffer, U., Online blood electrolyte monitoring with a chemFET microcell system, *Sensors Actuators,* B1, 477, 1990.

143. Pfeiffer, E. F., Meyerhoff, C., Bishof, F., Keck, F. S., and Kerner, W., On line continuous monitoring of subcutaneous tissue glucose is feasible by combining portable glucosensor with microdialysis, *Horm. Metab. Res.,* 25, 121, 1993

144. Laurell, T., Microdialysis implemented in the design of a system for continuous glucose monitoring, *Sensors Actuators,* B13-14, 323, 1993.

145. Clemens, J. A. and Phebus, L. A., Brain dialysis in conscious rat confirms *in vivo* electrochemical evidence that dopaminergic stimulation releases ascorbate, *Life Sci.,* 35, 671, 1984.

146. Flentge, F., Venema, K., Koch, T., and Korf, J., An enzyme-reactor for electrochemical monitoring of choline and acetylcholine: applications in high-performance liquid chromatography, brain tissue, microdialysis and cerebrospinal fluid, *Anal. Biochem.,* 204, 305, 1992.

147. Lunte, C. E., Scott, D. O., and Kissinger, P. T., Sampling living systems using microdialysis probes, *Anal. Chem.,* 63, 773A, 1991.

148. Arai, T., Tomita, Y., Kikuchi, M., and Negishi, N., Transcutaneous detection for blood glucose change by measurement of effusion fluid using suction ATR method, in *Proc. XIV ICMBE and VII ICMP,* Espoo, Finland, 1985, 303.

149. Kimura, J., Ito, N., Kuriyama, T., A novel blood glucose monitoring method and ISFET biosensor applied to transcutaneous effusion fluid, *J. Electrochem. Soc.,* 136, 1744, 1989.

150. Ito, N., Saito, A., Miyamoto, S., Shinohara, S., Kuriyama, T., Kimura, J., Arai, T., Kikuchi, M., Kayashima, S., Nagata, N., and Takatani, O., A novel blood glucose monitoring system based on an ISFET biosensor and its application to a human 75 g oral glucose tolerance test, *Sensors Actuators,* B1, 488, 1990.

151. Kayashima, S., Arai, T., Kikuchi, M., Sato, N., Nagata, N., Takatani, O., Ito, N., Kimura, J., Kuriyama, T., and Kaneyoshi, A., New noninvasive transcutaneous approach to blood glucose monitoring: successful glucose monitoring on human 75 g OGTT with novel sampling chamber, *IEEE Trans. Biomed. Eng.,* 38, 752, 1991.

152. Ito, N., Kayashima, S., Kimura, J., Kuriyama, T., Arai, T., Kikuchi, M., and Nagata, N., Development of a transcutaneous blood-constituent monitoring method using a suction effusion fluid collection technique and an ion-sensitive field-effect transistor glucose sensor, *Med. Biol. Eng. Comput.,* 32, 242, 1994.

153. Ito, N., Saito, A., Kayashima, S., Kimura, J., Kuriyama, T., Nagata, N., Arai, T., and Kikuchi, M., Transcutaneous blood glucose monitoring system based on an ISFET glucose sensor and studies on diabetic patients, *Frontiers Med. Biol. Eng.,* 6, 269, 1995.

154. Anderson, R. R. and Parrish, J. H., The optics of human skin, *J. Invest. Dermatol.,* 77, 13, 1981.

155. Tregear, R. I., *Physical Functions of Skin,* Academic Press, London, 1960.

156. Scheuplein, R. J. and Blank, I. H., Permeability of the skin, *Physiol. Rev.,* 51, 702, 1971.

157. Baumberger, J. P. and Goodfriend, R. B., Determination of arterial oxygen tension in man by equilibration through intact skin, *Fed. Proc.,* 10, 10, 1951.

158. Bruls, W. A. G. and van der Leun, J. C., Forward scattering properties of human epidermal layers, *Photochem. Photobiol.,* 40, 231, 1984.

159. Profio, A. E., Light transport in tissue, *Appl. Opt.,* 28, 2216, 1989.

160. Cheong, W. G., Prahl, S. A., and Welch, A. J., A review of the optical properties of biological tissues, *IEEE J. Q. Electron.,* 26, 2166, 1990.

161. Wilson, B. C. and Jacques, S. L., Optical reflectance and transmittance of tissues: principles and applications, *IEEE J. Q. Electron.,* 26, 2186, 1990.

162. Evans, N. T. S. and Naylor, P. F. D., The systemic oxygen supply to the surface of human skin, *Resp. Physiol.,* 3, 21, 1967.

163. Lübbers, D. W., Methods of measuring oxygen tensions of blood and organ surfaces, in *Oxygen Measurements in Blood and Tissues,* Payne, J. P., and Hill, D. W., Eds., Churchill, London, 1960, 103.

164. Huch, A., Huch, R., and Lübbers, D. W., Quantitative polarographische sauerstoff druckmessung auf der kopfhaut des neugeborenen, *Arch. Gynakol.,* 207, 443, 1969.

165. Eberhard, P., Hammacher, K., and Mindt, W., Methode zur kutanen messung des sauerstoffpartialdruckes, *Biomed. Tech.,* 6, 216, 1973.

166. Hutchison, D. S. C., Rocca, G., and Honeybourne, D., Estimation of arterial oxygen tension in adult subjects using a transcutaneous electrode, *Thorax,* 36, 473, 1981.

167. Huch, R., Huch, A., and Rolfe, P., Transcutaneous measurement of pO_2 using electrochemical analysis, in *Noninvasive Physiological Measurements,* Vol. 1, Rolfe, P., Ed., Academic Press, London, 1979.

168. Huch, R., Huch, A., and Lübbers, D.W., *Transcutaneous pO2,* Thieme-Stratton, New York, 1981.

169. Huch, R. and Huch, A., *Continuous Transcutaneous Blood Gas Monitoring,* Marcel Dekker, New York, 1983.

170. Rithalia, S. V. S., Developments in transcutaneous blood gas monitoring: a review, *J. Med. Eng. Tech.,* 15, 143, 1991.

171. Mindt, W., Eberhard, P., and Schäfer, R., Methodological factors affecting the relationship between transcutaneous and arterial pCO_2, in *Continuous Transcutaneous Blood Gas Monitoring,* Huch, A. and Huch, R., Eds., Marcel Dekker, New York, 1983, 199.

172. Lacy, J. F., Clinical uses of transcutenous blood gas measurements, *Adv. Pediatr.,* 28, 27, 1981.

173. Huch, A., Lübbers, D. W., and Huch, R., Patientübervachung durch transcutane pCO_2 messung bei gleichzeiligen kontrolle der relativen lokalen perfusion, *Anaesthetist. J.,* 22, 379, 1973.

174. Herrel, N., Martin, R. J., Pultasher, M., Lough, M., and Fanaraff, A., Optimal temperature for the measurement of transcutaneous carbon dioxide tension in the neonate, *J. Pediatr.,* 97, 114, 1980.

175. Severinghaus, J. W., A combined transcutaneous pO_2-pCO_2 electrode with electrochemical HCO_3^- stabilization, *J. Appl. Physiol.*, 51, 1027, 1981.

176. Delpy, D. and Parker, D., The application of mass spectrometry to transcutaneous blood gas analysis, in *Non-Invasive Physiological Measurements*, Rolfe, P., Ed., Academic Press, London, 1979, 334.

177. Behrens-Tepper, J. C., Massaro, T. A., Updike, S. J., and Folts, J. D., Non-polarographic blood gas analysis. II. *In vivo* evaluation of gas chromatograph system, Biomaterials, *Med. Dev. Artif. Organs*, 5, 193, 1977.

178. Mendelson, Y., Blood gas measurement, trancutaneous, in *Encyclopedia of Medical Devices and Instrumentions*, Webster, J., Ed., John Wiley & Sons, New York, 1988, 448.

179. Kramer, K., Ein verfahren zur fortlaufenden messung des sauerstoffgehaltes im stromenden blute an uneröffneten gefassen, *Z. Biol.*, 96, 61, 1935.

180. Matthes, K., Untersuchungen über die sauerstoffsättigungen des menschlichen arterienblutes, *Arc. Exp. Pathol. Parmakol.*, 179, 689, 1935.

181. Wood, E. and Geraci, J. E., Photoelectric determination of arterial oxygen satuaration in man, *J. Lab. Clin. Med.*, 34, 387, 1949

182. Aoyagi, T., Kishi, T., Yamaguchi, K., and Watanabe, S., Improvement of ear-piece oximator, in *Proc. 13th Jpn. Medical Electronics and Biological Engineering Conf., Osaka*, 1974, 90.

183. Mendelson, Y. and Ochs, B. D., Noninvasive pulse oximetry utilizing skin reflectance photoplethysmography, *IEEE Trans. Biomed. Eng.*, 10, 798, 1988.

184. Tur, E., Tur, M., Maibach, H. I., and Guy, H. R., Basal perfusion of the cutaneous microcirculation: measurements as a function of anatomic position, *J. Invest. Dermatol.*, 81, 442, 1983.

185. Olsson, S. G., Fletcher, R., Johnson, B., Nordström, L., and Prakash, O., Clinical studies of gas exchange during ventilatory support — a method using the Siemens-Elema CO_2 analyzer, *Br. J. Anaest.*, 52, 491, 1980.

186. Coombes, R. G. and Halsall, D., Carbon dioxide analyzers, in *Encyclopedia of Medical Devices and Instrumentations*, Webster, J. G., Ed., John Wiley & Sons, New York, 1988, 556.

187. Bhavani-Shankar, K., Mosley, H., Kumar, A. Y., and Delf, Y., Capnometry and anesthesia, *Can. J. Anaesth.*, 39, 617, 1992.

188. Arnoudse, P. B. and Pardue, H. L., Instrumentation for the breath-by-breath determination of oxygen and carbon dioxide based on nondispersive absorption measurements, *Anal. Chem.* 64, 200, 1992.

189. Riker, J. B., Arrt, B. H., Bourland, J. D., Miller, R., and Geddes, L. A., Expired gas monitoring by mass spectrometry in a respiratory intensive care unit, *Crit. Care Med.*, 4, 223, 1976.

190. Severinghaus, J. W. and Ozanne, G., Multioperating room monitoring with one mass spectrometer, *Acta Anaesth. Scand. (Suppl.)*, 70, 186, 1978.

191. Sodal, I. E., Clark, J. S., and Swanson, G. D., Mass spectrometers in medical monitoring, in *Encyclopedia of Medical Devices and Instrumentations*, Webster, J. G., Ed., John Wiley & Sons, New York, 1988, 1848.

192. McLean, J. A. and Tobin, G. *Animal and Human Calorimetry*, Cambridge University Press, London, 1987.

193. Humphrey, S. J. E. and Wolf, H. S., The Oxylog, *J. Physiol.*, 267, 12P, 1977.

194. Ballal, M. A. and MacDonald, I. A., An evaluation of the Oxylog as a portable device with which to measure oxygen consumption, *Clin. Phys. Physiol. Meas.*, 3, 57, 1982.

195. Kappagoda, C. T., Stoker, J. B., and Linden, R. J., A method for the continuous measurement of oxygen consumption, *J. Appl. Physiol.*, 37, 604, 1974.

196. Smales, O. R. C., Simple method for measuring oxygen consumption in babies, *Arch. Dis. Child*, 53, 53, 1978.

197. Newell, J. P., Kappagoda, C. T., and Linden, R. J., Method for continuous measurement of carbon dioxide output, *J. Appl. Physiol.*, 41, 810, 1976.

198. Webb, P. and Troutman, S. J., An instrument for continuous measurement of oxygen consumption, *J. Appl. Physiol.*, 28, 867, 1970.

199. Tamura, T., Sato, K., and Togawa, T., Ambulatory oxygen uptake measurement system, *IEEE Trans. Biomed. Eng.*, 39, 1274, 1992.

200. Gardner, J. W. and Bartlett, P. N., *Sensors and Sensory Systems for an Electronic Nose*, NATO ASI Series E, Vol. 212, Kluwer Academic, Dordrecht, The Netherlands, 1992.

201. Shurmer, H. V., Gardner, J. W., and Chan, H. T., The application of discrimination techniques to alcohols and tobaccos using tin oxide sensors, *Sensors Actuators*, 18, 361, 1989.

202. Shurmer, H. V., Gardner, J. W., and Corcoran, P., Intelligent vapour discrimination using a composite 12-element sensor array, *Sensors Actuators B*, 1, 256, 1990.

203. Nakamoto, T., Fukuhischi, K., and Moriizumi, T., Identification capacity of odour using quartz-resonator array and neural-network pattern recognitions, *Sensors Actuators B*, 1, 473, 1990.

204. Ödman, S., Personal communication, 1995.

205. Shurmer, H. V. and Gardner, J. W., Odour discrimination with an electronic nose, *Sensors Actuators B*, 8, 1, 1992.

206. Bartlett, P. N. and Ling Chang, S. K., Conducting polymer gas sensors. Part III. Results for four different polymers and five different vapours, *Sensors Actuators*, 20, 287, 1989.

207. Moriizumi, T., Langmuir-Blodgett films as chemical sensors, *Thin Solid Films*, 160, 413, 1988.

208. Lundström, I., Hedberg, E., Spetz, H., Sundgren, H., and Winqvist, F., Electronic noses based on field effect structures, in *Sensors and Sensory Systems for an Electronic Nose*, NATO ASI Series E, Vol. 212, Gardner, J. W. and Bartlett, P. N, Eds., Kluwer Academic, Dordrecht, The Netherlands, 1992.

INDEX

Pulse oximeter, 339–340
Pulsed wave excitation, 148, see also Nuclear magnetic
    resonance
Pulse-echo technique, 193–194, 196
PZT, see Lead zirconate titanate

Q

QCM, see Quartz crystal microbalance
Quadrature demodulation, 95, 96
Quadrupole analyzer, 315–316
Quantization error, 11
Quartz crystal microbalance (QCM), 305, 307
Quartz crystal tonometer, see Tonometer

R

Rabbits, 86
Radiation sources, 302, see also Chemical measurement
Radio pill, 243
Radioactive indicators, 131–132
Radio-iodinated serum albumin (RISA), 111
Radioisotopes, 111
Radiometers, 234–235
Raman effect, 301, 302
Random error, 11, see also Error
Random noise theory, 97, see also Noise
Random-signal Doppler system, 99, 100, 101
Range
    fundamental concepts in measurement, 6
    pressure and measurement sites, 14–16
Range-discriminating Doppler flowmeter, 99–102
Rapid injection method, 104
Rate gyro, 203
Rayleigh scattering, 94, 137, 302
Rayleigh wave, 306, see also Rayleigh scattering
Reaction force, 59–67
Real-time measurements, 280
Rebreathing method, 111
Recirculation component, 105–106
Reconstruction method, 151, see also Nuclear magnetic
    resonance
Rectal temperature, 239, see also Temperature/heat
    flow/evaporation measurement
Red blood cells, 118–120, 122, 146
Reduction reactions, 265, 288, see also Electrode theory
Reference electrode, 288–291
Reference points, 16–17, 18, 185
Reflected acoustic pathways, 93
Reflection, 302
Refractive index, 323
Relaxation time, see Longitudinal relaxation time
Remote-sensing aerials, 237, 238
Reproducibility, 6
Resistive potentiometers, 185–186, see also
    Potentiometric sensors
Resistivity, 115, 128, 223, 224
Resolution, 6
Resonant frequency
    chemical measurement, 305, 321

flow measurement, 89, 147–148, 149, 167
    motion/force measurement, 205
    pressure measurements, 24, 29, 34
    temperature measurement, 230
Respiration-measuring microbial sensors, 313, 314
Respiratory gas
    analysis, 340–343
    gas flow sensors
        hot-wire anemometer, 156–159
        pneumotachograph, 153–155
        rotameter, 152–153
        time-of-flight flowmeter, 159–160
        ultrasonic gas flowmeter, 161–162
        vortex flowmeter, 162–163
    lung plethysmography
        body plethysmography, 165–167
        impedance, 168–170
        inductance, 167–168
    requirements of flow measurement, 76
    units of flow measurement, 76
    volume-measuring spirometers, 163–65
Response time, 9, 159, 292, 294
Retina circulation, 116
Retro-reflecting markers, 196, 197
Reusable electrodes, 267
Reynold's number, 73, 78, 109, 156, 162
Right atrium, 17, see also Heart
Rigidity, 37, 39
RISA, see Radio-iodinated serum albumin
Rise time, 9
Rodents, 253
Rotameter, 122, 123, 152–153
Rotary photoencoder, see Photoencoder
Rotating polarizers, 198, 199
Rotation
    contact transducers in measurement
        body and extracted tissue, 185–192
        in vivo, 192–196
    moving coordinate systems, 185
    noncontact measurement
        magnetic methods, 199–200
        optical methods, 196–199
Rotational potentiometer, 186–187, see also
            Potentiometric sensors
Rotational viscometer, 322, 323

S

Safety, 31, 305
Saline, 111–112
Saturation effect, 148
Sauerbrey equation, 305
SAW, see Surface acoustic wave sensor
Sawtooth wave excitation, 80, 81
Scanning laser-Doppler device, 146, 147, see also
            Laser-Doppler flowmeters
Scanning system, 236
Scattering, 137, 302, see also Individual entries
Schmidt belt, 244, 245
Second-order system, 8, 9